For the last 30 years, international summer schools in plasma physics have been held at Culham Laboratory, in Oxfordshire. Culham is the United Kingdom centre for research in magnetically confined fusion plasmas, and is the site of the Joint European Torus. This book has been developed from lectures given at these schools, and provides a wide-ranging introduction to the theoretical and experimental study of plasmas and their applications.

The first few chapters deal with the fundamentals of plasma physics, covering such topics as particle dynamics, kinetic theory, waves and magnetohydrodynamics. Subsequent chapters describe the applications and properties of man-made and naturally occurring plasmas. These include magnetically confined fusion plasmas, industrial and laser-produced plasmas, as well as gravitational, solar, space and magnetospheric plasmas. In addition, there are chapters devoted to general phenomena such as turbulence and chaos. The computational techniques employed in modelling plasma behaviour are also described.

Since no prior knowledge of plasma physics is assumed, this book will act as an ideal introduction to the subject for final year undergraduates and graduate students in physics, astronomy, mathematics and engineering.

Plasma Physics: an Introductory Course

Plasma Physics: an Introductory Course

EDITED BY
RICHARD DENDY
Culham Laboratory
Abingdon, Oxfordshire

CAMBRIDGE
UNIVERSITY PRESS

Published by the Press Syndicate of the University of Cambridge
The Pitt Building, Trumpington Street, Cambridge CB2 1RP
40 West 20th Street, New York, NY10011–4211, USA
10 Stamford Road, Oakleigh, Melbourne 3166, Australia

First published 1993

Printed in Great Britain at the University Press, Cambridge

A catalogue record for this book is available from the British Library

Library of Congress cataloguing in publication data

Dendy, Richard.
Plasma Physics: an introductory course/Richard Dendy.
p. cm.
Includes index.
ISBN 0 521 43309 6
1. Plasma (Ionized gases) I. Title.
QC718.D386 1993
530.4′4–dc20 93–13675 CIP

ISBN 0 521 43309 6 hardback

VN

Contents

Contributors

A. R. Bell, Department of Physics, Imperial College of Science, Technology and Medicine, Prince Consort Road, London SW7 2BZ.

R. Bingham, Rutherford Appleton Laboratory, Chilton, Didcot, Oxfordshire OX11 0QX.

J. J. Binney, Department of Physics, University of Oxford, Keble Road, Oxford OX1 3NP.

D. A. Bryant, Rutherford Appleton Laboratory, Chilton, Didcot, Oxfordshire OX11 0QX.

R. A. Cairns, Department of Mathematical and Computational Sciences, University of St Andrews, St Andrews KY16 9SS, Fife, Scotland.

R. O. Dendy, AEA Technology, Culham Laboratory, Abingdon, Oxfordshire OX14 3DB.

J. P. Dougherty, Department of Applied Mathematics and Theoretical Physics, University of Cambridge, Silver Street, Cambridge CB3 9EW.

J. W. Eastwood, AEA Technology, Culham Laboratory, Abingdon, Oxfordshire OX14 3DB.

J. A. Elliott, Department of Physics, University of Manchester Institute of Science and Technology, PO Box 88, Manchester M60 1QD.

F. A. Haas, AEA Technology, Culham Laboratory, Abingdon, Oxfordshire OX14 3DB.

R. J. Hastie, AEA Technology, Culham Laboratory, Abingdon, Oxfordshire OX14 3DB.

A. W. Hood, Department of Mathematical and Computational Sciences, University of St Andrews, St Andrews KY16 9SS, Fife, Scotland.

K. I. Hopcraft, Department of Theoretical Mechanics, University of Nottingham, University Park, Nottingham NG7 2RD.

P. C. Johnson, AEA Technology, Culham Laboratory, Abingdon, Oxfordshire OX14 3DB.

G. McCracken, AEA Technology, Culham Laboratory, Abingdon, Oxfordshire OX14 3DB.

T. J. Mullin, Clarendon Laboratory, University of Oxford, Parks Road, Oxford OX1 3PU.

M. R. O'Brien, AEA Technology, Culham Laboratory, Abingdon, Oxfordshire OX14 3DB.

R. S. Pease, FRS, 'The Poplars', West Ilsley, Newbury, Berkshire RG16 0AW.

D. C. Robinson, AEA Technology, Culham Laboratory, Abingdon, Oxfordshire OX14 3DB.

T. E. Stringer, JET Joint Undertaking, Abingdon, Oxfordshire OX14 3DB.

T. N. Todd, AEA Technology, Culham Laboratory, Abingdon, Oxfordshire OX14 3DB

Preface

This book has been developed from lectures given in recent years at the Culham Summer School in Plasma Physics. Every summer since 1964, graduate-level students from throughout Europe, and beyond, have assembled at Culham Laboratory, near Oxford, for an intensive two-week introductory course in plasma physics. In the first week of the course, the initial objective is to build up a working knowledge of the fundamentals of plasma physics among students who may have no previous knowledge of the subject. The relevant material is contained in the first four chapters of this book, which cover plasma particle dynamics, kinetic theory, wave theory, and magnetohydrodynamics. Thereafter, the course follows three main themes. First, we consider certain phenomena and techniques which have applications in all fields of plasma physics: for example, we have chapters on turbulence, chaos, and computational physics. Secondly, we provide introductions to the main fields of research where a knowledge of plasma physics is essential: separate chapters cover magnetospheric and space plasmas, the sun, gravitational systems, laser-produced plasmas, industrial plasmas, and tokamak plasmas. The latter chapter also serves as an introduction to our third theme, the physics of fusion plasmas. Separate chapters address transport in magnetically confined plasmas, radio-frequency heating, boundary plasmas, and engineering physics issues. The conceptual development of fusion plasma physics is placed in its historical context in the final survey chapter.

Each chapter in this book is the work of a different author, who brings a personal perspective to the subject. I believe that the overall result has been to produce a book which is more diverse than a single-author volume of equivalent length. My primary objective has been to ensure that the coverage of subject matter is as comprehensive as possible. Finally, I should like to thank the lecturer-authors who have found the time to contribute chapters to this book, and also Conrad Guettler and Irene Pizzie of Cambridge University Press.

Richard Dendy
Culham Laboratory, January 1993

Symbols, notation, fundamental equations and numerical values

SI units are used throughout this book, and we have employed the following set of symbols and conventions for notation wherever possible.

Vector notation

vector quantities: bold face italic
tensor quantities: bold face italic
orthonormal unit vectors $\hat{\boldsymbol{e}}_x$, $\hat{\boldsymbol{e}}_y$, $\hat{\boldsymbol{e}}_z$
generalized spatial vector $\boldsymbol{x} = x\hat{\boldsymbol{e}}_x + y\hat{\boldsymbol{e}}_y + z\hat{\boldsymbol{e}}_z$
guiding-centre position $\boldsymbol{x}_c$
scalar product: raised bold face dot $\boldsymbol{a} \cdot \boldsymbol{b}$
vector product: bold face cross $\boldsymbol{a} \times \boldsymbol{b}$
gradient operator: $\boldsymbol{\nabla}$
divergence: $\boldsymbol{\nabla} \cdot$
curl: $\boldsymbol{\nabla} \times$
length element $\mathrm{d}\boldsymbol{l}$
surface elements $\mathrm{d}\boldsymbol{S}$
volume element $\mathrm{d}^3\boldsymbol{x}$
velocity space element $\mathrm{d}^3\boldsymbol{v}$

Length scales

Debye length λ_{D}
electron Debye length λ_{De}
Larmor radius r_{L}
electron Larmor radius r_{Le}

Frequencies and related quantities

plasma frequency ω_{p}
electron plasma frequency ω_{pe}
ion plasma frequency ω_{pi}
cyclotron frequency ω_{c}
electron cyclotron frequency ω_{ce}
ion cyclotron frequency ω_{ci}
collision frequency ν_{c}
electron–ion collision frequency ν_{ei}
Coulomb logarithm $\lambda_{\mathrm{C}} = \ln \Lambda$

Velocities

velocity $\boldsymbol{v}$
perpendicular velocity $v_\perp$
parallel velocity $v_\parallel$
thermal velocity v_{T}
electron thermal velocity v_{Te}
sound velocity V_{s}
Alfvén velocity V_{A}
phase velocity v_{ph}

Distributions and energy

Maxwellian $\sim \exp(-v^2/v_{\mathrm{T}}^2)$, hence $v_{\mathrm{T}} = (2k_{\mathrm{B}}T/m)^{1/2}$
distribution function $f(\boldsymbol{x}, \boldsymbol{v}, t)$
electron distribution function $f_{\mathrm{e}}(\boldsymbol{x}, \boldsymbol{v}, t)$
temperature T
electron temperature T_{e}
Boltzmann's constant k_{B}

Oscillation and waves

generalized oscillation, space and time sign conventions (except in Chapter 3): $\exp(\mathrm{i}\boldsymbol{k}\cdot\boldsymbol{x})$, $\exp(-\mathrm{i}\omega t)$, $\frac{1}{2}A\exp(\mathrm{i}\boldsymbol{k}\cdot\boldsymbol{x} - \mathrm{i}\omega t) + \mathrm{c.c.}$
refractive index $N = ck/\omega$

Density and mass

number density n
electron number density n_{e}
ion number density n_{i}
mass density ρ
electron mass m
proton mass m_{p}
ion mass M

Electric and magnetic quantities

charge density ρ_e
electron charge $-e$
proton charge e
ion charge Ze
electric and magnetic fields $\boldsymbol{E}$, $\boldsymbol{B}$
unit vector along magnetic field $\hat{\boldsymbol{b}}$
magnetic moment μ
current density $\boldsymbol{J}$

Fundamental equations of electricity and magnetism

The Coulomb field:

$$\boldsymbol{E} = \frac{1}{4\pi\varepsilon_0}\int \frac{\rho_e \boldsymbol{r}}{r^3}\,\mathrm{d}^3\boldsymbol{x}.$$

Lorentz force on an electron with charge $-e$:

$$\boldsymbol{F} = -e(\boldsymbol{E} + \boldsymbol{v} \times \boldsymbol{B}).$$

Maxwell's equations '*in vacuo*':

$$\nabla \cdot \boldsymbol{E} = \rho/\varepsilon_0,$$

$$\nabla \cdot \boldsymbol{B} = 0,$$

$$\nabla \times \boldsymbol{E} = -\partial \boldsymbol{B}/\partial t,$$

$$\nabla \times \boldsymbol{B} = \mu_0\left(\boldsymbol{J} + \varepsilon_0 \frac{\partial \boldsymbol{E}}{\partial t}\right).$$

Here $\mu_0\varepsilon_0 = c^{-2}$, and numerical values are given below.

Numerical values of constants

c	= speed of light in a vacuum	$= 2.998 \times 10^8\ \mathrm{m\,s^{-1}}$
ε_0	= permittivity of a vacuum	$= 8.854 \times 10^{-12}\ \mathrm{F\,m^{-1}}$
μ_0	= permeability of a vacuum	$= 4\pi \times 10^{-7}\ \mathrm{H\,m^{-1}}$
e	= charge of a proton	$= 1.602 \times 10^{-19}\ \mathrm{C}$
m	= mass of an electron	$= 9.109 \times 10^{-31}\ \mathrm{kg}$
m_p	= mass of a proton	$= 1.673 \times 10^{-27}\ \mathrm{kg}$
m_p/m	= proton–electron mass ratio	$= 1.836 \times 10^3$
k_B	= Boltzmann's constant	$= 1.381 \times 10^{-23}\ \mathrm{J\,K^{-1}}$
eV	= electron-volt	$= 1.602 \times 10^{-19}\ \mathrm{J}$
eV/k_B	= temperature equivalent to 1 eV	$= 1.160 \times 10^4\ \mathrm{K}$
G	= gravitational constant	$= 6.673 \times 10^{-11}\ \mathrm{N\,m^2\,kg^{-2}}$

Numerical values of plasma quantities

The plasma parameters in the following expressions are in SI units; hence, for example, n_e is in units of $\mathrm{m^{-3}}$, B is in units of tesla, $v_\perp$ is in units of $\mathrm{m\,s^{-1}}$, and T_e is in units of kelvin. Fundamental frequencies are:

ω_pe	= angular electron plasma frequency	$= 5.64 \times 10^{-1}\, n_\mathrm{e}^{1/2}\ \mathrm{rad\,s^{-1}}$,
ω_pp	= angular proton plasma frequency	$= 1.32 n_\mathrm{p}^{1/2}\ \mathrm{rad\,s^{-1}}$,
ω_ce	= angular electron cyclotron frequency	$= 1.76 \times 10^{11} B\ \mathrm{rad\,s^{-1}}$,
ω_cp	= angular proton cyclotron frequency	$= 9.58 \times 10^{7} B\ \mathrm{rad\,s^{-1}}$.

Note that frequencies in hertz are obtained by dividing the above expressions by 2π. To convert the expressions above which apply to protons into expressions for ions of charge Ze and mass M:

ω_pi, multiply ω_pp by $Z(m_\mathrm{p}/M)^{1/2}$,
ω_ci, multiply ω_cp by $Z(m_\mathrm{p}/M)$.

Fundamental velocities and length scales:

v_{Te} = thermal electron velocity, $(2k_B T_e/m)^{1/2} = 5.51 \times 10^3 T_e^{1/2}\ \mathrm{m\,s^{-1}}$
v_{Ti} = thermal ion velocity, $(2k_B T_i/M)^{1/2} = 1.29 \times 10^2 (m_p/M)^{1/2} T_i^{1/2}\ \mathrm{m\,s^{-1}}$
λ_{De} = electron Debye length, $(k_B T_e/m)^{1/2}/\omega_{pe} = 6.90 \times 10^1 T_e^{1/2} n_e^{-1/2}\ \mathrm{m}$
r_{Le} = electron Larmor radius, $v_\perp/\omega_{ce} = 5.68 \times 10^{-12} v_\perp B^{-1}\ \mathrm{m}$
r_{Lp} = proton Larmor radius, $v_\perp/\omega_{cp} = 1.04 \times 10^{-8} v_\perp B^{-1}\ \mathrm{m}$

Key vector identities

$$A + B = B + A$$

$$A \cdot B = B \cdot A$$

$$A \times B = -B \times A$$

$$A \cdot (B \times C) = C \cdot (A \times B) = B \cdot (C \times A)$$

$$A \times (B \times C) = B(A \cdot C) - C(A \cdot B)$$

$$(A \times B) \cdot (C \times D) = (A \cdot C)(B \cdot D) - (A \cdot D)(B \cdot C)$$

Key results from vector calculus

Gradient: $\nabla = \hat{e}_x \frac{\partial}{\partial x} + \hat{e}_y \frac{\partial}{\partial y} + \hat{e}_z \frac{\partial}{\partial z}$

For a vector field $A(x, y, z)$,

divergence: $\nabla \cdot A = \frac{\partial A_x}{\partial x} + \frac{\partial A_y}{\partial y} + \frac{\partial A_z}{\partial z}$

curl: $\nabla \times A = \begin{vmatrix} \hat{e}_x & \hat{e}_y & \hat{e}_z \\ \frac{\partial}{\partial x} & \frac{\partial}{\partial y} & \frac{\partial}{\partial z} \\ A_x & A_y & A_z \end{vmatrix}$

$$\nabla \cdot (\nabla \times A) = 0$$

$$\nabla \times (\nabla \times A) = \nabla(\nabla \cdot A) - \nabla^2 A.$$

For a scalar function $\phi(x, y, z)$,

$$\nabla \times \nabla\phi = 0$$

$$\nabla \cdot \nabla\phi = \nabla^2 \phi.$$

For a scalar function ϕ in combination with a vector field A,

$$\nabla \cdot (\phi A) = \phi \nabla \cdot A + A \cdot \nabla\phi$$

$$\nabla \times (\phi A) = \phi \nabla \times A + (\nabla\phi) \times A.$$

For two vector fields $A(x, y, z)$ and $B(x, y, z)$,

$$\nabla \cdot (A \times B) = B \cdot (\nabla \times A) - A \cdot (\nabla \times B)$$

$$\nabla \times (\boldsymbol{A} \times \boldsymbol{B}) = \boldsymbol{A}(\nabla \cdot \boldsymbol{B}) - \boldsymbol{B}(\nabla \cdot \boldsymbol{A}) + (\boldsymbol{B} \cdot \nabla)\boldsymbol{A} - (\boldsymbol{A} \cdot \nabla)\boldsymbol{B}.$$

Divergence theorem:

$$\int_V \nabla \cdot \boldsymbol{A} \, \mathrm{d}^3\boldsymbol{x} = \int_S \boldsymbol{A} \cdot \mathrm{d}\boldsymbol{S}.$$

Stokes's theorem:

$$\int_S (\nabla \times \boldsymbol{A}) \cdot \mathrm{d}\boldsymbol{S} = \oint_C \boldsymbol{A} \cdot \mathrm{d}\boldsymbol{l}.$$

Books by Culham Summer School lecturers

Several of the contributors to this volume are authors (or co-authors) of textbooks, and so are some of the other recent lecturers at the Culham Summer School. For convenience, we list brief publication details here.

Introductory texts

Cairns, R. A. (1985). *Plasma Physics*. Blackie, Glasgow.

Clemmow, P. C. and Dougherty, J. P. (1990). *Electrodynamics of Particles and Plasmas*. Addison-Wesley, London.

Dendy, R. O. (1990). *Plasma Dynamics*. Oxford University Press.

Specialized texts

Binney, J. J. and Tremaine, S. D. (1987). *Galactic Dynamics*. Princeton University Press, Guildford.

Cairns, R. A. (1991). *Radiofrequency Heating of Plasmas*. Adam Hilger, Bristol.

Hockney, R. W. and Eastwood, J. W. (1988). *Computer Simulation using Particles*. Adam Hilger, Bristol.

Hopcraft, K. I. and Smith, P. R. (1992). *An Introduction to Electromagnetic Inverse Scattering*. Kluwer Academic Publishers, London.

Manheimer, W. M. and Lashmore-Davies, C. N. (1989). *MHD and Micro-instabilities in Confined Plasmas*. Adam Hilger, Bristol.

Mullin, T. (1993). *The Nature of Chaos*. Oxford University Press.

Wesson, J. (1987). *Tokamaks*. Oxford University Press.

Introduction

R. O. DENDY

If we increase the temperature of a gas beyond a certain limit, it does not remain a gas: it enters a regime where the thermal energy of its constituent particles is so great that the electrostatic forces which ordinarily bind electrons to atomic nuclei are overcome. Instead of a hot gas composed of electrically neutral atoms, we have two commingled populations composed of oppositely charged particles – electrons and ionized nuclei. This is a plasma, and it is neither solid, liquid, nor gas. Its most notable feature is perhaps its high electrical conductivity; this is so great that externally applied electric fields are effectively cancelled, within the main body of the plasma, by the currents which they induce.

On Earth, plasmas occur naturally only in lightning and the aurora. Beyond our atmosphere, however, we are surrounded by the magnetosphere, which is a plasma system formed by the interaction of the Earth's magnetic field with the solar wind. The sun is made of plasma, and so are most stars, so that the plasma state can be said to dominate the visible universe. The practical terrestrial applications of man-made plasmas are also extensive. They range from the microfabrication of electronic components to demonstrations of substantial thermonuclear fusion power from magnetically confined plasmas. Furthermore, the concepts and techniques of plasma physics have found applications far beyond the original boundaries of the subject. For example, we describe in this book how the methods of plasma physics can be applied both to the etching of integrated circuits and to the elucidation of galactic spiral structure – thus we consider length scales from nanometres to kiloparsecs.

In order to study plasmas, physicists have developed a unified body of powerful theoretical techniques, which has been created in parallel with the discovery of many subtle experimental phenomena. The fundamental concepts and techniques of plasma physics now form a distinct branch of science, but it is interesting to note that many of them have evolved from concepts originally used to describe solids, liquids, or gases. For example, if one chooses to take as a starting point the fact that a plasma is a gas whose constituent particles are charged, it seems natural to try to extend the kinetic theory of gases so as to describe plasmas. This approach has been extremely successful. The need to include the distinctive features of plasmas – for example, in addition to being ionized, they frequently coexist with magnetic fields whose effects must be incorporated – has led to the development of plasma kinetic theory, which yields unique insights into phenomena such as the resonant transfer of energy between waves and particles. Many of the techniques of plasma kinetic theory apply also to the study of galactic dynamics; physically, this reflects the need in both cases to address the many-body problem for particles that interact through inverse square force fields – Coulomb for charged particles and gravitational for stars.

A different starting point for the study of plasmas – this time, one which is associated with the liquid state – is provided by fluid dynamics. If we knew nothing of microscopic physics, a plasma would present itself as a fluid with high electrical conductivity, in some ways analogous to a liquid metal. Such fluids interact strongly with magnetic fields within and around them, and the evolution of the combined fluid–field system is described by the equations of magnetohydrodynamics. The success of magnetohydrodynamics in dealing with the large scale equilibrium and stability of plasmas is remarkable given the fact that, in reality, a plasma is a far more complex system than the assumptions of magnetohydrodynamics might suggest. However, the kinetic approach provides a more fundamental level of description, and, indeed, the equations of magnetohydrodynamics can be derived from the kinetic equations by extracting appropriately averaged quantities from the latter. Much theoretical research is at present devoted to the problem of striking a useful balance that combines essential features of both the kinetic and fluid approaches – for example, in retaining kinetic information about selected populations of energetic ions so as to calculate their effect on the large scale magnetohydrodynamic stability of a plasma.

Having described how some aspects of plasma physics have conceptual origins in the study of liquids and of gases, let us turn briefly to the links between plasmas and solids. They are best illustrated by the wave modes that can arise in plasmas, whose description (essentially, in terms of dielectric tensors) and key properties (dispersion, polarization, and so on) strongly evoke the solid state. The link becomes closer still when one considers the wave modes supported by the conduction electrons within a metal, because the latter can in many senses be regarded as forming a plasma.

To summarize, the key concepts of plasma physics form a distinct intellectual structure: some of these concepts have evolved from progenitors which apply to solids, or to liquids, or to gases; others have arisen internally, within plasma physics. In principle, of course, in order to describe a plasma we require only a knowledge of the Lorentz force and of Maxwell's equations. The Lorentz force acting on each particle in the plasma depends only on its charge, its velocity, and the local values of the electric and magnetic fields, and, indeed, much can be learnt by studying single-particle orbits in fixed fields. However, as each particle in the plasma moves in response to the local fields, it contributes to the local charge density and electric current, and these are the sources of the electric and magnetic fields which affect the dynamics of all the other particles. The problem is to find a *self-consistent* description of the fields and particles in the plasma, a description which reflects the fact that the movement of each particle is both a consequence and a source of the electric and magnetic fields.

As an example, let us introduce the concept of the electron plasma frequency. We take a highly idealized model known as a 'cold plasma', in which equal numbers of electrons (charge $-e$) and ions (charge e) are initially motionless and uniformly distributed. For simplicity, we shall concentrate on electron dynamics, and assume that the mass of the ions is so much greater than that of the electrons that their inertia keeps them motionless on the short time scale of interest. The initial average charge density is zero everywhere, because the electrons and ions are uniformly distributed. Suppose that we now perturb this system by gathering together all the electrons from some small specific locality. This creates a local region of negative charge where the electrons are concentrated, with an adjoining positively charged region that has been depleted of electrons; beyond this region, the rest of the plasma is neutral. What happens next? The charge separation that we have created by perturbing the electrons is the source of an electric field, and, at the same time, this electric field will act on the electrons – everything that happens must be consistent at all times with both Poisson's equation and the Lorentz force equation. To begin with, the electric field accelerates the electrons back to their initial positions, so that momentarily the charge density is everywhere zero once more and the electric field vanishes. When this occurs, all the electrostatic potential energy associated with the initial perturbation has become transformed into the kinetic energy of the moving electrons – it cannot simply disappear. This kinetic energy is sufficient to carry the electrons on past their original positions. As they move away, they leave behind them an electron-depleted region which has positive charge. An electric field builds up, opposite in direction to that associated with the initial perturbation, which retards the electrons and eventually brings them to a standstill. At this stage, their kinetic energy has been completely transformed back into electrostatic potential energy, and the initial configuration has been reversed: the cloud of negative charge has moved to the opposite side of the region which is positively charged. This is the first half-cycle of an electron plasma oscillation, which is a form of simple harmonic motion that could, in principle, continue indefinitely, as the perturbed electrons move back and forth past their equilibrium position, converting electrostatic energy into kinetic energy and back again.

It is fairly easy to obtain an exact formula for the frequency of the oscillation that we have described – the electron plasma frequency – by solving Poisson's equation and the Lorentz force equation self-consistently as in Chapter 3. The point to note here, however, is that the electron plasma frequency is a concept which plasma physicists have created as an alternative to trying to calculate the behaviour of each individual particle. It describes a collective self-consistent phenomenon. The key role of such phenomena is perhaps the most characteristic and fundamental feature of plasma physics, and we shall meet it again in many contexts in this book: from the Debye length to flux-freezing, from the Grad–Shafranov equation to the Rayleigh–Taylor instability. Conversely, the rich variety of physical phenomena described in this book ultimately reflects the immense range and power of the concepts described by the Lorentz force equation and by Maxwell's equations.

1

Plasma particle dynamics

R. J. HASTIE

1.1 Introduction

In tenuous or high temperature plasmas, where collisional encounters between the constituent charged particles are rare, an understanding of plasma phenomena requires a knowledge of the individual particle trajectories in the self-consistent electromagnetic fields.

Only in highly symmetric fields can exact solutions of the equations of motion of a charged particle be obtained. In a constant and uniform magnetic field, for example, a charged particle trajectory is just a helix with its axis parallel (or antiparallel) to B. However, this simple orbit defines a fundamental time, the gyro-period, $2\pi/\omega_c$ and a fundamental length, the gyro- (or Larmor) radius r_L, and when considering trajectories in inhomogeneous and time dependent fields it is frequently the case that the characteristic time $(2\pi/\omega)$ and length (L) over which these fields vary are very large compared with the gyro-period and -radius of the orbiting particles. When this is the case, approximate, asymptotic, solutions to the orbit equations can be obtained by expansions in r_L/L and ω/ω_c. Such an approximation is referred to as the drift, guiding centre, or adiabatic approximation, and in it the charged particle motion is broken down into a local helical gyration, together with equations of motion for the instantaneous centre of this gyration, a point known as the guiding centre. A rigorous and systematic development of these guiding-centre equations can be obtained, but in this chapter we will first (Sections 1.1–1.4) give a heuristic derivation of the guiding-centre equations and discuss the importance of adiabatic invariants in determining guiding-centre trajectories in complicated electromagnetic fields. Alternative derivations can be found in the work of Northrop (1963) and Sivukhin (1965).

In Section 1.5 we investigate the properties of the adiabatic invariants of motion for a charged particle in more detail. In particular we show that these quantities are not exact constants of motion, and that the expressions which can be derived as infinite series in the expansion parameters, ω/ω_c and r_L/L, are not convergent.

In Sections 1.6 and 1.7 we examine the effects of interparticle collisions, and of emission of electromagnetic radiation, on particle trajectories, and adiabatic invariants.

Finally, in Section 1.8, we briefly discuss some consequences of collective behaviour which appear when many charged particles interact with each other as in a plasma. The key parameters distinguishing plasma behaviour are introduced.

1.2 Motion in constant uniform fields

The equations of motion for a charged particle with mass m and velocity $\boldsymbol{v}$ in fields $\boldsymbol{E}$ and $\boldsymbol{B}$ are

$$m\dot{\boldsymbol{v}} = e[\boldsymbol{E} + \boldsymbol{v} \times \boldsymbol{B}], \tag{1.2.1}$$

$$\dot{\boldsymbol{x}} = \boldsymbol{v}. \tag{1.2.2}$$

1.2.1 *Electric field only*

If $\boldsymbol{B} = 0$, $\boldsymbol{v}$ changes linearly with time. The particle accelerates freely in the direction of $\boldsymbol{E}$ (for positive charge).

1.2.2 *Magnetic field only*

Since we will frequently resolve vectors into components parallel to and perpendicular to $\boldsymbol{B}$, we introduce the following notation:

$$\boldsymbol{b} = \boldsymbol{B}/|\boldsymbol{B}|, \quad \boldsymbol{v}\cdot\boldsymbol{b} \equiv v_{\|}, \quad \boldsymbol{v}_{\perp} = \boldsymbol{v} - v_{\|}\boldsymbol{b}, \tag{1.2.3}$$

and similarly for other vectors.

Now taking the parallel component of Eq. (1.2.1) we find

$$\dot{v}_{\|} = 0; \quad v_{\|} = \text{constant}, \tag{1.2.4}$$

so that

$$\frac{m}{e}\dot{\boldsymbol{v}}_{\perp} = \boldsymbol{v}_{\perp} \times \boldsymbol{B}. \tag{1.2.5}$$

Taking the scalar product of Eq. (1.2.5) with $\boldsymbol{v}_{\perp}$, we have

$$\frac{m}{e}\boldsymbol{v}_{\perp}\cdot\dot{\boldsymbol{v}}_{\perp} = 0, \tag{1.2.6}$$

or

$$\frac{\mathrm{d}(v_{\perp}^2)}{\mathrm{d}t} = 0, \quad v_{\perp}^2 = \text{constant}. \tag{1.2.7}$$

Using Eq. (1.2.5) a second time, and recalling the vector identity $\boldsymbol{A} \times (\boldsymbol{B} \times \boldsymbol{C}) = \boldsymbol{B}(\boldsymbol{A}\cdot\boldsymbol{C}) - \boldsymbol{C}(\boldsymbol{A}\cdot\boldsymbol{B})$, we find

$$\left(\frac{m}{e}\right)^2 \ddot{\boldsymbol{v}}_{\perp} = \frac{m}{e}(\dot{\boldsymbol{v}}_{\perp} \times \boldsymbol{B}) = -B^2\boldsymbol{v}_{\perp}, \tag{1.2.8}$$

or

$$\ddot{\boldsymbol{v}}_{\perp} = -\omega_{\mathrm{c}}^2\boldsymbol{v}_{\perp}. \tag{1.2.9}$$

Thus each component of $\boldsymbol{v}_{\perp}$ satisfies the equation of the simple harmonic oscillator with frequency $\omega_{\mathrm{c}} = eB/m$. This is known as the gyro-frequency, Larmor frequency, or cyclotron frequency.

Introducing coordinates x, y, z with the z-axis in the direction of $\boldsymbol{B}$, we have

$$\begin{aligned} v_x &= v_{\perp}\cos(\omega_{\mathrm{c}}t + \phi), \\ v_y &= -v_{\perp}\sin(\omega_{\mathrm{c}}t + \phi), \end{aligned} \tag{1.2.10}$$

with the phase ϕ an arbitrary constant.

From Eq. (1.2.2) it now follows that $\boldsymbol{x}_\perp$ undergoes uniform circular motion perpendicular to $\boldsymbol{B}$, about a centre $\boldsymbol{x}_c$:

$$x = x_c + \frac{v_\perp}{\omega_c}\sin(\omega_c t + \phi),$$
$$y = y_c + \frac{v_\perp}{\omega_c}\cos(\omega_c t + \phi). \tag{1.2.11}$$

The frequency of gyration is again ω_c, and the radius of gyration, the Larmor radius, is

$$r_L \equiv |\boldsymbol{x}_\perp - \boldsymbol{x}_c| = \frac{v_\perp}{\omega_c}. \tag{1.2.12}$$

The position $\boldsymbol{x}_c$ of the centre about which the particle gyrates is known as the guiding centre. When considering particle motion in inhomogeneous fields, this becomes a useful concept whose motion (free from rapid gyration) can be studied. Defining $z_c \equiv z$, we see that in the present simple case, whereas the particle motion is helical (with clockwise rotation for negative charge), the guiding-centre motion is linear, parallel to $\boldsymbol{B}$ (Fig. 1.1).

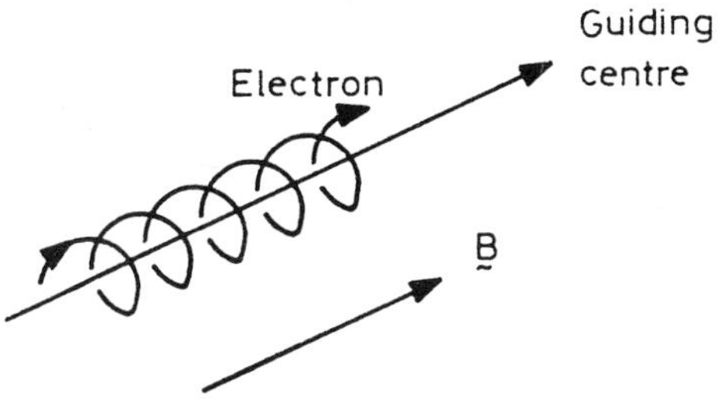

Fig. 1.1. Electron guiding-centre motion in the magnetic field $\boldsymbol{B}$.

The particle gyration constitutes a circulating electric current in a clockwise sense (for either sign of e). The magnetic moment associated with this current loop is given by

$$\begin{aligned}\mu &= \text{current} \times \text{area of loop}\\ &= \frac{e\omega_c}{2\pi}\pi r_L^2\\ &= \frac{mv_\perp^2}{2B},\end{aligned} \tag{1.2.13}$$

and the direction of the associated magnetic field is counter to that of the external field. Thus a collection of many charged particles is diamagnetic.

1.2.3 *Electric and magnetic fields*

Even with both $\boldsymbol{E}$ and $\boldsymbol{B}$ present, the parallel and perpendicular components of the motion decouple. Thus

$$\frac{m}{e}\dot{v}_{\parallel} = E_{\parallel}, \tag{1.2.14}$$

giving free acceleration parallel to $\boldsymbol{B}$, whereas

$$\frac{m}{e}\dot{\boldsymbol{v}}_{\perp} = (\boldsymbol{E}_{\perp} + \boldsymbol{v}_{\perp} \times \boldsymbol{B}). \tag{1.2.15}$$

Transforming to a reference frame moving with (constant) velocity $\boldsymbol{V}$, and writing

$$\boldsymbol{v}_{\perp} = \boldsymbol{c}_{\perp} + \boldsymbol{V}, \tag{1.2.16}$$

Eq. (1.2.15) becomes

$$\frac{m}{c}\dot{\boldsymbol{c}}_{\perp} = (\boldsymbol{E}_{\perp} + \boldsymbol{V} \times \boldsymbol{B}) + \boldsymbol{c}_{\perp} \times \boldsymbol{B}. \tag{1.2.17}$$

Then if $\boldsymbol{V}$ is chosen so that

$$\boldsymbol{V} = \frac{\boldsymbol{E} \times \boldsymbol{B}}{B^2}, \tag{1.2.18}$$

Eq. (1.2.17) reduces to the form of Eq. (1.2.5),

$$\frac{m}{e}\dot{\boldsymbol{c}}_{\perp} = \boldsymbol{c}_{\perp} \times \boldsymbol{B}, \tag{1.2.19}$$

which describes uniform rotation with frequency ω_c and radius $r_L = c_{\perp}/\omega_c$. Thus the particle motion is now composed of three parts:

$$\boldsymbol{v} = v_{\parallel}\boldsymbol{b} + \boldsymbol{V} + \boldsymbol{c}_{\perp}, \tag{1.2.20}$$

comprising motion along $\boldsymbol{B}$ (described by Eq. (1.2.14)), uniform drift velocity $\boldsymbol{V}$ perpendicular to $\boldsymbol{B}$, and Larmor gyration. Taking the time average of $\boldsymbol{v}$ over a gyro-period,

$$\frac{\omega_c}{2\pi}\int_0^{2\pi/\omega_c} \boldsymbol{v}\,\mathrm{d}t \equiv \langle \boldsymbol{v} \rangle = v_{\parallel}\boldsymbol{b} + \boldsymbol{V} \tag{1.2.21}$$

shows that $\boldsymbol{V} \equiv (\boldsymbol{E} \times \boldsymbol{B})/B^2$ is the average perpendicular velocity.

1.2.4 *Motion in a general force field*

If some additional force $\boldsymbol{F}$ acts on the particle, the foregoing analysis holds with $\boldsymbol{E}$ replaced by $(1/e)\boldsymbol{F}$. Thus the effect of such a force is to produce transverse drift with velocity

$$\frac{1}{e}(\boldsymbol{F} \times \boldsymbol{B})/B^2. \tag{1.2.22}$$

If $\boldsymbol{F}$ is charge-independent this drift is charge-dependent and induces charge separation.

1.3 Inhomogeneous and time-varying fields

The importance of understanding in detail the motion of a charged particle in constant uniform fields lies in the fact that, for most cases of interest, both in the plasmas encountered in fusion research and in natural plasmas, the scale of inhomogeneity of the fields is very large compared to a Larmor radius r_L, and the time scale is very long compared to a gyro-period $2\pi/\omega_c$. Under these conditions, the particle motion may be separated into a rapid gyration around its guiding centre, together with a slow drift of the guiding centre across the magnetic field. In the following we give an elementary discussion of the effects arising from nonuniformity of $\boldsymbol{B}$ and $\boldsymbol{E}$.

1.3.1 *Time-varying electric field*

We assume that $\boldsymbol{E}$ is perpendicular to $\boldsymbol{B}$ (a parallel component is easily handled) and spatially uniform. Transforming to a frame moving with velocity $\boldsymbol{V} = (\boldsymbol{E} \times \boldsymbol{B})/B^2$, we have $\boldsymbol{v}_\perp = \boldsymbol{V} + \boldsymbol{c}_\perp$, and

$$\frac{m}{e}\dot{\boldsymbol{c}}_\perp = -\frac{m}{e}\dot{\boldsymbol{V}} + \boldsymbol{c}_\perp \times \boldsymbol{B}. \tag{1.3.1}$$

This equation has a structure similar to that of Eq. (1.2.15), but with the property that the additional force term is of order $(1/\omega_c)(\mathrm{d}E/\mathrm{d}t)$ and is therefore small, $O(\omega/\omega_c)$, where ω characterizes the frequency of variation of $E(t)$.

Now introducing a second transformation (by analogy with Eq. (1.2.18)) to a frame moving with velocity

$$\begin{aligned}\boldsymbol{U} &= \frac{\left(-\dfrac{m}{e}\dot{\boldsymbol{V}}\right) \times \boldsymbol{B}}{B^2} \\ &= \frac{1}{\omega_c}\frac{\dot{\boldsymbol{E}}}{B},\end{aligned} \tag{1.3.2}$$

Eq. (1.3.1) becomes

$$\frac{\mathrm{d}}{\mathrm{d}t}\left[\frac{m}{e}(\boldsymbol{c}_\perp - \boldsymbol{U})\right] = (\boldsymbol{c}_\perp - \boldsymbol{U}) \times \boldsymbol{B} - \frac{\ddot{\boldsymbol{E}}}{\omega_c^2}. \tag{1.3.3}$$

Here the explicit E term is now of order $(\omega/\omega_c)^2$ and can be neglected as it is of second order in the small parameter. The equation for $(\boldsymbol{c}_\perp - \boldsymbol{U})$ then assumes the familiar form describing Larmor motion. Thus, averaging the total motion over a gyro-period (when $\langle(\boldsymbol{c}_\perp - \boldsymbol{U})\rangle \equiv 0$), we obtain the guiding-centre velocity,

$$\langle \boldsymbol{v} \rangle = v_\parallel \boldsymbol{b} + (\boldsymbol{E} \times \boldsymbol{B})/B^2 + \frac{m}{e}\frac{\dot{\boldsymbol{E}}}{B^2}. \tag{1.3.4}$$

The new drift velocity, $\boldsymbol{U}$, obtained here is charge-dependent, causing charge separation, and is known as the polarization drift.

Evidently the procedure adopted in the foregoing analysis could be continued indefinitely, resulting in a series of drift velocities of order $(\omega/\omega_c)^n$. This exemplifies the fact that the results to be obtained in varying $\boldsymbol{E}$ and $\boldsymbol{B}$ fields are approximate only, and represent the leading order effects in a systematic expansion. Higher order corrections are not usually as easily obtained as in this simple case.

1.3.2 *Time-varying magnetic field: constancy of magnetic moment*

Temporal variation of $\boldsymbol{B}$ is associated with spatial variation of $\boldsymbol{E}$ by Maxwell's equation

$$\nabla \times \boldsymbol{E} = -\frac{\partial \boldsymbol{B}}{\partial t}. \tag{1.3.5}$$

We assume $E_{\parallel} = 0$, and $E_{\perp} = 0$ at the guiding-centre position, so that there is no $(\boldsymbol{E} \times \boldsymbol{B})/B^2$ drift. From Eq. (1.2.1) we have

$$\frac{\mathrm{d}}{\mathrm{d}t}\left(\frac{m}{e}\frac{v_{\perp}^2}{2}\right) = \boldsymbol{E}\cdot\boldsymbol{v}_{\perp}. \tag{1.3.6}$$

Thus the change in perpendicular energy in one gyro-period is

$$\delta\left(\frac{1}{2}mv_{\perp}^2\right) = e\oint \boldsymbol{E}\cdot\boldsymbol{v}_{\perp}\,\mathrm{d}t = e\oint \boldsymbol{E}\cdot\mathrm{d}\boldsymbol{l}, \tag{1.3.7}$$

where the loop integral is taken around the gyration orbit. Using Stoke's theorem, and the Maxwell Eq. (1.3.5), this becomes

$$\delta\left(\frac{1}{2}mv_{\perp}^2\right) = e\iint(\nabla \times \boldsymbol{E})\cdot\mathrm{d}\boldsymbol{S} = -e\iint\frac{\partial \boldsymbol{B}}{\partial t}\cdot\mathrm{d}\boldsymbol{S} = e\frac{\partial B}{\partial t}\pi r_{\mathrm{L}}^2$$
$$= \left(\frac{1}{2}mv_{\perp}^2\right)\frac{2\pi}{\omega_{\mathrm{c}}}\frac{1}{B}\frac{\partial B}{\partial t}. \tag{1.3.8}$$

The change in B over one period of gyration is just

$$\delta B = \frac{\partial B}{\partial t}\delta t = \frac{\partial B}{\partial t}\frac{2\pi}{\omega_{\mathrm{c}}}, \tag{1.3.9}$$

so that we finally obtain

$$\delta\left(\frac{1}{2}mv_{\perp}^2\right) = \frac{1}{2}\,mv_{\perp}^2\frac{\delta B}{B}, \tag{1.3.10}$$

showing that

$$\delta\mu = 0 \quad \text{or} \quad \mu \equiv \frac{mv_{\perp}^2}{2B} = \text{constant}, \tag{1.3.11}$$

where μ is the magnetic moment introduced at Eq. (1.2.13).

1.3.3 *Inhomogeneous magnetic field*

Here we assume that $\boldsymbol{E} = 0$ and take a coordinate system (x, y, z) with the z-direction along $\boldsymbol{B}$ at the particle guiding centre. The various spatial gradients appear in the dyadic $\nabla\boldsymbol{B}$:

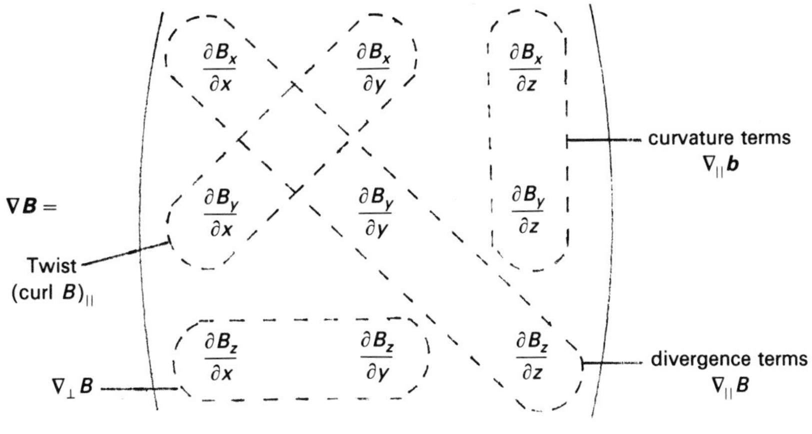

Because of the small Larmor orbit assumption ($r_L \ll$ spatial variation of $\boldsymbol{B}$), the fields experienced by the particle are dominated by B_z. Thus $|\boldsymbol{B}| \simeq B_z$, and $\boldsymbol{b}\cdot\nabla \simeq \partial/\partial z$. Thus the terms $\partial B_z/\partial x$, $\partial B_z/\partial y$ represent transverse gradients of $|\boldsymbol{B}|$, $(\nabla_\perp B)$; the term $\partial B_z/\partial z$ represents a longitudinal gradient of $|\boldsymbol{B}|$, $(\nabla_\| B)$; and the terms $\partial B_x/\partial z$, $\partial B_y/\partial z$ represent change of direction of $\boldsymbol{B}$, curvature. The remaining terms represent twisting, but are unimportant for particle motion.

1.3.4 *Divergence terms*

Taking the z-component (the component parallel to $\boldsymbol{B}$ at the guiding centre) of Eq. (1.2.1) gives

$$\begin{aligned}\frac{m}{e}\dot{v}_z &= v_x B_y - v_y B_x - v_x\frac{\partial B_y}{\partial y}y - v_y\frac{\partial B_x}{\partial x}x \\ &= \frac{v_\perp^2}{\omega_c}\cos^2(\omega_c t + \phi)\frac{\partial B_y}{\partial y} + \frac{v_\perp^2}{\omega_c^2}\sin^2(\omega_c t + \phi)\frac{\partial B_x}{\partial x}\end{aligned} \tag{1.3.12}$$

on using Eq. (1.2.10), and assuming that all other components of $\nabla\boldsymbol{B}$ (except $\partial B_z/\partial z$) are zero. Now averaging over a gyro-period, and using $\nabla\cdot\boldsymbol{B} = 0$,

$$\frac{m}{e}\dot{v}_z = -\frac{K_\perp}{eB}\frac{\partial B_z}{\partial z} \tag{1.3.13}$$

where $K_\perp \equiv \frac{1}{2}mv_\perp^2$ is the perpendicular kinetic energy of the particle. In the absence of an electric field, total energy is conserved so that we also have

$$\begin{aligned}\dot{K}_\perp &= \frac{\mathrm{d}}{\mathrm{d}t}\langle\tfrac{1}{2}mv_\perp^2\rangle = -mv_z\langle\dot{v}_z\rangle \\ &= \frac{K_\perp}{B}\frac{\partial B_z}{\partial z}v_z \simeq \frac{K_\perp}{B}\frac{\mathrm{d}B}{\mathrm{d}t}.\end{aligned} \tag{1.3.14}$$

Hence,

$$\frac{\mathrm{d}}{\mathrm{d}t}\left(\frac{K_\perp}{B}\right) \equiv \frac{\mathrm{d}\mu}{\mathrm{d}t} = 0, \tag{1.3.15}$$

so that once again the magnetic moment remains constant.

As a consequence of this, and the constancy of $K = \frac{1}{2}mv^2$, we have

$$v_\parallel = \left\{\frac{2}{m}[K - \mu B]\right\}^{1/2} \tag{1.3.16}$$

with K, μ constants of the motion. Thus as a charged particle moves along the magnetic field into regions of stronger field B, its longitudinal velocity diminishes, and reflection occurs at a point where $B(\boldsymbol{x}) = (K/\mu)$. This is the magnetic mirror effect.

1.3.5 *Magnetic curvature*

As the guiding centre moves along a curved field line, it experiences a centrifugal force (see Fig. 1.2)

$$\boldsymbol{F} = \frac{mv_\parallel^2}{R_c^2}\boldsymbol{R}_c. \tag{1.3.17}$$

This force is perpendicular to $\boldsymbol{B}$ and can be imagined as an extraneous force as discussed earlier. It therefore gives rise to a drift with velocity

$$\boldsymbol{v}_d = \frac{\boldsymbol{F}\times\boldsymbol{B}}{eB^2} = \frac{mv_\parallel^2}{\omega_c}\frac{\boldsymbol{R}_c\times\boldsymbol{B}}{R_c^2}. \tag{1.3.18}$$

Note that this velocity, known as the curvature drift velocity, depends on e through ω_c, so that it gives rise to charge separation.

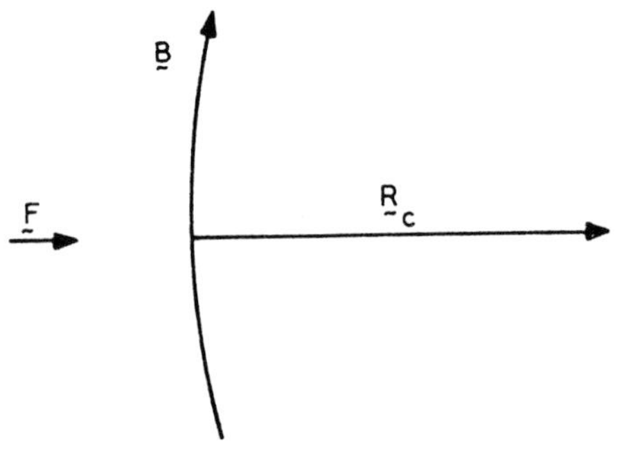

Fig. 1.2. Centrifugal force due to motion along a curved magnetic field line.

1.3.6 *Transverse gradient of field strength*

If $|\boldsymbol{B}|$ varies in the x-direction, the Larmor radius $r_L = \left(\frac{v_\perp}{B}\right)\left(\frac{m}{e}\right)$ varies as the particle gyrates (see Fig. 1.3), being consistently smaller on the right than on the left. The result is a drift in the y-direction, which is the direction of $\boldsymbol{B}\times\nabla B$.

To obtain the magnitude of this drift quantitatively, we return to Eq. (1.2.1) and transform to a frame of reference moving with unknown velocity $\boldsymbol{v}_d$ so that $\boldsymbol{v}_\perp = \boldsymbol{c}_\perp + \boldsymbol{v}_d$, and Taylor-expand $\boldsymbol{B}$ about $x = 0$ to obtain

$$\dot{\boldsymbol{c}}_\perp + \dot{\boldsymbol{v}}_d = \frac{e}{m}\boldsymbol{c}_\perp\times\boldsymbol{B}(0) + \frac{e}{m}\boldsymbol{c}_\perp\times\boldsymbol{B}(0)\left(\frac{1}{B}\frac{\partial B}{\partial x}\right)x + \frac{e}{m}\boldsymbol{v}_d\times\boldsymbol{B}. \tag{1.3.19}$$

Now dropping $(m/e)\dot{\boldsymbol{v}}_d$, which will be of second order in the smallness parameter $(r_L(\partial/\partial x) \ll 1)$, and anticipating the 'gyrating' solution for $\boldsymbol{c}_\perp$ by writing

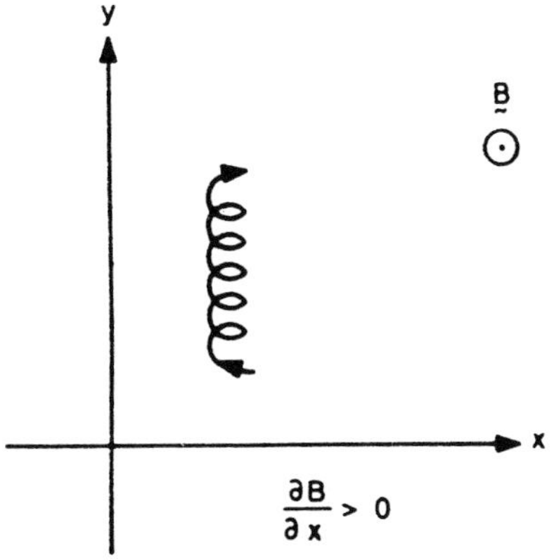

Fig. 1.3. Particle drift in an inhomogeneous magnetic field.

$$x = \frac{c_\perp}{\omega_c}\sin(\omega_c t + \phi), \quad c_x = c_\perp \cos(\omega_c t + \phi), \quad c_y = -c_\perp \sin(\omega_c t + \phi),$$

we find

$$\dot{\boldsymbol{c}}_\perp = \frac{e}{m}\boldsymbol{c}_\perp \times \boldsymbol{B}(0) + \omega_c\left\{-\frac{c_\perp^2}{\omega_c}\frac{1}{B}\frac{\partial B}{\partial x}\sin^2(\omega_c t + \phi) + v_{dy}\right\}\hat{\boldsymbol{e}}_x$$
$$+ \omega_c\left\{-\frac{c_\perp^2}{2\omega_c}\frac{1}{B}\frac{\partial B}{\partial x}\sin 2(\omega_c t + \phi) - v_{dx}\right\}\hat{\boldsymbol{e}}_y. \quad (1.3.20)$$

Because of the rapid variation in the sine terms it is not possible to remove these terms completely by suitable choice of $\boldsymbol{v}_d$ and to consistently neglect $\dot{\boldsymbol{v}}_d$ terms. However, it is the secular nature of the $\sin^2(\omega_c t + \phi)$ which must be removed to preserve the basic oscillatory nature of $\boldsymbol{c}_\perp$. Thus we must take

$$\boldsymbol{v}_d = \frac{1}{2}\frac{c_\perp^2}{\omega_c}\frac{1}{B}\frac{\partial B}{\partial x}\hat{\boldsymbol{e}}_y = \frac{1}{2}\frac{c_\perp^2}{\omega_c}\frac{\boldsymbol{B} \times \nabla B}{B^2} \quad (1.3.21)$$

The remaining small oscillatory terms are then minor perturbations to the gyration.

Both the grad B drift described by Eq. (1.3.21) and the curvature drift are small relative to the basic particle velocity either along $\boldsymbol{B}$, or around $\boldsymbol{B}$ (gyration velocity).

Thus $v_d/v \sim O(r_L/L)$ where L typifies the scale of variation of $B(\boldsymbol{x})$.

1.3.7 *Summary*

In the foregoing, we have seen that the motion of charged particles orbiting in slowly varying electric and magnetic fields may be broken down into two parts, rapid gyration around the guiding centre, and a guiding-centre motion. In this, the adiabatic approximation, the guiding-centre trajectory is essentially along the magnetic field, with a slow drift velocity transverse to $\boldsymbol{B}$. In the absence of electric fields, the appropriate equations are

$$\dot{\boldsymbol{X}}_c = v_\parallel \boldsymbol{b} + v_\parallel^2\frac{\boldsymbol{b} \times \boldsymbol{R}_c}{\omega_c R_c^2} + \mu\frac{\boldsymbol{b} \times \nabla B}{\omega_c}, \quad (1.3.22)$$

$$v_\parallel = \left\{\frac{2}{m}(K - \mu B)\right\}^{1/2}, \quad (1.3.23)$$

where the energy K and the magnetic moment μ are to be held constant in integrating Eq. (1.3.22), and where $\boldsymbol{b} \equiv \boldsymbol{B}/B$. Although Eqs (1.3.22) and (1.3.23) are more complicated than the original equations of motion, their advantage lies in their having averaged out the gyro-motion. For numerical computation they are much more convenient.

1.4 Particle trajectories in tokamak devices

As an elementary example of the application of the foregoing orbit theory, we investigate the nature of electron and ion orbits in a tokamak. The magnetic field of a tokamak is composed of two, axisymmetric, components: a strong toroidal magnetic field, generated by external coils, and a much weaker poloidal magnetic field generated by the toroidal current which flows in the plasma.

The geometry of a typical tokamak is shown in Fig. 1.4, which defines the

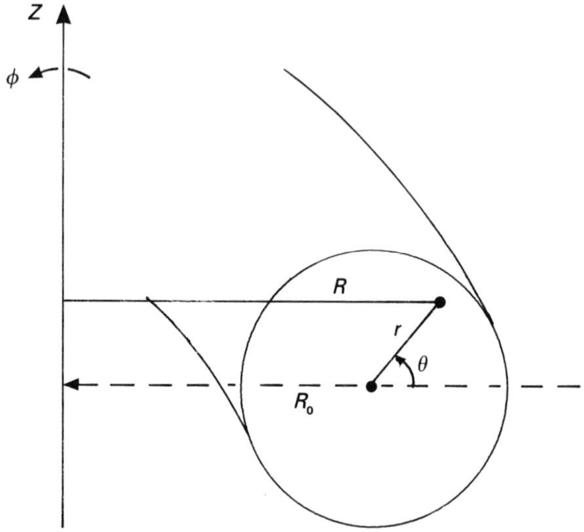

Fig. 1.4. Geometry and coordinates (R, ϕ, Z), (r, θ, ϕ) of a tokamak.

cylindrical coordinates R, ϕ, Z, and the toroidal coordinates (r, θ, ϕ). In these latter coordinates, the magnetic field has the approximate form

$$\boldsymbol{B} = (0, B_\theta(r), B_0)\left(1 - \frac{r}{R_0}\cos\theta\right). \tag{1.4.1}$$

The magnetic field lines lie on the nested surfaces of constant r, and spiral slowly around these surfaces. The rate of spiralling is obtained from

$$\frac{r\,\mathrm{d}\theta}{B_\theta} = \frac{R_0\,\mathrm{d}\phi}{B_0},$$

giving $\mathrm{d}\phi/\mathrm{d}\theta = rB_0/R_0B_\theta(r) = q(r)$, where $q(r)$, known as the winding number (or safety factor), typically lies in the range $0.7 < q(r) < 3$ or 4.

In such magnetic fields, electrons and ions will, to leading order in r_L/R, follow magnetic fields, gyrating around them, with constant energy K and magnetic moment μ. Their guiding-centre motion is described by Eqs (1.3.22) and (1.3.23). Expressing Eq. (1.3.22) in terms of $\dot{r}$, $\dot{\theta}$, $\dot{\phi}$, we find

$$R\dot{\phi} = v_{\parallel}\frac{B_0}{|B|}, \tag{1.4.2}$$

$$r\dot{\theta} = v_{\parallel}\frac{B_\theta}{|B|} + v_{\parallel}\frac{\partial}{\partial r}\left(\frac{v_{\parallel}}{\omega_c}\right), \tag{1.4.3}$$

$$\dot{r} = \frac{v_{\parallel}}{r}\frac{\partial}{\partial \theta}\left(\frac{v_{\parallel}}{\omega_c}\right). \tag{1.4.4}$$

First we note that magnetic mirroring must occur for those particles for which $K/B_0(1 + r/R_0) < \mu < K/B_0(1 - r/R_0)$, since for such particles $v_{\parallel}$ vanishes for some value of θ. Such particles are referred to as trapped particles, and on any given magnetic surface r they constitute a fraction $\sim(2r/R)^{1/2}$ of all the particles in an isotropic distribution. They have a most important influence on many phenomena observed in tokamaks. The remaining particles circulate around the minor and major dimensions with values of $v_{\parallel}$ which are modulated by the magnetic field variations, but which do not change sign.

Next we consider the nature of the guiding-centre trajectories of both trapped and passing particles, projected onto the (r, θ)-plane. Because the fields are axisymmetric, toroidal motion does not affect this projection. Making use of the small Larmor radius expansion r_L/R, and defining $\omega_{c\theta} = eB_\theta/m$ we obtain

$$\frac{dr}{d\theta} = \frac{\partial}{\partial \theta}\left(\frac{v_{\parallel}}{\omega_{c\theta}}\right), \tag{1.4.5}$$

which integrates to give

$$r = r_0 + \left[\frac{v_{\parallel}}{\omega_{c\theta}}(r, \theta) - \frac{v_{\parallel}}{\omega_{c\theta}}(r, 0)\right], \tag{1.4.6}$$

where r_0 is the radial position of the guiding centre as it crosses the mid-plane $(\theta = 0)$. Eq. (1.4.6) shows that, for the bulk of particles, the width of the drift orbit is given by

$$\delta r \sim \frac{\delta v_{\parallel}}{\omega_{c\theta}} \sim q r_L \tag{1.4.7}$$

since $\delta v_{\parallel} \sim \varepsilon v$, where $\varepsilon \equiv r/L \ll 1$. However, for trapped particles (which have $v_{\parallel}(r, 0) \sim \varepsilon^{1/2} v$, and hence $\delta v_{\parallel} \sim \varepsilon^{1/2} v$), we obtain

$$\Delta r \sim q r_L/\varepsilon^{1/2}. \tag{1.4.8}$$

The r, θ projections of these drift orbits are shown in Fig. 1.5(a), whereas Fig. 1.5(b) shows the projection of the equivalent particle motion.

One other feature of the trapped orbits (known, for obvious reasons, as banana orbits) is worthy of note. This is the phenomenon of toroidal precession. Since these particles are confined on the outboard side of the cross-section ($\theta = 0$ side) they experience a cumulative cross-field drift $\dot{\theta}$ ($\propto \cos\theta$) as well as their oscillatory radial drift ($\propto \sin\theta$). Reflection nevertheless always occurs at the same field strength, and therefore poloidal angle $\theta = \theta_c$. The result (shown schematically in Fig. 1.6) is that mirror reflection points of successive reflections advance around the torus, and the whole banana orbit precesses toroidally.

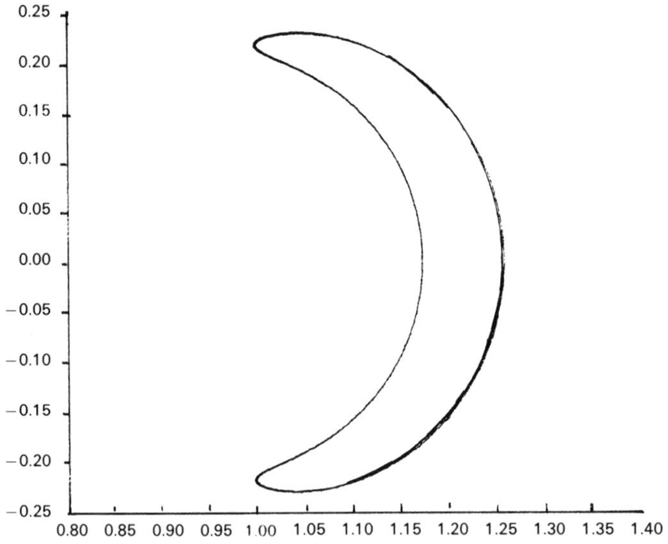

(*a*)

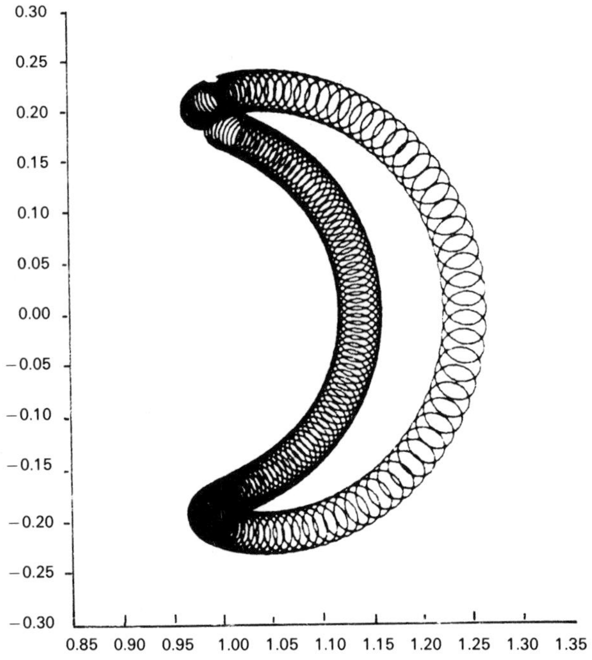

(*b*)

Fig. 1.5. (*a*) Projection of guiding-centre motion of a trapped particle in a simple tokamak field. (*b*) Projection of gyrating motion of a trapped particle in a simple tokamak field. Both the projections are in the (R,Z)-plane.

The effect provides an important mechanism for the resonant release of energy by energetic particles to drive certain low frequency waves which can resonate with the precessional motion. It has been identified as a driving mechanism for Alfvénic fluctuations in tokamaks, and possibly also in the magnetosphere.

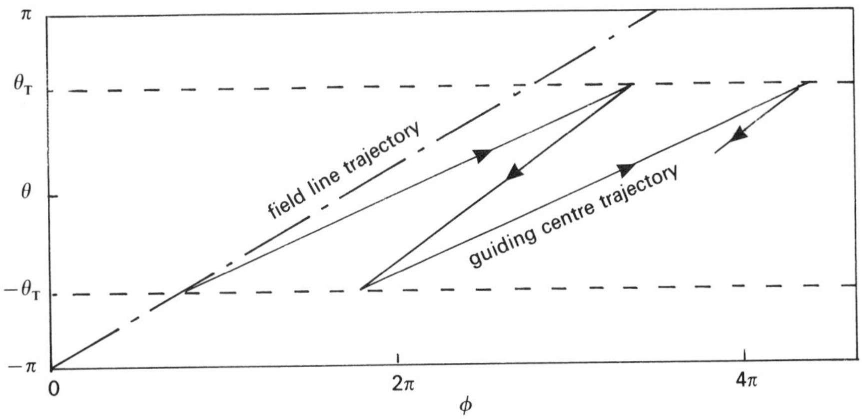

Fig. 1.6. Guiding-centre trajectory showing toroidal precession of a trapped particle reflecting at $\pm\theta_T$.

1.5 Adiabatic invariants

In the previous sections on orbit theory it was shown that the magnetic moment μ is 'approximately invariant' for particles orbiting in slowly varying electric and magnetic fields, but what this means was not made clear. Does μ retain the same value over long times, oscillate about a constant value, or drift slowly away from its initial value? In this section we will examine more precisely the meaning of this question, establish what an adiabatic invariant is, and in what manner exact constancy is violated. The higher order invariants of single-particle motion, the longitudinal invariant J and the flux invariant Φ, will also be investigated and the consequences of their existence outlined.

Returning to the magnetic moment μ, it has been shown (Berkowitz and Gardner, 1959; see also Gardner, 1959, and Kruskal, 1962) that the magnetic moment is the leading term of an asymptotic series in the adiabatic parameter $\varepsilon = r_L/L$, and that this series

$$\hat{\mu}(\varepsilon) = \mu + \varepsilon\mu_1 + \varepsilon^2\mu_2 + \cdots \tag{1.5.1}$$

is constant to all orders in ε. This does not mean that $\hat{\mu}$ is an exact constant but that

$$\lim_{\varepsilon\to 0}\{(\hat{\mu} - \text{const})/\varepsilon^N\} = 0 \quad \text{for any } N. \tag{1.5.2}$$

Hence the deviation from constancy, $\Delta\mu = (\hat{\mu} - \text{const})$, goes to zero faster than any power of ε as $\varepsilon \to 0$.

We will outline in the following a method for calculating the higher order corrections $\mu_1(r, v)$, $\mu_2(r, v)$, in the invariant series, and discover the existence of second and third invariants, but at this stage it is instructive to examine the results of numerical integration of exact particle orbits.

We consider proton trajectories in a linear quadrupole magnetic field, generated by two current carrying wires (see Fig. 1.7).

Fig. 1.8 shows plots of μ, $\hat{\mu}_1 \equiv \mu + \varepsilon\mu_1$ and $\hat{\mu}_2 = \mu + \varepsilon\mu_1 + \varepsilon^2\mu_2$, against time for a charged particle in a linear multipole magnetic field (Hastie, Hobbs and Taylor, 1969). The oscillations of μ increase in amplitude as the particle passes

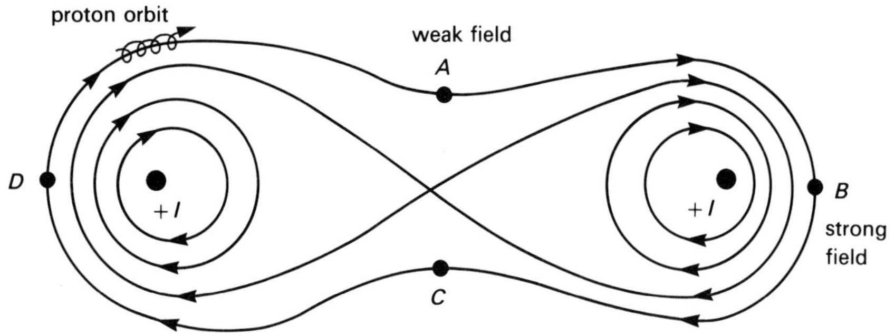

Fig. 1.7. Quadrupole magnetic field lines, generated by two linear conductors carrying current I.

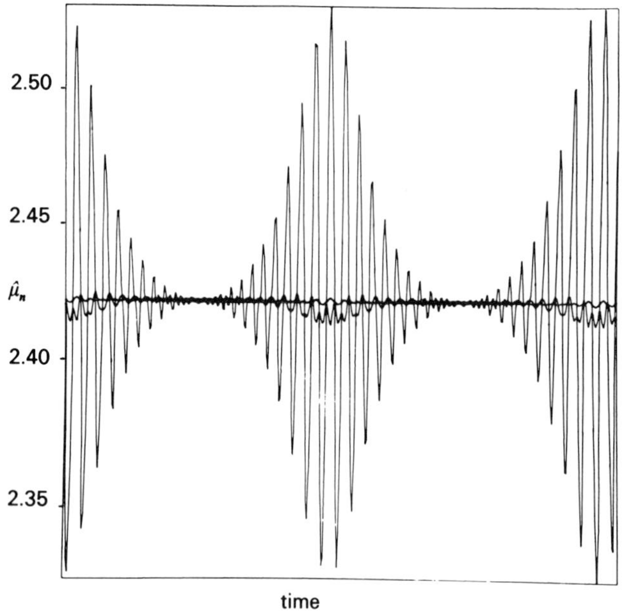

Fig. 1.8. Variation of $\hat{\mu}_0 \equiv v_\perp^2/2B$, $\hat{\mu}_1 \equiv \mu_0 + \varepsilon\mu_1$, and $\hat{\mu}_2 \equiv \mu_0 + \varepsilon\mu_1 + \varepsilon^2\mu_2$ as a function of time for charged particles in a quadrupole magnetic field. The large oscillations are those of μ_0. (From Hastie, Hobbs and Taylor, 1969.)

through the weak field region (where its Larmor radius and value of ε are largest) but the quantities $\hat{\mu}_1$ and $\hat{\mu}_2$ appear to be effectively constant.

In Fig. 1.9 a similar plot of $\hat{\mu}_N$ against time is shown, but for a particle of higher energy. Now, although $\hat{\mu}$ appears to be effectively constant over most of the trajectory, as the particle passes through the weak field region an easily measured jump $\Delta\mu$ takes place. Fig. 1.10 shows the dependence of this jump $\Delta\mu(\varepsilon)$ on ε, with $\ln \Delta\mu$ being a linear function of ε; thus

$$\Delta\mu \propto \exp(-\alpha/\varepsilon), \tag{1.5.3}$$

where α may be a function of the initial value of μ, and will depend on the structure of the magnetic field. Note that this form for $\Delta\mu \equiv (\hat{\mu} - \text{const})$ is entirely consistent with Eq. (1.5.2) and the theorem of Berkowitz and Gardner (1959), since $\exp(-\alpha/\varepsilon) \to 0$ as $\varepsilon \to 0$ faster than any power of ε.

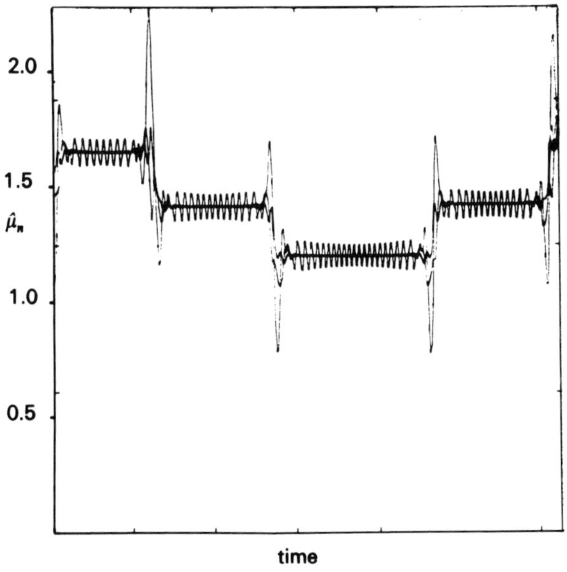

Fig. 1.9. Variation of $\hat{\mu}_0$, $\hat{\mu}_1$ and $\hat{\mu}_2$ for a particle of higher energy in the same magnetic field. The distinct jumps $\Delta\mu$ occur as the particle passes the weaker magnetic field. (From Hastie, Hobbs and Taylor, 1969.)

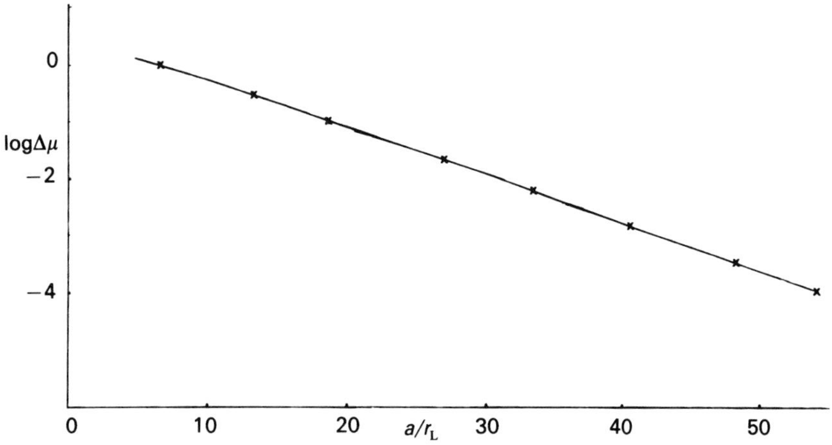

Fig. 1.10. Dependence of the nonadiabatic jumps, $\Delta\mu$, on adiabaticity parameter $\varepsilon = r_L/R$. (From Hastie, Hobbs and Taylor, 1969.)

1.5.1 *Consequences of nonadiabatic behaviour*

In Fig. 1.11 we follow a particle with significant nonadiabatic behaviour through many transits of the weak field region. Initially the particle is mirror-trapped with $\mu(t = 0) > K/B_{max}$, but its magnetic moment is 'scattered' on each transit and appears to perform a random walk until $\mu B_{max} < K$, at which point the particle has entered the so-called loss cone, and now passes around the closed magnetic field lines.

Few containment devices in thermonuclear fusion research are likely to experience nonadiabatic phenomena. However, reactor designs are required to confine the 3.5 MeV alpha particles produced by deuterium–tritium fusion for long enough to allow them to give up their energy to the reacting plasma, and

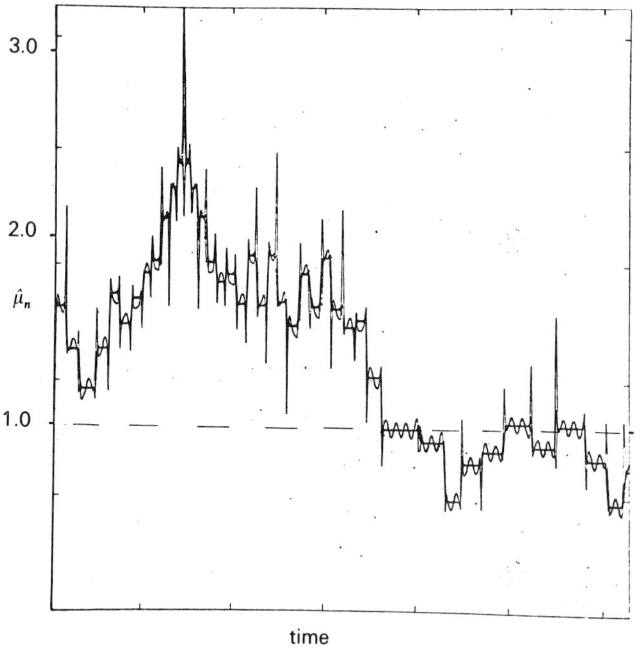

Fig. 1.11. Variation of μ_n for a nonadiabatic particle in a quadrupole magnetic field. For $\mu_0 > 1$ the particle is mirrored between regions of strong field. For $\mu_0 < 1$ it passes around the complete closed magnetic field line. Transitions between the trapped and circulating states occur. (From Hastie, Hobbs and Taylor, 1969.)

nonadiabaticity of the second invariant J provides a possible loss mechanism at such high energies.

1.5.2 *Invariant series: the higher order corrections*

Having examined briefly the way in which the magnetic moment invariant breaks down, we return to the calculation of the higher order corrections to μ, and of the other invariants in general. The method we use is important since it also has application in the analysis of micro-instabilities.

To obtain the most general invariant, I, together with its higher order corrections we solve

$$\frac{\mathrm{d}I}{\mathrm{d}t} = 0,$$

or equivalently

$$\boldsymbol{v}\cdot\nabla I + \frac{e}{m}(\boldsymbol{v}\times\boldsymbol{B})\cdot\frac{\partial I}{\partial \boldsymbol{v}} = 0, \tag{1.5.4}$$

by expansion in powers of ε

$$I = I_0 + \varepsilon I_1 + \varepsilon^2 I_2 + \cdots \tag{1.5.5}$$

noting that the second term of Eq. (1.5.4) is $\mathrm{O}(1/\varepsilon)$ relative to the first.

We note first that Eq. (1.5.4) is just the Vlasov equation or collisionless kinetic equation for the distribution function $f(\boldsymbol{x}, \boldsymbol{v})$ for charged particles in the given fields, so that solutions of Eq. (1.5.4) will have obvious application in kinetic theory.

Since we consider the simplest case of no electric field, the energy $K = \frac{1}{2}mv^2$ is an exact constant, and it is convenient to transform to the velocity variables (μ, K, ϕ), where ϕ is the phase angle of the gyro-motion. Thus

$$v = v_{\parallel} \boldsymbol{b} + v_{\perp} \cos \phi \boldsymbol{e} + v_{\perp} \sin \phi \boldsymbol{e}' \tag{1.5.6}$$

with

$$v_{\parallel} \equiv \left[\frac{2}{m}(K - \mu B)\right]^{1/2}, \quad v_{\perp} = \left[\frac{2\mu B}{m}\right]^{1/2} \tag{1.5.7}$$

and $\boldsymbol{e}$ and $\boldsymbol{e}'$ are unit vectors normal to $\boldsymbol{b}$ and to each other.

In real space it is convenient to use a coordinate system associated with the magnetic field structure. Thus we take (α, β, l) where

$$\boldsymbol{B} = \nabla\alpha \times \nabla\beta \tag{1.5.8}$$

and l measures arc length along $\boldsymbol{B}$.

In these coordinates Eq. (1.5.4) takes the form

$$\omega_c \frac{\partial I}{\partial \phi} = \mathscr{D} I, \tag{1.5.9}$$

where $\mathscr{D}$ is a complicated differential operator (Hastie, Taylor and Haas, 1967) with the properties that $\oint \mathrm{d}\phi \mathscr{D} \equiv \langle \mathscr{D} \rangle = v_{\parallel} (\partial/\partial l)$, and that $\mathscr{D}$ is $O(\varepsilon)$ relative to $\omega_c(\partial/\partial\phi)$.

Thus inserting the expansion Eq. (1.5.5) into Eq. (1.5.9), we obtain in lowest order

$$\omega_c \frac{\partial I_0}{\partial \phi} = 0, \tag{1.5.10}$$

with solution $I_0 = I_0(\mu, K, \alpha, \beta, l)$. In effect this equation states that any function of the variables μ, K, α, β, l remains constant on the time scale of ω_c^{-1}.

In next order one obtains

$$\omega_c \frac{\partial I_1}{\partial \phi} = \mathscr{D} I_0. \tag{1.5.11}$$

Now since I_1 must be a periodic function of ϕ, there is an integrability condition associated with Eq. (1.5.11). This is obtained by annihilating the $\partial I_1/\partial\phi$ by integrating over a period of ϕ:

$$\langle \mathscr{D} I_0 \rangle = 0,$$

hence

$$v_{\parallel} \frac{\partial I_0}{\partial l} = 0. \tag{1.5.12}$$

Thus although Eq. (1.5.11) appears to be an equation for I_1, it also contains further information on the functional dependence of I_0, which must now be of the form

$$I_0 = I_0(\mu, K, \alpha, \beta).$$

Solving Eq. (1.5.11) for I_1 now gives

$$I_1 = \frac{1}{\omega_c}\int^{\phi} \mathscr{D}I_0\,\mathrm{d}\phi + \bar{I}_1(\mu, K, \alpha, \beta, l), \tag{1.5.13}$$

where $\bar{I}_1$ is, as yet, unknown.

Thus if we had chosen I_0 to be the magnetic moment, μ, Eq. (1.5.13) gives the oscillatory (in ϕ) part of the first correction μ_1 explicitly, although the complete first correction μ_1 is not yet fully determined.

Proceeding to the next order, we find the equation

$$\omega_c \frac{\partial I_2}{\partial \phi} = \mathscr{D}I_1. \tag{1.5.14}$$

Clearly in this order we begin to construct I_2, the second order correction to the invariant I_0, but as before there is an integrability condition obtained by annihilating the left hand side:

$$\langle \mathscr{D}I_1 \rangle \equiv \langle \mathscr{D}\bar{I}_1 \rangle + \left\langle \mathscr{D}\frac{1}{\omega_c}\int^{\phi} \mathscr{D}I_0 \right\rangle = 0. \tag{1.5.15}$$

Explicit evaluation of the I_0 term in Eq. (1.5.15) gives the following result:

$$v_{\parallel}\frac{\partial \bar{I}_1}{\partial l} + \boldsymbol{v}_d \cdot \nabla I_0 = 0 \tag{1.5.16}$$

with the velocity $\boldsymbol{v}_d$ being precisely the sum of the two transverse guiding-centre drift velocities, the $\nabla \boldsymbol{B}$ drift and the curvature drift velocity,

$$\boldsymbol{v}_d = v_{\parallel}^2 \frac{\boldsymbol{b} \times \boldsymbol{R}_c}{\omega_c R_c^2} + \mu B \frac{\boldsymbol{b} \times \nabla B}{\omega_c B}. \tag{1.5.17}$$

Eq. (1.5.16) now determines the hitherto undetermined part of I_1. However, if the longitudinal motion of the particle is periodic, Eq. (1.5.16) also has an integrability condition, since the $\bar{I}_1$ term can be annihilated by integrating $\oint \mathrm{d}l/v_{\parallel}$ round the closed orbit. Periodicity in l may arise in two ways; either (i) because the magnetic field lines are themselves closed loops, or (ii) because mirror reflection returns the particle to its initial position. In either case the constraint equation takes the form

$$\oint \frac{\mathrm{d}l}{v_{\parallel}} v_d \cdot \nabla I_0 = 0, \tag{1.5.18}$$

where I_0 has already been constrained by Eqs (1.5.16) and (1.5.12) to have the functional form $I_0 = I_0(\mu, K, \alpha, \beta)$.

To see the meaning of Eq. (1.5.18), we consider the particular case of a vacuum magnetic field. In this case $\boldsymbol{B}$ may be written as the gradient of a scalar potential

$$\boldsymbol{B} = \nabla \chi,$$

where the surfaces of constant χ are orthogonal to the surfaces of α and β ($\nabla\alpha \cdot \nabla\chi = \nabla\beta \cdot \nabla\chi = 0$), while the curvature is

$$\frac{\boldsymbol{R}_c}{R_c^2} = \frac{\nabla_{\perp} B}{B}.$$

Using these results Eq. (1.5.18) takes the form

$$\oint \frac{\mathrm{d}\chi}{B^2} \boldsymbol{B} \times \nabla\left(\frac{v_{\|}}{B}\right) \cdot \nabla I_0 = 0, \tag{1.5.19}$$

with $v_{\|}$ given by Eq. (1.5.7). Now expressing the gradients as

$$\nabla = \nabla\alpha \frac{\partial}{\partial \alpha} + \nabla\beta \frac{\partial}{\partial \beta} + \nabla\chi \frac{\partial}{\partial \chi} \tag{1.5.20}$$

using $\boldsymbol{B} = \nabla\alpha \times \nabla\beta$, and the orthogonality of the χ surfaces, this becomes

$$\frac{\partial I_0}{\partial \alpha}\frac{\partial J}{\partial \beta} - \frac{\partial I_0}{\partial \beta}\frac{\partial J}{\partial \alpha} = 0, \tag{1.5.21}$$

where

$$J \equiv \oint \mathrm{d}\chi \frac{v_{\|}}{B} \equiv \oint v_{\|}\, \mathrm{d}l. \tag{1.5.22}$$

The constraint equation Eq. (1.5.21) therefore requires that I_0 be of the form

$$I_0 = I_0(\mu, K, J), \tag{1.5.23}$$

where I_0 is the leading term of the most general form of invariant. Proceeding to higher order, it transpires that no new constraints on I_0 appear. Now, by taking I_0 identically equal to μ or J, and evaluating the higher order terms I_1, I_2, we may construct the asymptotic series for the adiabatic invariants. The corrections μ_1, μ_2, and J_1 have all been constructed by this method. In addition, higher order guiding-centre velocities can be obtained from equations analogous to Eq. (1.5.16), i.e.

$$\sum_{j=0}^{l} \boldsymbol{v}_j \cdot \nabla I_{l-j} = 0. \tag{1.5.24}$$

1.5.3 *Importance of the longitudinal invariant J*

The importance of the longitudinal invariant is that, when it exists, the particle is constrained to drift on the surface described by

$$J(\mu, K, \alpha, \beta) = \text{const.}, \tag{1.5.25}$$

so that much is learned about the particle trajectory over very long times without the need to integrate the equations of motion either of the gyrating particle or of its guiding centre.

1.5.4 *The flux invariant Φ*

If the drift surfaces described by Eq. (1.5.21) are closed, the slow drift motion is also periodic, and yet another invariant exists. For the stationary fields we have been considering, this additional invariant is just the particle energy, but if slow time dependence of the fields is introduced so that energy is not conserved, it is the flux invariant $\boldsymbol{\Phi}$ (the flux contained within the drift surface), which is an adiabatic invariant of the motion.

1.6 The effect of Coulomb collisions on charged particle trajectories

The effect of a Coulomb collisional encounter resulting in a large-angle scatter of the gyro-angle of a charged particle is shown in Fig. 1.12. The centre of gyration, or guiding centre, of the particle trajectory experiences a discontinuous jump (from A to A' in Fig. 1.12) a distance of order r_L.

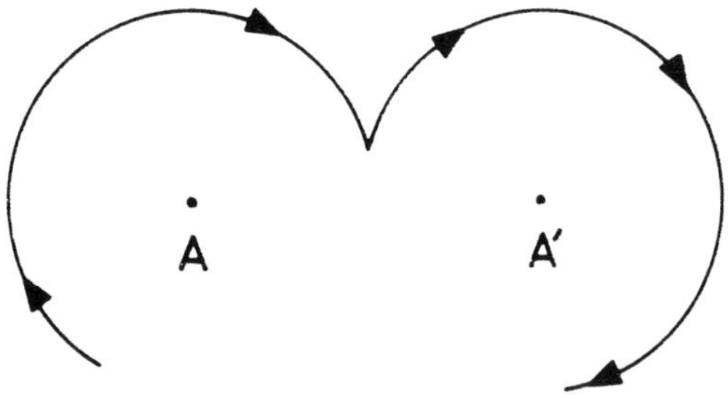

Fig. 1.12. Effect of large-angle scattering on the motion of a charged particle in a magnetic field.

Now considering the cumulative effect of such collisions on a charged test particle, one can see that the particle experiences a random walk in which characteristic displacements of order r_L occur with frequency ν_c, the collision frequency. This results in spatial diffusion carrying the particle across the magnetic field with diffusion coefficient $D_\perp$ such that

$$D_\perp \sim \nu_c r_L^2. \tag{1.6.1}$$

This diffusion, which scales as B^{-2}, is known as classical diffusion. Because the characteristic collision frequency for electrons is only $(m_i/m_e)^{1/2}$ larger than for ions, it follows that $D_{\perp e} \sim D_{\perp i}(m_e/m_i)^{1/2} \ll D_{\perp i}$. Since unequal diffusion rates would generate large electric fields, in practice charge separation induces electric fields to develop which restrain the ions, and both charged species diffuse across the magnetic field at the slower rate characteristic of electrons.

An interesting phenomenon occurs at the very low collision frequencies found in very hot or tenuous plasmas. We have seen that, in addition to their Larmor gyration, electrons and ions perform a slow drift motion (the guiding-centre drift) across the magnetic field. The trajectories of this drift motion may also be closed, so that a particle may drift around some mean position in much the same way as it gyrates (on a smaller scale and faster time scale) around its guiding centre. (The banana orbits of magnetically trapped particles in tokamaks provide one example of such behaviour.) However, particles of different magnetic moment μ may have quite different drift orbits, and so a Coulomb collisional encounter which significantly alters μ may result in the guiding centre setting off along a totally new drift orbit. This is shown schematically in Fig. 1.13, where $|A - A'| = r_D$ may greatly exceed r_L. The appropriate diffusion coefficient for this process is now

$$D'_\perp = \nu' r_L^2. \tag{1.6.2}$$

In tokamaks the so-called neoclassical diffusion is of this type, and at low collision frequencies, $\nu' \sim R/r\nu_c$ and $r_D \sim qr_L(R/r)^{1/2} \gg r_L$ (where R and r refer to the major and minor radii of the toroidal surfaces, respectively). Evidently neoclassical transport in tokamaks exceeds the classical value in magnitude, but still scales as B^{-2}.

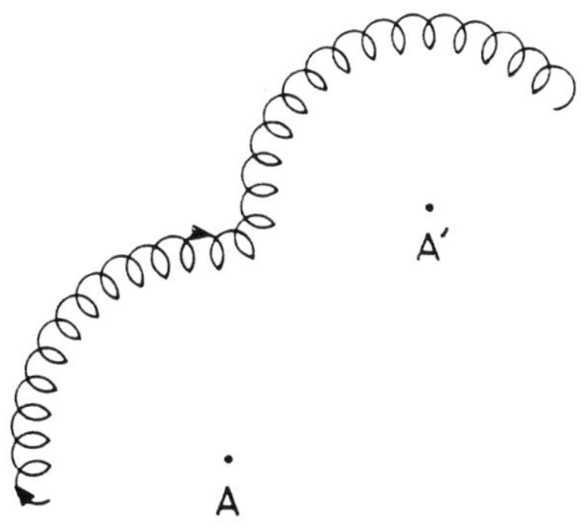

Fig. 1.13. Effect of large-angle scattering in the guiding-centre motion.

In practice, the diffusion observed in tokamaks is much larger (by two orders of magnitude) than the neoclassical value. The reason for this has not yet been resolved, but one possible mechanism for the anomalously fast diffusion may be visualized in terms of orbit theory, as follows. Tokamak plasmas are not completely quiescent. A low level of electrostatic activity is generally present. These fluctuations typically are of low frequency (small compared to the transit frequency of electrons between magnetic mirrors), long parallel wavelength (of the order of the connection length along a field line from weak to strong field), and have perpendicular wavelengths of the order of the ion Larmor radius.

Consequently the fluctuations appear relatively static to the rapidly moving electrons, and the magnetically trapped electrons may be trapped in a spatial region where the phase of the wave does not change. This is represented schematically in Fig. 1.14. The figure shows how a trapped electron, with mirror points at l_1 and l_2, is confined to a region in which the guiding-centre drift induced by an electrostatic wave

$$V = \frac{\boldsymbol{E} \times \boldsymbol{B}}{B^2} \sim -(\boldsymbol{k} \times \boldsymbol{B})\frac{\phi}{B^2} \tag{1.6.3}$$

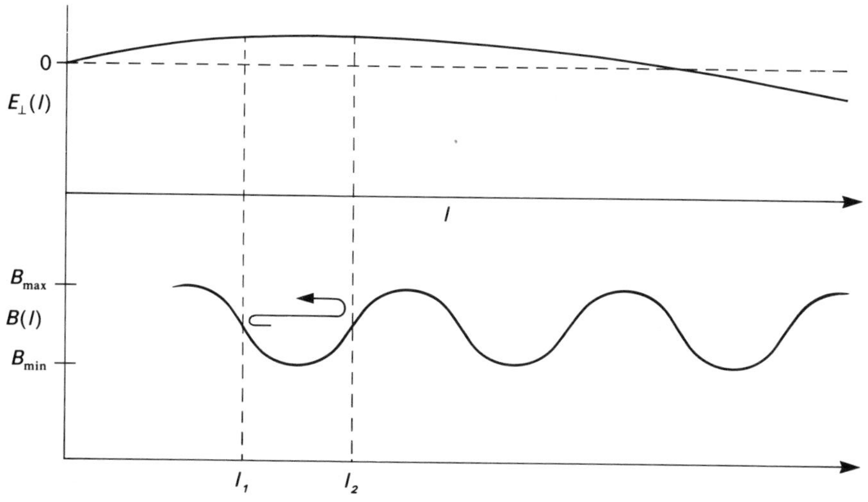

Fig. 1.14. Effect of magnetic trapping (between l_1 and l_2) in the presence of a quasistationary, long wavelength electric field. $V_E = \boldsymbol{E} \times \boldsymbol{B}/B^2$ does not oscillate so $\int V_E \, dt$ accumulates.

remains in the same direction for a time of order ω^{-1}. The distance covered by an electron in this time is therefore

$$\Delta r \sim \frac{k\phi}{\omega B} \sim \frac{e\phi}{T}\frac{T}{m_i\omega_{ci}}\frac{k}{\omega} \simeq \frac{e\phi}{T}\left(\frac{v_{Ti}}{\omega}\right)(kr_{Li}), \tag{1.6.4}$$

This distance may greatly exceed both the electron Larmor radius r_{Le} and the electron Banana width $qr_{Le}(R/r)^{1/2}$. Thus trapped electrons in tokamaks should be thought of as guiding centres attempting to execute quite large drift orbit excursions under the influence of low frequency electrostatic oscillations. In practice, the time for which such cross-field excursions continue may be determined by Coulomb collisions (which detrap the electrons), by the decorrelation time of the wave, or by the ∇B and curvature drifts of the electron (which can carry it through a perpendicular wavelength of the electrostatic wave).

Of course other mechanisms may be more important, but this example serves to illustrate the usefulness of simple particle trajectory considerations when seeking a physical picture for some familiar observations in plasma physics.

1.7 The effect of electromagnetic radiation on charged particle trajectories

Throughout the foregoing discussion, we have ignored all radiation processes. As a consequence the particle energy was found to be an exact constant of particle motion for electrons and ions moving in static fields if $\boldsymbol{E}\cdot\boldsymbol{B} = 0$, and the magnetic moment was found to be an approximate constant of motion. However, since an accelerating charge radiates energy in the form of electromagnetic waves, it loses energy at the rate (see Bekefi, 1966)

$$\frac{\mathrm{d}K}{\mathrm{d}t} = \frac{e^2}{6\pi\varepsilon_0 mc^3}a^2, \tag{1.7.1}$$

where a is the acceleration. As we have seen, all charged particles are being continuously accelerated by the magnetic field with $\boldsymbol{a} = (e/m)\boldsymbol{v}\times\boldsymbol{B}$.

The resulting radiation is termed cyclotron radiation (sometimes referred to as magnetic bremsstrahlung, and as synchrotron radiation if relativistic energies are involved), and causes a gradual energy loss. Thus, integrating Eq. (1.7.1) we obtain the result

$$K_\perp = K_{\perp 0}\exp(-t/\tau_R), \tag{1.7.2}$$

where $K_\perp = \frac{1}{2}mv_\perp^2$ is the energy of gyration of the particle and the characteristic time τ_R for radiative energy loss is

$$\tau_R = 3\pi\varepsilon_0 mc^3/e^2\omega_c^2. \tag{1.7.3}$$

Since τ_R scales as m^3, this process can only be of significance for electrons. For the plasmas of typical fusion devices τ_R lies in the range 1–10 s, and is therefore considerably larger than other characteristic times associated with the particle orbit.

In addition, these expressions overestimate the rate of energy loss of an electron, since local absorption of the radiation from neighbouring electrons also takes place.

This inverse process is exploited in radio-frequency heating schemes in fusion devices, where intense radio-frequency waves at the cyclotron frequency of electrons or ions, or at a low harmonic of this frequency, are launched into the plasma and are resonantly absorbed by the gyrating particles.

1.8 Collective behaviour and Debye screening

Up to this point we have considered the motion of a single particle moving in known electromagnetic fields. In discussing collisions, we noted that each individual particle interacts with other charged particles, and this was discussed in terms of a binary collisional event resulting in a deflection of the particle orbit. We also considered, briefly, the role of fluctuating electric fields which might, for example, be generated by the collective motion of electrons and ions in an unstable plasma. However, there exists another type of collective behaviour which is of fundamental importance in a plasma. This is known as Debye screening. If a positive test charge of magnitude Ze is placed in a plasma, it attracts electons and repels ions in such a way that its Coulomb electrostatic potential $\phi_c \approx Ze/4\pi\varepsilon_0 r$ is attenuated at distances beyond a Debye length. To calculate this effect, we solve for the potential $\phi(r)$ generated by such a test charge. Assuming the plasma to be in thermal equilibrium, the distribution functions of electrons and ions are of the Maxwell–Boltzmann form

$$f(\boldsymbol{x}, \boldsymbol{v}) = n_0 \exp\left(-\frac{mv^2}{2k_B T} + \frac{e_j\phi}{k_B T}\right), \tag{1.8.1}$$

and the densities are $n_j(r) = n_0 \exp(e_j\phi(r)/k_B T)$. Here $\phi(r)$ is the potential generated by the test charge, which is as yet unknown. Since this potential must satisfy Poisson's equation

$$\nabla^2\phi = \frac{1}{\varepsilon_0}\rho(r), \tag{1.8.2}$$

with the charge density $\rho(r) = \sum_j e_j n_j(r)$, it follows that, assuming spherical symmetry, ϕ satisfies the equation

$$\frac{1}{r^2}\frac{\mathrm{d}}{\mathrm{d}r} r^2 \frac{\mathrm{d}\phi}{\mathrm{d}r} = \frac{2n_0 e^2}{\varepsilon_0 k_B T}\phi; \tag{1.8.3}$$

here we have assumed that the potential is small enough that $e\phi/k_B T \ll 1$.

Taking the solution of Eq. (1.8.3) which vanishes as $r \to \infty$, we obtain

$$\phi = \frac{A}{r}\exp(-r/\lambda_D), \tag{1.8.4}$$

where $\lambda_D \equiv (\varepsilon_0 kT/2n_0 e^2)^{1/2}$ is known as the Debye length, and A is not yet determined. To evaluate the constant A, we must match the potential to the 'bare' Coulomb potential of the test charge, $\phi_c = Ze/4\pi\varepsilon_0 r$, at a distance r from the charge which is small compared to the average interparticle distance $n_0^{-1/3}$. The result is that $A = Ze/4\pi\varepsilon_0$, provided that $n_0^{-1/3} \ll \lambda_D$. Eq. (1.8.4) then shows that, at distances greater than a Debye length, the potential of a test charge in a plasma is exponentially attenuated below the value it would have in a vacuum. This cutoff of the potential has important implications for the collisional events in a plasma, limiting them to interactions between charged particles which approach within a

Debye length of each other. It also follows that each charged particle in a plasma is simultaneously undergoing many collisional encounters with all the particles in its Debye sphere (the sphere of radius λ_D surrounding the particle), a number of order $n_0\lambda_D^3$. For most laboratory and space plasmas, this is a very large number. In typical tokamaks, for example, λ_D is of similar magnitude to the electron gyro-radius $\simeq 4 \times 10^{-5}$ m, while the average interparticle distance is $n_e^{-1/3} \simeq 3 \times 10^{-7}$ m.

A modification to Debye screening might be expected in the limit of very strong magnetic fields, when the electron gyro-radius $r_{Le} \ll \lambda_D$, so that electrons behave as guiding centres tied to the magnetic field, rather than as 'free' particles moving in the potential of neighbouring charges. Since guiding centres are not free to 'cluster around' a test charge, it might be conjectured that Debye screening is modified or impaired in this limit. However, screening of a test charge still occurs. This problem provides an instructive application for guiding-centre theory.

1.9 Summary

In this chapter we have outlined the theory of charged particle motion in electromagnetic fields and shown how, when the inhomogeneities in these fields are relatively weak, different characteristic times can be distinguished naturally, and approximate constants of motion (the adiabatic invariants) can be constructed. The importance of these adiabatic invariants lies in the fact that, when use is made of them, predictions can be made about the nature of particle trajectories over very long periods of time.

The way in which these adiabatic invariants vary in time was investigated, as was the effect upon them of Coulomb collisions and radiative processes, and a heuristic description of the mechanism underlying collisional transport across magnetic fields was presented. One possible mechanism for enhanced transport in fluctuating electrostatic fields was described. Finally we described, briefly, the most fundamental aspect of the collective behaviour of the charged particles comprising a plasma, which is the phenomenon of Debye screening.

Many complex phenomena can be approached using this basic picture of a plasma, based on particle dynamics.

References and bibliography

BEKEFI, G. (1966). *Radiation Processes in Plasmas.* Wiley, New York.

BERKOWITZ, J. and GARDNER, C. (1959). *Comm. Pure Appl. Math.*, **12**, 501.

BOGOLYUBOV, N. N. and MITROPOLSKII, Y. A. (1961). *Asymptotic Methods in the Theory of Nonlinear Oscillations.* Gordon and Breach, New York.

BOGOLYUBOV, N. N. and ZUBAREV, D. N. (1955). *Ukrainian Math. J.*, **7**, 5.

BRAGINSKII, S. I. (1956). *Ukrainian Math. J.*, **8**, 119.

GARDNER, C. (1959). *Phys. Rev.*, **115**, 791.

HASTIE, R. J., HOBBS, G. D. and TAYLOR, J. B. (1969). *3rd Int. Conf. on Plasma Physics and Contr. Nucl. Fus. Research*, Novosibirsk, paper CN-24 # C6, p. 389. IAEA, Vienna.

HASTIE, R. J., TAYLOR, J. B. and HAAS, F. A. (1967). *Ann. Phys. (U.S.A.)*, **41**, 302.

KRUSKAL, M. (1962). *J. Math. Phys.*, **3**, 806.

NORTHROP, T. G. (1963). *The Adiabatic Motion of Charged Particles.* Interscience, New York.

SIVUKHIN, D. V. (1965). In *Reviews of Plasma Physics*, vol. 1, p. 1. (M. A. Leontovich, ed.). Consultants Bureau, New York.

2

Plasma kinetic theory

J. A. ELLIOTT

2.1 Introduction

There are three basic approaches to the theory of plasmas. The first one met by most students of plasma physics is the description in terms of particle orbit theory, and this has been set out in Chapter 1. Each particle is considered individually, and its motion followed under the influence of the Lorentz force

$$\boldsymbol{F} = m\frac{\mathrm{d}\boldsymbol{v}}{\mathrm{d}t} = q(\boldsymbol{E} + (\boldsymbol{v} \times \boldsymbol{B})). \tag{2.1.1}$$

This approach is very valuable for gaining physical insight into plasma behaviour. We discover that particles move on helical trajectories, along field lines; we discover a wide variety of mechanisms causing particles to drift across the magnetic field, and the physical origin is revealed of currents which appear in magnetically confined plasma. But the model tacitly assumes the presence of, for example, an electrical field $\boldsymbol{E}$. It offers no explanation of the origin of such a field, and indeed we have seen in the physics of Debye shielding that such a field cannot simply be imposed by external means. It often arises self-consistently from the cooperative motion of all the plasma charges, and cannot be described by the single-particle approach. An alternative way of presenting the limitations of the single-particle model is to observe that the variables are particle position and velocity: useful concepts, but not measurable or even knowable, and therefore not relatable to experiment. We wish to deal with measurable variables such as fluid velocity and particle density, and these cannot be found from a single-particle model, since they depend in a very complex manner on the motion of the individual particles.

For these reasons, most plasma physics problems are handled using a fluid model, in which the plasma is described by a modified set of fluid equations, incorporating the electromagnetic forces which are so central to plasma behaviour. These fluid equations are not self-evident, however, and must be derived on some firm and reliable basis. That basis is the subject of this chapter.

The kinetic theory of plasma is the most fundamental description of the plasma state; we shall develop the essential core of the approach and show how the fluid equations can be derived, and, very importantly, their reliability assessed. We shall define distribution functions for the particle species, and show how measurable variables like density, velocity and temperature can be obtained by averaging over the distribution.

In case these objectives appear a little dull, however virtuous, we shall then proceed to show that the kinetic theory is much more than a necessary foundation for the practical, and useful, fluid approach. It assumes very great significance and importance due to the low collisionality of many plasmas. Indeed, most of the major phenomena characterizing the plasma state appear in collisionless plasmas, where the cooperative motion causing fluid-like behaviour is not due to collisions (as in a gas) but due to electromagnetic coupling of the particles. It is quite common, therefore, for particle species in a plasma to have distinctly non-Maxwellian distribution functions, the physical effects of which cannot be described using a fluid theory. We shall give some examples of this important feature later in this chapter, and reveal some physics which, when first discovered, was quite startlingly new.

2.2 The distribution function

How do we begin to describe a plasma in detail? If there are N particles, and each particle has a position and velocity $\boldsymbol{x}_i$ and $\boldsymbol{v}_i$ $(i = 0, \ldots, N)$, then the total number of coordinates is $6N$. The actual configuration of the system at any given time is represented by a single point in the $6N$-dimensional space $(\boldsymbol{x}_1 \ldots \boldsymbol{x}_N, \boldsymbol{v}_1 \ldots \boldsymbol{v}_N)$. Some advanced texts (for example, Clemmow and Dougherty, 1969) adopt this as the way into the problem.

The decription has, of course, to be simplified at some stage, and we shall do that now. We may reduce the description down to six dimensions $(\boldsymbol{x}, \boldsymbol{v})$, the coordinate space for a single particle, and define a distribution function $f(\boldsymbol{x}, \boldsymbol{v}, t)$. Using this, we find that

$$f(\boldsymbol{x}, \boldsymbol{v}, t)\,\mathrm{d}x\,\mathrm{d}y\,\mathrm{d}z\,\mathrm{d}v_x\,\mathrm{d}v_y\,\mathrm{d}v_z \tag{2.2.1}$$

is then the number of particles in the volume element $\mathrm{d}x\,\mathrm{d}y\,\mathrm{d}z$ at position $\boldsymbol{x}$, and the element $\mathrm{d}v_x\,\mathrm{d}v_y\,\mathrm{d}v_z$ in velocity space with velocity $\boldsymbol{v}$, at time t. The $(\boldsymbol{x}, \boldsymbol{v})$ space is called phase space.

The spatial density of particles, $n(\boldsymbol{x},t)$, can now be obtained by integrating Eq. (2.2.1) over all velocities:

$$n(\boldsymbol{x}, t) = \int_{-\infty}^{\infty} f(\boldsymbol{x}, \boldsymbol{v}, t)\,\mathrm{d}^3\boldsymbol{v}, \tag{2.2.2}$$

where $\mathrm{d}^3\boldsymbol{v}$ means $\mathrm{d}v_x\,\mathrm{d}v_y\,\mathrm{d}v_z$. The velocity distribution function, normalized to unity, is thus

$$f'(\boldsymbol{x}, \boldsymbol{v}, t) = \frac{f(\boldsymbol{x}, \boldsymbol{v}, t)}{n(\boldsymbol{x}, t)}. \tag{2.2.3}$$

If the system is collisional, with collision frequency ν, then after a time which is long compared to the collision time $1/\nu$, equipartition of energy by collisions (i.e. thermalization) will always cause the system to move towards a Maxwellian velocity distribution:

$$f'_{\mathrm{M}} = \left(\frac{m}{2\pi k_{\mathrm{B}} T}\right)^{3/2} \exp\left(-v^2/v_{\mathrm{T}}^2\right), \tag{2.2.4}$$

where $v_{\mathrm{T}} = (2k_{\mathrm{B}}T/m)^{1/2}$ is the thermal velocity characterizing the distribution. For any given species, this distribution is defined by one parameter only, the temperature T. It is the only distribution for which a temperature can be properly defined.

of v_x; a_x is thus independent of v_x, and similarly for the y- and z-components. Since the only derivatives of $\boldsymbol{a}$ appearing in the expansion of Eq. (2.6.5) are of the form $\partial a_x/\partial v_x$ and so on, the introduction of the Lorentz force is, happily, quite in order.

2.7 The Vlasov equation

Inserting the Lorentz force explicitly into Eq. (2.6.7), we obtain the Vlasov equation:

$$\frac{\partial f}{\partial t} + \boldsymbol{v}\cdot\frac{\partial f}{\partial \boldsymbol{x}} + \frac{q}{m}(\boldsymbol{E} + \boldsymbol{v} \times \boldsymbol{B})\cdot\frac{\partial f}{\partial \boldsymbol{v}} = 0, \tag{2.7.1}$$

which lies at the heart of plasma physics.

We must now address the problem we created when we assumed that the particles were all noninteracting, when we know that interactions are the very essence of plasma behaviour. In fact this is not much of a problem. We can justify using the Vlasov equation in the following way.

First, we assume no collisions – no short range, local interactions. These are certainly not described by the Vlasov equation as presented. In collisionless plasmas, each particle moves in the average Coulomb field due to thousands of others, and this is in fact our second assumption – that the fields $\boldsymbol{E}$ and $\boldsymbol{B}$ in the equation are fields due to the rest of the plasma, and *they* describe the interaction of the particles. They are often called self-consistent fields. Any externally applied fields can also be included, of course. This separation of the collisional and long range interactions is valid only if the Debye sphere contains a large number of particles, and the plasma is therefore a truly cooperative system, but this is usually the case.

The field $\boldsymbol{E}$, and in general also the field $\boldsymbol{B}$, which are determined by the distribution in the rest of the plasma, both depend on the distribution function f. Thus the Vlasov equation is nonlinear, and analytic solutions are in general not possible.

We should also add that a Vlasov equation is needed for each separate species of particle in the plasma, with different mass, charge, and distribution function f.

In a few specialized applications, where temperatures are very high, relativistic effects may need to be taken into account, and the relativistic Vlasov equation is then needed. To obtain this, only a small modification to Eq. (2.7.1) is needed: we put

$$\boldsymbol{p} = m\boldsymbol{v} = \gamma m_0 \boldsymbol{v},$$

where m_0 is the rest mass. The Vlasov equation then becomes

$$\frac{\partial f}{\partial t} + \frac{\boldsymbol{p}}{m}\cdot\frac{\partial f}{\partial \boldsymbol{x}} + \boldsymbol{F}\cdot\frac{\partial f}{\partial \boldsymbol{p}} = 0, \tag{2.7.2}$$

where

$$\boldsymbol{F} = \boldsymbol{F}_{\text{ext}} + q(\boldsymbol{E} + \boldsymbol{v} \times \boldsymbol{B}).$$

2.8 The effect of collisions

We must now include the effect of collisions of the type which we have so far excluded from the description: short range interactions not included in the effect of the fields $\boldsymbol{E}$ and $\boldsymbol{B}$.

If the plasma is only partially ionized, collisions with neutrals will occur. In a fully ionized plasma, as well as the cooperative effects described by the fields $\boldsymbol{E}$ and $\boldsymbol{B}$ ($\boldsymbol{B}$ will often contain an externally applied component), particles do in fact undergo microcollisions, i.e. they are gradually deflected by large numbers of small deflections due to local Coulomb interactions. This is the cause of the ultimate thermalization. These collisions are not described by the Vlasov equation, and are treated by the inclusion of a collision term $(\partial f/\partial t)_{\mathrm{c}}$:

$$\frac{\partial f}{\partial t} + \boldsymbol{v}\cdot\frac{\partial f}{\partial \boldsymbol{x}} + \frac{q}{m}(\boldsymbol{E} + \boldsymbol{v} \times \boldsymbol{B})\cdot\frac{\partial f}{\partial \boldsymbol{v}} = \left(\frac{\partial f}{\partial t}\right)_{\mathrm{c}}. \tag{2.8.1}$$

One has to choose a suitable representation for the collision term. The simplest is the Krook collision term (Bhatnagar, Gross and Krook, 1954):

$$\left(\frac{\partial f}{\partial t}\right)_{\mathrm{c}} = -\left(\frac{f - f_{\mathrm{m}}}{\tau}\right), \tag{2.8.2}$$

where f_{m} is the Maxwellian equilibrium distribution function, to which the system is tending, and τ is the mean (constant) collision time. Eq. (2.8.2) integrates to yield

$$f(t) = f_{\mathrm{m}} + (f(0) - f_{\mathrm{m}})\exp(-t/\tau). \tag{2.8.3}$$

The model is not very good if the masses of the colliding species are very different, such as in electron-neutral collisions. Under such circumstances, the collision time τ should be replaced by $(m_{\mathrm{n}}/2m_{\mathrm{e}})\tau$. Discussions of the Krook collision approximation are given by Boyd and Sanderson (1969) and Clemmow and Dougherty (1969).

2.9 The Fokker–Planck equation

In fully ionized plasmas, the collisionality is the effect of many small Coulomb collisions, and the usual procedure is to use the Fokker–Planck equation for the collision term. It has its origins in the study of Brownian motion, a similar phenomenon in that there also the deflections of the large suspended particles are due to the accumulation of many microdeflections. We may derive it in the following way.

We define the probability $\psi(\boldsymbol{v}, \Delta\boldsymbol{v})$ that a particle initially with velocity $\boldsymbol{v}$, undergoing many microcollisions, acquires an increment of velocity $\Delta\boldsymbol{v}$ in a time Δt. We remember that $f(\boldsymbol{x}, \boldsymbol{v}, t)$ is itself a probability distribution, and its form at time t can be written as a product of the distribution at a time Δt earlier, multiplied by the probability of change in the time Δt, and integrated over the possible $\Delta\boldsymbol{v}$:

$$f(\boldsymbol{x}, \boldsymbol{v}, t) = \int f(\boldsymbol{x}, \boldsymbol{v} - \Delta\boldsymbol{v}, t - \Delta t)\psi(\boldsymbol{v} - \Delta\boldsymbol{v}, \Delta\boldsymbol{v})\, \mathrm{d}^3(\Delta\boldsymbol{v}). \tag{2.9.1}$$

Using Taylor's theorem, we can expand the product $f\psi$ inside the integral to second order:

$$f(\boldsymbol{x},\boldsymbol{v},t) = \int \mathrm{d}^3(\Delta\boldsymbol{v})\left[f(\boldsymbol{x},\boldsymbol{v},t-\Delta t)\psi(\boldsymbol{v},\Delta\boldsymbol{v}) - \Delta\boldsymbol{v}\cdot\left(\frac{\partial}{\partial\boldsymbol{v}}(f\psi)\right) + \tfrac{1}{2}\Delta\boldsymbol{v}\Delta\boldsymbol{v}:\left(\frac{\partial^2}{\partial\boldsymbol{v}\,\partial\boldsymbol{v}}(f\psi)\right)\right]. \tag{2.9.2}$$

(We are using standard tensor notation here. A brief summary will be found in Appendix 2.17.)

We now work through the integrals in Eq. (2.9.2), remembering that $\int\psi \mathrm{d}^3(\Delta\boldsymbol{v}) = 1$, by definition:

$$f(\boldsymbol{x},\boldsymbol{v},t) = f(\boldsymbol{x},\boldsymbol{v},t-\Delta t) - \frac{\partial}{\partial\boldsymbol{v}}\cdot(f\langle\Delta\boldsymbol{v}\rangle) + \frac{1}{2}\frac{\partial^2}{\partial\boldsymbol{v}\,\partial\boldsymbol{v}}:(f\langle\Delta\boldsymbol{v}\Delta\boldsymbol{v}\rangle) \tag{2.9.3}$$

where

$$\langle\Delta\boldsymbol{v}\rangle = \int\psi\Delta\boldsymbol{v}\,\mathrm{d}^3(\Delta\boldsymbol{v})$$

and

$$\langle\Delta\boldsymbol{v}\Delta\boldsymbol{v}\rangle = \int\psi\Delta\boldsymbol{v}\Delta\boldsymbol{v}\,\mathrm{d}^3(\Delta\boldsymbol{v}).$$

Now again by definition,

$$\left(\frac{\partial f}{\partial t}\right)_{\mathrm{c}} = \frac{f(\boldsymbol{x},\boldsymbol{v},t) - f(\boldsymbol{x},\boldsymbol{v},t-\Delta t)}{\Delta t},$$

and hence

$$\left(\frac{\partial f}{\partial t}\right)_{\mathrm{c}}\Delta t = -\frac{\partial}{\partial\boldsymbol{v}}\cdot(f\langle\Delta\boldsymbol{v}\rangle) + \frac{1}{2}\frac{\partial^2}{\partial\boldsymbol{v}\,\partial\boldsymbol{v}}:(f\langle\Delta\boldsymbol{v}\Delta\boldsymbol{v}\rangle), \tag{2.9.4}$$

which is the Fokker–Planck equation.

The physical meaning of the first term on the right of Eq. (2.9.4) can be understood by noting that $\langle\Delta\boldsymbol{v}\rangle/\Delta t$ is an acceleration, or force/unit mass. This term therefore describes the frictional force slowing down fast particles and accelerating slow ones. The negative divergence in velocity space describes a narrowing of the distribution.

In the second term, $\langle\Delta\boldsymbol{v}\,\Delta\boldsymbol{v}\rangle/\Delta t$ is a coefficient of diffusion in velocity space. This term then describes the fact that a narrow velocity distribution (e.g. a beam) will broaden as a result of collisions. The two terms thus operate in opposite senses, and are in balance for an equilibrium (Maxwellian) distribution.

The actual physics of the collisional process is contained in the function $\psi(\boldsymbol{v},\Delta\boldsymbol{v})$, and this function is commonly derived using a Rutherford scattering model (see, for example, Krall and Trivelpiece, 1973). We shall not delve further into the details of collisional processes in this chapter, but we will include the effects of collisions in the later development of the fluid equations.

2.10 The equivalence of kinetic theory and orbit theory

In this section, we follow the useful approach given by Boyd and Sanderson (1969) and Clemmow and Dougherty (1969).

The collisionless Boltzmann equation is the root of kinetic theory:

$$\frac{\partial f}{\partial t} + \boldsymbol{v}\cdot\frac{\partial f}{\partial \boldsymbol{x}} + \frac{\boldsymbol{F}}{m}\cdot\frac{\partial f}{\partial \boldsymbol{v}} = 0. \qquad (2.10.1)$$

The base equation of orbit theory is Newton's law: $m\,\mathrm{d}^2\boldsymbol{x}/\mathrm{d}t^2 = \boldsymbol{F}$. Being a second order differential equation in three dimensions, the general solution of Newton's equation must contain six constants of integration, $\alpha_1, \ldots, \alpha_6$. We write the solution

$$\left.\begin{aligned} \boldsymbol{x} &= \boldsymbol{x}(\alpha_1, \ldots, \alpha_6, t) \\ \boldsymbol{v} &= \boldsymbol{v}(\alpha_1, \ldots, \alpha_6, t). \end{aligned}\right\} \qquad (2.10.2)$$

In principle, we can formally solve these six scalar equations for the α_i:

$$\alpha_i = \alpha_i(\boldsymbol{x}, \boldsymbol{v}, t), \quad i = 1\text{–}6. \qquad (2.10.3)$$

Now any arbitrary function of the α_i, $f = f(\alpha_i, \ldots, \alpha_6)$, is a solution of the Boltzmann equation above; we can show this by substituting the function f directly into Eq. (2.10.1):

$$\sum_i \left(\frac{\partial \alpha_i}{\partial t} + \boldsymbol{v}\cdot\frac{\partial \alpha_i}{\partial \boldsymbol{x}} + \frac{\boldsymbol{F}}{m}\cdot\frac{\partial \alpha_i}{\partial \boldsymbol{v}} \right) = \sum_i \frac{\partial f}{\partial \alpha_i}\frac{\mathrm{d}\alpha_i}{\mathrm{d}t} \equiv 0. \qquad (2.10.4)$$

The result is identically zero because the α_i's are constants.

Thus the general solution of the Vlasov equation is an arbitrary function of the integrals of Newton's law, and the two approaches are equivalent. This was first shown by Jeans in connection with stellar dynamics, and is sometimes called Jeans's theorem (Chandrasekhar, 1960).

2.11 The fluid equations

We have seen in Section 2.5 that real physical quantities are derived as moments of the distribution function. We can therefore expect to obtain physical equations relating macroscopic variables such as density, fluid velocity and pressure by taking moments of the Vlasov equation, Eq. (2.7.1). This is how we proceed.

We will first derive the equation of continuity, describing conservation of particles for each species. This equation follows from taking the zeroth order moment. The first order moment will give us the force balance equation, describing conservation of momentum. The second order moment gives us an equation describing conservation of energy; however, the derivation is quite lengthy and the reader will be referred to specialist texts for a full treatment. We shall also discover that this sequence of equations obtained from moments of the Vlasov equation is in principle not closed: each equation contains quantities which have to be derived from the next higher order equation. An approximation has to be found at some stage, which allows the set of equations to be closed in practice.

2.11.1 *The zeroth order moment*

We derive first, then, the continuity equation describing particle conservation. We could write this down as self-evident, but its derivation is straightforward and a very good illustration of the method.

Taking the zeroth order moment, equivalent to a straight integration of the equation, we obtain:

$$\int \frac{\partial f}{\partial t} \mathrm{d}^3 \boldsymbol{v} + \int \boldsymbol{v} \cdot \frac{\partial f}{\partial \boldsymbol{x}} \mathrm{d}^3 \boldsymbol{v} + \frac{q}{m} \int (\boldsymbol{E} + \boldsymbol{v} \times \boldsymbol{B}) \cdot \frac{\partial f}{\partial \boldsymbol{v}} \mathrm{d}^3 \boldsymbol{v} = \int \left(\frac{\partial f}{\partial t}\right)_{\mathrm{c}} \mathrm{d}^3 \boldsymbol{v}. \tag{2.11.1}$$

The first term gives

$$\int \frac{\partial f}{\partial t} \mathrm{d}^3 \boldsymbol{v} = \frac{\partial}{\partial t} \int f \, \mathrm{d}^3 \boldsymbol{v} = \frac{\partial n}{\partial t}. \tag{2.11.2}$$

The second term integrates as follows:

$$\int \boldsymbol{v} \cdot \frac{\partial f}{\partial \boldsymbol{x}} \mathrm{d}^3 \boldsymbol{v} = \frac{\partial}{\partial \boldsymbol{x}} \cdot \int \boldsymbol{v} f \, \mathrm{d}^3 \boldsymbol{v} = \frac{\partial}{\partial \boldsymbol{x}} \cdot (n \bar{\boldsymbol{v}}) \equiv \frac{\partial}{\partial \boldsymbol{x}} \cdot (n \boldsymbol{u}). \tag{2.11.3}$$

We have used the fact that $\boldsymbol{x}$ and $\boldsymbol{v}$ are independent variables. In Eq. (2.11.3), $\boldsymbol{u}$ is the 'fluid' velocity.

The term in $\boldsymbol{E}$ in fact vanishes. We show this by rewriting it using Gauss's theorem in velocity space:

$$\int \boldsymbol{E} \cdot \frac{\partial f}{\partial \boldsymbol{v}} \mathrm{d}^3 \boldsymbol{v} = \int \frac{\partial}{\partial \boldsymbol{v}} \cdot (f \boldsymbol{E}) \, \mathrm{d}^3 \boldsymbol{v} = \int_{\mathrm{s}} f \boldsymbol{E} \cdot \mathrm{d} \boldsymbol{S}_{\mathrm{v}} = 0. \tag{2.11.4}$$

The surface integral vanishes if we take the surface S to infinity: the surface area increases as v^2, but a real distribution $f(\boldsymbol{v})$ (for example, a Maxwellian) will in practice always go to zero much faster, typically like $\exp - (v^2/v_{\mathrm{T}}^2)$. The B term also vanishes:

$$\int (\boldsymbol{v} \times \boldsymbol{B}) \cdot \frac{\partial f}{\partial \boldsymbol{v}} \mathrm{d}^3 \boldsymbol{v} = \int \frac{\partial}{\partial \boldsymbol{v}} \cdot [f (\boldsymbol{v} \times \boldsymbol{B})] \, \mathrm{d}^3 \boldsymbol{v} - \int f \frac{\partial}{\partial \boldsymbol{v}} \cdot (\boldsymbol{v} \times \boldsymbol{B}) \, \mathrm{d}^3 \boldsymbol{v} = 0. \tag{2.11.5}$$

The first term again reduces to a vanishing surface integral, whereas the second is zero because $\boldsymbol{v} \times \boldsymbol{B}$ is perpendicular to $\partial/\partial \boldsymbol{v}$ (this point was discussed in Section 2.6).

The collision term also vanishes, because

$$\int \left(\frac{\partial f}{\partial t}\right)_{\mathrm{c}} \mathrm{d}^3 \boldsymbol{v} = \left[\frac{\partial}{\partial t} \int f \, \mathrm{d}^3 \boldsymbol{v}\right]_{\mathrm{c}} = 0, \tag{2.11.6}$$

since the total number of particles of the species considered must remain constant as collisions proceed. (We do not consider recombination events in this context. Such events would of course remove particles from the distribution, but they are extremely rare, being essentially a three-particle collision.)

All that remains from Eq. (2.11.1) is

$$\frac{\partial n}{\partial t} + \frac{\partial}{\partial \boldsymbol{x}} \cdot (n \boldsymbol{u}) = 0, \tag{2.11.7}$$

which expresses conservation of particles. It is only a matter of multiplying by m or q to convert this to an equation describing conservation of mass, or charge.

2.11.2 *The closure problem*

This result immediately illustrates the closure problem associated with this procedure: taking the zeroth moment of Eq. (2.7.1) has introduced a higher order moment, the first velocity moment of f, which is $\boldsymbol{u}$. This will have to be obtained by taking the first moment of the equation, but this in turn will introduce a second moment of f, etc. The sequence will have to be terminated by some justifiable procedure. For most purposes, the closure is effected by setting the third velocity moment of f, describing thermal conductivity, to zero. We shall not in fact pursue the details in this treatment, but shall quote results.

2.11.3 *The first order moment*

To take the first order moment of the Vlasov equation, we multiply Eq. (2.7.1) by $m\boldsymbol{v}$ and integrate. This will yield a macroscopic equation describing conservation of momentum. We have

$$m\int \boldsymbol{v}\frac{\partial f}{\partial t}\mathrm{d}^3\boldsymbol{v} + m\int \boldsymbol{v}\left(\boldsymbol{v}\cdot\frac{\partial}{\partial \boldsymbol{x}}\right) f\,\mathrm{d}^3\boldsymbol{v} + q\int \boldsymbol{v}(\boldsymbol{E} + \boldsymbol{v}\times\boldsymbol{B})\cdot\frac{\partial f}{\partial \boldsymbol{v}}\mathrm{d}^3\boldsymbol{v} = \int m\boldsymbol{v}\left(\frac{\partial f}{\partial t}\right)\mathrm{d}^3\boldsymbol{v}. \tag{2.11.8}$$

The first term of Eq. (2.11.8) gives

$$m\int \boldsymbol{v}\frac{\partial f}{\partial t}\mathrm{d}^3\boldsymbol{v} = m\frac{\partial}{\partial t}\int \boldsymbol{v}f\,\mathrm{d}^3\boldsymbol{v} = m\frac{\partial}{\partial t}(n\boldsymbol{u}). \tag{2.11.9}$$

Taking the third term next, it may be expanded to become:

$$\int \boldsymbol{v}(\boldsymbol{E} + \boldsymbol{v}\times\boldsymbol{B})\cdot\frac{\partial f}{\partial \boldsymbol{v}}\mathrm{d}^3\boldsymbol{v} = \int \frac{\partial}{\partial \boldsymbol{v}}\cdot[f\boldsymbol{v}(\boldsymbol{E} + \boldsymbol{v}\times\boldsymbol{B})]\,\mathrm{d}^3\boldsymbol{v} - \int f\boldsymbol{v}\frac{\partial}{\partial \boldsymbol{v}}\cdot(\boldsymbol{E} + \boldsymbol{v}\times\boldsymbol{B})\,\mathrm{d}^3\boldsymbol{v}$$
$$- \int f(\boldsymbol{E} + \boldsymbol{v}\times\boldsymbol{B})\cdot\frac{\partial \boldsymbol{v}}{\partial \boldsymbol{v}}\mathrm{d}^3\boldsymbol{v}. \tag{2.11.10}$$

(Note: products like $\boldsymbol{ab}$ are called tensor products, or dyads: see the appendix, Section 2.17.)

The first integral in Eq. (2.11.10) vanishes, as can be seen by applying Gauss's theorem as we did in Section 2.11.1 earlier; the second integral is also zero, because $\boldsymbol{E}$ is not a function of $\boldsymbol{v}$, and $(\boldsymbol{v}\times\boldsymbol{B})$ is perpendicular to $\partial/\partial\boldsymbol{v}$; and $\partial\boldsymbol{v}/\partial\boldsymbol{v}$ is the identity tensor $\boldsymbol{I}$. This third term therefore reduces to

$$-q\int(\boldsymbol{E} + \boldsymbol{v}\times\boldsymbol{B})f\,\mathrm{d}^3\boldsymbol{v} = -qn(\boldsymbol{E} + \boldsymbol{u}\times\boldsymbol{B}). \tag{2.11.11}$$

Let us now consider the second term in Eq. (2.11.8). Remembering that $\boldsymbol{x}$ and $\boldsymbol{v}$ are independent variables, we write

$$\int \boldsymbol{v}\left(\boldsymbol{v}\cdot\frac{\partial}{\partial \boldsymbol{x}}\right)f\,\mathrm{d}^3\boldsymbol{v} = \int \frac{\partial}{\partial \boldsymbol{x}}\cdot(f\boldsymbol{vv})\,\mathrm{d}^3\boldsymbol{v} = \frac{\partial}{\partial \boldsymbol{x}}\cdot\int f\boldsymbol{vv}\,\mathrm{d}^3\boldsymbol{v} = \frac{\partial}{\partial \boldsymbol{x}}\cdot(n\overline{\boldsymbol{vv}}). \tag{2.11.12}$$

At this point, it is helpful to separate the velocity $\boldsymbol{v}$ into a mean (fluid) velocity $\boldsymbol{u}$ and a random (thermal) velocity $\boldsymbol{w}$: $\boldsymbol{v} = \boldsymbol{u} + \boldsymbol{w}$. Then

$$\frac{\partial}{\partial \boldsymbol{x}} \cdot (n\overline{\boldsymbol{v}\boldsymbol{v}}) = \nabla \cdot (n\overline{\boldsymbol{v}\boldsymbol{v}}) = \nabla \cdot (n\boldsymbol{u}\boldsymbol{u}) + \nabla \cdot (n\overline{\boldsymbol{w}\boldsymbol{w}}) + \nabla \cdot n(\boldsymbol{u}\bar{\boldsymbol{w}} + \bar{\boldsymbol{w}}\boldsymbol{u}). \tag{2.11.13}$$

The final term here is zero, since $\bar{\boldsymbol{w}} \equiv 0$ by definition. The first term can be written

$$\nabla \cdot (n\boldsymbol{u}\boldsymbol{u}) = \boldsymbol{u}\nabla \cdot (n\boldsymbol{u}) + n(\boldsymbol{u} \cdot \nabla)\boldsymbol{u}. \tag{2.11.14}$$

In the second term, the quantity $mn\overline{\boldsymbol{w}\boldsymbol{w}}$ appears. This clearly has the dimensions of energy density, and is called the stress tensor or pressure tensor $\boldsymbol{P}$. We will discuss the interpretation of this quantity later in this section.

There remains the collision term, which we shall represent as

$$\int m\boldsymbol{v}\left(\frac{\partial f}{\partial t}\right)_{\mathrm{c}} \mathrm{d}^3\boldsymbol{v} = \left[\frac{\partial}{\partial t}\int m\boldsymbol{v} f\, \mathrm{d}^3\boldsymbol{v}\right]_{\mathrm{c}} = \boldsymbol{P}_{ij}. \tag{2.11.15}$$

It represents the rate of change of momentum density due to collisions between different species i and j. Collisions between like particles cannot produce a net change of momentum of that species.

Collecting all the terms together, we have for Eq. (2.11.8), the first moment equation,

$$m\frac{\partial}{\partial t}(n\boldsymbol{u}) + m\boldsymbol{u}\nabla \cdot (n\boldsymbol{u}) + mn(\boldsymbol{u} \cdot \nabla)\boldsymbol{u} + \nabla \cdot \boldsymbol{P} - qn(\boldsymbol{E} + \boldsymbol{u} \times \boldsymbol{B}) = \boldsymbol{P}_{ij}. \tag{2.11.16}$$

Finally we combine the first two terms using the continuity equation above, Eq. (2.11.7), giving

$$mn\left[\frac{\partial \boldsymbol{u}}{\partial t} + (\boldsymbol{u} \cdot \nabla)\boldsymbol{u}\right] = qn(\boldsymbol{E} + \boldsymbol{u} \times \boldsymbol{B}) - \nabla \cdot \boldsymbol{P} + \boldsymbol{P}_{ij}. \tag{2.11.17}$$

2.11.4 *The pressure tensor*

Let us now examine the meaning of the pressure tensor $\boldsymbol{P}$, defined in Section 2.11.3 above as $\boldsymbol{P} = nm\overline{\boldsymbol{w}\boldsymbol{w}}$. It can be written

$$\boldsymbol{P} = \begin{pmatrix} p_{xx} & p_{xy} & p_{xz} \\ p_{yx} & p_{yy} & p_{yz} \\ p_{zx} & p_{zy} & p_{zz} \end{pmatrix}. \tag{2.11.18}$$

$\boldsymbol{P}$ can be interpreted in the following way. Consider a closed surface S in the plasma, and consider a small element $\mathrm{d}S$ of this, with normal vector $\hat{\boldsymbol{n}}$. The force/unit area on $\hat{\boldsymbol{n}}\,\mathrm{d}S$ is

$$-\boldsymbol{P} \cdot \hat{\boldsymbol{n}} = -nm\langle \boldsymbol{w}(\boldsymbol{w} \cdot \hat{\boldsymbol{n}})\rangle, \tag{2.11.19}$$

where $\langle\ \rangle$ signifies an average. This result follows from the following observations: (1) $n(\boldsymbol{w} \cdot \hat{\boldsymbol{n}})$ is the *outward* flux/unit area of particles from the volume enclosed by S; (2) $nm\boldsymbol{w}(\boldsymbol{w} \cdot \hat{\boldsymbol{n}})$ is then the outward flux density in the $\hat{\boldsymbol{n}}$-direction of momentum in the $\boldsymbol{w}$-direction. The minus sign appears because an outward flux gives rise to a *lower* pressure.

Now consider a special case, where $\hat{\boldsymbol{n}}$ is in the x-direction: $\hat{\boldsymbol{n}} = (1, 0, 0)$. Then $-\boldsymbol{P}\cdot\hat{\boldsymbol{n}} = -(p_{xx}, p_{yx}, p_{zx})$. Fig. 2.2 shows that p_{xx} is a *pressure* force, whereas p_{yx} and p_{zx} are *shear* forces. Clearly, $p_{xy} = p_{yx}$, etc. These off-diagonal elements describe viscosity in the plasma: transfer of momentum in directions perpendicular to the particle motion. This may often be ignored in plasmas. In conventional fluids viscous effects are most prominent in the interaction of the fluid with boundaries, such as walls of pipes or confining vessels. In the situations in which plasmas are normally considered, there are no material confining vessels, and no walls. Viscosity therefore has very little importance for plasmas, but when needed it can be included in this term.

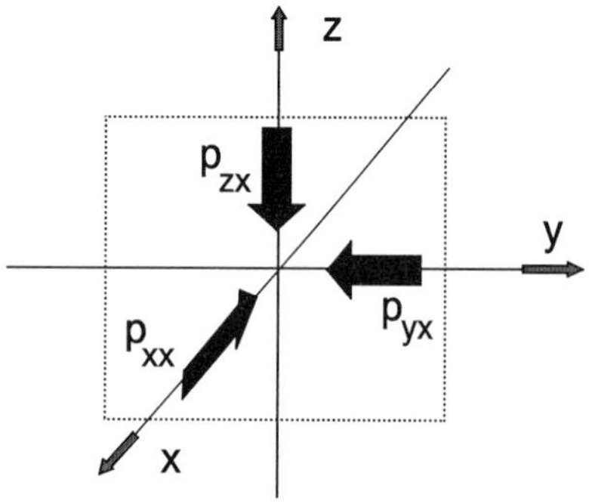

Fig. 2.2. The presssure tensor $\boldsymbol{P}$: three components are shown; p_{yx} and p_{zx} represent shear forces, and p_{xx} represents a pressure force.

The diagonal elements represent normal hydrostatic pressure. In an isotropic plasma, $p_{xx} = p_{yy} = p_{zz} = p$, the scalar pressure. Thus for most purposes, the pressure tensor can be written $\boldsymbol{P} = p\boldsymbol{I}$, where $\boldsymbol{I}$ is the unit tensor, and $\boldsymbol{\nabla}\cdot\boldsymbol{P}$ becomes $\boldsymbol{\nabla}p$, the pressure gradient.

The square bracket in Eq. (2.11.17) is the convective derivative $\mathrm{d}\boldsymbol{u}/\mathrm{d}t$. The equation is the *fluid equation of motion*, and describes the force balance in this component of the plasma. There will be such an equation for each species, but it is often useful to combine them into single-fluid equations describing the neutral plasma. We will develop this in Section 2.12.

2.11.5 *The second order moment*

The second moment is obtained by multiplying Eq. (2.7.1) by $m\boldsymbol{v}\boldsymbol{v}/2$ and integrating. We shall not pursue it here; the procedure is well presented in several texts (for example Bittencourt, 1986). The second order equations, one for each particle species, describe energy conservation within and between the species: when reduced to a single equation for the neutral fluid (a procedure we shall follow in the next section), it becomes

$$p/\rho^{\gamma} = \text{constant}. \tag{2.11.20}$$

the adiabatic equation of state.

2.12 The single-fluid equations

We first define the appropriate parameters for a neutral plasma: mass density

$\rho = n_{\mathrm{i}}M + n_{\mathrm{e}}m \approx nM,$

mass velocity

$$\boldsymbol{v} = \frac{1}{\rho}(n_i M \boldsymbol{u}_i + n_e m \boldsymbol{u}_e) \approx \frac{M\boldsymbol{u}_i + m\boldsymbol{u}_e}{M + m} \approx \boldsymbol{u}_i,$$

current density

$$\boldsymbol{J} = e(n_i \boldsymbol{u}_i - n_e \boldsymbol{u}_e) \approx ne(\boldsymbol{u}_i - \boldsymbol{u}_e). \qquad (2.12.1)$$

These approximations indicate that the mass and velocity of the single fluid are essentially provided by the ions. We can therefore write the continuity equation, describing conservation of mass in the neutral plasma, using Eq. (2.11.7) for the ions:

$$\frac{\partial \rho}{\partial t} + \boldsymbol{\nabla}\cdot(\rho \boldsymbol{v}) = 0. \qquad (2.12.2)$$

We will also need a useful form for the electron–ion collision term $\boldsymbol{P}_{ei}$ in Eq. (2.11.17). Since it describes momentum exchange between species, it is this term which describes resistive effects in the plasma. We present here a simple argument, starting with the resistive Ohm's law $\boldsymbol{E} = \eta \boldsymbol{J}$. Though simple, the procedure allows us to develop an approximate form for the resistive term which is both valid and useful. The accurate treatment of this problem in plasma physics is a difficult challenge.

We write:

$$\text{force/unit volume on the electrons} = ne\boldsymbol{E} = ne\eta\boldsymbol{J} = \mathrm{d}\boldsymbol{p}/\mathrm{d}t, \qquad (2.12.3)$$

where $\boldsymbol{p}$ is the momentum density. We equate this rate of change of momentum density to the collision term $\boldsymbol{P}_{ei}$. Thus, using the expression from Eq. (2.12.1) for $\boldsymbol{J}$, we have

$$\boldsymbol{P}_{ei} = n^2 e^2 \eta(\boldsymbol{u}_i - \boldsymbol{u}_e). \qquad (2.12.4)$$

The ion–electron collision term $\boldsymbol{P}_{ie}$ is clearly given by $-\boldsymbol{P}_{ei}$: total momentum must be conserved in the collisions between the two fluids. This simple model gives the result that momentum exchange is proportional to the difference in the fluid velocities, and is approximately correct if this difference is fairly small and the distributions are Maxwellian. Assuming the validity of this velocity dependence, we can also write for the rate of change of momentum density

$$\boldsymbol{P}_{ei} = nm\nu_{ei}(\boldsymbol{u}_i - \boldsymbol{u}_e), \qquad (2.12.5)$$

where ν_{ei} is the electron–ion collision frequency. Comparing with the previous expression for $\boldsymbol{P}_{ei}$, we obtain the relation between resistivity and collision frequency:

$$\eta = \frac{m\nu_{ei}}{ne^2}. \qquad (2.12.6)$$

Finally we find that, since $\boldsymbol{P}_{ei} = -\boldsymbol{P}_{ie}$,

$$m\nu_{ei} = M\nu_{ie}. \qquad (2.12.7)$$

It may be a surprise to find that ν_{ei} and ν_{ie} are not equal. This relationship makes clear that it is the mass difference which causes that inequality: if we consider $1/\nu_{ei}$

to be the mean time for an electron to be deviated by $\pi/2$ rad, due to a process involving many small interactions, it is clear that it will take a much larger number of such interactions, and hence a longer time, to deviate an ion by the same amount.

The two fluid equations, from Eq. (2.11.17), are:

$$Mn\left(\frac{\partial \boldsymbol{u}_\mathrm{i}}{\partial t} + (\boldsymbol{u}_\mathrm{i}\cdot\nabla)\boldsymbol{u}_\mathrm{i}\right) = en(\boldsymbol{E} + \boldsymbol{u}_\mathrm{i} \times \boldsymbol{B}) - \nabla p_i + \boldsymbol{P}_\mathrm{ie}, \tag{2.12.8}$$

$$mn\left(\frac{\partial \boldsymbol{u}_\mathrm{e}}{\partial t} + (\boldsymbol{u}_\mathrm{e}\cdot\nabla)\boldsymbol{u}_\mathrm{e}\right) = -en(\boldsymbol{E} + \boldsymbol{u}_\mathrm{e} \times \boldsymbol{B}) - \nabla p_\mathrm{e} + \boldsymbol{P}_\mathrm{ei}. \tag{2.12.9}$$

We can obtain two powerful equations by taking different linear combinations of these two fluid equations. The first is obtained by adding, and yields the force balance equation for the neutral plasma, and the second is obtained by taking a different linear combination to yield the effective Ohm's law for the system.

Adding the equations, we obtain

$$n\left(\frac{\partial}{\partial t}(M\boldsymbol{u}_\mathrm{i} + m\boldsymbol{u}_\mathrm{e}) + M(\boldsymbol{u}_\mathrm{i}\cdot\nabla)\boldsymbol{u}_\mathrm{i} + m(\boldsymbol{u}_\mathrm{e}\cdot\nabla)\boldsymbol{u}_\mathrm{e}\right)$$
$$= en(\boldsymbol{u}_\mathrm{i} - \boldsymbol{u}_\mathrm{e}) \times \boldsymbol{B} - \nabla(p_i + p_\mathrm{e}) \tag{2.12.10}$$

(note $\boldsymbol{P}_\mathrm{ei} = -\boldsymbol{P}_\mathrm{ie}$). Since $p = p_i + p_\mathrm{e}$, and using the above definitions for $\boldsymbol{v}$, $\boldsymbol{J}$, and the approximation $m \ll M$, we obtain

$$\rho\left(\frac{\partial \boldsymbol{v}}{\partial t} + (\boldsymbol{v}\cdot\nabla)\boldsymbol{v}\right) = \boldsymbol{J} \times \boldsymbol{B} - \nabla p + \boldsymbol{F}, \tag{2.12.11}$$

where we have generalized the equation by including a term $\boldsymbol{F}$ representing any external force, such as gravity. This important equation describes momentum conservation in the neutral plasma, or force balance.

The second equation, known as the *generalized Ohm's law*, is obtained by taking a particular linear combination of the fluid equations. We multiply the ion equation, Eq. (2.12.8), by m, and the electron equation, Eq. (2.12.9), by M, and subtract:

$$Mmn\left(\frac{\partial}{\partial t}(\boldsymbol{u}_\mathrm{i} - \boldsymbol{u}_\mathrm{e}) + (\boldsymbol{u}_\mathrm{i}\cdot\nabla)\boldsymbol{u}_\mathrm{i} - (\boldsymbol{u}_\mathrm{e}\cdot\nabla)\boldsymbol{u}_\mathrm{e}\right)$$
$$= en(M + m)\boldsymbol{E} + en(m\boldsymbol{u}_\mathrm{i} + M\boldsymbol{u}_\mathrm{e}) \times \boldsymbol{B} - m\nabla p_\mathrm{i} + M\nabla p_\mathrm{e} - (M + m)\boldsymbol{P}_\mathrm{ei}. \tag{2.12.12}$$

Let us first deal with the convective terms on the left-hand side. For many purposes, they may be neglected, and this is the common case. The simplest justification for this is to restrict ourselves to small velocities, in which case these terms, being essentially quadratic in $\boldsymbol{u}_\mathrm{i}$ and $\boldsymbol{u}_\mathrm{e}$, may be neglected. This, however, is not very satisfactory. A better insight can be gained by roughly assessing the relative magnitude of the terms by a dimensional argument, and we shall do this later in this section. For the moment, we shall neglect the convective terms.

Using our earlier definitions for $\boldsymbol{J}$, Eq. (2.12.1) and $\boldsymbol{P}_\mathrm{ei}$, Eq. (2.12.4), the above equation reduces, in the limit $m \ll M$, to

$$\frac{Mmn}{e}\frac{\partial}{\partial t}\left(\frac{\boldsymbol{J}}{n}\right) = e\rho\boldsymbol{E} - Mne\eta\boldsymbol{J} + M\nabla p_e + en(m\boldsymbol{u}_i + M\boldsymbol{u}_e)\times\boldsymbol{B}. \tag{2.12.13}$$

The last term can be simplified by the ingenious device of writing

$$\begin{aligned} m\boldsymbol{u}_i + M\boldsymbol{u}_e &= M\boldsymbol{u}_i + m\boldsymbol{u}_e - (M-m)(\boldsymbol{u}_i - \boldsymbol{u}_e) \\ &\simeq \frac{\rho}{n}\boldsymbol{v} - \frac{M}{ne}\boldsymbol{J}. \end{aligned} \tag{2.12.14}$$

Dividing by $e\rho$ and rearranging terms we finally obtain

$$\frac{m}{ne^2}\frac{\partial \boldsymbol{J}}{\partial t} = \boldsymbol{E} + (\boldsymbol{v} + \boldsymbol{B}) - \frac{1}{ne}(\boldsymbol{J}\times\boldsymbol{B}) + \frac{1}{ne}\nabla p_e - \eta\boldsymbol{J}, \tag{2.12.15}$$

which is indeed a generalized Ohm's law. Each term is an emf, measured in volts/metre: the left-hand side describes the effect of electron inertia, and only matters for very high frequency phenomena. In the special case of uniform, collisionless plasma without magnetic field, Eq. (2.12.15) reduces to

$$\frac{m}{ne^2}\frac{\partial \boldsymbol{J}}{\partial t} = \boldsymbol{E},$$

and putting $\boldsymbol{J} = ne\boldsymbol{v}$ this in turn reduces to Newton's law:

$$m\frac{\partial \boldsymbol{v}}{\partial t} = e\boldsymbol{E}.$$

If the electron mass is ignored, and collisions included, Eq. (2.12.15) becomes $\boldsymbol{E} = \eta$, the simple Ohm's law. The $(\boldsymbol{J}\times\boldsymbol{B})$ term describes the Hall effect, and the ∇p_e term describes the fact that electron pressure gradients will drive currents.

It is often the case that some of the terms in the generalized Ohm's law Eq. (2.12.15) can be neglected, and this can be illustrated by the dimensional analysis mentioned earlier. It is a rough-and-ready argument, but useful. It should be noted that the argument follows a slightly different course depending on whether we are considering electrostatic phenomena (where the field $\boldsymbol{B}$ is essentially a vacuum field, or zero) or true magnetohydrodynamic (MHD) phenomena, where $\boldsymbol{B}$ is determined by the current density $\boldsymbol{J}$ within the plasma.

We use the rough dimensional relationhip $v = \omega L$, where $1/\omega$ and L are characteristic time and length scales. For MHD, we use Maxwell's equation reduced to a scalar form: $\nabla\times\boldsymbol{B} \approx B/L \approx \mu_0\boldsymbol{J}$. We also use the relationships $\omega_c = eB/m$, $\omega_p = ne^2/m\varepsilon_0$, $c^2 = 1/\varepsilon_0\mu_0$, $V_s^2 \approx p/\rho$.

Take the first two terms on the right of Eq. (2.12.15). It is generally the case that the velocity v is of the order E/B, hence these terms are the same order of magnitude. Now (for electrostatic cases) compare the left-hand side term with the $\boldsymbol{E}$ term; it helps to take the divergence of both. The ratio of the terms is:

$$\frac{m}{ne^2}\frac{\partial}{\partial t}(\nabla\cdot\boldsymbol{J}):\nabla\cdot\boldsymbol{E} = \frac{m}{ne^2}\frac{\partial^2\rho}{\partial t^2}:\frac{\rho}{\varepsilon_0}.$$

Putting $\partial/\partial t = \mathrm{i}\omega$, the ratio may be written

$$\frac{\varepsilon_0 m\omega^2}{ne^2} = \frac{\omega^2}{\omega_p^2},$$

demonstrating that the left-hand side term matters only at frequencies approaching the plasma frequency. For the MHD case we compare with the $\boldsymbol{v} \times \boldsymbol{B}$ term, and use $B/L \approx \mu_0 J$, to obtain the ratio

$$\left(\frac{\omega}{\omega_\mathrm{p}} \frac{c}{v}\right)^2.$$

Putting $J = nev_\mathrm{c}$, where v_c represents the velocity of the charge carriers producing the current, the ratio of the Hall term to the $(\boldsymbol{v} \times \boldsymbol{B})$ term is of the order v_c/v, and the Hall term may be neglected for small currents and high densities. Alternatively, using the relation $B/L \approx \mu_0 J$ one can show that the ratio is of the order

$$\left(\frac{\omega\omega_\mathrm{ce}}{\omega_p^2}\right)\left(\frac{c^2}{v^2}\right),$$

again showing that the term may be neglected for sufficiently low frequencies.

The ratio of the ∇p_e term to the $(\boldsymbol{v} \times \boldsymbol{B})$ term for MHD is of the order

$$\frac{\omega}{\omega_\mathrm{ci}}\left(\frac{V_\mathrm{s}}{v}\right)^2.$$

Finally, the ratio of the resistive term to the $\boldsymbol{E}$ term for electrostatic cases is $\omega\nu/\omega_\mathrm{p}^2$, where ν is the collision frequency, and is $(\omega\nu/\omega_\mathrm{p}^2)(c/v)^2$ for MHD.

Now we must return to consider the convective terms in Eq. (2.12.12), which we so glibly neglected earlier in this section. The ratio of the $\boldsymbol{u}_i$ term to the $(\boldsymbol{v} \times \boldsymbol{B})$ term is $\omega/\omega_\mathrm{ce}$, and is thus negligible for frequencies small compared to the electron cyclotron frequency. For the $\boldsymbol{u}_\mathrm{e}$ term, the ratio is

$$\frac{\frac{m}{e}(\boldsymbol{u}_\mathrm{e}\cdot\nabla)\boldsymbol{u}_\mathrm{e}}{vB} \approx \frac{vu_\mathrm{e}^2}{Lv^2\omega_\mathrm{ce}} \approx \frac{\omega}{\omega_\mathrm{ce}}\left(\frac{u_\mathrm{e}}{v}\right)^2.$$

giving the additional condition that the electron fluid velocity should be not large compared with the global velocity.

These arguments are not totally reliable: we have ignored all vector relationships, which may not always be admissible. We do see that it may well be necessary to consider the convective terms in some circumstances. An interesting example where their effect is important has been highlighted by Jones and Hugrass (1981) and Hugrass (1982), who point out that for frequencies ω intermediate between ω_ci and ω_ce, the magnetic field is tied to the electrons, but not to the ions. In effect, the field is frozen into the species with the smallest Larmor radius; the field is even tied preferentially to the lower energy electrons in the distribution. A practical application of the effect is seen in the rotamak (Hugrass *et al.*, 1980), where a rotating magnetic field is used to drive a continuous unidirectional electron current.

In conclusion to this section, the two single-fluid equations we have just derived are the very powerful starting point for the wide subject of magnetohydrodynamics (MHD).

2.13 The MHD equations

For low frequency phenomena, the MHD equations are now found, from the results obtained in Sections 2.11 and 2.12 as Eqs (2.11.20), (2.12.2), (2.12.11), and (2.12.15), to be:

$$\frac{\partial \rho}{\partial t} + \nabla \cdot (\rho \boldsymbol{v}) = 0. \tag{2.13.1}$$

$$\rho \frac{\mathrm{d}\boldsymbol{v}}{\mathrm{d}t} = (\boldsymbol{J} \times \boldsymbol{B}) - \nabla p. \tag{2.13.2}$$

$$\frac{p}{\rho^{\gamma}} = \text{const.}, \tag{2.13.3}$$

$$\boldsymbol{E} + (\boldsymbol{v} \times \boldsymbol{B}) = \eta \boldsymbol{J}, \tag{2.13.4}$$

and we include from Maxwell's equations

$$\nabla \times \boldsymbol{E} = -\frac{\partial \boldsymbol{B}}{\partial t}, \tag{2.13.5}$$

$$\nabla \times \boldsymbol{B} = \mu_0 \boldsymbol{J}. \tag{2.13.6}$$

There are 14 scalar variables $(\rho, p, \boldsymbol{v}, \boldsymbol{J}, \boldsymbol{E}, \boldsymbol{B})$ and 14 equations, which is an encouraging outcome from our analysis, showing that the set is complete and in principle soluble.

If the resistivity η is neglected, we find that $\boldsymbol{E}$ can be eliminated, giving

$$\nabla \times (\boldsymbol{v} \times \boldsymbol{B}) = \frac{\partial \boldsymbol{B}}{\partial t}. \tag{2.13.7}$$

This situation, in which the effects of resistivity and electron inertia are neglected, is commonly known as *ideal MHD*.

2.14 An application of kinetic theory: electron plasma waves

These are waves at very high frequency, close to the plasma frequency, and are in essence compressional waves in the electron fluid. The ions remain virtually fixed at these frequencies, and may be treated merely as a constant background. Electron plasma waves (Langmuir waves) are an excellent example of the power of the kinetic theory approach. The dispersion relation can be derived from the fluid equations (see Chapter 3), which yield

$$\omega^2 = \omega_{\mathrm{p}}^2 + \gamma v_{\mathrm{T}}^2 k^2, \tag{2.14.1}$$

where v_{T} is the electron thermal speed. However, some very important and significant physics has been lost in this derivation. The fluid equations, as we have seen, are obtained by averaging over the distribution functions, and thus any effects due to nonequilibrium (non-Maxwellian) distribution functions have been irretrievably lost. When the same phenomenon – electron plasma waves – is approached using the kinetic equations, we find a new, interesting and important phenomenon, known as Landau damping after the Soviet physicist Lev D. Landau who first predicted it mathematically (Landau, 1946).

The essence of Landau damping is that the electron plasma waves are strongly damped, even in the complete absence of resistivity, and when there is therefore absolutely no dissipative mechanism in the conventional sense. The phenomenon is not restricted to electron plasma waves, and similar damping is widely observed in other circumsances. We shall present a simplified derivation of the effect, and then discuss the physical interpretation of this strange behaviour.

2.14.1 *The dispersion relation*

We start by going right back to the Vlasov equation for the electron gas. We are dealing with a high frequency phenomenon, and so the ions can be considered to be a stationary background system taking no part in the oscillations. We assume no magnetic field, and we shall work in a one-dimensional system for simplicity. We have

$$\frac{\partial f}{\partial t} + v\frac{\partial f}{\partial x} + \frac{qE}{m}\frac{\partial f}{\partial v} = 0. \tag{2.14.2}$$

When a wave is propagating through the electron gas, the local distribution function, which in the absence of the wave is a Maxwellian $f_0(x, v, t)$, will be perturbed by an amount $\tilde{f}(x, v, t)$. We write the new distribution function f as $f = f_0 + \tilde{f}$. Let $E = \tilde{E}$, a small field due to the wave in the plasma. Making these substitutions, the Vlasov equation may be linearized by retaining only first order terms:

$$\frac{\partial \tilde{f}}{\partial t} + v\frac{\partial \tilde{f}}{\partial x} + \frac{q\tilde{E}}{m}\frac{\partial f_0}{\partial v} = 0. \tag{2.14.3}$$

We now assume that the perturbations are due to plane waves of the form $\exp(ikx - i\omega t)$. We can put $\partial/\partial t \to -i\omega$, $\partial/\partial x \to ik$. We obtain

$$-i\omega\tilde{f} + ikv\tilde{f} + \frac{q\tilde{E}}{m}\frac{\partial f_0}{\partial v} = 0; \tag{2.14.4}$$

therefore

$$\tilde{f} = -\frac{q\tilde{E}}{m}\frac{\partial f_0/\partial v}{i(kv - \omega)}. \tag{2.14.5}$$

The density perturbation can be found as $\tilde{n} = \int_{-\infty}^{+\infty} \tilde{f}\,dv$; therefore

$$\tilde{n} = -\frac{q\tilde{E}}{im}\int_{-\infty}^{+\infty} \frac{\partial f_0/\partial v}{kv - \omega}\,dv. \tag{2.14.6}$$

Let $\tilde{E} = -\partial\tilde{\phi}/\partial x = -ik\tilde{\phi}$. Poisson's equation gives $\partial\tilde{E}/\partial x = \tilde{\rho}/\varepsilon_0$, from which $k^2\tilde{\phi} = \tilde{n}q/\varepsilon_0$, and we have

$$k^2 = \frac{q^2}{m\varepsilon_0}\int_{-\infty}^{+\infty} \frac{\partial f_0/\partial v}{v - \omega/k}\,dv. \tag{2.14.7}$$

This is, in fact, a dispersion relation, although not yet in a comprehensible form. We observe that the integrand contains a singularity at $v = \omega/k$. The proper way to handle this is to use contour integration in the complex plane: the solution can be found in standard texts (for example Bittencourt, 1986). However, we can

occurs if $T_i \geqslant T_e$ and the particle velocity is comparable with the wave velocity. It is observed that ion acoustic waves only propagate in plasmas for which $T_e \gg T_i$. Such conditions are, however, quite commonly encountered in laboratory plasmas.

There are very interesting examples of Landau damping, and other aspects related to plasma kinetic theory, in other areas of physics. For example, an assembly of particles which are gravitationally coupled, (i.e. stars in galaxies), is subject to much the same physics as a plasma with Coulomb coupling. Some of the techniques and results described in this chapter will be encountered again in Chapter 11, when we consider gravitational plasmas.

2.17 Appendix: Tensor basics

We work in two dimensions, for simplicity of expression.

The *tensor product* or *dyad* $\boldsymbol{ab}$ of two vectors $\boldsymbol{a}$ and $\boldsymbol{b}$ is defined by

$$\boldsymbol{T} = \boldsymbol{ab} = \begin{pmatrix} a_x b_x & a_x b_y \\ a_y b_x & a_y b_y \end{pmatrix}.$$

The *tensor dot product* is itself a vector, and is defined as

$$\boldsymbol{T}\cdot\boldsymbol{c} = \begin{pmatrix} a_x b_x & a_x b_y \\ a_y b_x & a_y b_y \end{pmatrix}\begin{pmatrix} c_x \\ c_y \end{pmatrix}; \quad \boldsymbol{c}\cdot\boldsymbol{T} = (c_x \quad c_y)\begin{pmatrix} a_x b_x & a_x b_y \\ a_y b_x & a_y b_y \end{pmatrix}.$$

From these definitions we can obtain the following relations:

$$(\boldsymbol{ab})\cdot\boldsymbol{c} = \boldsymbol{a}(\boldsymbol{b}\cdot\boldsymbol{c}) = (\boldsymbol{c}\cdot\boldsymbol{b})\boldsymbol{a},$$

$$\boldsymbol{c}\cdot(\boldsymbol{ab}) = (\boldsymbol{c}\cdot\boldsymbol{a})\boldsymbol{b}$$

and

$$\nabla\cdot(\boldsymbol{ab}) = \boldsymbol{b}(\nabla\cdot\boldsymbol{a}) + (\boldsymbol{a}\cdot\nabla)\boldsymbol{b}.$$

The product $\boldsymbol{S}:\boldsymbol{T}$ of two tensors $\boldsymbol{S}$ and $\boldsymbol{T}$ follows the rules of matrix multiplication, and is itself a tensor:

$$\boldsymbol{S}:\boldsymbol{T} = \begin{pmatrix} S_{xx} & S_{xy} \\ S_{yx} & S_{yy} \end{pmatrix}\begin{pmatrix} T_{xx} & T_{xy} \\ T_{yx} & T_{yy} \end{pmatrix}.$$

References

BHATNAGAR, P. L., GROSS, E. P. and KROOK, M. (1954). *Phys. Rev.*, **94**, 511.

BITTENCOURT, J. A. (1986). *Fundamentals of Plasma Physics*. Pergamon, Oxford.

BOYD, T. J. M. and SANDERSON, J. J. (1969). *Plasma Dynamics*. Nelson, London.

CHANDRASEKHAR, S. (1960). *Principles of Stellar Dynamics*. University of Chicago Press.

CHEN, F. F. (1984). *Introduction to Plasma Physics and Controlled Fusion*. Plenum, New York.

CLEMMOW, P. C. and DOUGHERTY, J. P. (1969). *Electrodynamics of Particles and Plasmas*. Addison-Wesley, USA.

HUGRASS, W. N. (1982) *J. Plasma Phys.*, **28**, 369.

HUGRASS, W. N., JONES, I. R., McKENNA, K. F., PHILLIPS, M. G. R., STORER, R. G. and TUCZEK, H. (1980). *Phys. Rev. Lett.*, **44**, 1676.

JONES, I. R. and HUGRASS, W. N. (1981). *J. Plasma Phys.*, **26**, 441.

KRALL, N. A. and TRIVELPIECE, A. W. (1973). *Principles of Plasma Physics*. McGraw-Hill, USA.

LANDAU, L. D. (1946). *J. Phys. (USSR)*, **10**, 25.

3

Waves in plasmas

J. P. DOUGHERTY

3.1 Introduction

For any dynamical system, much can be learnt by investigating the possible modes of small-amplitude oscillations, or waves. In the case of a continuous medium, one can begin by simplifying the problem even further, and consider plane waves propagating in an infinite homogeneous and time-independent medium; in later work some of these idealizations can be removed. For example, in an ideal, neutral gas, this procedure leads to the theory of sound waves and the well-known fact that the speed of (small-amplitude) plane sound waves is independent of frequency. By introducing more physics (for example, heat conduction and viscosity) we could enquire about the decay of sound waves.

A plasma is physically much more complicated than an ideal gas. This is especially true if there is an externally applied magnetic field. As a result, the properties of plane waves show considerable variety. Their investigation reveals many aspects of the physics of plasmas. In spite of the idealized nature of the work, the results have been found useful in interpreting observation, both in laboratory experiments and in natural plasmas. Radio-wave propagation in the ionosphere was indeed an early stimulus for the development of the theory.

In this chapter, we shall study plane waves in a homogeneous plasma from the continuum point of view, dealing mostly with the 'cold' plasma. Some properties of waves in a 'hot' plasma are also found by adding a pressure term (just as in elementary sound waves); but some phenomena are only revealed by means of a kinetic treatment, for which the reader is referred to Chapter 2.

3.2 Equations of motion

For a single species in a plasma, with number density n, velocity $\boldsymbol{u}$, pressure p, we have Eulerian equations of motion in electric and magnetic fields $\boldsymbol{E}$, $\boldsymbol{B}$,

$$\dot{n} + \nabla \cdot (n\boldsymbol{u}) = 0, \tag{3.2.1}$$

$$nm\frac{\mathrm{D}\boldsymbol{u}}{\mathrm{D}t} = qn(\boldsymbol{E} + \boldsymbol{u} \times \boldsymbol{B}) - \nabla p, \tag{3.2.2}$$

$$\frac{\mathrm{D}}{\mathrm{D}t}(pn^{-\gamma}) = 0, \tag{3.2.3}$$

where the particles have mass m, charge q, γ is the usual specific heat ratio, and $\mathrm{D}/\mathrm{D}t$ is the convective rate of change $\partial/\partial t + \boldsymbol{u} \cdot \nabla$. We have already omitted collisions, viscosity and heat conduction, which are outside the scope of this chapter.

An actual plasma must have at least two species since it is on average electrically neutral. So we need a suffix 's' on the fluid variables (e.g. n_s) to label the species, and Eqs (3.2.1)–(3.2.3) hold for each species. In what follows we shall omit the suffix s in passages where we are only concerned with one species at a time. Our practical calculations will be restricted to the case where there are just two species, namely positive ions with charge $+e$ and electrons with charge $-e$. The corresponding suffixes s will then be denoted by i and e. However, the generalization to the case of several species of ions, possibly not singly charged, is straightforward although cumbersome; therefore some equations will appear with the label s so that the reader can easily imagine what would be involved in extending the theory.

The electric charge density and current are given by

$$\rho = \sum_s n_s q_s, \quad \boldsymbol{J} = \sum_s n_s q_s \boldsymbol{u}_s. \tag{3.2.4}$$

These quantities are the source terms for Maxwell's equations

$$\nabla \cdot \boldsymbol{B} = 0 \tag{3.2.5}$$

$$\nabla \cdot \boldsymbol{E} = \rho/\varepsilon_0 \tag{3.2.6}$$

$$\nabla \times \boldsymbol{B} = \mu_0 \boldsymbol{J} + \varepsilon_0 \mu_0 \dot{\boldsymbol{E}} \tag{3.2.7}$$

$$\nabla \times \boldsymbol{E} = -\dot{\boldsymbol{B}} \tag{3.2.8}$$

(of course $\varepsilon_0 \mu_0 = c^{-2}$).

To deal with small-amplitude waves, we assume a 'background' situation representing a uniform infinite plasma. The values of n, $\boldsymbol{u}$, p, $\boldsymbol{E}$, $\boldsymbol{B}$ for this will be denoted by n_0, etc.; however, here we shall take $\boldsymbol{u} = \boldsymbol{E} = 0$ in the unperturbed state. Clearly then $\boldsymbol{J} = 0$, and all of Eqs (3.2.1)–(3.2.8) are satisfied except possibly Eq. (3.2.6), which requires $\rho = 0$, whence

$$\sum_s n_s q_s = 0, \tag{3.2.9}$$

the condition of charge neutrality. For our simple two-species plasma this becomes

$$n_{0\mathrm{i}} = n_{0\mathrm{e}} \quad (= n_0, \text{ say}). \tag{3.2.10}$$

We now introduce perturbations, denoted by the suffix 1, namely

$$n = n_0 + n_1, \quad p = p_0 + p_1, \quad \boldsymbol{B} = \boldsymbol{B}_0 + \boldsymbol{B}_1. \tag{3.2.11}$$

For the other variables, which vanish in the unperturbed state, the labels 0 and 1 are unnecessary. We insert the perturbed values into Eqs (3.2.1)–(3.2.3), neglect all second order terms, and obtain

$$\dot{n}_1 + n_0 \nabla \cdot \boldsymbol{u} = 0, \tag{3.2.12}$$

$$n_0 m \dot{\boldsymbol{u}} = q n_0 (\boldsymbol{E} + \boldsymbol{u} \times \boldsymbol{B}_0) - \nabla p_1, \tag{3.2.13}$$

$$\frac{p_1}{p_0} = \gamma \frac{n_1}{n_0}. \tag{3.2.14}$$

In dealing with Eq. (3.2.3) it has been assumed that n_1 and p_1 do not include constant parts that could have been absorbed into n_0 and p_0. Taking each species to be a perfect gas, with unperturbed temperature T (which could be different for each species) we have $p_0 = n_0 k_B T$ (k_B = Boltzmann's constant) so Eq. (3.2.14) becomes

$$p_1 = \gamma k_B T n_1. \tag{3.2.15}$$

As Maxwell's equations are already linear, Eqs (3.2.5)–(3.2.8) can be simply rewritten with the perturbation quantities replacing the general ones.

To develop the theory of plane waves, we search for solutions of the linearized equations in which all perturbation quantities are proportional to*

$$\exp[\mathrm{i}(\omega t - \boldsymbol{k}\cdot\boldsymbol{x})]. \tag{3.2.16}$$

Each variable $n_1, \ldots, \boldsymbol{E}, \ldots$ is represented in magnitude and phase by a complex amplitude, to which the same symbol is ascribed. The physical value of the variable is obtained by inserting the factor (3.2.16) and taking the real part. With this convention, $\partial/\partial t$ is replaced by $\mathrm{i}\omega$, and the operator $\boldsymbol{\nabla}$ by $-\mathrm{i}\boldsymbol{k}$.

Before proceeding in detail with such calculations, let us point out some special features of Maxwell's equations that result from (3.2.16). As is well known, the divergence operator, applied to Eqs (3.2.7) and (3.2.8) yields the time derivatives of Eqs (3.2.6) and (3.2.5), respectively, after taking account of the equation of conservation of charge

$$\dot{\rho} + \boldsymbol{\nabla}\cdot\boldsymbol{J} = 0 \tag{3.2.17}$$

(and we note that Eq. (3.2.17) is not a new equation, as Eq. (3.2.1) implies conservation of each species). Thus Eqs (3.2.5) and (3.2.6) themselves are not independent of Eqs (3.2.7) and (3.2.8) in the case of harmonic fields. So it is sufficient to use only Eqs (3.2.7) and (3.2.8), and this is what we shall usually do.

As a slight exception, let us examine Eqs (3.2.5)–(3.2.8) in the case of longitudinal waves, that is waves in which the vector wave fields $\boldsymbol{E}, \boldsymbol{B}, \boldsymbol{u}, \boldsymbol{J}$ are all parallel to the wave vector $\boldsymbol{k}$. Since $\nabla \times \boldsymbol{E} = 0$ becomes $-\mathrm{i}\boldsymbol{k} \times \boldsymbol{E} = 0$, we find $\boldsymbol{B}_1 = 0$, so Eq. (3.2.5) is redundant. With $\nabla \times \boldsymbol{B}_1 = 0$, Eq. (3.2.7) becomes equivalent to Eq. (3.2.6) since $\boldsymbol{J}$ and $\boldsymbol{E}$ are parallel to $\boldsymbol{k}$. One may therefore use either Eq. (3.2.6) or Eq. (3.2.7), and it seems physically more natural to use Eq. (3.2.6), since this can be interpreted as saying that Maxwell's equations have been reduced to Poisson's equation. This electric field thus behaves as if it were a static one (even though it is in fact time-dependent).

Transverse waves are those having the vector fields perpendicular to $\boldsymbol{k}$ (of course $\boldsymbol{B}_1$ is always so, by Eq. (3.2.5)). By Eq. (3.2.6), ρ must vanish for such waves. Transverse waves are rather similar to vacuum electromagnetic waves.

Commonly, in plasma physics, waves are neither longitudinal nor transverse, but have a more complicated structure.

* Although this sign convention is the opposite of that used in other chapters of this book, it is the standard one in texts on this subject.

3.3 Waves in an unmagnetized plasma

3.3.1 *Plasma oscillations*

In this section we will take $\boldsymbol{B}_0 = 0$ throughout. To begin with, let us also take $T = 0$, and consider the motion of electrons only, with the ions staying at rest as a uniform background (formally, take $m_i \to \infty$). Eq. (3.2.13) for the electrons, is then

$$m_e i\omega \boldsymbol{u} = -e\boldsymbol{E}. \tag{3.3.1}$$

By (3.2.12)

$$n_1 = -\frac{n_0 e}{m_e \omega^2} \nabla \cdot \boldsymbol{E}.$$

Thus

$$\rho = \frac{n_0 e^2}{m_e \omega^2} \nabla \cdot \boldsymbol{E}. \tag{3.3.2}$$

At this point we can choose to seek a longitudinal wave, so that $\nabla \cdot \boldsymbol{E} \neq 0$. To do this we need only combine Eq. (3.3.2) with Eq. (3.2.6), and we find that

$$\omega^2 = \frac{n_0 e^2}{m_e \varepsilon_0} = \omega_{pe}^2 \tag{3.3.3}$$

is required; here ω_{pe} is the electron plasma frequency. This solution describes plasma oscillations. The electrons are compressed and rarefied in the manner of sound waves but the restoring force is the field $\boldsymbol{E}$ set up by charge separation, rather than pressure. Eq. (3.3.3) is our first example of a dispersion relation, however it is conspicuously different from the corresponding result for sound waves ($\omega^2 = V_s^2 k^2$ where V_s is the speed of sound). In plasma oscillations, it appears that only one frequency is permitted, namely ω_{pe}, irrespective of the wave number. Indeed, the wave number did not appear in the derivation, and it is not necessary to assume any particular spatial form for the disturbance. Related to this is the fact that the group velocity $\partial\omega/\partial k$ vanishes identically. This suggests that a wave packet made of plasma oscillations would not propagate, and in fact it is easy to see that if only a part of the medium is disturbed initially, the rest of it remains undisturbed. The word 'oscillation' may be preferred to 'wave' for this reason.

A straighforward exercise for the reader is to reintroduce the ion dynamics, by writing an analogue of Eq. (3.3.1) for the ions and computing the total electrical charge. One finds that, since $m_i \gg m_e$, the ions oscillate at a much lower amplitude than the electrons, and the outcome is merely a slight correction to the frequency, namely

$$\omega^2 = \omega_p^2 = \omega_{pe}^2 + \omega_{pi}^2 = \omega_{pe}^2\left(1 + \frac{m_e}{m_i}\right). \tag{3.3.4}$$

3.3.2 *Transverse waves*

The step in which we combined Eqs (3.3.2) and (3.2.6) becomes meaningless if $\nabla \cdot \boldsymbol{E}$ should happen to vanish; this indicates the possibility of the existence of transverse waves (which would not satisfy Eq. (3.3.4)). To find them, we return to Maxwell's equations, and follow the more usual course of using the two curl equations, as mentioned earlier. We shall need the current density, and for the electrons the contribution is $-n_0 e\boldsymbol{u}$, where $\boldsymbol{u}$ is given by Eq. (3.3.1), leading to

$$\boldsymbol{J} = \frac{n_0 e^2}{m_e \mathrm{i}\omega}\boldsymbol{E} = \sigma \boldsymbol{E}, \tag{3.3.5}$$

say. Here σ is the 'AC conductivity'. More generally, σ would be a sum of terms of this type (note that the sign of the charge is irrelevant).

It is convenient to define a vector $\boldsymbol{D}$ so that the right-hand side of Eq. (3.2.7) is $\mu_0 \dot{\boldsymbol{D}}$, just as for simple dielectric media. Thus

$$\mathrm{i}\omega \boldsymbol{D} = \boldsymbol{J} + \varepsilon_0 \mathrm{i}\omega \boldsymbol{E} \tag{3.3.6}$$

or

$$\boldsymbol{D} = \varepsilon_0 \varepsilon \boldsymbol{E}, \tag{3.3.7}$$

where ε is the (dimensionless) dielectric constant:

$$\varepsilon = 1 + \frac{\sigma}{\mathrm{i}\omega\varepsilon_0} = 1 - \frac{n_0 e^2}{\varepsilon_0 m_e \omega^2} = 1 - \frac{\omega_{pe}^2}{\omega^2}. \tag{3.3.8}$$

We may then write Eqs (3.2.7) and (3.2.8) as

$$\nabla \times \boldsymbol{B}_1 = \frac{1}{c^2}\varepsilon \mathrm{i}\omega \boldsymbol{E}, \quad \nabla \times \boldsymbol{E} = -\mathrm{i}\omega \boldsymbol{B}_1. \tag{3.3.9}$$

These equations are identical with those encountered in the elementary development of electromagnetic waves in a simple dielectric, and (assuming $\varepsilon \neq 0$) the solutions have $\boldsymbol{E}$ and $\boldsymbol{B}$ both perpendicular to $\boldsymbol{k}$. The refractive index, N, is just $\varepsilon^{\frac{1}{2}}$ so we have

$$N = \frac{ck}{\omega} = \left(1 - \frac{\omega_{pe}^2}{\omega^2}\right)^{1/2} \tag{3.3.10}$$

($k = |\boldsymbol{k}|$). This can also be written

$$\omega^2 = \omega_{pe}^2 + c^2 k^2. \tag{3.3.11}$$

As with electromagnetic waves in vacuo, two independent polarizations are possible, but both satisfy Eq. (3.3.10): one could choose right-handed and left-handed circularly polarized waves, for example, or superpose them to make plane polarized waves.

An unfamiliar aspect of Eq. (3.3.10) is that, if $\omega > \omega_{pe}$, $N < 1$, implying that the phase speed exceeds c. However, from Eq. (3.3.11), the group speed (which is vectorially parallel to $\boldsymbol{k}$) is $\partial\omega/\partial k = c^2 k/\omega = cN < c$. Therefore, signalling takes place at below the speed of light, in accordance with the special theory of relativity. Also, for $\omega < \omega_{pe}$, N becomes imaginary, and a glance at Eq. (3.3.11) shows that $k = \pm \mathrm{i}(\omega_{pe}^2 - \omega^2)^{1/2}/c$ in this case. Propagation in the ordinary sense is not then possible since the exponent $-\mathrm{i}\boldsymbol{k}\cdot\boldsymbol{x}$ is real. To understand this physically, consider an interface such that in $x < 0$ we have a vacuum, in $x > 0$ a plasma of density n_0. An electromagnetic wave with $\omega < \omega_{pe}$ is incident normally from $x < 0$. It is easy to show that the reflection problem is solved by having a totally reflected wave in $x < 0$, whereas in $x > 0$ the wave field penetrates only slightly, being proportional to

$$\exp[-(\omega_{pe}^2 - \omega^2)^{1/2} x/c]. \tag{3.3.12}$$

Such a wave is known as *evanescent*, and the medium is described as *over-dense*. If $\omega \ll \omega_{pe}$, the depth of penetration is of order c/ω_{pe}, the plasma wavelength.

In Fig. 3.1 we give two plots in common use to describe Eqs (3.3.10) and (3.3.11), namely N^2 as a function of ω, and ω as a function of k. The latter is particularly suitable for describing the phase and group velocities since these are given by ω/k and $\partial\omega/\partial k$, respectively.

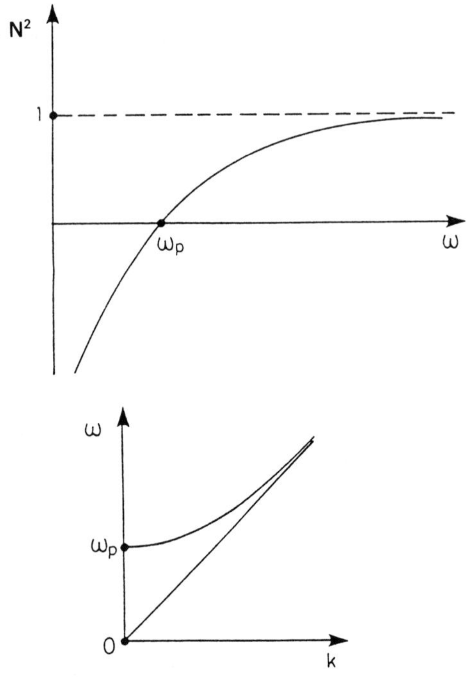

Fig. 3.1. Propagation in an unmagnetized plasma. N^2 as a function of ω, and ω as a function of k.

It may be asked how the plasma oscillations can be obtained from the curl equations (3.3.9), since those equations appear to demand that $\boldsymbol{E}$ and $\boldsymbol{B}_1$ be perpendicular to $\boldsymbol{k}$. The answer is that there is a singular solution with $\boldsymbol{B}_1 = 0$ and $\boldsymbol{E}$ parallel to $\boldsymbol{k}$, *provided* $\varepsilon = 0$; that condition is, by Eq. (3.3.8), just Eq. (3.3.3).

It is easy to show that inclusion of ion dynamics makes exactly the same (uninteresting) correction as we noted when dealing with plasma oscillations.

3.3.3 *Effect of pressure*

Here we investigate the effect of pressure on the dynamics of the ions and electrons. This means we must readmit the ∇p_1 term in Eq. (3.2.13) (while still omitting the $\boldsymbol{B}_0$ term). Using Eqs (3.2.12) and (3.2.15) to eliminate p_1 and n_1 we find

however, the dispersion curves can readily be computed numerically for any given situation.

In practice, the sound speed (for each species) is much less than c, and usually the Alfvén speed is much less than c also. These inequalities permit approximations which deal fairly well with the new features, without recourse to the involved analysis which the exact treatment requires.

We note first that the pressure term in Eq. (3.5.1) has no effect on purely transverse waves, as $\boldsymbol{k} \cdot \boldsymbol{u} = 0$. So the circularly polarized modes for $\theta = 0$, and the ordinary mode for $\theta = \frac{1}{2}\pi$, remain as exact solutions. So do our plasma and ion sound waves of Section 3.3.3 if $\theta = 0$, since $\boldsymbol{B}_0$ has no effect on them.

Next we ask whether the pressure term in Eq. (3.5.1) is negligible owing to V_s being small. Comparing the term with the left-hand side, we can expect this to be so provided $\omega^2/|k|^2 \gg V_s^2$, that is the phase speed should be much greater than the sound speed. But many of our results in Section 3.4 described waves whose phase speeds were of order c; for them it is legitimate to ignore pressure effects. So we need only seek corrections where $N \gg 1$. Fig. 3.2 discloses two areas for study in this respect.

The first concerns the 'resonances' where $N \to \infty$ (see Eq. (3.4.32) and the ensuing discussion). The pressure term modifies the resonances so that N, although large, does not become infinite. The curves $N(\omega)$ are therefore qualitatively altered in those regions, and similarly the refractive index surfaces are modified. But since N is large, simplification of the analysis can be obtained by replacing the full set of Maxwell's equations by electrostatics, as described earlier. The resulting approximate dispersion relation really does relate ω and $\boldsymbol{k}$ rather than merely pick out the resonance frequencies, as did Eq. (3.4.32); this dispersion relation gives the form of the modifications needed near the resonant frequencies.

The second area for study is the subject of the next section.

3.5.2 *Waves at low frequency*

For $\omega \ll \omega_{ci}$ our 'pressureless' theory described Alfvén waves, and since usually $V_A \ll c$ we may expect to find modifications due to the pressure term. But if $\theta = 0$ the waves are transverse and there is no effect (as already remarked). For general θ, the shear Alfvén wave has $\boldsymbol{k} \cdot \boldsymbol{u} \approx 0$, so the pressureless theory should once more be adequate, but for the compressional Alfvén wave a fresh calculation is indicated. Even this is quite lengthy, and the results will only be summarized here. For details, the reader is referred to Stringer (1963) and Clemmow and Dougherty (1990).

In the limit $\omega \to 0$, with k/ω finite, there are three modes of propagation for a given direction θ; for example if $\theta = 0$ we have the two Alfvén waves and the ion sound wave. For general θ one of the modes is the shear Alfvén wave, with phase speed $V_A \cos\theta$, and the other two are given by the quadratic for N^2

$$V_{cs}^2 \cos^2\theta N^4 - \left(\frac{V_{cs}^2 \omega_p^2}{\omega_{ce}\omega_{ci}} + c^2 + V_{cs}^2 \cos^2\theta\right) N^2 + c^2 \left(1 + \frac{\omega_p^2}{\omega_{ce}\omega_{ci}}\right) = 0, \quad (3.5.2)$$

where V_{cs} is the composite sound speed already encountered at Eq. (3.3.21):

$$V_{cs}^2 = \frac{\gamma k_B (T_i + T_e)}{m_i + m_e}. \quad (3.5.3)$$

We recall that the Alfvén speed $V_A = c(\omega_{ci}\omega_{ce})^{1/2}/\omega_p$ (cf. Eq. (3.4.19)), and if $V_A \ll c$, Eq. (3.5.2) gives approximately

$$\frac{\omega^2}{k^2} = \tfrac{1}{2}\{V_{cs}^2 + V_A^2 \pm [(V_{cs}^2 + V_A^2)^2 - 4V_{cs}^2 V_A^2 \cos^2\theta]^{1/2}\}. \tag{3.5.4}$$

The two modes given by Eq. (3.5.4) are called the *fast* and *slow* magnetosonic waves (for upper and lower signs, respectively); they are characterized by

$$\frac{\omega}{k} \gtrless \text{both } V_{cs}, V_A,$$

respectively. In the limit $\theta \to \frac{1}{2}\pi$, the phase speeds are $(V_{cs}^2 + V_A^2)^{1/2}$ and 0, respectively. The fast wave in this case has a clear physical interpretation: the restoring force is a combination of gas pressure and magnetic pressure. Note that the ratio V_{cs}^2/V_A^2 is of course just the plasma parameter commonly denoted by β.

3.6 Waves on plasma streams

3.6.1 *The dispersion relation*

Another way in which our work can be generalized arises from permitting the unperturbed state to involve steady motion of the plasma particles; for instance each species s could have an unperturbed flow velocity $\boldsymbol{u}_{0s}$. Still more generally there could be a number of streams, labelled by s, having different velocities $\boldsymbol{u}_{0s}$, but the particles need not all be of different species. Such a state of affairs may arise in beam experiments. Of course nothing new occurs if all the $\boldsymbol{u}_{0s}$ are the same, but if they differ the propagation of waves is modified, and in particular there are instabilities. We investigate this briefly, confining ourselves to the case $\boldsymbol{B}_0 = 0$, $T = 0$.

First consider the unperturbed situation. Since $\boldsymbol{B}_0 = 0$, any constant values of $\boldsymbol{u}_{0s}$ may be adopted, but we must have $\boldsymbol{E} = 0$; the momentum equation for each species is then satisfied. The streams must be such that the total charge and current densities vanish in Eq. (3.2.4). (Actually the latter condition is sometimes ignored, since for short wavelengths and a plasma of finite extent, the resulting magnetic field generated by the plasma may be neglected elsewhere in the calculation.)

In the perturbed state, with $\boldsymbol{u}_s = \boldsymbol{u}_{0s} + \boldsymbol{u}_{1s}$, the linearized momentum equation is

$$m\left(\frac{\partial \boldsymbol{u}_1}{\partial t} + \boldsymbol{u}_0 \cdot \nabla \boldsymbol{u}_1\right) = q\boldsymbol{E} + q\boldsymbol{u}_0 \times \boldsymbol{B}_1. \tag{3.6.1}$$

If $\boldsymbol{k}$, $\boldsymbol{u}_0$ and $\boldsymbol{E}$ are in general directions, the last term of Eq. (3.6.1) must be retained; however we will omit it since it disappears in the only application that we will consider; nevertheless we will retain general vector notation a little longer. Using our usual harmonic factor, Eq. (3.6.1) becomes

$$m[\mathrm{i}\omega \boldsymbol{u}_1 - \mathrm{i}(\boldsymbol{k} \cdot \boldsymbol{u}_0)\boldsymbol{u}_1] = q\boldsymbol{E},$$

whence

$$\boldsymbol{u}_1 = \frac{q\boldsymbol{E}}{\mathrm{i}m(\omega - \boldsymbol{k}\cdot\boldsymbol{u}_0)}. \tag{3.6.2}$$

The equation of continuity reads

$$\dot{n}_1 = -\boldsymbol{\nabla}\cdot(n\boldsymbol{u}) = -\boldsymbol{\nabla}\cdot(n_0\boldsymbol{u}_1 + n_1\boldsymbol{u}_0)$$

to the first order, and combining with Eq. (3.6.2) one finds,

$$n_1 = \frac{n_0 q}{\mathrm{i}m(\omega - \boldsymbol{k}\cdot\boldsymbol{u}_0)^2\, \boldsymbol{k}\cdot\boldsymbol{E}} \tag{3.6.3}$$

with resulting charge density $n_1 q$.

Let us use this result to consider plasma oscillations in a medium composed of such streams, with $\boldsymbol{u}_{0\mathrm{s}}$ (for all s), $\boldsymbol{k}$, and $\boldsymbol{E}$ all parallel. (Note that the 'omitted' term of Eq. (3.6.1) then vanishes as $\boldsymbol{B}_1 = 0$.) For this problem, only Poisson's equation

$$-\mathrm{i}\boldsymbol{k}\cdot\boldsymbol{E} = \varepsilon_0^{-1}\sum_{\mathrm{s}} q_{\mathrm{s}} n_{1\mathrm{s}}$$

is needed, and (dropping vector notation) we find

$$\sum_{\mathrm{s}} \frac{\omega_{\mathrm{ps}}^2}{(\omega - k u_{0\mathrm{s}})^2} = 1. \tag{3.6.4}$$

This is the generalization of the results of Section 3.3.1. The effect of the streaming velocities $\boldsymbol{u}_{0\mathrm{s}}$ is simply to impose a Doppler shift on the frequency ω.

3.6.2 *The two-stream instability*

Just two streams are sufficient to exhibit the possibility of instability contained in Eq. (3.6.4). Fig 3.6 shows the left-hand side plotted as a function of ω. Depending on the circumstances, the horizontal line representing the right-hand side may meet the curve either two or four times. If it meets it only twice, there must be a complex conjugate pair of complex roots in ω; one of these implies growth, and hence instability. The instability is helped by higher density; however, for given stream densities it can always be obtained by choosing a low enough k, i.e. long enough wavelength.

Fig. 3.6. The two-stream instability.

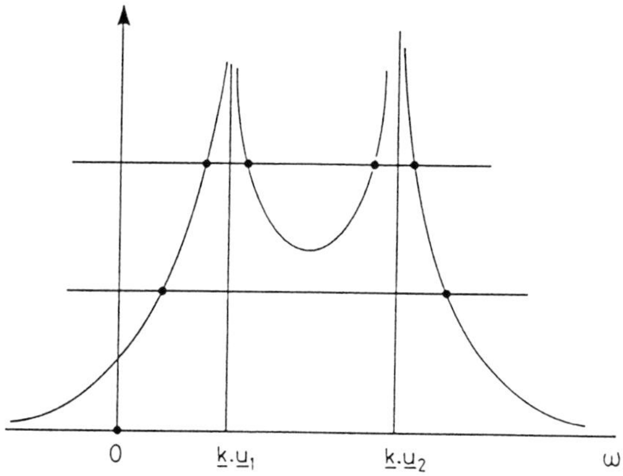

As a worked example, we may take a plasma with infinitely massive ions, and two streams of electrons with density $\frac{1}{2}n$ and velocity $\pm\frac{1}{2}v$. The dispersion relation (3.6.4) becomes

$$\frac{\frac{1}{2}}{\left(\dfrac{\omega}{\omega_{\mathrm{p}}}-\dfrac{kv}{2\omega_{\mathrm{p}}}\right)^2}+\frac{\frac{1}{2}}{\left(\dfrac{\omega}{\omega_{\mathrm{p}}}+\dfrac{kv}{2\omega_{\mathrm{p}}}\right)^2}=1, \qquad (3.6.5)$$

where ω_{p} is the plasma frequency corresponding to density n. The solution is

$$\frac{2\omega^2}{\omega_{\mathrm{p}}^2}=1+\frac{k^2v^2}{2\omega_{\mathrm{p}}^2}\pm\left(1+\frac{2k^2v^2}{\omega_{\mathrm{p}}^2}\right)^{1/2}, \qquad (3.6.6)$$

and although ω^2 is always real, it is negative (for the lower sign) if $kv < 2\omega_{\mathrm{p}}$. By varying k (ω_{p} and v being fixed) we can select the maximum growth rate. This occurs when $k^2 = 3\omega_{\mathrm{p}}^2/2v^2$, giving $\omega^2 = -\omega_{\mathrm{p}}^2/8$. We note that the growth time for such waves is of order ω_{p}^{-1}.

For cold interpenetrating streams, it appears there is always an instability. However, this is misleading. If the streams are warm, so that in kinetic theory terminology they have a velocity distribution, the situation is different. One finds there is a threshold relative velocity below which there is no instability, for any k.

References

CLEMMOW, P. C. and DOUGHERTY, J. P. (1990). *The Electrodynamics of Particles and Plasmas*. Addison-Wesley, Reading, Mass.
STRINGER, T. E. (1963). *Plasma Phys.*, **5**, 89.
WALKER, A. D. M. (1977a). *J. Plasma Phys.*, **17**, 467.
WALKER, A. D. M. (1977b). *J. Plasma Phys.*, **18**, 339.

4
Magnetohydrodynamics

K. I. HOPCRAFT

4.1 Introduction

The topic of magnetohydrodynamics (MHD) is ubiquitous in plasma physics. Examples where the theory has been used with success range from explaining the spontaneous generation and subsequent evolution of magnetic fields within stellar and planetary interiors, to accounting for the gross stability of magnetically confined thermonuclear plasmas. Often, it is found that the scale length of many instabilities and waves which are able to grow or propagate in a system are comparable with the plasma size. It transpires that MHD is capable of providing a good description of such large scale disturbances, indicating that the MHD account of plasma behaviour is necessarily a macroscopic one. In essence, the theory is a marriage between fluid mechanics and electromagnetism. When compared with the extremely detailed particle and kinetic theories described in Chapters 1 and 2, MHD is a relatively simple theory, perhaps even crude! Despite its apparent simplicity, MHD describes a remarkably rich and varied mix of phenomena and the subject is one whose development continues to flourish.

The relevant figure of merit for determining the conditions for break-even in thermonuclear plasmas is given by the Lawson parameter $n\tau_E$. Here n is the particle density and τ_E the energy confinement time. Roughly speaking, τ_E tends to be affected by microscopic processes occurring in the plasma. These processes can be described by two entirely equivalent methodologies: first there is the particle dynamics approach, leading naturally via statistical mechanics to the alternative kinetic description. Eventually both lead to a macroscopic account of transport processes and, as a consequence, to those important mechanisms responsible for energy distribution and deposition in the plasma. In this way the microscopic behaviour ultimately determines the plasma temperature upon which the Lawson product $n\tau_E$ implicitly depends. Thus a feedback mechanism is established, as illustrated in Fig. 4.1. This figure also serves to orientate the reader to the overall significance of MHD in thermonuclear research. By way of contrast, the particle density tends to be influenced by the macroscopic properties of the plasma, and this regime is within the ambit of MHD. Three general problems of particular importance to which answers may be sought using MHD are:

- finding magnetic field configurations capable of confining a plasma in equilibrium;
- the linear stability properties of such equilibria; and
- the nonlinear development of instabilities and their consequences.

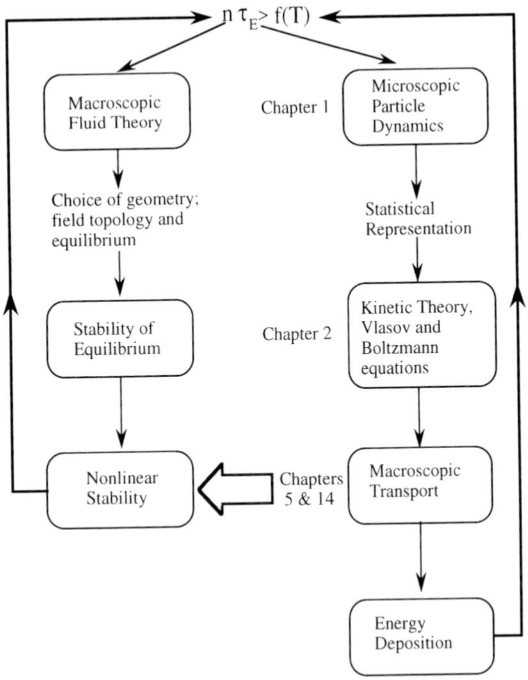

Fig. 4.1. The respective roles of macroscopic and microscopic theories and the relevance of MHD.

Constraints imposed by the solution to these problems can act on n either individually or severally to provide an alternative route for the feedback mechanism, also shown in Fig. 4.1. Of course, this overview of the respective roles of microscopic and macroscopic phenomena is a gross simplification of the true state of affairs. The microscopic description is pre-eminent and necessarily provides considerable input to the macroscopic picture, especially when the nonlinear consequences of MHD are considered.

The purpose of this introduction to MHD is to cover some topics which arise from those points listed above. We will first discuss the MHD model itself, paying attention to the relevant time and length scales for which the theory is valid. This serves to provide clear indications of what the theory can and cannot do. We then distinguish between 'ideal' and 'nonideal' MHD, before examining some important consequences of the theory. In particular, we deduce the constraints on the allowable class of fluid displacements, and discuss what bearing these have on the topology of the magnetic field. The properties of MHD equilibria are then examined in a variety of configurations, culminating in the derivation of the equation governing axisymmetric two-dimensional equilibria in a torus. The important topic of linear stability will then be addressed, and the role played by the magnetic field in the context of stability will be illustrated. A nonideal instability is then examined, the consequence of which is that the field topology is able to change. The behaviour of this instability is then pursued into its nonlinear regime.

4.2 What is MHD?

MHD provides a macroscopic dynamical description of an electrically conducting fluid in the presence of magnetic fields. The separate identities of the positively and negatively charged species do not feature in the formulation, so the conducting medium is a continuum or fluid through which are threaded lines of magnetic field. This field may be externally applied, produced by current flowing in the fluid, or a combination of both.

Such a single-fluid description will only be valid provided the plasma is collision dominated. One condition for collision dominance is that the distribution functions of the particle species are locally Maxwellian. If this condition is to pertain, then the MHD time scale t must be sufficiently long for there to be adequately many collisions between particle species. Should this be the case, t will certainly be much longer than the time required for light to traverse the plasma. This allows the displacement current to be neglected from Maxwell's equations. A concomitant condition for collision dominance pertains to the MHD scale length a, which must be large compared to the mean free path between collisions. The scale length a should also be large compared to the Debye length. If this is so, the fluid will be quasineutral; consequently the number densities of the electrons and protons will be approximately equal. If, in addition, a is large compared with the ion Larmor radius r_{Li}, then the electron diamagnetic and Hall effects which appear in the electron momentum equation can be neglected. This allows the electron inertia to be incorporated into the ion momentum equation to yield a single-fluid momentum equation. Our first constraint summarizes the condition for collision dominance as being:

$$\frac{r_{\mathrm{Li}}}{a} \ll 1.$$

Another stipulation for the single-fluid picture to be valid is that the charged species should have identical temperatures, a condition which when combined with quasineutrality implies that the pressures are also the same. This requires the energy equilibrium time to be short compared with the characteristic time t, a constraint which is rather more restrictive than requiring collision dominance. The relevant condition is

$$\left(\frac{m_{\mathrm{i}}}{m}\right)^{1/2} \frac{v_{\mathrm{Ti}}\tau_{\mathrm{ii}}}{a} \ll 1,$$

where m_{i} and m are the ion and electron masses, respectively, v_{Ti} is the ion thermal speed and τ_{ii} is the collision time between ions. If both these conditions are satisfied simultaneously, the equations of MHD then become:

$$\frac{\partial \rho}{\partial t} + \nabla \cdot (\rho,\boldsymbol{v}) = 0 \qquad \text{(mass continuity)}, \qquad (4.2.1)$$

$$\frac{\mathrm{d}}{\mathrm{d}t}\left(\frac{p}{\rho^{\gamma}}\right) = 0 \qquad \text{(adiabatic equation of state)}, \qquad (4.2.2)$$

$$\rho\frac{\mathrm{d}\boldsymbol{v}}{\mathrm{d}t} = \boldsymbol{J} \times \boldsymbol{B} - \nabla p \qquad \text{(momentum equation)}, \qquad (4.2.3)$$

$$\nabla \times \boldsymbol{B} = \mu_0 \boldsymbol{J} \qquad \text{(Ampère's law)}, \qquad (4.2.4)$$

$$\nabla \times \boldsymbol{E} = -\frac{\partial \boldsymbol{B}}{\partial t} \qquad \text{(Faraday's law)}, \qquad (4.2.5)$$

$$\nabla \cdot \boldsymbol{B} = 0, \tag{4.2.6}$$

$$\boldsymbol{E} + \boldsymbol{v} \times \boldsymbol{B} = \eta \boldsymbol{J} \qquad \text{(resistive Ohm's law)}, \tag{4.2.7}$$

where each of the symbols has its customary meaning, with γ being the ratio of specific heats, η the electrical resistivity of the plasma and $\mathrm{d}/\mathrm{d}t$ the convective derivative

$$\frac{\mathrm{d}}{\mathrm{d}t} \equiv \frac{\partial}{\partial t} + \boldsymbol{v} \cdot \nabla.$$

It may be the case that the plasma is sufficiently collision dominated for the fluid picture to apply, yet sufficiently 'collisionless' for the resistivity appearing in Eq. (4.2.7) to be negligible. Alternatively, the plasma scale length may be sufficiently large so that the resistive diffusion time $\eta/a^2 \ll t$. The conflation of these rather delicate conditions can be represented by

$$\left(\frac{m}{m_{\mathrm{i}}}\right)^{1/2} \frac{r_{\mathrm{Li}}^2}{v_{\mathrm{Ti}} \tau_{\mathrm{ii}} a} \ll 1,$$

in which case the right-hand side of Eq. (2.4.7) may be neglected to yield:

$$\boldsymbol{E} + \boldsymbol{v} \times \boldsymbol{B} = 0 \qquad (\textit{ideal} \text{ Ohm's law}). \tag{4.2.8}$$

This states that there is no electric field in the rest frame of the fluid. Eqs (4.2.1)–(4.2.6) with Eq. (4.2.8) constitute the *ideal* MHD equations, which is usually contracted to *MHD*. The inclusion of Eq. (4.2.7) is described as resistive MHD.

An enormous amount of physics has been discarded in obtaining the MHD equations, so it is pertinent to enquire what precisely has been retained. It can be seen that the equations are essentially an amalgam of fluid mechanics and 'pre-Maxwell' electromagnetism. The fluid inertia is affected by forces due to the fluid pressure gradients and to the $\boldsymbol{J} \times \boldsymbol{B}$ term, which is the Lorentz force in continuum form. It will be noted however that the Ohm's law couples the fluid to the fields. If the magnetic flux is conserved, then this equation provides constraints on the allowable class of fluid displacements described by the theory, and this in turn has implications for the topology of the magnetic fields, as will be discussed in Section 4.3.

MHD possesses those conservation properties enjoyed by fluid mechanics and electromagnetism, namely:

- conservation of mass;
- conservation of momentum;
- conservation of energy (both mechanical and electromagnetic);
- conservation of magnetic flux.

The physical effects described by MHD are remarkably diverse, there being a plethora of unstable motions which manifest themselves according to the particular geometry and magnetic field topology under consideration. There are also three classes of waves which are capable of propagating in the plasma:

4.4.3 *The z-pinch*

The z-pinch confines the plasma in cylindrical geometry by using an axial current and its associated poloidal magnetic field. So

$$\boldsymbol{B} = B_\theta(r)\hat{\boldsymbol{e}}_\theta, \quad \boldsymbol{J} = \frac{1}{\mu_0}\frac{\mathrm{d}B_\theta(r)}{\mathrm{d}r}\hat{\boldsymbol{e}}_z,$$

and the equilibrium equation becomes

$$\frac{\mathrm{d}}{\mathrm{d}r}\left(\frac{B_\theta^2}{2\mu_0} + p\right) + \frac{B_\theta^2}{\mu_0 r} = 0. \tag{4.4.6}$$

The last term on the left can be attributed to the tension experienced by the field lines in being 'wrapped' over the surface of a cylinder. A typical z-pinch profile is shown in Fig. 4.4(a). Again one field variable must be prescribed, after which the entire equilibrium configuration may be determined.

4.4.4 *Axisymmetric toroidal equilibria*

The determination of two-dimensional equilibria is considerably more complex than the one-dimensional counterparts, requiring two functions to be prescribed. We shall derive the equilibrium equation which must be satisfied by an axisymmetric toroidal plasma. The resulting second order, nonlinear partial differential equation is known as the 'Grad–Shafranov' equation. The basic strategy of the derivation is first to exploit the axisymmetry in order to deduce the general structure of the magnetic field via the equation $\nabla \cdot \boldsymbol{B} = 0$. This enables the field to be expressed in terms of a gradient of a scalar flux function. This representation of the field is then substituted into the equilibrium equation $\boldsymbol{J} \times \boldsymbol{B} = \nabla p$. The two prescribed arbitrary functions correspond to the pressure gradient and a function which is a measure of the total plasma current.

The cylindrical coordinate system (R, ϕ, Z) we shall adopt is shown in Fig. 4.5. Writing out Eq. (4.2.6) in these coordinates gives

$$\nabla \cdot \boldsymbol{B} = \frac{1}{R}\frac{\partial(RB_R)}{\partial R} + \frac{\partial B_Z}{\partial Z} + \frac{1}{R}\frac{\partial B_\phi}{\partial \phi} = 0.$$

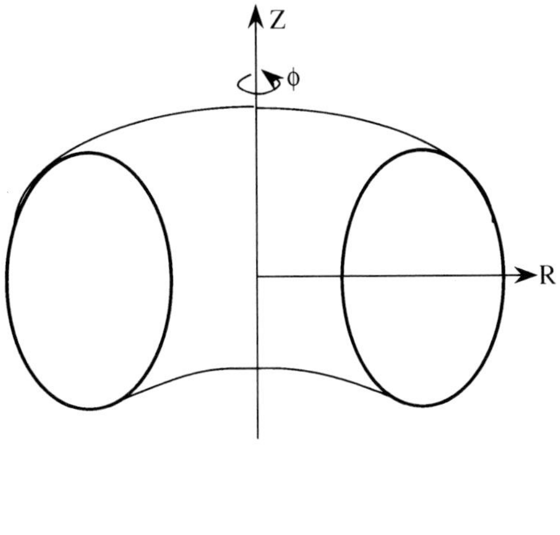

Fig. 4.5. The coordinate system for an axisymmetric toroidal plasma.

By the assumption of axisymmetry, we have $\partial/\partial\phi = 0$, and so $\boldsymbol{B} = \boldsymbol{B}(R, Z)$ only. This allows us to introduce a flux function to represent the field components in the poloidal plane,

$$RB_R = \frac{\partial\psi}{\partial Z} \quad \text{and} \quad RB_Z = -\frac{\partial\psi}{\partial R};$$

or, more succinctly,

$$\boldsymbol{B} = \frac{1}{R}\boldsymbol{\nabla}\psi \times \hat{\boldsymbol{e}}_\phi + \mathrm{B}_\phi \hat{\boldsymbol{e}}_\phi. \tag{4.4.7}$$

Having obtained the field, the plasma current can now be calculated using Ampére's law Eq. (4.2.4). The components of the current are

$$\mu_0 J_R = \frac{\partial B_\phi}{\partial Z}, \quad \mu_0 J_Z = -\frac{1}{R}\frac{\partial(RB_\phi)}{\partial R}, \quad \mu_0 J_\phi = -\frac{\partial}{\partial R}\left(\frac{1}{R}\frac{\partial\psi}{\partial R}\right) - \frac{1}{R}\frac{\partial^2\psi}{\partial Z^2}.$$

On grouping the poloidal and toroidal components of current, this may be written compactly as

$$\mu_0 \boldsymbol{J} = \frac{1}{R}\boldsymbol{\nabla}(RB_\phi) \times \hat{\boldsymbol{e}}_\phi - \frac{1}{\mathrm{R}}\Delta^*\psi\hat{\boldsymbol{e}}_\phi, \tag{4.4.8}$$

with Δ^* representing the Laplacian operator in this coordinate system,

$$\Delta^* \equiv R^2\boldsymbol{\nabla}\cdot\left\{\frac{1}{R^2}\nabla\right\} \equiv R\frac{\partial}{\partial R}\left\{\frac{1}{R}\frac{\partial}{\partial R}\right\} + \frac{\partial^2}{\partial Z^2}. \tag{4.4.9}$$

Now substitute the expression for the magnetic field given by Eq. (4.4.7) and the current Eq. (4.4.8) into the equilibrium Eq. (4.4.1). Specifically, if we examine the ϕ-component of the equilibrium equation, and once more appeal to axisymmetry,

$$(\boldsymbol{J} \times \boldsymbol{B})\cdot\hat{\boldsymbol{e}}_\phi = \boldsymbol{\nabla}p\cdot\hat{\boldsymbol{e}}_\phi = 0$$

or

$$\left\{\frac{1}{\mu_0 R}\frac{\partial(RB_\phi)}{\partial R}\right\}\left\{\frac{1}{R}\frac{\partial\psi}{\partial Z}\right\} - \left\{\frac{1}{\mu_0 R}\frac{\partial(RB_\phi)}{\partial Z}\right\}\left\{\frac{1}{R}\frac{\partial\psi}{\partial R}\right\} = 0.$$

This states that the Jacobian of the functions RB_ϕ/μ_0 and ψ is zero. The vanishing of the Jacobian implies that $RB_\phi/\mu_0 = F(\psi)$. Now use the fact that the field lines lie in isobaric surfaces:

$$\boldsymbol{B}\cdot\boldsymbol{\nabla}p = \frac{1}{R}\boldsymbol{\nabla}\psi \times \hat{\boldsymbol{e}}_\phi\cdot(\boldsymbol{\nabla}p)_{\text{pol}} = \frac{1}{R}\frac{\partial\psi}{\partial R}\frac{\partial p}{\partial Z} - \frac{1}{R}\frac{\partial\psi}{\partial Z}\frac{\partial p}{\partial R} = 0,$$

which is the Jacobian of ψ and p, hence $p = p(\psi)$. If we now write down the R-component of the equilibrium equation:

$$(\boldsymbol{J} \times \boldsymbol{B})\cdot\hat{\boldsymbol{e}}_R = \frac{\partial p}{\partial R} = -\left\{\frac{1}{\mu_0 R}\frac{\partial(RB_\phi)}{\partial R}\right\}B_\phi - \frac{1}{\mu_0 R}\Delta^*\psi\left\{\frac{1}{R}\frac{\partial\psi}{\partial R}\right\}.$$

Using the fact that both p and RB_ϕ are functions of ψ,

$$\frac{\mathrm{d}p(\psi)}{\mathrm{d}\psi}\frac{\partial\psi}{\partial R} = -\frac{1}{R}\frac{\mathrm{d}F(\psi)}{\mathrm{d}\psi}\frac{\partial\psi}{\partial R}\left(\frac{\mu_0 F(\psi)}{R}\right) - \frac{1}{\mu_0 R}\Delta^*\psi\frac{1}{R}\frac{\partial\psi}{\partial R},$$

hence we may write

$$\Delta^*\psi = -\mu_0 R^2\frac{\mathrm{d}p(\psi)}{\mathrm{d}\psi} - \mu_0^2 F(\psi)\frac{\mathrm{d}F(\psi)}{d\psi}. \tag{4.4.10}$$

This is a nonlinear second order elliptic partial differential equation for calculating the equilibrium in terms of the flux, and is usually known as the Grad–Shafranov equation. The procedure for finding its solution is first to prescribe the pressure and current functions p and F as some physically reasonable distribution of the flux ψ. The Grad–Shafranov equation is then solved, given appropriate boundary conditions satisfied by ψ, to obtain ψ as a function of R and Z. Invariably the solution is found computationally, due to the complexity of the prescribed functions and boundary conditions which must be imposed. Having determined the spatial distribution of the flux, the poloidal field, toroidal field, current and pressure can all be found as functions of R and Z. In Fig. 4.6 are shown the contours of ψ drawn in the poloidal plane for a tokamak device with noncircular cross-section. It can be noted that the surfaces of constant ψ form a nested set, rather like the layers within an onion. Also shown are the toroidal magnetic field and current distributions, together with the pressure across the toroidal and poloidal mid-planes.

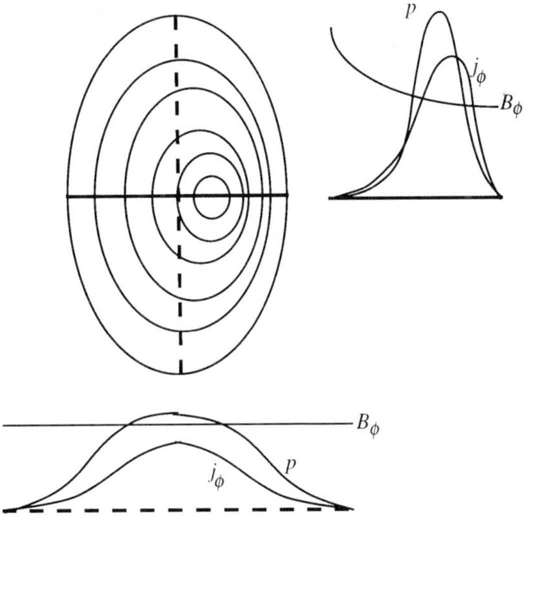

Fig. 4.6. The nested flux surfaces which typically result from solving the Grad–Shafranov equation for a tokamak with noncircular cross-section. Also shown are the toroidal field and current distributions and the pressure distributions across the major (dotted line) and minor (full line) mid-planes.

4.5 Linear stability

We have seen how equilibria can be calculated for a variety of configurations. The next task is to develop techniques for testing whether such equilibria are stable to small perturbations, and hence to determine if the equilibria will persist or ultimately be destroyed. If the perturbations do not grow, they still constitute a disturbance representing a stable wave propagating through the plasma. MHD permits only instabilities or stable waves and not a mixture. This is a consequence of the nature of the MHD force operator, as we shall later show.

To study the stability properties of equilibria, we need to first linearize the MHD equations about a stationary equilibrium. To facilitate this we shall adopt a common convention whereby the equilibrium variables are denoted with the subscript '0', and perturbed variables are identified using lower-case symbols. One way of viewing the linearization process is to write all the variables (be they vector or scalar) as

$$Q(\boldsymbol{x},t) = Q_0(\boldsymbol{x},t) + \varepsilon q(\boldsymbol{x},t), \tag{4.5.1}$$

and then order the equations in terms of the smallness parameter ε. At this stage, it is customary to introduce the fluid displacement vector

$$\boldsymbol{\xi}(\boldsymbol{x},t) = \int \boldsymbol{v}(\boldsymbol{x},t)\,\mathrm{d}t$$

rather than the velocity $\boldsymbol{v}$ itself. Because the equilibria are static, $\boldsymbol{v}_0(\boldsymbol{x},t) = 0$ and clearly $\boldsymbol{\xi}$ is a first order quantity.

To order $\varepsilon^{(0)}$ substitution of Eq. (4.5.1) into the MHD equations gives precisely the magnetostatic equilibrium equations

$$(\nabla \times \boldsymbol{B}_0) \times \boldsymbol{B}_0 = \mu_0 \nabla p_0, \tag{4.5.2}$$

$$\nabla \cdot \boldsymbol{B}_0 = 0. \tag{4.5.3}$$

To first order $\varepsilon^{(1)}$ the linearized momentum balance equation is

$$\rho_0 \frac{\partial^2 \boldsymbol{\xi}}{\partial t^2} = -\nabla p + \boldsymbol{j} \times \boldsymbol{B}_0 + \boldsymbol{J} \times \boldsymbol{b},$$

with

$$\boldsymbol{j} = \frac{1}{\mu_0} \nabla \times \boldsymbol{b}.$$

There is only a single term representing the fluid inertia on the left-hand side because the fluid displacement is a first order quantity. The other term appearing in the convective derivative is the advection term $\boldsymbol{v} \cdot \nabla \boldsymbol{v}$, which is second order in ε. The linearized pressure can be found by combining the conservation of mass with the adiabatic equation of state:

$$\frac{\partial p}{\partial t} + \gamma p_0 \nabla \cdot \left(\frac{\partial \boldsymbol{\xi}}{\partial t} \right) + \frac{\partial \boldsymbol{\xi}}{\partial t} \cdot \nabla p_0 = 0,$$

which on integrating and using the fact that there is no initial displacement at $t = 0$ gives

$$p = -\gamma p_0 \nabla \cdot \boldsymbol{\xi} - \boldsymbol{\xi} \cdot \nabla p_0.$$

The linearized magnetic field $\boldsymbol{b}$ can be calculated by combining Faraday's law with the ideal Ohm's law,

$$\rho = -\xi_y \frac{d\rho_0}{dy}. \tag{4.5.13}$$

Substituting Eqs (4.5.11) and (4.5.12) into Eq. (4.5.9) gives the eigenvalue equation

$$\frac{d}{dy}\left\{\rho_0 \frac{d\xi_y}{dy}\right\} - k^2\rho_0\xi_y - \left\{\frac{k}{\omega}\right\}^2 g\frac{d\rho_0}{dy}\xi_y - \frac{1}{\rho_0}\left\{\frac{kB_0}{\omega}\right\}^2\left\{\frac{d\xi_y}{dy^2} - k^2\xi_y\right\} = 0.$$

This equation can be integrated with respect to y across the interface separating the two regions shown in Fig. 4.8 whilst applying the boundary conditions that the fluid displacement is zero as $|y| \to \infty$ and that ξ_y is continuous at $y = 0$. The values of ω^2 which permit these boundary conditions to be implemented are given by the dispersion relation

$$\omega^2 = kg\left(\frac{\rho_2 - \rho_1}{\rho_2 - \rho_1}\right) + \frac{2}{\mu_0}\left(\frac{k^2B_0^2}{\rho_2 + \rho_1}\right). \tag{4.5.14}$$

The first term on the right-hand side can be positive or negative, depending on the relative densities of the two fluids. In the absence of a magnetic field, $\omega^2 < 0$ if $\rho_2 < \rho_1$ giving the same condition for instability as before. Because the second term on the right is always positive, the presence of a magnetic field has a stabilizing effect. This is because work is expended by the fluid in bending the field lines. Moreover, because the field-line bending term is proportional to k^2, this stabilization is most effective for short wavelength modes. Indeed one may write down a necessary stability criterion for the critical wavelength of the modes:

$$k_{\text{crit}} > \frac{g\mu_0}{2B_0^2}(\rho_1 - \rho_2).$$

The situation which has just been described is a rather special one, for which the perturbation is directed parallel to the field. For the sake of completeness, we should examine the case when the perturbation is aligned at some arbitrary angle θ to the field. This necessitates redoing the calculation for the more general perturbation

$$\xi(\boldsymbol{x}, t) = \xi(y)\exp[i(k(z\cos\theta + x\sin\theta) - \omega t)].$$

In this case the dispersion relation becomes

$$\omega^2 = kg\left(\frac{\rho_2 - \rho_1}{\rho_2 - \rho_1}\right) + \frac{2}{\mu_0}\frac{(\boldsymbol{k}\cdot\boldsymbol{B}_0)^2}{\rho_2 + \rho_1}, \tag{4.5.15}$$

which is clearly identical to Eq. (4.5.14) when k is parallel to $\boldsymbol{B}_0$. Note, however, that when k is perpendicular to $\boldsymbol{B}_0$, the stabilizing effect of the field is entirely nullified, so that the mode with $\theta = \pi/2$ will always be unstable everywhere within the plasma volume if $\rho_1 > \rho_2$. This undesirable situation can be ameliorated by introducing a *sheared* magnetic field, by which is meant a field whose direction changes as a function of position. Fig. 4.9 shows such a field configuration, comprising a stack of sinusoidally perturbed surfaces in the plasma on which are drawn the sheared magnetic field lines. The shear is such that the field rotates through an angle π so that the direction of the field reverses, between the two extremes in height. At the bottom of the stack, the perturbation

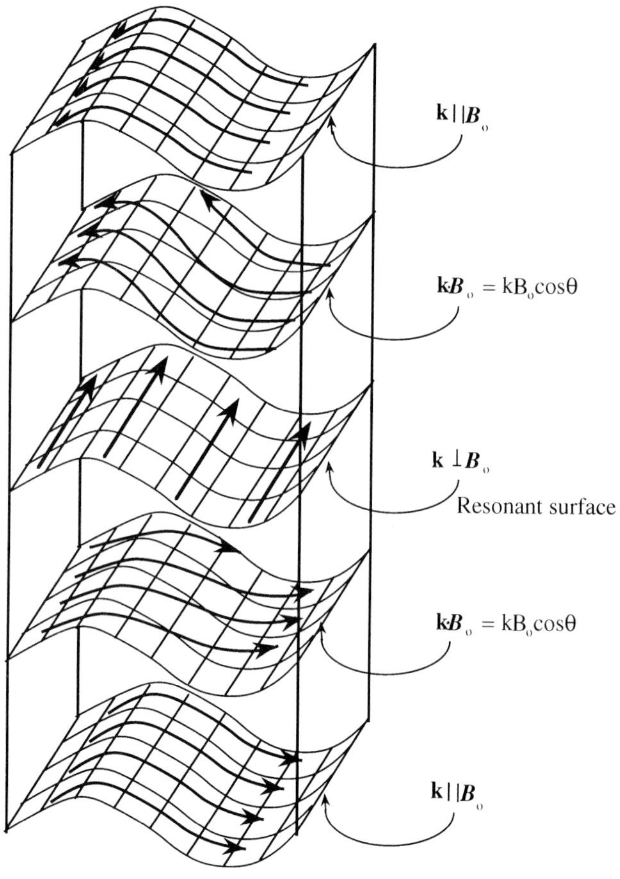

Fig. 4.9. The effect of introducing shear into the magnetic field. The perturbation is perpendicular to the field on only one surface within the plasma, thereby spatially localizing the region of unconditional instability if the fluid is top heavy.

is parallel to the field, so this surface would be the most stable to the Rayleigh–Taylor instability. Moving to the next surface up, the angle between the field and perturbation is θ, and so it follows from Eq. (4.5.15), that the stabilizing effect of the field has been reduced by a factor $\cos\theta$. There is only one surface in the plasma, where $\theta = \pi/2$, where the plasma is unconditionally unstable if $\rho_1 > \rho_2$. Hence it may be seen that the introduction of a sheared magnetic field effectively localizes the region of unconditional instability to the surface within the plasma where $\boldsymbol{k} \cdot \boldsymbol{B}_0 = 0$, leaving the rest of the plasma conditionally unstable. The surface where $\boldsymbol{k} \cdot \boldsymbol{B}_0 = 0$ is known as the resonant surface.

4.5.4 *Cylindrical stability*

In cylindrical geometry, a sheared equilibrium magnetic field also leads to localization of unstable MHD modes. We may consider a periodic cylinder to be an approximation to the tokamak configuration. We may define an effective major radius R of the device as being a periodic length along the axis of a cylinder through $z = 2\pi R$. The equilibrium has the form $\boldsymbol{B}_0 = (0, B_\theta(r), B_\phi)$, which is perturbed by a mode with the structure

$$\xi(\boldsymbol{x},t) = \xi(r)\exp[\mathrm{i}(m\theta + n\phi - \omega t)];$$

here m and n are the poloidal and toroidal mode numbers which give the pitch of a helical deformation. Recall that the safety factor $q(r)$ is a measure of the pitch of the field line on a magnetic surface at radius r. In this instance the $\boldsymbol{k}\cdot\boldsymbol{B}_0$ term becomes

$$\frac{B_\theta}{r}(m - nq(r)),$$

and so the resonant surface will occur at radius r_s defined by $q(r_s) = m/n$.

The energy integral δW for such a plasma column of minor radius a, surrounded by a perfectly conducting shell of radius $b \geqslant a$, is given by the following expression, which is accurate to second order in the inverse aspect ratio $\varepsilon = a/R$:

$$\delta W = \frac{\pi^2 B_\phi^2}{R\mu_0}\int_0^a \left\{\frac{n}{m} - \frac{1}{q}\right\}^2 \left\{\left(r\frac{\mathrm{d}\xi_r}{\mathrm{d}r}\right)^2 + (m^2 - 1)^2\xi_r^2\right\} r\,\mathrm{d}r$$
$$+ \frac{(\pi B_\phi a\xi_r(a))^2}{R\mu_0}\left\{\frac{2}{q(a)}\left(\frac{n}{m} - \frac{1}{q(a)}\right) + (1 + m\lambda)\left(\frac{n}{m} - \frac{1}{q(a)}\right)^2\right\}, \qquad (4.5.16)$$

where

$$\lambda = \frac{1 + (a/b)^{2m}}{1 - (a/b)^{2m}} \quad \text{and} \quad m > 1.$$

If the conducting shell sits on the plasma surface, then the boundary condition for the displacement is $\xi_r(a) = 0$, in which case the second term on the right is identically zero, and δW is positive definite for all displacements. Hence in this instance the plasma is unconditionally stable to ideal displacements. If the conducting wall is situated at $b > a$, the surface term can provide a destabilizing contribution to the energy if $q(a) < m/n$. Therefore a sufficient condition for the stability of a cylindrical plasma is that $q(a) > m/n$ (for $m > 1$), which is equivalent to demanding that all the resonant surfaces lie *within* the plasma volume. If a particular resonant surface lies in the vacuum region, then there is energy to drive the instabilities, which are known by the generic term *kink modes* because of the way they buckle the magnetic surfaces and plasma column into a helix. Whilst this stability criterion is revealing and places effective limits on the plasma current, it is not the end of the story. It must be borne in mind that this analysis is only true for ideal MHD displacements. We shall find that if a resonant surface exists within the plasma, then the resistive MHD model predicts the growth of a new class of instability, albeit with a slower growth rate.

4.5.5 *The $m = 1$ internal kink mode*

When dealing with the kink mode we specifically excluded the possibility of $m = 1$. We shall now give due consideration to this case, which has a rather different structure from the other kink modes, corresponding to a rigid displacement of the plasma column rather than a kinking of the plasma.

The most important $m = 1$ mode is that with toroidal mode number $n = 1$. From the previous discussion, this mode will be stable provided there is no $q = 1$

surface in the plasma. Hence, provided the smallest value of the safety factor q_0 is greater than unity, the plasma will be stable. This leads to the important Kruskal–Shafranov stability condition for the internal kink mode, namely that $q_0 > 1$. Note however that if the plasma is bounded by a perfectly conducting wall, a rigid displacement in the radial direction, ξ_r = constant, reduces the integral appearing in Eq. (4.5.16) to zero. Although ξ_r = constant minimizes the energy to zero, it is manifestly inconsistent with the perfectly conducting wall boundary condition that $\xi_r(a) = 0$. Hence ξ_r must necessarily fall to zero over a short distance l near to the plasma edge, as shown in Fig. 4.10. It transpires that the energy due to the change in ξ_r over this region is vanishingly small in the limit as $l \to 0$. Hence the $m/n = 1$ mode is marginally stable to order ε^2. The nonvanishing contribution to the energy which drives the mode unstable does not appear in the calculation until terms of order ε^4 are considered. The physics which is described to this order in the energy can be attributed to the effects of toroidicity in the equilibrium. The mode is actually driven by a combination of gradients in both pressure and parallel current profiles, so that its designation as a kink mode is rather a misnomer. It is an important and potentially limiting instability for plasmas whose ratio of kinetic to magnetic energy (the plasma β) is comparatively high. Detailed and rather technical calculation shows that the mode is unstable if:

$$\frac{20R^2\mu_0}{3r_1B_\phi^2}\int_0^{r_1}\left\{-\frac{dp_0}{dr}\right\}r^2\,dr > 1,$$

where r_1 is the radius of the $q = 1$ surface.

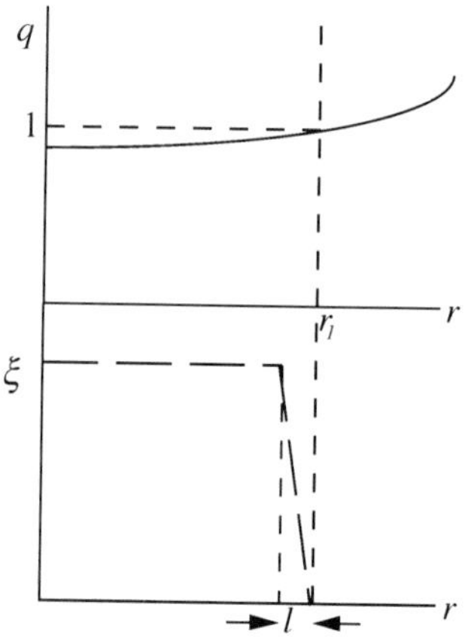

Fig. 4.10. A q-profile which is unstable to the internal $m = 1$ kink mode and the resulting eigenfunction. The displacement, which corresponds to a rigid shift of the plasma column, is constant over most of the plasma radius, falling to zero over a short distance l near the periphery.

4.5.6 *The tearing mode*

Use of the energy principle shows that kink modes are stable in ideal MHD if the resonant surface is inside the plasma. Whilst an identical outcome results if the normal mode method is used, a potential pitfall becomes highlighted by using this analysis. The difficulty arises because the differential equation for the eigenfunction of the stable mode is singular at the resonant surface, indicating that the ideal MHD model has broken down. The singularity can be resolved by including more physics in the model, in particular by using the resistive Ohm's law Eq. (4.2.7) instead of its ideal counterpart Eq. (4.2.8).

The resistivity of a fusion plasma is very small, being typically $\sim 10^{-8}\,\Omega$m. On first consideration, one might expect that the diffusion of field lines resulting from this resistivity to occur on a time scale $\tau_R \sim (\eta/L^2)^{-1}$, where L is the size of the plasma. This is very long compared with the characteristic time scale of ideal MHD (the magnetic Lundquist number $S = \tau_R/\tau_{MHD} \sim 10^8$). However, at a resonant surface where ideal MHD breaks down, resistive effects are all that remain and therefore they dominate the dynamics. The size of the singular layer is not the plasma size L but a much smaller region $w \ll L$. In this case, it can be seen that the diffusion of the field lines can be significantly rapid over a small region of the plasma since $\tau_R \sim (\eta/w^2)^{-1}$. The analysis of the tearing mode proceeds by using the techniques of boundary layer theory. Far away from the singular surface, ideal MHD provides an adequate description of the physical processes, whilst near to the singular surface, resistivity must be included in the prescription and the solutions to the equations in either region matched together. This matching procedure serves to determine the eigenvalue (which is the growth rate) of the resulting instability.

Because the growth rate of the tearing mode is long compared to ideal MHD time scales, the plasma exists in a state of stable equilibrium away from the resonant surface, so that $\boldsymbol{J} \times \boldsymbol{B} = \nabla p$, and it therefore follows that $\nabla \times (\boldsymbol{J} \times \boldsymbol{B}) = 0$. The region where ideal MHD is still valid is commonly referred to as the outer region. Assuming that the inverse aspect ratio $\varepsilon \ll 1$, this equation can be written in cylindrical coordinates, combined with Ampère's law, and then linearized to yield the following equation for B_r which is the perturbed component of the magnetic field in the radial direction:

$$\frac{d}{dr}\left(r\frac{d}{dr}(rB_r)\right) - m^2 B_r - \frac{mr^2\mu_0(dJ_{0\phi}/dr)}{B_{0\theta}(m - nq(r))}B_r = 0. \tag{4.5.17}$$

From this equation, it may be seen at once that the resonant surface where $q(r) = m/n$ is also a singular point of the equation. The last term on the left of this equation can drive the instability, depending on the sign of the current gradient. This destabilization is dominant at the resonant surface where the stabilizing effect from the bending of field lines (recall the discussion of the Rayleigh–Taylor instability) is nullified because both field and perturbation possess the same pitch.

Within the narrow layer where the resistivity is important, the plasma is no longer in equilibrium as was the case in the outer region, beyond the layer. Within the layer, referred to as the *inner region*, the effects of resistivity and of plasma inertia must be included in the analysis. An equation can be derived for the perturbed radial field inside the layer by elimination of the fluid displacement. This equation is fourth order and has the growth rate γ^2 as eigenvalue (where $\gamma = -i\omega$). Since the perturbed field must be continuous, the only way in which the second order solution in the outer region is able to match smoothly to the fourth order solution in the inner region is if the derivative of B_r is discontinuous across the layer.

The growth rate of the instability is found by matching the inner solution with the outer solution, using the following prescription. By first integrating Eq. (4.5.17) from the resonant surface r_s to the origin, where $B_r \sim r^{m-1}$, and then integrating from the resonant surface to the plasma boundary at $r = a$, where (say) $B_r = 0$, the quantity which measures the finite jump in the derivative of B_r can be determined from

$$\Delta' = \lim_{\varepsilon \to 0} \frac{1}{B_r(r_s)} \frac{dB_r}{dr}\bigg|_{r_s-\varepsilon}^{r_s+\varepsilon},$$

where ε is a measure of the width of the inner region. If the fourth order equation for B_r is solved in the inner region, and its solution matched at the extremities of the layer to the solution in the outer region, there will be an analogous jump δ in the derivative of B_r across the layer width:

$$\Delta'_{\text{inner}} = \frac{1}{B_r(r_s)} \delta \left\{ \frac{dB_r}{dr} \right\}.$$

In this equation, B_r is an explicit function of the growth rate γ. The growth rate is extracted by solving the dispersion relation

$$\Delta' = \Delta'_{\text{inner}}(\gamma),$$

and if $\Delta' > 0$, the growth rate is given by

$$\gamma = 0.55 \left(\frac{a^2 n}{R} \left(\frac{d \ln q}{dr} \right)_{r_s} \right)^{2/5} (a\Delta')^{4/5} \frac{1}{\tau_R^{3/5} \tau_{\text{MHD}}^{2/5}}. \tag{4.5.18}$$

Note first that the characteristic time of the instability occurs on a hybrid time scale, being neither purely resistive nor purely the ideal MHD or Alfvén transit time. This is because the region over which the diffusion is important is small compared with the plasma size. Secondly, the instability depends on the shear at the resonant surface, which in turn depends on the equilibrium toroidal current gradient. The current gradient also determines the stability parameter Δ', which, it will be noted, depends only on the solution in the outer region. Thus the stability properties are entirely governed by the equilibrium magnetic field structure outside the layer.

Although the growth rate of this instability is much slower than the ideal kink modes described previously, they are in certain respects more insidious and potentially more damaging, especially if the plasma discharge persists for a long period. The reason for this is that the resistive diffusion at the resonant surface removes the frozen field constraint of ideal MHD and allows the field lines to slip through the fluid, thereby enabling the plasma to change its field line topology. The magnetic surfaces which were merely distorted and advected by the fluid motion can now 'tear' apart and rejoin in a different configuration, hence the name *tearing mode*. As this instability grows, and the perturbed fields and displacements increase in magnitude, the linear approximation used to analyse its behaviour breaks down so that the nonlinear behaviour of the mode requires investigation.

4.6 Nonlinear instability

The nonlinear consequence of the tearing mode is that the resistive layer centred at the resonant surface grows to form a magnetic island, this being a region with different magnetic field topology from that outside the layer. The island can grow to cover a significant portion of the plasma, and the transport of heat is enhanced across the island, so that the hot central plasma can lose energy to the cooler periphery and thereby significantly reduce the energy confinement time.

Before examining the nonlinear tearing mode, it is important to consider the change in the magnetic field line topology caused by the presence of a magnetic island. This is accomplished by perturbing the equilibrium flux surfaces with the radially directed magnetic field introduced by the tearing mode. The tearing layer is centred on the resonant surface, where both field line and perturbation have the same helical structure, represented by the angle $\zeta = m\theta - n\phi$. The equilibrium magnetic field when projected in the ζ-direction is given by

$$\frac{\boldsymbol{k}\cdot\boldsymbol{B}}{|\boldsymbol{k}|} \equiv B^* = B_\theta\left(1 - \frac{nq(r)}{m}\right),$$

which reverses sign as the resonant surface is traversed, as depicted in the upper panel of Fig. 4.11. The trajectory of a field line near to the resonant surface is found by solving

$$\frac{\mathrm{d}r}{\mathrm{d}\zeta} = r_\mathrm{s}\frac{B_\mathrm{r}}{B_\theta}\left(1 - \frac{nq(r)}{m}\right)^{-1}.$$

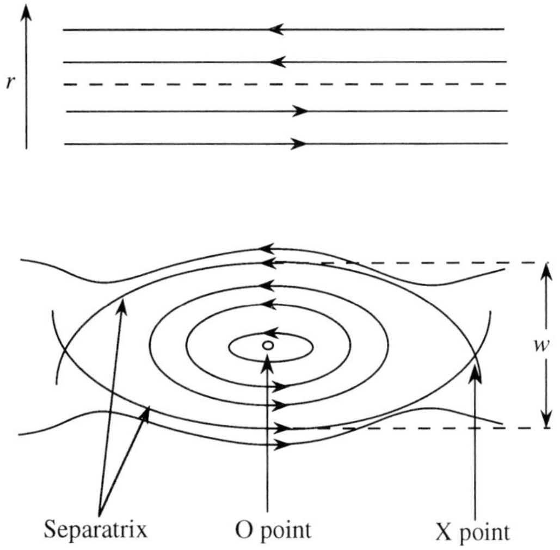

Fig. 4.11. The change in magnetic field topology due to the nonlinear tearing mode. The top panel shows $\boldsymbol{k}\cdot\boldsymbol{B}$ close to the resonant surface. The tearing mode produces a perturbed radial component of $\boldsymbol{B}$ which generates a magnetic island of width w with distinct nested flux surfaces.

It can be shown that the radial field has the form $B_\mathrm{r} \sim B_{\mathrm{r}0}\sin\zeta$, with $B_{\mathrm{r}0}$ approximately constant across the island. On Taylor expanding about the resonant surface, the field line equation integrates to give

$$(r - r_\mathrm{s})^2 = 2w^2(\cos\zeta - \cos\zeta_0), \quad \text{where } w^2 = \left(\frac{mrB_{\mathrm{r}0}}{nq'B_\theta}\right)_{r_\mathrm{s}}$$

is the width of the magnetic island which scales with the square root of the radial field. The magnetic surfaces inside the island are centred on the 'O point', whilst the line marking the boundary of the island also marks the change in the field topology and is known as the separatrix. The separatrix is a multiple valued curve, and the branches cross at the 'X point' as shown in the lower panel of Fig. 4.11.

Detailed analysis of the nonlinear tearing mode shows that the presence of a small magnetic island creates fluid vortices whose inertia acts in such way as to retard further growth. Any subsequent change of the island then occurs on the purely resistive time scale τ_R, so the island changes according to a diffusion process. We will show this by the following heuristic argument.

Close to an island, the evolution of the magnetic field can be found from combining Ampère's law with Faraday's law and the resistive Ohm's law $\boldsymbol{E} = \eta \boldsymbol{J}$ to give the following diffusion equation for the radial component of the field:

$$\frac{\partial B_r}{\partial t} = \frac{\eta}{\mu_0} \frac{\partial^2 B_r}{\partial r^2}.$$

Since B_r is approximately constant across the island, this equation can be integrated across the island

$$\int_{r_s - w/2}^{r_s + w/2} \frac{\partial B_r}{\partial t} \mathrm{d}r = \frac{\eta}{\mu_0} \int_{r_s - w/2}^{r_s + w/2} \frac{\partial^2 B_r}{\partial r^2} \mathrm{d}r,$$

so that

$$w \frac{\partial B_r}{\partial t} \approx \frac{\eta}{\mu_0} \frac{\partial B_r}{\partial r}\bigg|_{r_s - w/2}^{r_s + w/2}.$$

On recalling that $w \propto B_r^{1/2}$ this equation can be written as

$$\frac{\mathrm{d}w}{\mathrm{d}t} \approx \frac{\eta}{2\mu_0} \Delta'(w) \quad \text{where } \Delta'(w) = \frac{\eta}{\mu_0} \frac{1}{B_r} \frac{\partial B_r}{\partial r}\bigg|_{r_s - w/2}^{r_s + w/2}. \tag{4.6.1}$$

This is the nonlinear analogue of the Δ' parameter encountered in the linear theory of the tearing mode, which is now an implicit function of the island width. From Eq. (4.6.1), it can be seen that the evolution of the island depends linearly on the resistivity, which is very small for plasmas of interest to fusion. This may lead one to suspect that the island would grow slowly, but this is not necessarily the case because of the $\Delta'(w)$ term. This term can be very large and of either sign, depending on the shape of the equilibrium toroidal current profile, and despite the smallness of η, $\mathrm{d}w/\mathrm{d}t$ can still be large. Hence for certain profiles, the nonlinear growth of the tearing mode may lead to a large island inside the plasma volume. Since islands are capable of growing at any resonant surface, it is possible that islands with different pitches may interact with each other. This can lead to a randomization of the field lines, so that the field structure becomes ergodic or chaotic within their vicinity. Particles gyrating around such field lines are therefore essentially unconfined, are able to traverse large tracts of the plasma, and so lead to poor energy confinement times.

Bibliography

Two excellent texts which deal with MHD in the context of fusion research are:

FREIDBERG, J. P. (1987). *Ideal Magnetohydrodynamics*. Plenum Press, New York.
WESSON, J. A. (1987). *Tokamaks*. Oxford University Press.

Fig. 5.10. $\tilde{B}_\theta/B_\theta$ versus E_z in the steady state. Bars show fluctuation amplitudes. $S = 10^3$, η = flat.

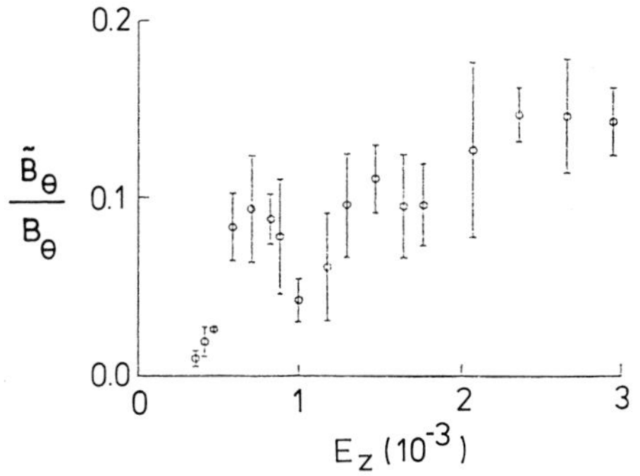

Fig. 5.11. $J_\parallel/B$ versus radius (solid line) and comparison with experiment (points). $S = 10^3$, η = flat.

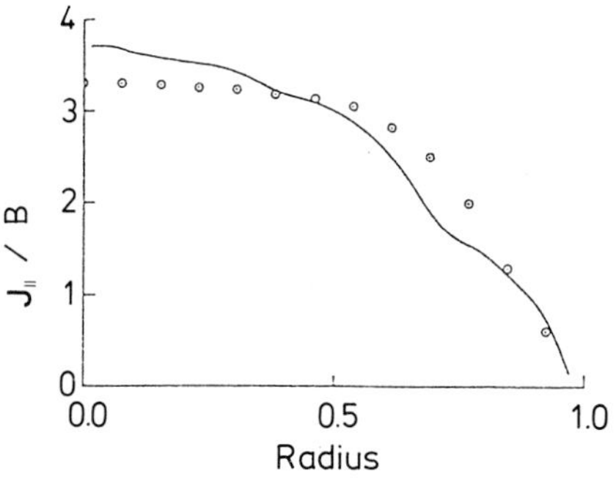

Fig. 5.12. Energy spectrum of B_r versus time, $S = 10^3$, η = flat. (*a*) corresponds to $m = 1$, $n = -2$; (*b*) $m = 1$, $n = -3$.

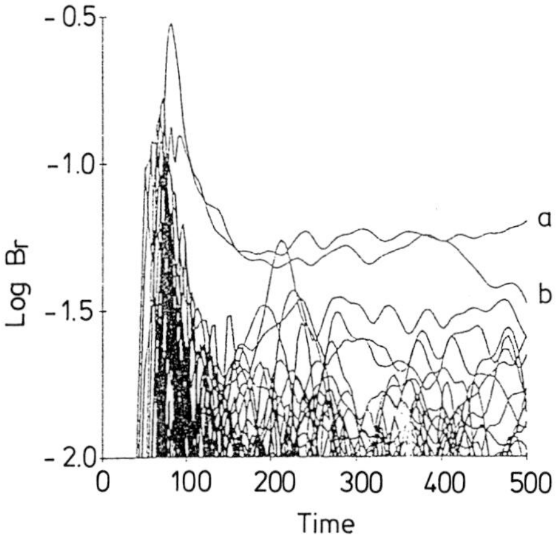

Fig. 5.13. Energy spectrum of B_r at time $t = 500$. $S = 10^3$, η = flat.

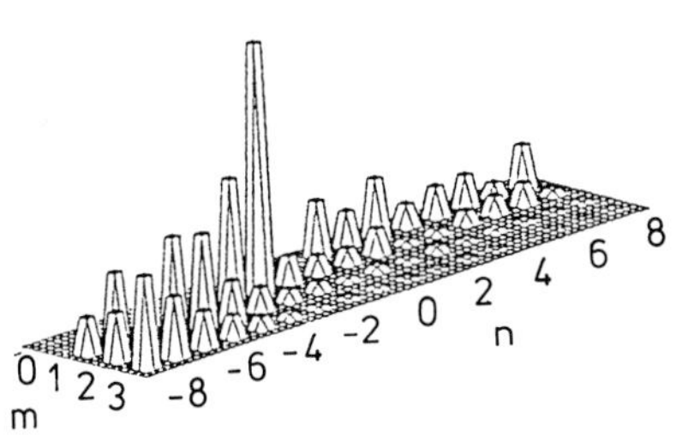

In conclusion, we see that, despite the relative simplicity of the model, a three-dimensional numerical simulation of an RFP based on an incompressible single fluid leads to a good description of the basic experimental properties. In particular, field reversal is demonstrated, and the properties of the $F-\theta$ curve are exhibited. These show good qualitative agreement, as does the $J_{\|}/B$ profile. Furthermore, the magnitude of the fluctuation levels, and some spectral details, also show reasonable agreement with experiment. However, the principal result, namely field reversal due to single-fluid MHD dynamo action, is clearly demonstrated. The calculation also demonstrates the essential three-dimensional nature of the turbulence. In some sense, the phenomenon of reversal must be expremely rugged as, first, the precribed values for S are much smaller than experiment suggests, and, secondly, some of the model assumptions are definitely unrealistic – for example, the simple forms for η.

5.3.4 *Sawteeth in tokamaks*

In our discussion of the pinch, we considered the way in which gross modes lead (through turbulence) to a gross effect – field reversal through dynamo action. Here, we again consider the solution of the MHD equations, but turbulence is now introduced into the transport processes. This influences the forms of the calculated profiles of current and temperature.

The sawtooth is a well-attested phenomenon in tokamaks, and can occur in a wide variety of circumstances. It gets its name from the time-dependent form of the X-ray intensity emitted from the central region of the tokamak (see Fig. 5.14). Since the X-ray intensity is a function of the plasma electron temperature, the observed signal can be interpreted as a measure of the electron temperature at the magnetic axis. In JET, for example, the sawtooth period τ_s is of order 0.1 s, while the crash time, τ_{crash}, is of order 100 μs. Thus the phenomenon is cyclical, and has the characteristics of a 'relaxation' oscillation. The first theoretical interpretation of this phenomenon was put forward by Kadomtsev, and we now give a brief outline of this mechanism, which is generally referred to as the reconnection model. An important concept here is the safety factor, $q = rB_{tor}(r)/RB_{pol}(r)$, and in particular, its value on the magnetic axis, namely $q(0)$. The process proceeds as follows: (a) starting from a situation in which $q > 1$ on axis, as the electron temperature increases (ramp phase), the current density J also rises and hence q falls; (b) when $q(0)$ becomes less than unity the configuration becomes unstable to an MHD mode, namely, $m = 1$; (c) the cycle is now completed by the plasma returning to the condition in which $q(0) > 1$. The stages of the whole cycle are shown in Fig. 5.15; a key feature is the resistively growing $m = 1$, $n = 1$ island.

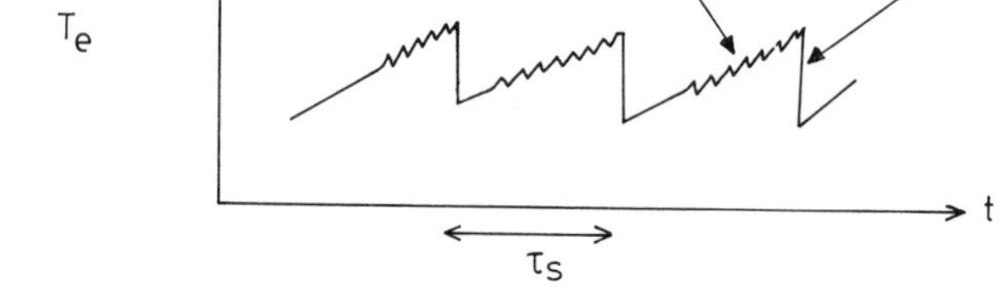

Fig. 5.14. Typical plot of X-ray intensity (or central electron temperature) as a function of time in a tokamak.

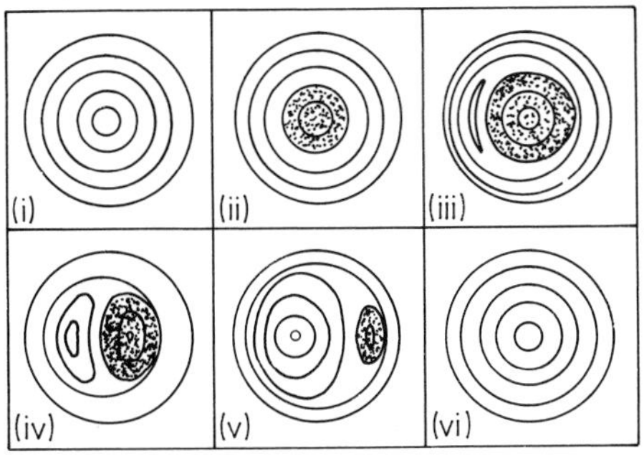

Fig. 5.15. Possible development of the magnetic field structure during the rapid fall stage of the sawtooth instability. The $m = 1$ instability displaces the $q < 1$ region (shaded) and restores $q(0)$ to a value larger than unity.

However, diagnostic measurement of the safety factor on axis suggest that – in most experiments at least – the Kadomtsev (reconnection) model is not correct. In fact, $q(0)$ is always less than unity. In JET, for example, $q(0) \sim 0.8$ and remains fairly steady, the change in the safety factor Δq being less than or of order 5% during the sawtooth cycle. Using the MHD equations Aydemir *et al.* (1989) have attempted to simulate the phenomenon. In their calculation the resistivity aids both reconnection and the annihilation of the magnetic field. An energy balance equation is used which includes both radial thermal diffusivity $\chi_\perp$ and parallel thermal diffusivity $\chi_\parallel$, with the resistivity and electron temperature related through the Spitzer formula ($\eta \sim T_e^{-3/2}$). Now Aydemir *et al.* find that their simulation recovers the Kadomtsev reconnection model, which, as we have already mentioned, is not borne out in most experiments.

In an attempt to interpret experiment, Goedheer and Westerhof (1988) have recently carried out a simulation in which they introduce a turbulence 'recipe' due to Dubois and Samain. When the resistivity growing $m = 1$, $n = 1$ island reaches a critical size, turbulence is switched into the calculation – the intention being to increase the radial transport. This ensures that sawtooth collapse occurs, but with $q(0)$ always below unity.

More recently, Haas and Thyagaraja (1992) have given an alternative interpretation of sawteeth in tokamaks. In this, taking partial reconnection ($q(0) < 1$) as an experimental fact and assuming certain turbulent constitutive relations for the anomalous particle and energy transport, nonlinear equations for the electron temperature (or β-poloidal) and the magnetic turbulence energy are derived. The solutions for these variables show periodic behaviour, with that for the electron temperature exhibiting typical sawtooth oscillations.

These different investigations demonstrate that turbulence may well play a crucial role in the appropriate interpretation of some of the global properties of tokamaks.

5.4 Turbulence theories of transport

So far, we have discussed the gross modes and the influence of turbulence on the global dynamics. We now consider the relationship of particle and energy transport to microscopic turbulence and fluctuations. This entails the investigation of the two-fluid equations. These are derived from the Boltzmann equations for the electrons and ions, which in their general form are

$$\frac{\partial f_e}{\partial t} + \boldsymbol{v}\cdot\frac{\partial f_e}{\partial \boldsymbol{r}} - \frac{e}{m_e}(\boldsymbol{E} + \boldsymbol{v}\times\boldsymbol{B})\cdot\frac{\partial f_e}{\partial \boldsymbol{v}_e} = C(f_e, f_e) + C(f_e, f_i) + S_e,$$

$$\frac{\partial f_i}{\partial t} + \boldsymbol{v}\cdot\frac{\partial f_i}{\partial \boldsymbol{r}} + \frac{e}{m_i}(\boldsymbol{E} + \boldsymbol{v}\times\boldsymbol{B})\cdot\frac{\partial f_i}{\partial \boldsymbol{v}_i} = C(f_i, f_i) + C(f_i, f_e) + S_i,$$

where the distribution functions f_e, f_i are functions of $\boldsymbol{r}$, $\boldsymbol{v}$ at t. The collision terms, C, take account of all possible interactions, and S_e, S_i denote external particle sources. The collision terms in the Boltzmann equations need to be treated with care, and although these may be represented by suitable model operators, it is essential that they conserve number density, momentum and energy as accurately as possible.

Taking moments of the Boltzmann equations leads to the continuity, momentum and energy equations for each of the particle species, namely electrons and ions. As far as the energy balance equations are concerned, there is the complication we mentioned earlier. The energy equations involve parallel thermal diffusivities, that is thermal transport along the equilibrium magnetic field. Now under typical tokamak conditions the particle mean free path is long compared to the major circumference $2\pi R$. Equivalently, we can write $L_{mfp} \gg 2\pi R \sim qR$, where qR denotes the so-called connection length, q being the safety factor; the connection length is a measure of the distance between the stable and unstable curvature regions of the equilibrium magnetic field.

Under these conditions, the form of the parallel electron thermal diffusivity, for example, needs careful consideration. The difficulty lies in the fact that the mean free path loses its significance as a step length wherever it exceeds the periodicity length of the system. Dimensional considerations suggest that the mean free path be replaced by the major circumference $2\pi R$ so that $\chi_{\parallel e} \sim (2\pi R)^2/\tau_e$, where τ_e is the electron collision time. This form can be described as the Knudsen corrected parallel electron thermal diffusivity. A thorough investigation of the appropriate Boltzmann equation confirms the proposed form, and of course determines the value of the order-one constant of proportionality required in the actual form. In principle, given the Knudsen corrected forms for the parallel transport coefficients, a turbulence investigation of the two-fluid equations can now be carried out.

Turbulence theories of transport can be viewed as having two distinct parts:

(a) *Mechanisms or dynamics.* Theoretical investigations reveal a plethora of linear (small-amplitude) instabilities to which a plasma may be subject. In principle, these can grow and evolve towards a nonlinearly saturated state, say. Alternatively, the initial plasma state may be linearly stable, but nonetheless be unstable to a nonlinear mode of instability, which can also lead on to saturation. In either case, the final saturated state carries information on the turbulence through the spectrum of the fluctuations, thus providing, in principle, information on phases and amplitudes.

(b) *Energetics.* This refers to the energy transport arising from the power spectrum (fluctuations). Clearly, the energetics mediate in the saturation process referred to in the dynamics.

The two parts are obviously complementary, and, in general, an investigation can become prohibitively complicated.

5.4.1 *Mechanisms*

Plasmas show a very wide range of instabilities. These have been investigated – mostly in the linear phase – but a number of nonlinear studies have also been carried out. Three possible sources of free energy are available for driving an instability, namely the density, temperature, and current profiles. Thus, for example, gradients in the particle densities and temperature can be shown to lead theoretically to the so-called η_i-mode, the collisional drift wave, and so on. These may be described as micro-instabilities, as distinct from the macroscopic or gross modes. On the other hand, the current density, $\boldsymbol{J}$, can drive macroscopic (MHD) modes, such as the kink (ideal) or tearing (resistive) modes. The current density can also provide a source of free energy for the microtearing mode, which is a high m, n analogue of the tearing mode – which arises from single-fluid resistive MHD. Some or all of these instabilities could be responsible for the fluctuations and turbulence observed in magnetic confinement experiments. It is widely believed that an understanding of plasma turbulence would, in principle, lead to a proper understanding of the dynamics and transport in these devices. We continue by discussing two basic theorems in particle transport.

5.4.2 *Role of compressibility and nonlinearity*

The significance of compressibility and nonlinearity in particle transport has been discussed by Haas and Thyagaraja (1986). Consider the plasma to be described by the electron and ion fluid equations, and assume the existence of a nested set of closed flux surfaces ψ_0 in a toroidal system (see Fig. 5.16). For such a system, the equilibrium (or 'mean' state in a turbulence sense) magnetic field $\boldsymbol{B}_0$ is related to ψ_0 through $\boldsymbol{B}_0 \cdot \nabla\psi_0 = 0$. The electron continuity equation is given by

$$\frac{\partial n_e}{\partial t} + \nabla \cdot (n_e \boldsymbol{v}_e) = S_e(\boldsymbol{r}),$$

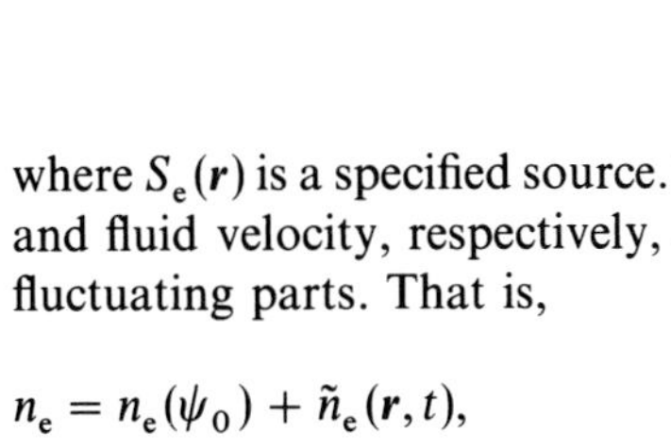

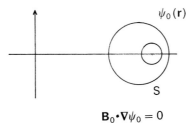

Fig. 5.16. Nested set of flux surfaces $\psi_0(\boldsymbol{r})$ in an axisymmetric toroidal system.

where $S_e(\boldsymbol{r})$ is a specified source. The quantities n_e and $\boldsymbol{v}_e$ are the electron density and fluid velocity, respectively, and can be expressed as the sum of mean and fluctuating parts. That is,

$$n_e = n_e(\psi_0) + \tilde{n}_e(\boldsymbol{r}, t),$$

$$\boldsymbol{v}_e = \boldsymbol{v}_{e0}(\psi_0) + \tilde{n}_e(\boldsymbol{r}, t).$$

It should be noted that the fluctuating parts are not necessarily small. We now introduce the standard averaging procedure of ordinary fluid turbulence. Thus the average of a fluid quantity $Q(\boldsymbol{r},t)$ is defined to be

$$\langle Q(\boldsymbol{r},t)\rangle = \lim_{T\to\infty} \frac{1}{T}\int_0^T Q(\boldsymbol{r},t)\,\mathrm{d}t,$$

where T is a macroscopic time scale, such as the pulse length of an experiment, or the time over which the source of particles is held constant, say. Put another way, this definition will be appropriate for all turbulence frequencies ω such that $\omega \gg 1/T$.

Now let us asume the electron fluid to be incompressible, that is $\nabla\cdot\boldsymbol{v}_\mathrm{e} = 0$. Then it follows that

$$\nabla\cdot\boldsymbol{v}_{\mathrm{e}0} = 0 \quad\text{and}\quad \nabla\cdot\tilde{\boldsymbol{v}}_\mathrm{e} = 0,$$

and some straightforward analysis yields the result

$$-\frac{1}{2}\int_S \frac{1}{|\nabla n_0|}\nabla\cdot\langle\tilde{\boldsymbol{v}}_\mathrm{e}\tilde{n}_\mathrm{e}^2\rangle\,\mathrm{d}S = \int_V S_\mathrm{e}(\boldsymbol{r})\,\mathrm{d}V.$$

We immediately observe that the turbulent particle flux is at least cubic in the fluctuation amplitudes. This is a general result which is independent of the mechanism, and depends only on the existence of flux surfaces in the mean state, and the existence of the average. This suggests that electron compressibility is essential if a particle flux is to be obtained which depends quadratically on the amplitude. A more detailed examination confirms this conclusion, it being found that both electron and ion compressibility are required. In practice, amplitudes are found to be small, and so this becomes an important result.

Apart from its significance with regard to particle transport, it is also a means of ruling out all those mechanisms which are dependent on incompressibility. It should be noted that in the TEXT tokamak, for example, the measured fluctuations at the edge suggest that a quadratic dependence on the amplitude is consistent with the observed particle loss.

5.4.3 *Adiabatic response*

In many treatments of drift wave turbulence, it is assumed that the density fluctuations are related to the electrostatic potential fluctuations, $\tilde{\phi}$ through the Boltzmann form, that is,

$$\frac{\tilde{n}}{n_0} = \frac{e\tilde{\phi}}{T_\mathrm{e}}.$$

An argument due to Rosenbluth and Rutherford (see Haas and Thyagaraja, 1986) indicates that any physical process in which this condition holds cannot lead to significant particle transport. We now demonstrate this with the following argument. For the confinement zone in a tokamak, the density fluctuations $\tilde{n}/n_0$ are typically of order 10^{-2}. In this circumstance, the fluctuating electron velocity perpendicular to the magnetic field is given by

$$\tilde{\boldsymbol{v}}_{\mathrm{e}\perp} = \frac{\nabla\phi\times\boldsymbol{B}}{B^2} + \text{nonlinear corrections.}$$

The electrostatic potential fluctuations follow the adiabatic response referred to above, together with additional 'nonadiabatic' corrections. That is,

$$\tilde{\phi} = \frac{T_e}{e} \cdot \frac{\tilde{n}}{n_0} + \text{'nonadiabatic' corrections.}$$

The additional terms can arise through effects such as Landau damping, but they will be small if $\omega \lesssim 100\,\text{kHz}$. Now the 'radial' particle flux Γ_e is given by

$$\Gamma_e = \langle \tilde{v}_{\perp e} \tilde{n} \rangle,$$

where the brackets denote the averaging process defined earlier. Since the linear terms are $\pi/2$ out of phase, they will make no contribution to the particle flux. This means that the particle flux will – at most – be cubic in the amplitude, or alternatively the particle flux will be due to the 'nonadiabatic' corrections which will lead to an even smaller contribution. Thus for significant turbulent particle transport to occur it is essential that the underlying processes be sufficiently nonadiabatic. It should be noted that experimental results from the TEXT and ASDEX tokamaks strongly suggest that the fundamental mechanisms – whatever they are – are nonadiabatic.

5.4.4 *'Energetics' method*

It should be apparent that to solve the full nonlinear two-fluid equations – continuity, momentum, energy and Maxwell's equations – is prohibitively difficult. One practical approach which has been used by a number of authors (see for example Kadomtsev and Pogutse, 1979, and Thyagaraja and Haas, 1989), is the so-called 'energetics' approach; an outline of the method is shown below:

prescribed power spectrum (guided by experiment) → measured $T_e(r)$

prescribed power spectrum (guided by experiment)		measured $T_e(r)$
↓		↑
energy flux		compare
↓		↑
radial energy balance	→	calculate $T_e(r)$

In a typical tokamak experiment, TEXT for example, the electron temperature profile is readily measured. On the other hand, only limited information is available on the details of the fluctuations in density, electrostatic potential and magnetic field. However, guided by this fragmentary information, a power spectrum is constructed for one of the fluctuating quantities which is consistent with the known facts. In practice, it is most straightforward to assume a form for the magnetic fluctuation spectrum. The procedure is then to use this form through linking relations to derive the forms of all the other spectra, the amplitude of the spectra being set by a comparison with experiment at a single radial point. This spectrum is used to evaluate the energy flux, which is then substituted into the radial energy balance equation. The latter equation is solved to obtain the electron temperature profile which can then be compared directly with the observed T_e profile. Within its limitations, this approach gives good qualitative agreement, together with some quantitative agreement. In the case of

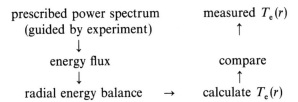

TEXT, the theory definitely demonstrates that the electrons respond nonadiabatically to the electrostatic potential – as indicated by experiment.

Although this method can only be regarded as a 'halfway' house to the complete problem, the resulting simplifications do have certain advantages. First, the energetics method is independent of the actual physical mechanism, or mechanisms. Secondly, the derivation of the spectral details, that is the nonlinear part of the problem, is bypassed. Finally, the energetics part of the calculation involves only a linear or quasilinear analysis, which is relatively straightforward to carry out.

We now give a brief overview of the way in which this procedure has been applied to TEXT by Thyagaraja and Haas. This is based on nonlinear kinetic equations together with model collision terms. It should be stressed that it is most important that the model collision terms adequately conserve particle number, momentum and energy. Any failure here will ultimately lead to the incorrect phases between fluctuating quantities, and hence errors in the particle and energy fluxes. As mentioned earlier, the $\tilde{B}$ spectrum is prescribed – a typical approach would be to choose a spectrum comprising m, n modes up to 50×50. Now, it can be shown that through the constraint of quasineutrality, the theory provides linking relations between $\tilde{B}$, $\tilde{n}$ and $\tilde{\phi}$. Hence the magnetic spectrum is adjusted until the magnitude and radial variations of the observed density fluctuation spectrum is achieved. Three pertinent questions can now be asked. Will this spectrum lead to:

(1) a nonadiabatic or adiabatic response for the electrons (more specifically, does it lead to the observed electrostatic potential fluctuations)?
(2) the observed electron heat flux, $Q_e(r)$?
(3) the observed ion heat flux, $Q_i(r)$?

Fig. 5.17 shows the variation of the magnitude of the radial magnetic fluctuations with radius for the modes which comprise the complete spectrum, namely the continuum and the (2, 1), (3, 1) and (5, 2) discrete modes.

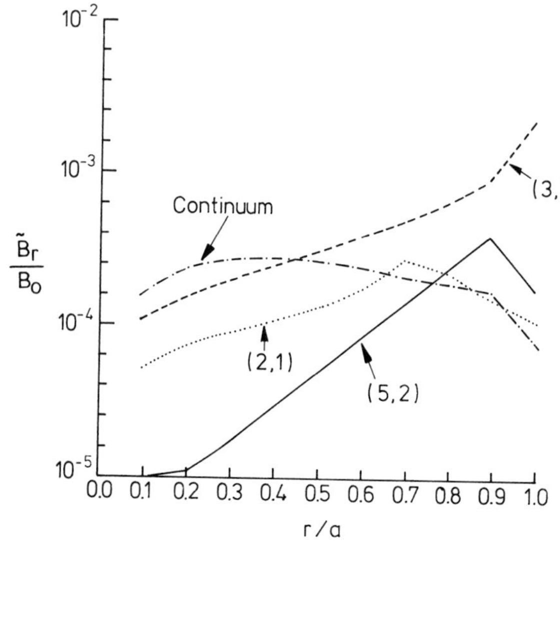

Fig. 5.17. Variation of the magnitude of the magnetic fluctuations with radius for the modes comprising the complete spectrum, namely the continuum and the (2, 1), (3, 1) and (5, 2) discrete modes.

Fig. 5.18 shows the corresponding density and electrostatic potential fluctuations as obtained from the theory (solid triangles and squares, respectively). The figure also shows the equivalent experimental results (open triangles and squares). Overall, good qualitative agreement is obtained. Figs 5.19 and 5.20 show the calculated electron and ion heat fluxes, respectively (indicated by the squares). The curves show the total heat fluxes as obtained from experiment; it will be seen that the experimental error bars are wide, but nevertheless the theory does produce at least qualitative agreement.

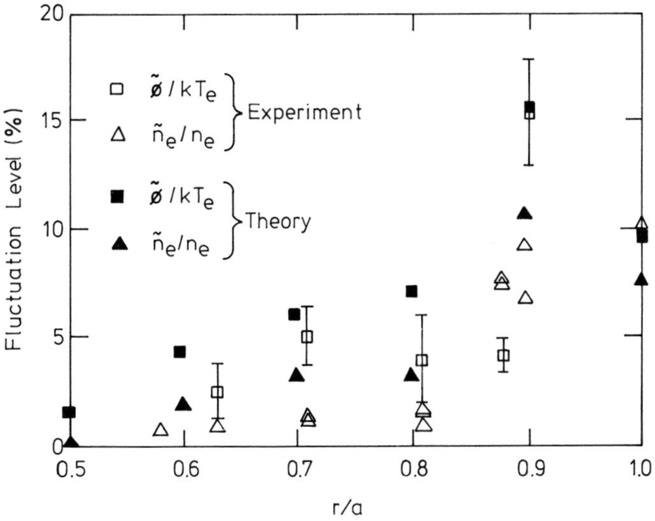

Fig. 5.18. The density and electrostatic potential fluctuations (solid triangles and squares, respectively) as derived from the spectrum of Fig. 5.17, with the equivalent experimental results (open triangles and squares).

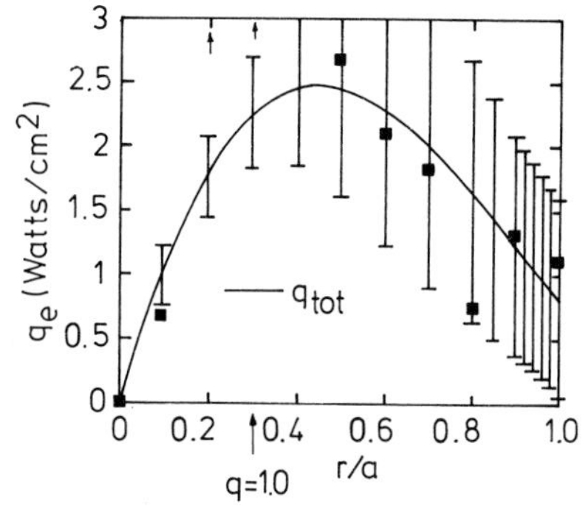

Fig. 5.19. Calculated electron heat flux (squares). The curve shows the total heat flux as obtained from experiment.

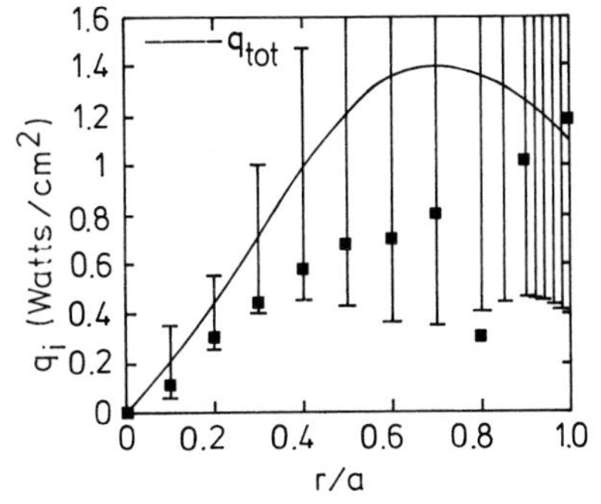

Fig. 5.20. Calculated ion heat flux (squares). The curve shows the total heat flux as obtained from experiment.

In conclusion, the prescribed magnetic spectrum gives the density and electrostatic potential fluctuations which are in good agreement with experiment, and demonstrate nonadiabaticity. The heat fluxes evaluated only agree qualitatively with observation. An important question arises as to whether the prescribed spectrum is unique or not. In practice, it has proved very difficult to construct an alternative spectrum which can match experiment to the same degree. Nonetheless, strictly speaking, the question of uniqueness is an open one.

5.4.5 *Collisional drift wave in a shearless magnetic field*

We now consider a specific physical mechanism and its consequences for turbulence. The collisional drift wave is anticipated to be relevant to turbulence at the edge of a small tokamak. In order to simplify the problem, we consider slab geometry (see Fig. 5.21). In this we treat the case of a uniform magnetic field B_0 in the z-direction, the x- and y-directions representing the radial and poloidal directions, respectively. The equilibrium density profile n_0 is assumed to have an inverse exponential dependence on x, namely, $n_0 \propto \exp(-\bar{K}x)$, where $\bar{K} = 1/L_n$, L_n being the density scale length. It is this density profile which provides the source of free energy for the instability. The problem is treated as two-dimensional in the variables x, y. The ions are described by warm fluid equations with a kinematic ion viscosity $\mu = 3T_i\nu_{ii}/10m_i\omega_{ci}^2$, where T_i and ν_{ii} are the ion temperature and collision frequency, respectively. The electrons are assumed to be isothermal along B_0, since the electron thermal conductivity is taken to be very large along the magnetic field. The model is, of course, resistive and the perturbations are purely electrostatic. The analysis leads to a pair of coupled nonlinear equations in the fluctuating electrostatic potential $\tilde{\phi}$ and density $\tilde{n}$. For the particular choice of mean state the coefficients in the equation are constant and the solutions for $\tilde{\phi}$ and $\tilde{n}$ can be developed as Fourier expansions. The wave numbers in the x- and y-directions are expressed as multiples of the basic wave numbers k_{x0}, k_{y0}, that is

$$k_x = lk_{x0} \quad \text{and} \quad k_y = mk_{y0},$$

where, l, m each range from 1 to 24. The initial turbulence is prescribed and the resistivity η and viscosity μ are fixed. Varying the parameter $\bar{K}$ – that is the free energy available for instability – a linear instability is observed to grow and finally nonlinear saturation is achieved. Fig. 5.22 shows the time evolution of four

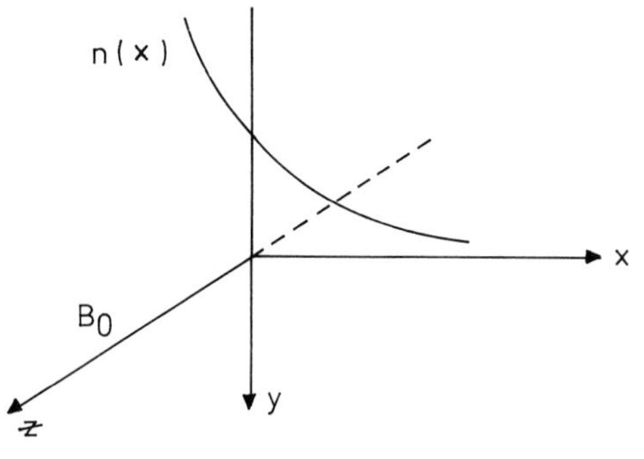

Fig. 5.21. Simple configuration for investigation of a collisional drift wave in a uniform magnetic field.

separate modes. It will be noted that the total energy of the modes saturates after sufficient time has elapsed. Fig. 5.23 shows how the magnitude of the energy of saturation increases linearly with $\bar{K}$, that is, increasing density gradient in the mean state. It will also be noted that the dependence is stronger if ion Landau damping is included (open circles). Finally, Fig. 5.24 gives the energy spectrum. It can be seen that the spectral dependence on k_x (that is, $E \approx k_x^{-3}$), is reminiscent of the 'Kolmogoroff–Kraichnan' spectrum, $E_k \simeq k^{-3}$. In conclusion the results show that the turbulence is nonadiabatic, that is $\tilde{n}/n_0 \neq e\tilde{\phi}/T_e$ – as suggested by experiment. Furthermore, the turbulence-driven particle flux is proportional to $\bar{K}$, that is $<\tilde{n}\tilde{v}_x> \propto \bar{K}$, and $E_{sat} \propto \bar{K}$; these results would appear to be intuitively reasonable.

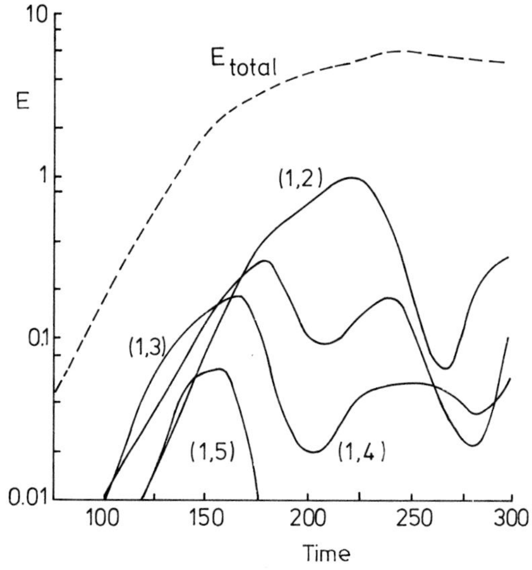

Fig. 5.22. Time evolution of energy in four separate modes. $\bar{K} = -0.3927$, $C_2 = 0.005$, $n = \phi = 0$ for $k_y < 0$ at $t = 0$.

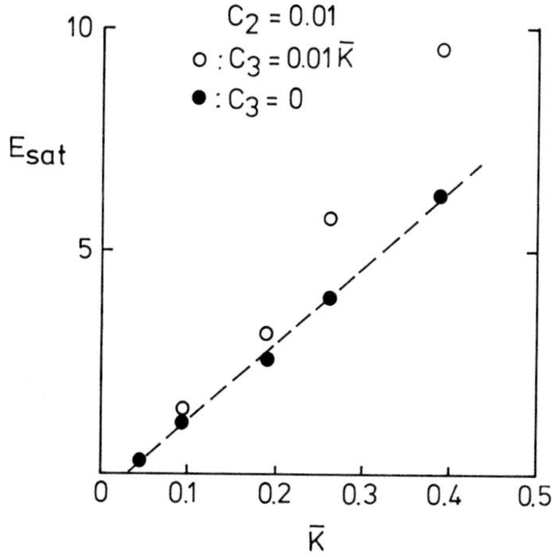

Fig. 5.23. Magnitude of the energy of saturation as a function of density gradient in the mean state.

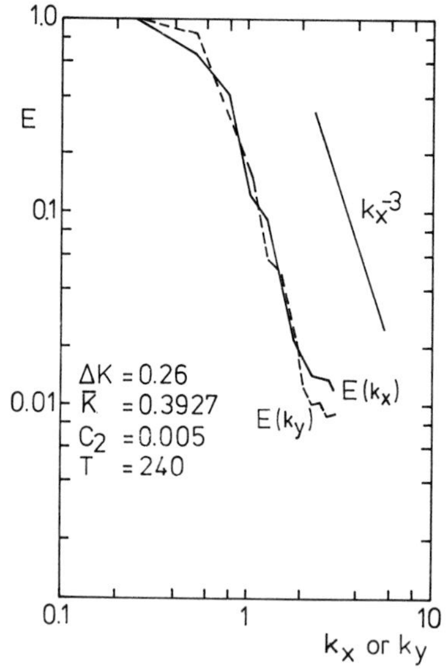

Fig. 5.24. Energy spectrum.

5.4.6 *A collisional drift wave in a sheared magnetic field*

In the model just described, the magnetic field is taken to be uniform. In practice, however, magnetic fields used for the confinement of plasmas are invariably sheared, and this can have significant theoretical consequences. Recently, Scott has investigated a slab model in which a sheared magnetic field is represented by the form

$$\boldsymbol{B} = B_0\left(\hat{\boldsymbol{e}}_z + \frac{x\hat{\boldsymbol{e}}_y}{L_s}\right),$$

where $\hat{\boldsymbol{e}}_y$, $\hat{\boldsymbol{e}}_z$ are unit vectors and L_s is the so-called shear length (see Fig. 5.25). The density and temperature are both assumed to have a gradient in the 'radial', that is x-direction. Collisions are included, and the study is restricted to electrostatic turbulence in ohmically heated discharges. The physical conditions are expected to be typical of those believed to occur at the edge of a tokamak like ASDEX, say. Energy transport is classical and includes thermal diffusivity χ, resistivity η and ion viscosity μ. Cold ions provide the inertia in the calculation and the electron dynamics are described by fluid equations. The calculation is restricted to a single resonant surface ($\boldsymbol{k}\cdot\boldsymbol{B} = 0$), as well as being two-dimensional (x, y). The resulting analysis leads to a coupled nonlinear set of equations in the variables $\tilde{n}$, $\tilde{T}$, $\tilde{\phi}$ and $\tilde{v}_\parallel$. Now it can be shown that this model is linearly stable: the aim of the investigation is to demonstrate the existence of a nonlinear (finite amplitude) instability, which subsequently goes to a saturated ('self-sustained') steady state. The initial turbulence is prescribed, with $\tilde{\phi}$ having a broadband structure, and being related to $\tilde{n}$ through adiabaticity, that is, $\tilde{n}/n_0 = e\tilde{\phi}/T_e$, with $\tilde{T} = \tilde{v}_\parallel = 0$. Varying the starting conditions, two types of solution are discovered: (a) a continuous decay, (b) steady self-sustainment. An example of the second type is shown in Fig. 5.26.

We now review the principal conclusions of this investigation. Despite the fact

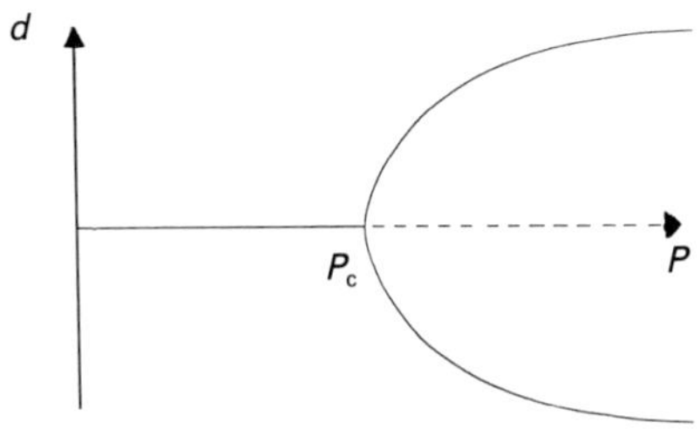

Fig. 6.2. Schematic plot of the symmetry-breaking pitchfork bifurcation in the Euler strut. The functional chosen to distinguish between the solutions is the mid-point displacement, which is shown plotted against the load.

deflection at the mid-point of the strut as the functional which distinguishes between the solution types. It may be seen in Fig. 6.2 that a change in the number of solutions from one to three occurs at P_c with increase of load. Thus, above this critical load there are two stable buckled states separated by the unstable straight solution. The buckled states have lost the up–down symmetry of the trivial state, and so we refer to this as a symmetry-breaking bifurcation. Algebraic forms for bifurcations of this type may be reduced from equations such as Eq. (6.2.1) by use of a technique called Lyapunov–Schmidt reduction. The form for the present case is:

$$G(x, \lambda) = x^3 - \lambda x, \tag{6.2.2}$$

where $\lambda = P/B$. If we set $G(x, \lambda) = 0$, we obtain the equilibrium solution which shows that the amplitude of the buckled states grows quadratically with increase in load.

One clear criticism of the analysis thus far is the assumed perfect symmetry of the situation. It is obvious even in the crude demonstration of the buckling of a plastic ruler that there will always be a preferred direction of bending of the ruler and that it is never convincingly straight, even at small loads. The perfect symmetry assumed in the mathematical model can never be realized in practice, even if we use an elaborate experimental arrangement. In other words, perfect symmetry is a non-robust mathematical abstraction which can only be approximated in the real world. However, the physical situation can be modelled by adding an imperfection term to the algebraic form as follows:

$$G(x, \lambda) = x^3 - \lambda x + I, \tag{6.2.3}$$

where I is a constant and is known as the imperfection term. This has the effect of disconnecting the pitchfork bifurcation as shown in Fig. 6.3. Thus there will now be the smooth evolution of one state with increase of load together with a

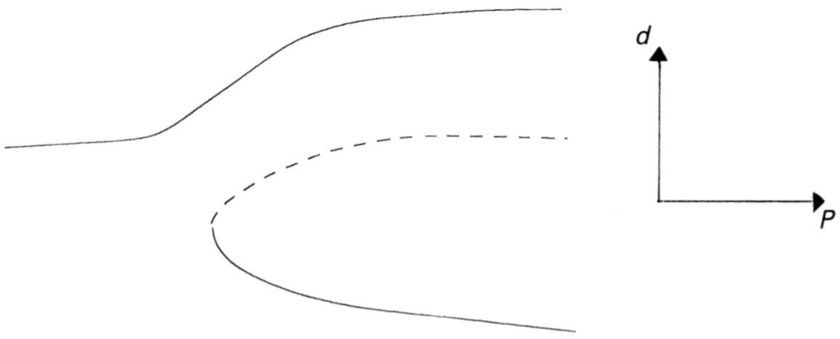

Fig. 6.3. The disconnected pitchfork for the Euler strut found when imperfections are included.

disconnected solution above a critical load. The disconnected state can only be reached by a discontinuous jump in load and it will snap back to the preferred state below a critical value. This disconnected form of the pitchfork is a good model of the physical system, and a lucid accound of the Euler strut problem is given by Zeeman (1976). We will now discuss the application of these ideas to a fluid flow problem.

6.3 Flow through a sudden expansion

An example of a pitchfork symmetry-breaking bifurcation in a fluid mechanical system is that of flow through a two-dimensional sudden symmetric expansion. A schematic diagram of the flow configuration is given in Fig. 6.4. Details of a numerical and experimental study can be found in Fearn, Mullin and Cliffe (1990). This particular example was chosen for detailed investigation because previous observations by Durst, Melling and Whitelaw (1974) had shown the existence of flows which are asymmetric about the centre plane of the channel. In addition, it is also known that the flow is symmetric about the same plane when the Reynolds number is small, and thus it must lose symmetry when this parameter is increased.

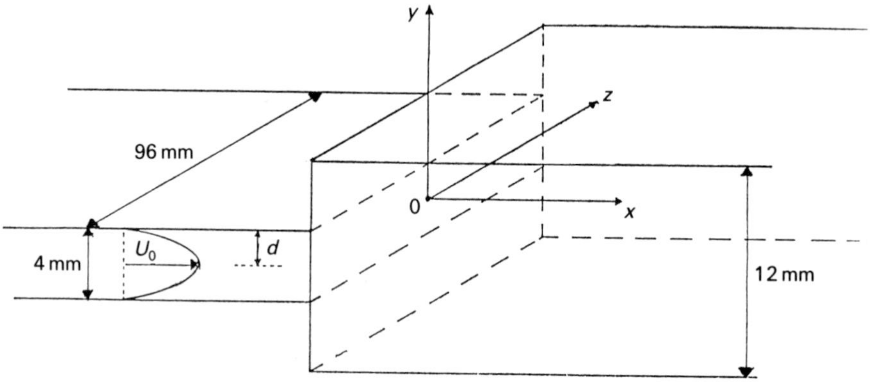

Fig. 6.4. Sketch of the test section used in the study of flow through a sudden expansion.

Numerical investigations by Cliffe and Greenfield (1982), and Sobey and Drazin (1986), of the related problem of flow through a smooth two-dimensional expansion had shown that the asymmetric states arose through a bifurcation. Cliffe and Greenfield used finite element approximations in their study and found a simple symmetry-breaking pitchfork, whereas Sobey and Drazin used finite differences and found a much more complicated bifurcation structure including multiple solutions and hysteresis phenomena. However, it has recently been shown by Sobey and Mullin (1992) that the numerical study of bifurcation phenomena using finite difference techniques can give misleading results. It is not yet clear if this completely accounts for the differences in the two sets of numerical results. As described in Section 6.1, the steady solution structure can underpin the time dependent phenomena in nonlinear problems. Therefore, before proceeding to the more complicated dynamical behaviour which arises at higher Reynolds numbers, it is clearly desirable to establish the exact nature of basic bifurcation structure.

Finally, the asymmetric states appear to persist into the turbulent regime as shown by Restivo and Whitelaw (1978). Thus the multiplicity of states is not restricted to laminar or weakly time dependent flows but can be a feature of 'real world' flows. It is clear, therefore, that an understanding of their origins and possible effects on the global dynamical behaviour is highly desirable.

6.3.1 *Pitchfork symmetry-breaking bifurcation*

In the experiment, a carefully controlled recirculating water flow channel was used to produce a steady flow with a parabolic profile in a nominally two-dimensional plane channel with aspect ratio of 24:1. Although a purely two-dimensional flow cannot be produced in the laboratory, it is found in practice that three-dimensional effects are confined to thin layers along the side boundaries at the low Reynolds numbers used here. Thus, from now on, the steady flows will be discussed as if they were ideally two-dimensional. The flow then passes through a sudden expansion of ratio 1:3. Observations were made on the flow by either flow visualization techniques or using laser Doppler velocimetry which gives an accurate measure of a velocity component at preselected points in the flow field.

At small Reynolds numbers (less than ≈ 35), the flow is symmetric about the mid-plane and it can be seen that the recirculation regions behind each step in the flow visualization photograph shown in Fig. 6.5 are of the same extent. In

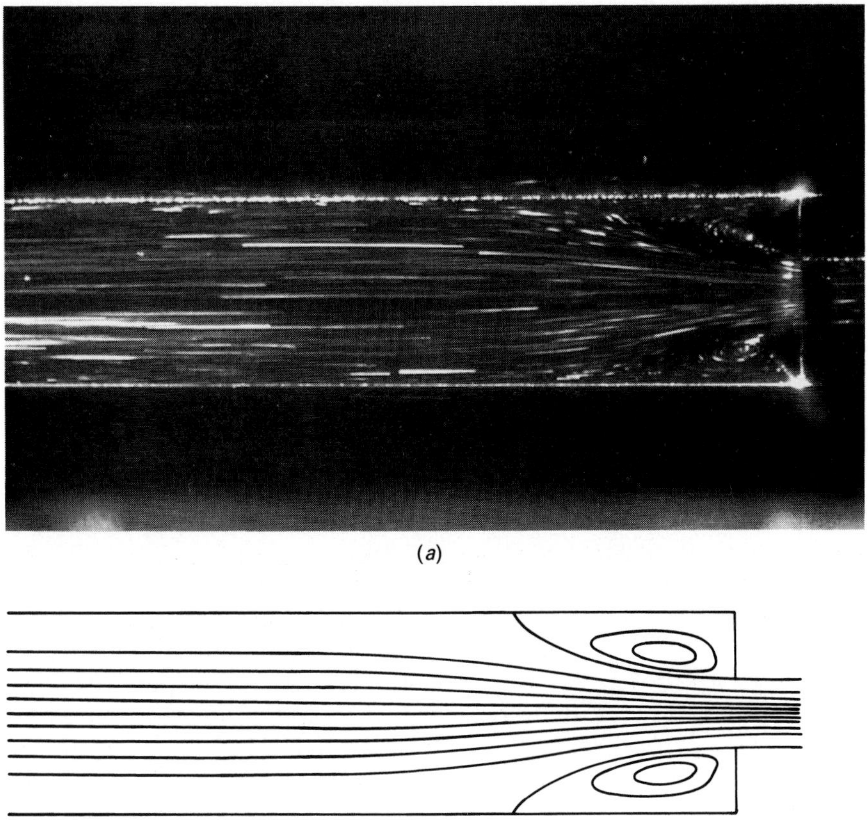

(*a*)

(*b*)

Fig. 6.5. (*a*) Flow visualization photograph and (*b*) calculated streamline plot for the symmetric flow at $R = 25.0$.

addition, we have shown the computed streamlines for comparison and they also show a symmetric form. (An outline of the numerical methods used in these calculations and the Taylor–Couette problem are given in Appendix A.) This is the only state which can be realized at these Reynolds numbers and thus the flow is uniquely defined at small R and is symmetric.

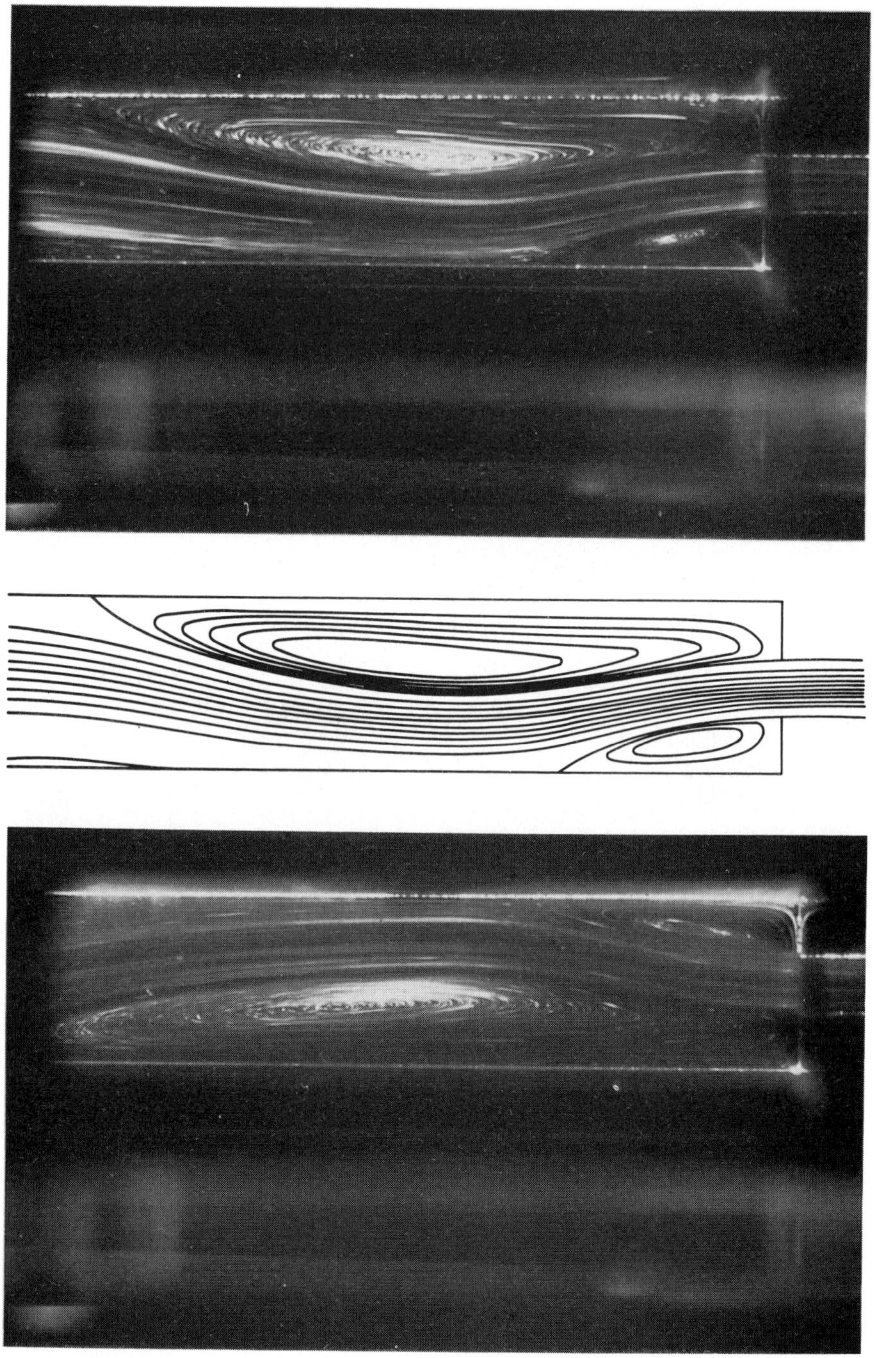

Fig. 6.6. Pair of flow visualization photographs of asymmetric flows at $R = 80.0$ shown with one of the streamline plots.

When the Reynolds number is increased above this range, the flow loses this symmetry such that one of the recirculating regions becomes longer than the other. There must now be two stable flow states if the boundary conditions are symmetric and each flow will be a mirror image of the other. We show in Fig. 6.6 flow visualization photographs of both flows along with one of the respective streamline plots at $R = 80$. The asymmetry is clear and continues to grow at higher Reynolds numbers but then saturates when other asymmetric recirculations grow downstream.

The asymmetric states have arisen though a bifurcation, and its form was calculated using the numerical bifurcation techniques which are outlined in Appendix A. We show the results of these calculations in Fig. 6.7, where it can be seen that the bifurcation has the form of a simple pitchfork. At low R there is a unique trivial solution which corresponds to the symmetric flow state. There is then an exchange of stability between this and a pair of asymmetric flows through the pitchfork bifurcation at $R_{crit} = 40.5$. Thus, above this value of Reynolds number the symmetric flow is unstable and the stable (and therefore observable) flows are the asymmetric states.

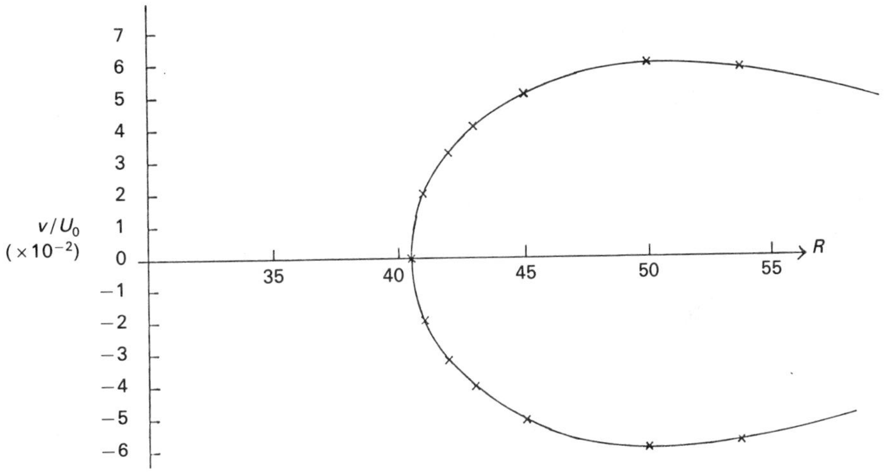

Fig. 6.7. Numerically determined bifurcation curve for the sudden expansion flow assuming perfect symmetry. The chosen functional is the vertical velocity component, shown plotted versus the Reynolds number.

Experimental estimates of the bifurcation curve can be made by measuring some feature of the flow which distinguishes between the different flow states. Here we choose the vertical velocity on the mid-plane as the functional which discriminates between flow types. For a symmetric flow the vertical velocity will therefore be zero and correspondingly positive or negative for the asymmetric states. The results of the measurements are shown superposed on the numerical results in Fig. 6.8. There is clearly some agreement far away from the bifurcation point but obvious disagreement close to the numerically determined critical point. The reason for the disagreement is the effect of imperfections in the experiment, which disconnect the bifurcation. This effect is the same as previously discussed for the Euler strut, where we saw that any small departure from symmetry gives rise to a disconnection. In the present case the numerical

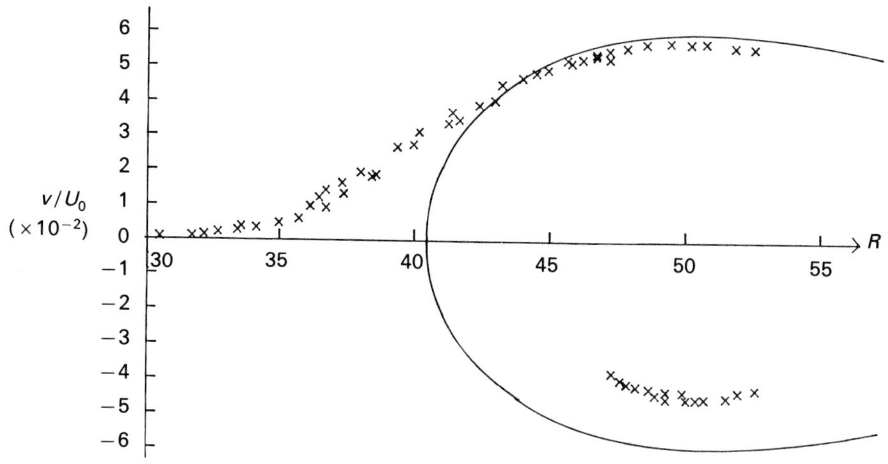

Fig. 6.8. Comparison of numerical and experimental results for the sudden expansion flow assuming perfect symmetry in the numerics.

calculations are carried out on perfectly symmetric boundary conditions which cannot be achieved in the physical experiment.

The imperfection in the experiment can be modelled by adding a small offset to the inflow boundary in the numerical scheme. This was done for the results shown in Fig. 6.9, and the overall agreement between calculation and experiment is now greatly improved. The actual offset is equivalent to $\approx 1/20$ mm in the experiment, which is on the limit of the accuracy of the construction of the apparatus. The disconnection is $\approx 13\%$ in R, which is an order of magnitude larger than the equivalent effects in Taylor–Couette flow. The difference between the two situations seems to be that the present case is an example of an open flow where disturbances may grow as they progress downstream, whereas the Taylor–Couette problem involves a closed recirculating flow.

As a result of the presence of this imperfection, only one of the two flow states can be realized by a continuous increase in R. The other state can only be reached by a sudden start of the experiment above $R \approx 50$. Thus after the flow-control valve is opened suddenly above this value there is a 50/50 chance of achieving the

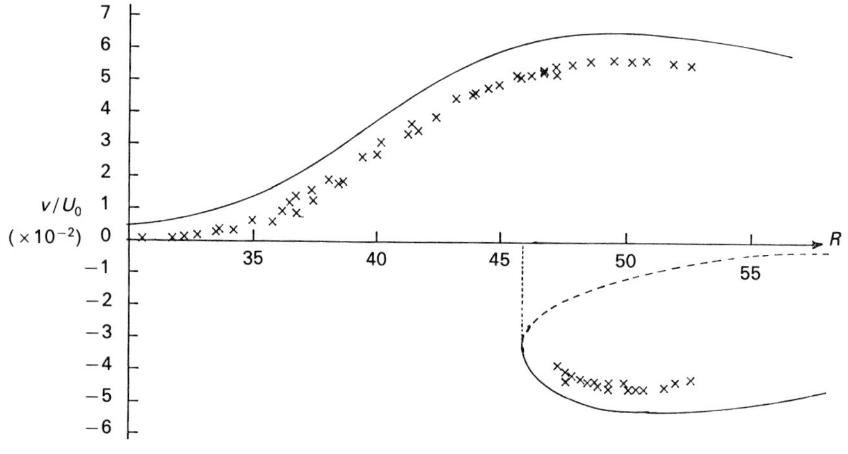

Fig. 6.9. Comparison of numerical and experimental results for the sudden expansion flow with a small imperfection included in the numerical model.

disconnected state after a transient. Once the disconnected solution has been reached, the Reynolds number can be reduced and an estimate of the limit point of the fold made. Thus we have conclusively shown that there is a pitchfork bifurcation in this open flow system.

If the Reynolds number is now increased further, time dependent motion sets in above $R \approx 152$ as shown by Fearn, Mullin and Cliffe (1990). Just prior to the onset of the time dependence the three-dimensional aspects of the laboratory flow become more prominent. In addition, there is conflicting numerical evidence for the presence of a Hopf bifurcation in the equations. In general, Hopf bifurcations give rise to singly periodic time dependent motion from a steady state, and in the present case would correspond to the sudden appearance of an oscillation in the flow as the Reynolds number is increased. However, it has not yet been proven if the time dependent flow in the experiment arises through such a mechanism. The experimentally observed motion is not strictly periodic, as can be seen from its power spectrum presented in Fig. 6.10. There is some width to the spectral peak indicating a variation in the frequency about a mean value.

Fig. 6.10. Power spectrum for the first observed time dependence in the flow through a sudden expansion.

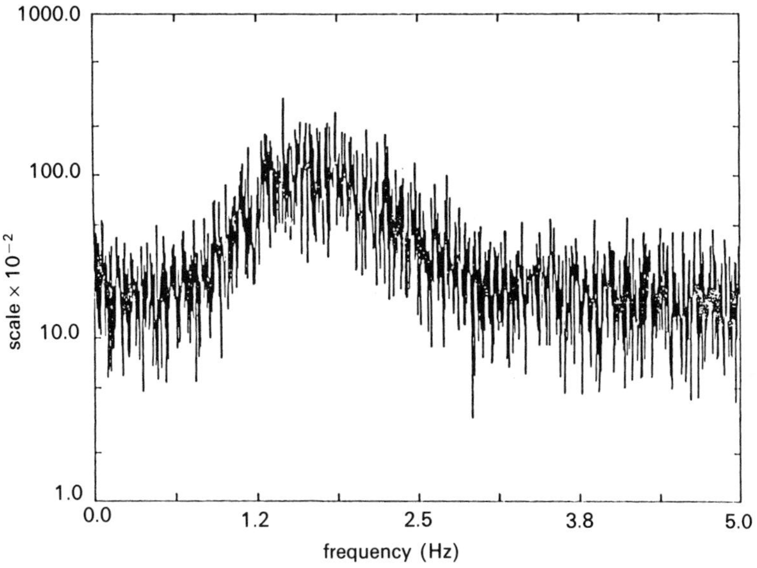

Recent developments in nonlinear signal processing techniques now permit us to reconstruct the attractor from the experimental time series as described in Appendix B. In the present case, a simple periodic oscillation would have the form of a connected loop or limit cycle in a reconstructed phase space, and the period of the oscillation would just be the time to orbit the loop. In addition, if there were two incommensurate frequencies present in the flow field the attractor would have the form of a torus or doughnut. In general, low-dimensional dynamical behaviour would be seen in the phase space representation as an attractor which has some geometrical structure. If we now return to the simplest dynamical behaviour observed in the sudden expansion flow and reconstruct the attractor corresponding to Fig. 6.10, we obtain the result shown in Fig. 6.11. There is no obvious structure in the reconstructed phase portrait, which ought to

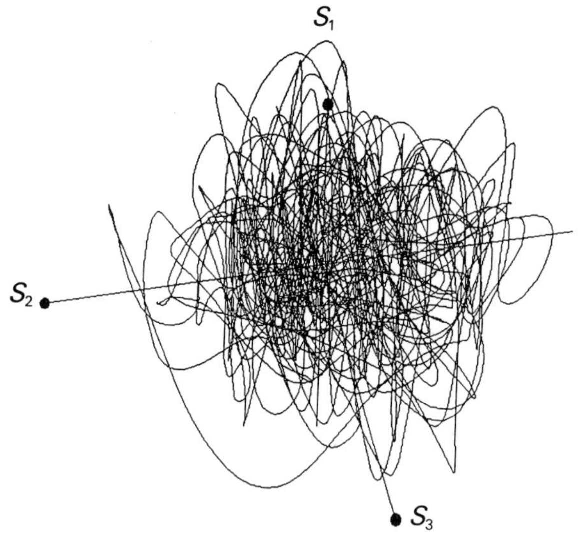

Fig. 6.11. Reconstructed attractor for the sudden expansion flow.

be present if low-dimensional behaviour is relevant. Thus we might conclude that low-dimensional dynamical behaviour is not present here when the system undergoes 'natural' transition to turbulence. However, we will return to this experiment after a discussion of Taylor–Couette flows and see that there is in fact evidence for low-dimensional chaos in a slightly modified version of the system.

6.4 Taylor–Couette flow

The Taylor–Couette flow problem is concerned with fluid motion in the gap between concentric rotating cylinders, and we will consider the case where the inner cylinder alone rotates. The situation involving the rotation of both cylinders is richer in phenomena, as shown by Andereck, Liu and Swinney (1986), but the theoretical understanding of that case is not yet as well developed as for the standard Taylor–Couette system. The interested reader is referred to Langford *et al.* (1988), for a discussion of the application of ideas from singularity theory to the double rotating cylinder problem. We will concentrate on the apparently simpler problem of the flow generated when the inner cylinder rotates and the outer remains stationary.

As discussed in the previous section, this is an example of a 'closed' flow system, unlike the fluid motion through the sudden expansion, which is an 'open' flow, where disturbances may grow as they are carried off downstream. One feature of the Taylor–Couette system is that it exhibits a rich variety of hydrodynamic instabilities. The lowest order instability of the mathematical model was first connected with the formation of cellular structure in the experimental flow by Taylor (1923). A photograph of a cellular flow is given in Fig. 6.13, and it is discussed in detail in the next section. Taylor's work was an important development in hydrodynamic stability theory since it provided the first quantitative agreement between theory and experiment for the appearance of an instability. Since that time there have been many papers written on the topic, and a good review of the subject is to be found in DiPrima and Swinney (1981). In more recent years, the problem has been extensively studied in relation to the application of the ideas of finite-dimensional dynamics to fluid flows. It is this latter issue that we will address here, but before proceeding with discussion of these complicated flows we will first review the fundamental problem of the steady bifurcation set associated with the cellular structure.

6.4.1 *The appearance of cells*

There are three control parameters which govern the Taylor–Couette system when the inner cylinder rotates. These are the radius ratio $\eta = r_1/r_2$, the aspect ratio $\Gamma = l/d$, where l is the length of the annular domain and d the gap width, and the Reynolds number which is defined as follows

$$R = \frac{\Omega r_1 d}{\nu},$$

where Ω is the angular speed of the inner cylinder, radius r_1, d is the width of the cylindrical gap and $\nu = \mu/\rho$ is the kinematic viscosity of the fluid. A schematic diagram of the experimental arrangement is presented in Fig. 6.12. The radius ratio is fixed in any experiment, and the latter two parameters may then be used to define a control plane. They can be defined with high precision in the experiment by accurate construction of the apparatus, fine control of the motor speed and close regulation of the temperature, which affects the viscosity of the fluid. These controls are important because critical bifurcation sequences can be covered by very narrow spans of parameter space. Observations can either be made using flow visualization techniques to view the flow structure, or a point measurement of a velocity component obtained using a laser Doppler velocimeter.

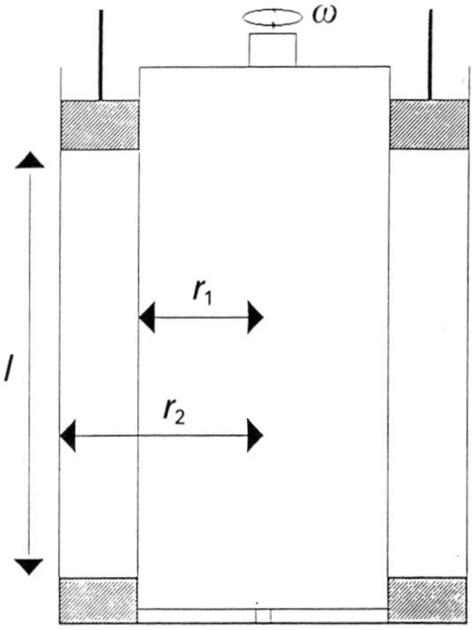

Fig. 6.12. Schematic diagram of the Taylor–Couette flow apparatus. $R = \omega r_1(r_2 - r_1)/\nu$; $\Gamma = l/(r_2 - r_1)$; $\eta = r_1/r_2$.

At small R the fluid is dragged around by the motion of the inner cylinder so that there is a shear across the fluid layer. If particles which reflect the incident light are added to the flow, then this flow regime will give an almost uniform reflectance except for some three-dimensional effects at the ends of the cylinders. When R is increased above a critical range, a banded or cellular structure appears along the length of the cylinder as shown in Fig. 6.13, and the flow has the appearance of a series of doughnuts stacked upon each other.

There is now a small secondary circular motion superimposed upon the main azimuthal component so that the particles follow spiral paths around the inner cylinder. If we now take a sectional view through the fluid as shown in Fig. 6.14 then the circular motion may be clearly seen. The photograph shows a view across the cylindrical gap illuminated by a planar sheet of light at right angles to the camera. We also show alongside, for comparison, the calculated streamline pattern which was obtained by K. A. Cliffe using the numerical methods outlined

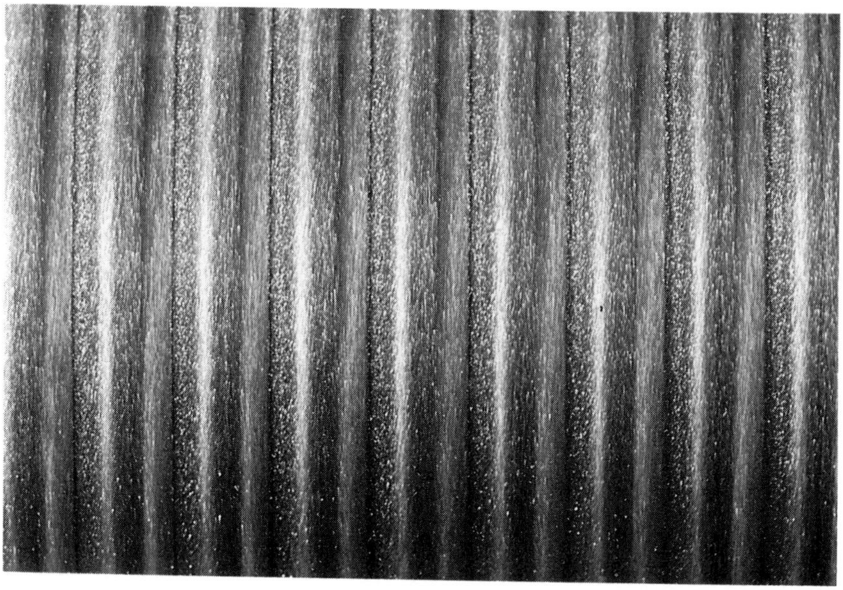

Fig. 6.13. Flow visualization of Taylor cells seen from the front of the apparatus.

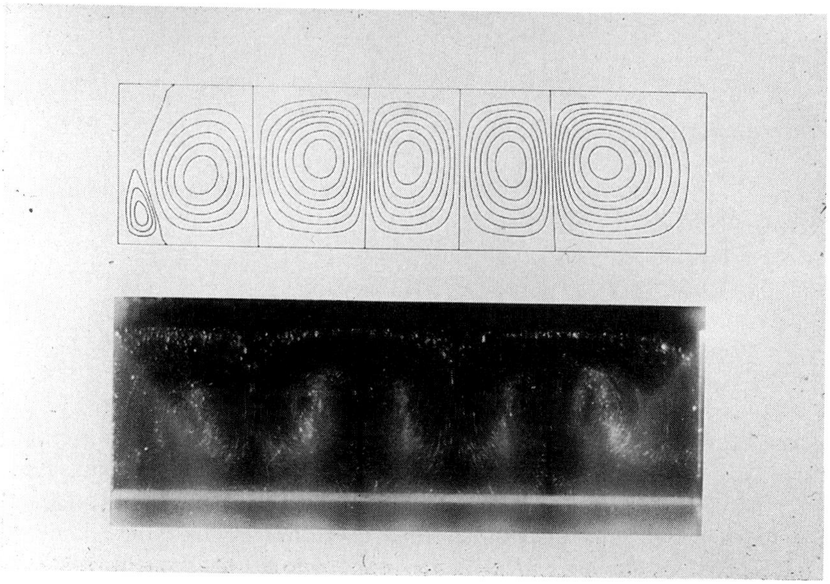

Fig. 6.14. Sectional flow visualization picture of cell structure shown in comparison with calculated streamlines.

in Appendix A. The reason we see the banded structure in Fig. 6.13 is that the particles are anisotropic and have the shape of small platelets. They are aligned by the shear in the flow and in the present regime follow spiral paths which are a combination of the main azimuthal and secondary circular flows. The incident light is reflected when they are face-on, but not when they are edge-on to the field of view. Thus dark bands correspond to regions where the flow is directed towards or away from the inner cylinder and the platelets are aligned by the action of the shear.

On symmetric boundary conditions, an even number of cells will appear with a smooth increase in R. The end cells will have a direction of rotation such that there is inflow along the stationary end wall. A physical argument which is often used to explain this preferred direction of rotation is that the centrifugal force gradient which drives the secondary circulation falls off towards the stationary end wall, and this determines the direction of rotation. However, as we shall see below, cells with the opposite sense of rotation are also observed, and so this physical reasoning can be challenged.

The number of cells which appear depends on the aspect ratio or size of the system. In general, n cells will be accommodated in an aspect ratio of size n, so that the cells tend to be square in cross-section. The mechanism for the exchange between different cellular states as the aspect ratio is varied involves a hysteretic interaction and will be discussed below. Even in this steady flow regime, the particular state which can be realized on the same boundary conditions is not unique. Thus other cell sizes and directions of rotation are stable as well as the 'preferred' or 'primary' mode, which we will define to be the state which is achieved by smooth increase in R. The primary state is equivalent to the connected buckled state in the Euler strut problem which was obtained by smoothly increasing the load. We call the disconnected states 'secondary' modes and they can generally only be created by discontinuous increases in R from rest into a range above the first onset of the primary flow. Nonuniqueness or multiplicity of states will be discussed later, as it has a direct bearing on the observed dynamical behaviour.

In a brilliant combination of classical analysis on the Navier–Stokes equations and fluid experiment, Taylor (1923) gave an explanation for the first appearance of the cells in terms of a hydrodynamic instability. In his model he considered the cylinders to be infinitely long so that periodic boundary conditions could be applied. Then the cellular instability can be represented by simple trigonometric functions if the additional assumption of a narrow gap between the cylinders is used. The mathematical problem then becomes amenable to direct methods of analysis.

In modern bifurcation parlance, the situation can be thought of as a pitchfork bifurcation from the trivial rotary shear state to an axisymmetric cellular flow, as shown in Fig. 6.15. Each branch of the pitchfork would correspond to a halfwavelength shift with respect to the other along the length of the cylinder, as indicated. This is of course easily achieved on the periodic boundary conditions of the model, where there is translational invariance in the assumed periodicity, but it is a situation which cannot be realized in an experiment. Thus, the central portion of the laboratory flow may appear to have an approximately periodic structure, but its phase is fixed by the ends of the physical system, no matter how large the apparatus.

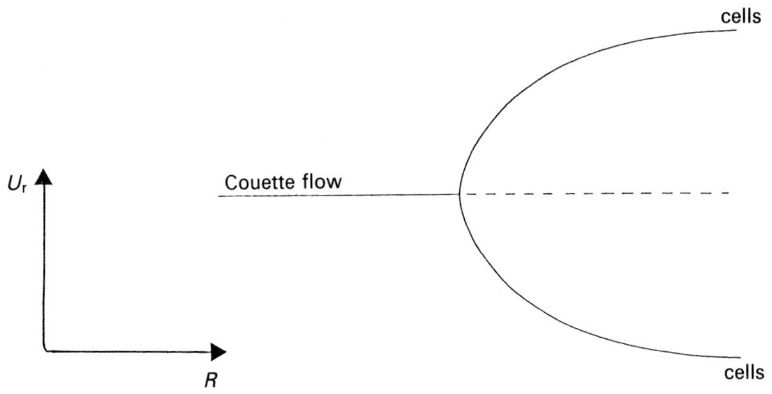

Fig. 6.15. Schematic pitchfork bifurcation for the appearance of steady cells in the Taylor–Couette model problem with nonflux boundary conditions.

We have already seen how the presence of physical imperfections can be accommodated in bifurcation theory in the examples of the Euler strut and flow through the sudden expansion. Indeed, there are three-dimensional flow effects at the ends of the flow domain which are precursors to the appearance of cells. Thus it might seem that we could represent the effects of the ends as a simple softening of the original pitchfork, but this is too simplistic a view. We can in fact measure the effects of the imperfection by obtaining estimates of the lower limits of stability of the disconnected portion of the pitchfork. This was first carried out by Benjamin (1978) and later extended both numerically and experimentally in a number of investigations. (See Mullin, 1991, for a review of this work.)

The central result of all of these studies is that the bifurcation is disconnected by a factor of ≈ 2.5 when the radius ratio of the cylinders is 0.6, as shown schematically in Fig. 6.16. A recent study by Cliffe, Kobine and Mullin (1992) has shown that, as the gap between the cylinders is narrowed, the disconnection becomes even greater. Thus, as the model conditions used by Taylor are approached, the quantitative difference between theory and experiment grows. This is an intriguing result since the experimental estimate for the appearance of cells is sharpened in this limit, and excellent agreement for this feature is obtained between theory and experiment as first shown by Taylor (1923). It also helps explain why these disconnected states went unnoticed in over 50 years of study of the problem until the work of Benjamin, since most experiments were focused on the narrow gap limit. Here the disconnected states only exist at Reynolds

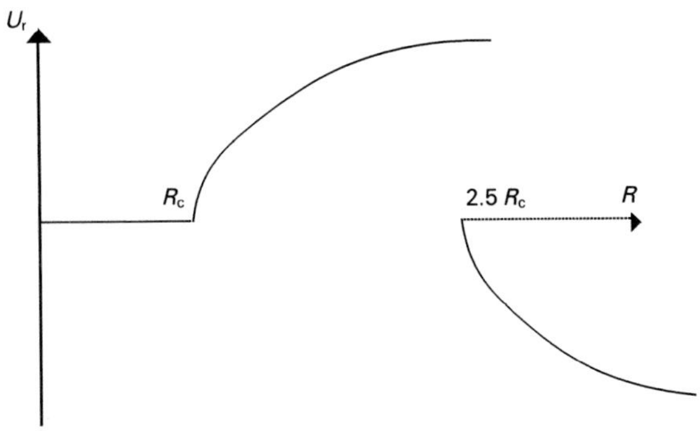

Fig. 6.16. Disconnection of model pitchfork bifurcation in the Taylor problem when realistic boundary conditions are applied.

numbers which are far into the time dependent regime and their existence can only be confirmed numerically by calculations of the steady flow equations. Nevertheless, they form an essential part of the solution set and since they had never previously been recorded, Benjamin called them 'anomalous modes'.

In summary, this is not an example of the simple softening of a pitchfork such as we saw earlier for the examples of the Euler strut and sudden expansion flow. The reason is that the essential symmetry of the model Taylor–Couette problem which has been lost in the experiment is that of translational invariance, whereas in the previous examples it is the loss of geometrical mirror-plane symmetry that is important. In other words, in the Taylor–Couette problem, the physical experiment is a major perturbation on the pitchfork in the model.

We stress the importance of this fundamental issue here, because it has very important consequences for the dynamical behaviour. It is well known that low-dimensional dynamical behaviour is often organized by the presence of double bifurcation points in a system (see e.g. Guckenheimer and Holmes, 1983). These can occur when lines of bifurcation points meet as the system's control parameters are changed. Thus, for example, we could have a line of Hopf bifurcations meeting a line of pitchforks in the present problem as the Reynolds number and aspect ratio are varied. The global dynamical behaviour which is born at interactions of this type persists for small disconnections of the pitchfork. In the present problem, if the Reynolds number is increased beyond the appearance of cells, then a time dependent instability arises through a Hopf bifurcation as first shown by Davey, DiPrima and Stuart (1968). This has the physical form of a simple travelling wave which travels at some fraction of the speed of the inner cylinder. Therefore, one might expect an interaction between the disconnected pitchfork and the Hopf as the control parameters are varied in the experiment, which could in turn give rise to low-dimensional dynamical behaviour.

However, since the pitchfork bifurcation of the model Taylor–Couette problem has to all intents and purposes been destroyed in the physical realization, we might expect that any associated finite-dimensional dynamics would also disappear. However, as we shall see below, there is in fact a simple geometrical pitchfork symmetry-breaking bifurcation in the physical system which helps organize the dynamics, and gives rise to low-dimensional dynamical behaviour. It cannot be accommodated in the periodic model which is often used to interpret the Taylor–Couette experiment, and requires a proper account of the realistic boundary conditions in any numerical investigation. Yet despite its nontrivial and nonlinear nature, the consideration of this pitchfork bifurcation is crucial in understanding the low-dimensional dynamical behaviour. In addition, if we wish to make mathematical progress in establishing a connection between the Navier–Stokes equations and low-dimensional dynamical systems, then we need to seek situations where weakly nonlinear methods are applicable. In the present problem these would be provided by multiple bifurcation points. Before we move on to a discussion of these, we must first discuss the interaction of steady states.

6.4.2 *The cellular exchange process*

A particular n-cell state will smoothly evolve with increase of R in a system with aspect ratio Γ as described above. Here n is an even integer, and the question we wish to consider now is how the exchange takes place between an n-cell state and its neighbouring $n \pm 2$ as the aspect ratio is varied. The problem was first investigated by Benjamin (1978) and we present in Fig. 6.17 a schematic diagram of the exchange mechanism he proposed. In Fig. 6.17(*a*) the aspect ratio is such that the n-cell state develops smoothly and the $n + 2$ state is a disconnected solution. The $+1$ and -1 give an indication of stability where the -1 signifies instability. As the aspect ratio is changed, there is an interaction between the two solutions so that the smooth curve develops a fold as shown in Fig. 6.17(*b*). The primary flow branch now has two stable parts which are separated by an unstable and therefore unobservable portion. The upper and lower fold points can be observed in the experiment as there is a definite jump in the intensity of the motion at these points, which can be precisely determined. Accurate estimates of the fold points, and thus the extent of the hysteresis, can be made in the experiment. Further change of the aspect ratio in the same direction leads to the two branches touching through the transcritical bifurcation in Fig. 6.17(*c*), and then the solution curves separate as shown in Fig. 6.17(*d*). Thus there is a smooth exchange between the two solutions, which involves hysteresis in the folded solution of Fig. 6.17(*b*).

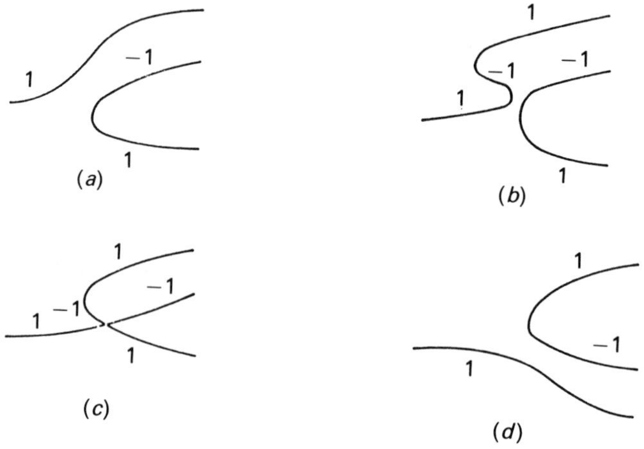

Fig. 6.17. Sequence of bifurcation diagrams showing the exchange between the n and $n + 2$ primary states when the aspect ratio is varied.

Benjamin confirmed this picture with experimental observations of the exchange between two and four cells. We show in Fig. 6.18 the experimental bifurcation set obtained by Mullin (1982) for the exchange between four and six cells. In order to relate these results to the sequence of bifurcations shown in Fig. 6.17, it must be remembered that only critical points can be determined conclusively in an experiment. Thus it is not possible to define the point at which cells appear on the smooth branches without adopting some arbitrary criterion. Therefore, the results shown in Fig. 6.18 are loci of limit points for the folds of the disconnected solutions and folded primary branch in the interaction region. The experiment is performed at various fixed values of aspect ratio and then the Reynolds number is varied. The line AB is the locus of limit points of the six-cell flow and CD that for the four cell when they are the respective secondary modes. Thus above B the six-cell flow will evolve smoothly with increase of R, and the four cell is a secondary mode. C is therefore the transcritical bifurcation point and

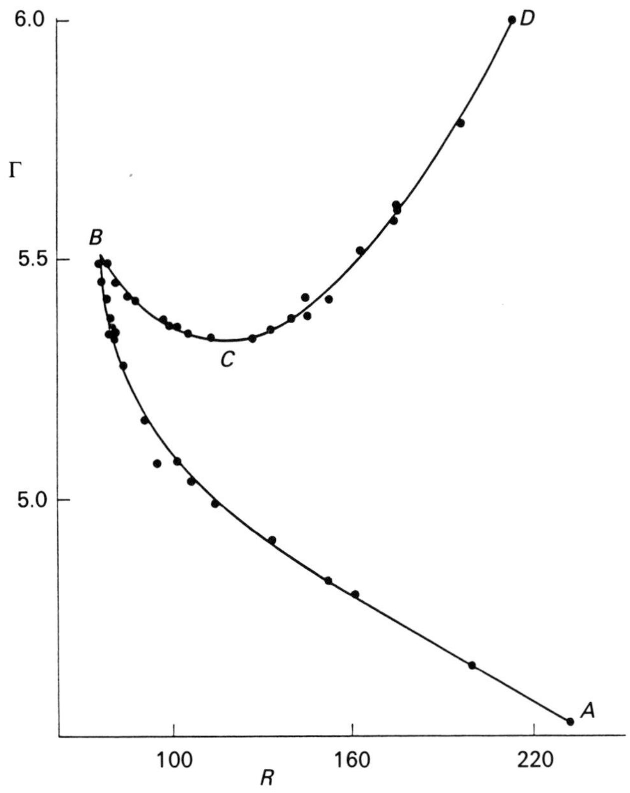

Fig. 6.18. Experimentally determined bifurcation set for the four- to six-cell exchange. *AB* and *CD* show the lower limits of the six- and four-cell states, respectively. *BC* is the upper hysteresis limit for the folded six-cell branch.

CB is the upper limit of the folded six-cell branch. The diagram can be related to Fig. 6.17 by choosing the sequence (*d*), (*c*), (*b*), (*a*) as the aspect ratio is increased.

Next we present in Fig. 6.19 the numerical results of Cliffe (1988) shown in comparison with the experimental points. The solid lines were computed on a grid which covered half the cylinder height and the solution is then reflected about the mid-plane. This assumption of symmetry of the solution was imposed for computational economy. It can be seen that, whilst there is good agreement between calculation and experiment along *BCD*, there is very poor correspondence along most of *AB*. The dashed line which begins at *E* was then calculated on a full grid without the symmetry assumption. It shows the locus of pitchfork symmetry-breaking bifurcations, and it is the path along which the six-cell flow loses stability to an unstable asymmetric state before it reaches the limit point, as proposed in the model shown in Fig. 6.17. These new solutions are asymmetric about the mid-plane of the flow and are equivalent to those discussed for the sudden expansion and Euler strut. An example of a pair of asymmetric four-cell states is shown in Fig. 6.23 for a slightly different problem, but they are qualitatively the same as in the present case. They are stable in some parameter ranges, as we shall see, but the main point we wish to emphasize here is that they can affect the stability of the symmetric solutions and need to be accounted for. The path of pitchfork bifurcations passes onto the unstable part of the solution surface just below *B* and thus has no effect on the experimental observations. However, it re-emerges on the stable surface above *D*, which is the point we will now describe.

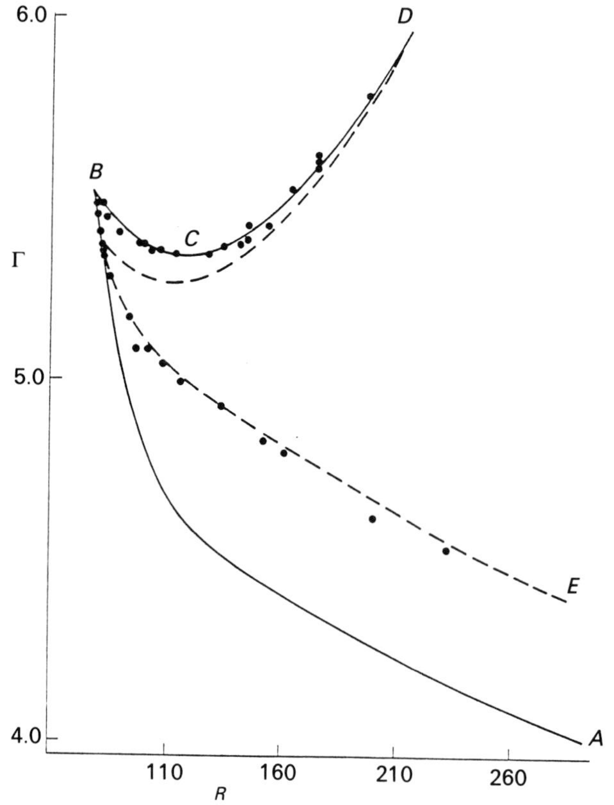

Fig. 6.19. Comprison between the numerical and experimental results for the four- to six-cell exchange. The solid lines are calculated assuming symmetric solutions and the dashed curves show the loci of symmetry-breaking bifurcations.

At the point D the path of symmetry-breaking pitchfork bifurcations crosses that for the folds of the secondary four-cell state. Guckenheimer (1981) has shown that double singular points of this type can have a line of Hopf bifurcations originating from them under certain circumstances. The situation is shown schematically in Fig. 6.20. Figure 6.20(a) shows the case when the symmetry breaking intersects the stable part of the fold. The symmetric and antisymmetric eigenvalues are labelled A and S, respectively, and a negative sign again indicates instability. The situation depicted in Fig. 6.20(b) arises when the symmetry-breaking bifurcation has moved around to the unstable surface which will be the case for aspect ratios below those corresponding to D. The branches with broken symmetry will initially be doubly unstable both to antisymmetric and symmetric perturbations by continuity from the unstable part of the fold. If on the other hand we started with situation (a) and continuously changed the aspect ratio to situation (b) then at the special point of coincidence of the two bifurcation points, the entire stability of broken-symmetry branch would have to change from $(+,+)$ to $(-,-)$. This is a highly unlikely scenario and so the dilemma is resolved by a line of Hopf bifurcations arising at D which act as secondary bifurcations to aid the exchange of stability. The two negative eigenvalues coalesce on the real axis, complexify and cross to the right-hand side through the imaginary axis as a complex conjugate pair in the Argand diagram. This is the condition for a Hopf bifurcation. In addition, the eigenvalues cross the imaginary axis close to zero,

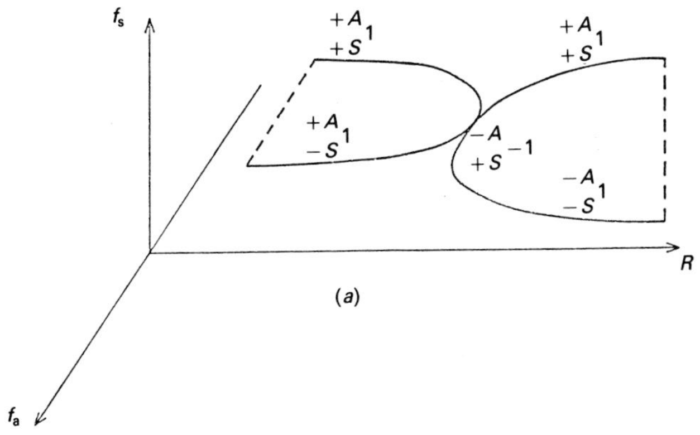

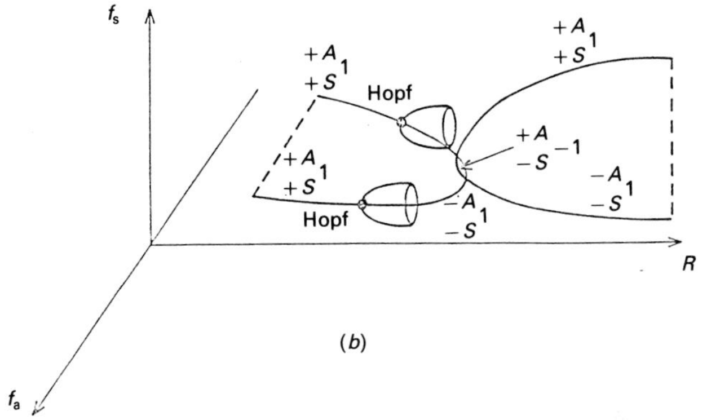

Fig. 6.20. Schematic diagram of the symmetric–asymmetric four-cell steady state interaction near point D in Fig. 6.19 showing the origin of the Hopf bifurcations.

which implies that the oscillatory states will have long periods.

We show the comparison between experimental and numerical results for the parameter range near the double singular point in Fig. 6.21. A full discussion of these results is given in Mullin, Cliffe and Pfister (1987). The line ADE originates at the double singular point A, and is the locus of Hopf bifurcations to oscillatory motion which is axisymmetric around the cylindrical gap between the two cylinders. The time signature of the flow can now be represented as a limit cycle using the methods outlined in Appendix B. AB is the locus of fold points and AC that for the symmetry-breaking bifurcations. The numerical results give the location of the Hopf points and indicate whether it is sub- or supercritical, but, as indicated in Appendix A, we cannot yet compute the periodic orbits which arise there so that the bifurcating branch cannot be followed numerically. The Hopf bifurcation was found, numerically, to be supercritical along AD and subcritical along DE, so that the axisymmetric oscillation exists for parameter values outside the region enclosed by ADE. The oscillation to the left of AD was confirmed to be axisymmetric in the experiment by splitting the laser beam and using two detectors to determine the phase relationship around the cylindrical gap. Each bifurcation point in the experiment was determined at a fixed value of Γ by varying R as before.

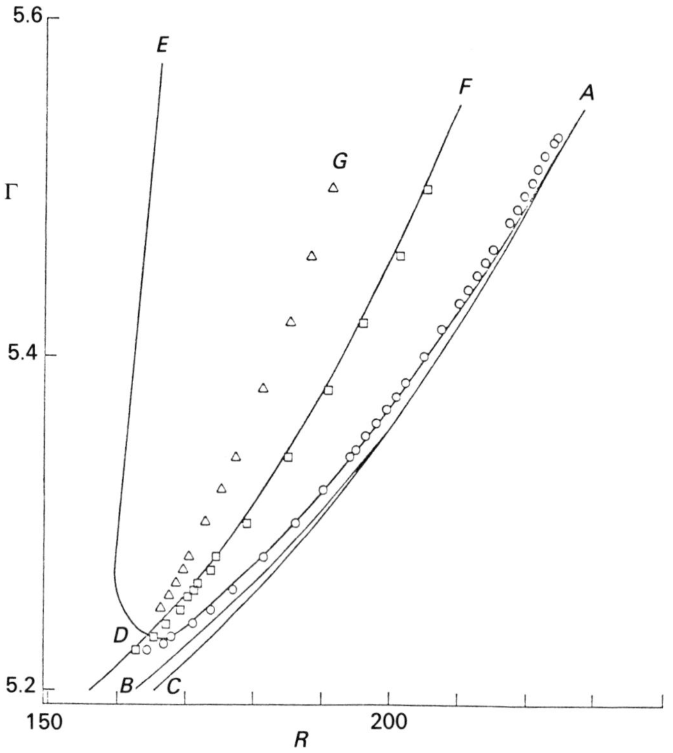

Fig. 6.21. Experimental and numerical results for steady and time dependent bifurcation set near the double singular point. *AC* is the line of symmetry-breaking bifurcations; *AB* is the fold line for the symmetric four-cell state; *ADE* is the locus of axisymmetric Hopf bifurcations; *FD* is the path of Hopf bifurcations which give an antisymmetric wave; and *GD* is the experimentally determined line of torus bifurcations.

The line *DF* in Fig. 6.21 is the locus of Hopf bifurcations to an azimuthal travelling wave with a wave number of one and the bifurcations are subcritical so that the onset of the wave occurs with decrease in *R*. This wave has the form of a tilt which has a 180° phase difference on opposite sides of the cylinder and travels at $\approx 1/10$ the speed of the inner cylinder. *DG* is the experimentally determined line of secondary Hopf bifurcations where there is an interaction between the two modes. To the left of this curve the time dependent motion is thus quasiperiodic as there are two incommensurate frequencies present. We can therefore represent this as motion on a torus in a reconstructed phase space.

Finally, we show in Fig. 6.22 a comparison between the numerical and experimental estimates of the period of the axisymmetric oscillation as the double singular point is approached in the (R,Γ)-plane. As discussed above, the frequency goes to zero as the double singular point is neared. In addition, the experimental observations show more scatter for the lower frequencies because the periods of the oscillations now correspond to approximately 100 inner cylinder rotations and seem to be sensitively dependent on the Reynolds number. There is now some evidence for irregularity in the periods and perhaps even finite-dimensional chaos. This is the topic for discussion in the next section where we use a more clear cut example which involves a variant of the Taylor–Couette system.

6.5 Low-dimensional chaos

Now we will consider some of the events which are observed in a modified version of the Taylor–Couette system. Here we rotate the ends of the fluid column with

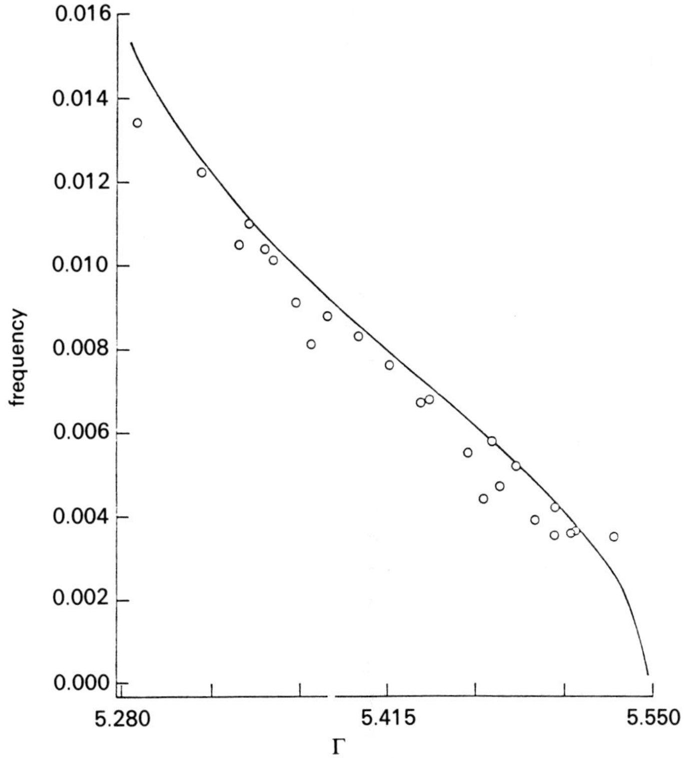

Fig. 6.22. Numerical and experimental comparison for the frequency of the axisymmetric oscillation near the double singular point.

the inner cylinder in an attempt to impose a strong symmetric forcing on the flow. The reason for the additional forcing is that, as we have just discussed, the standard system with stationary ends gives rise to symmetry-breaking bifurcations. The flows which arise at these points break the up–down symmetry of the flow domain as shown in Fig. 6.23. They must necessarily exist in pairs and are akin to the buckled beam states in the Euler strut. The idea behind the new system was that symmetric forcing would suppress the symmetry-breaking bifurcations, but in fact it turned out that their occurrence was made more likely. It would therefore appear that they are an essential feature of recirculating flow fields and, as we shall see, they form organizing centres for low-dimensional dynamical behaviour through the interaction of the asymmetric states with the symmetric flows.

We now wish to discuss results which came from a study whose primary aim was to obtain a better understanding of the finite-dimensional dynamics arising out of symmetric–antisymmetric interactions. The full details are given in Tavener, Mullin and Cliffe (1991). We show in Fig. 6.24 the numerically determined bifurcation set in the (R,Γ)-plane for a particular range of aspect ratio. It is clearly a very complicated diagram and so we will only give a brief overview of the features which are essential to understanding the organizing centres for the low-dimensional chaos. Inside the region *EQDK* there exist pairs of asymmetric four-cell flows and there is some hysteresis in the development of the asymmetry between *GQ* and *QE*. We need only discuss the results for one of the pair of solution surfaces here.

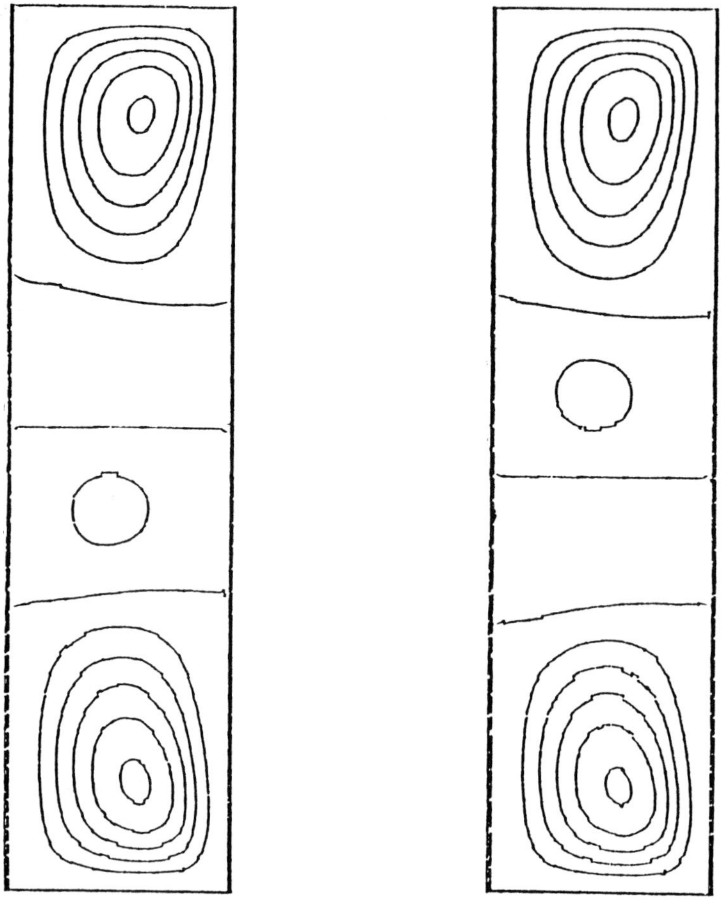

Fig. 6.23. Numerically determined streamline plots for a pair of asymmetric states in the rotating-ends Taylor system.

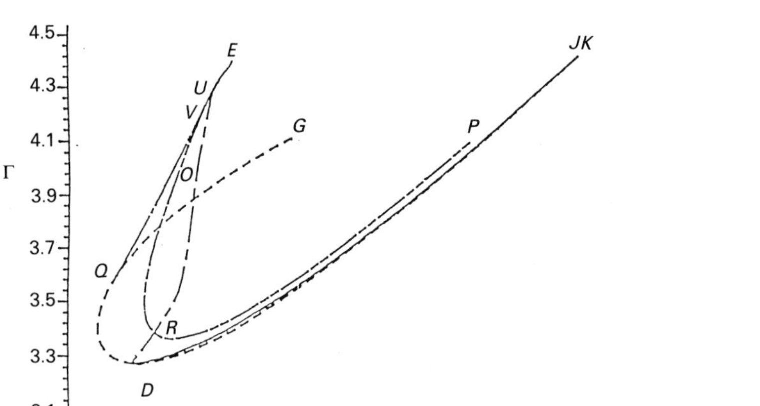

Fig. 6.24. Computed loci of bifurcation points for both steady and time dependent flows for the symmetric and asymmetric four-cell state. PRV is the locus of Hopf bifurcation to a wave number one flow which crosses the fold line QE at V. RU is the locus of Hopf bifurcations to an axisymmetric wave which terminates on QE in a Takens–Bogdanov point U.

Fig. 6.27. (*a*) Velocity time series for the radial velocity component measured at mid-gap with aspect ratio 4.25 and $R = 127.9$. (*b*) Reconstructed attractor from the experiment. (*c*) Model attractor with $p = 0.7$.

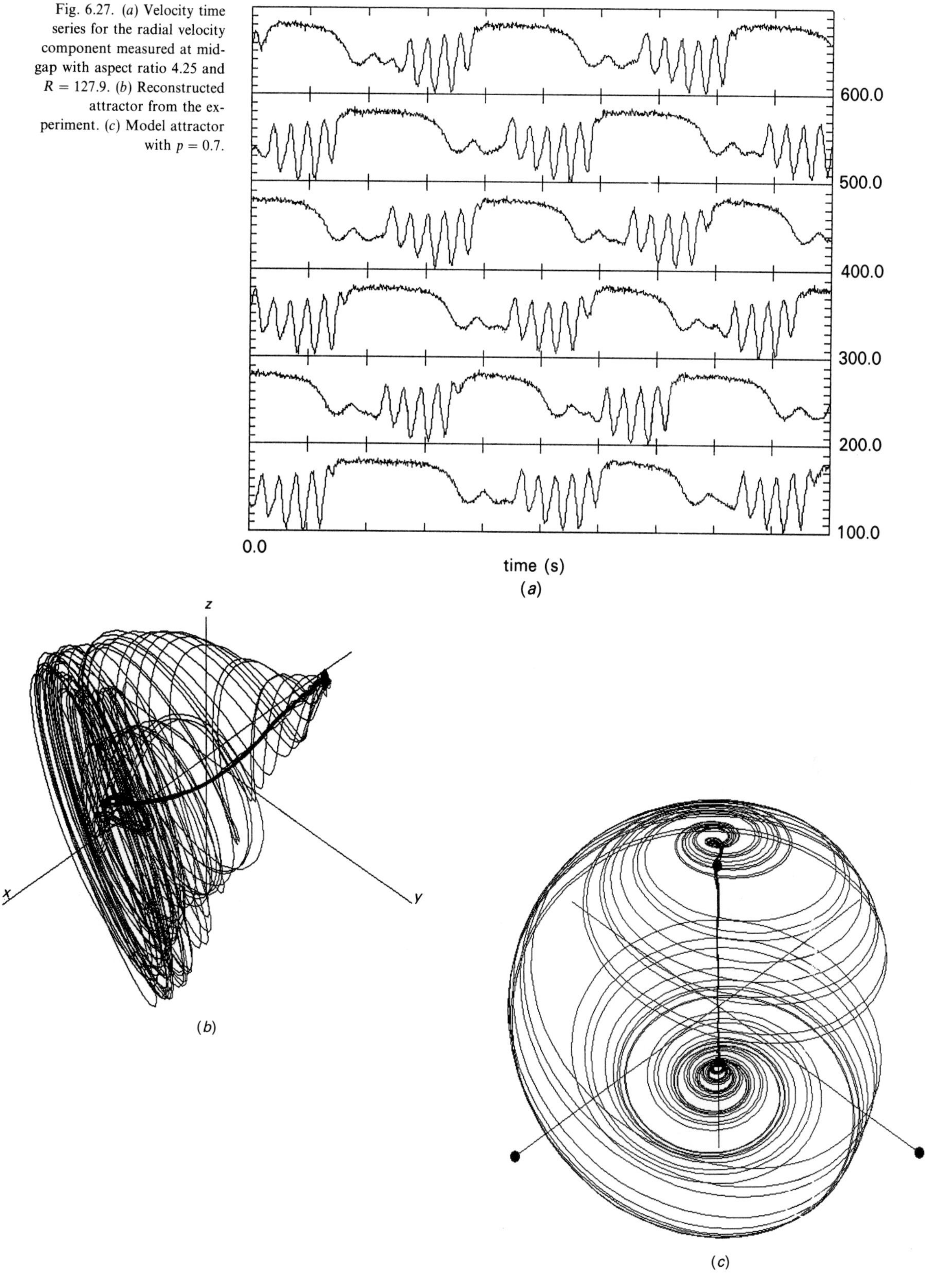

in terms of local critical events in parameter space. This observation makes the possibility of establishing a connection between the governing equations of motion and the considerably simplified dynamical models realistic.

Despite the considerable successes of the present studies of the application of modern ideas on chaos to well-controlled fluid flows, they appear to have little relevance when applied to the more general problem of fluid flow turbulence. One reason for this is that turbulence is almost certainly an infinite-dimensional phenomenon and so is not understandable directly in terms of the drastic restrictions imposed in the small scale experiments of the type discussed above. We have only really concerned ourselves with weakly irregular motion around well defined spatial states. Thus these ideas may have practical relevance when describing the large scale coherent motions which are often observed in flows such as atmospheric waves and large eddy flows in rivers and shear layers. It is perhaps even more important to note that the ideas of multiplicity of states and symmetry-breaking bifurcations do have direct practical value.

We end on a note of optimism for the future of the application of the ideas of low-dimensional dynamical behaviour to fluid flows. The open flow we discussed previously, which was concerned with flow through a sudden expansion, appeared to show no evidence for finite-dimensional behaviour in the dynamical regime. The observed steady laminar flow was replaced by a complicated three-dimensional time dependent state as the Reynolds number was increased. However, recent work by Madden and Mullin (1993) has shown that finite-dimensional dynamics including period doubling, homoclinicity and intermittency can be induced by adding a small periodic disturbance to the inflow. We show an example of this in Fig. 6.28 in the form of the reconstructed phase portrait of a torus. The two frequencies on the torus can be made to lock by changing either the frequency of the applied oscillation or the Reynolds number. In either case, finite-dimensional chaos results, which we believe is one of the first experimental observations of its kind. Thus low-dimensional dynamical behaviour may well be relevant to the understanding of a much wider class of weakly disordered flows than the perhaps rather special case of flow between rotating cylinders.

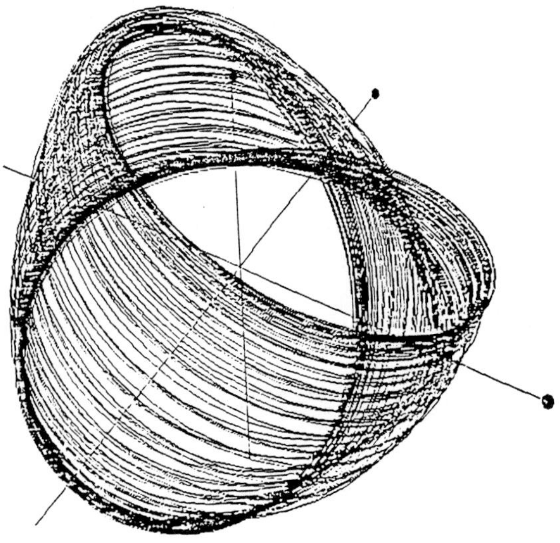

Fig. 6.28. Reconstructed experimental phase portrait for the modulated flow through a sudden expansion. The attractor has the form of a torus showing the presence of two incommensurate frequencies.

6.7 Appendix A

6.7.1 *Introduction*

We present in this section a brief description of the numerical methods which are used to compute bifurcations in the discrete two-dimensional Navier–Stokes equations. We will concentrate here on the problem of the flow through a sudden symmetric expansion although the techniques have also been used to explore the bifurcation structure of the Taylor–Couette flow, which is the main topic of discussion on low-dimensional chaos in fluid mechanics. The calculations in all of these studies were performed on the CRAY at AERE Harwell, using the ENTWIFE package in collaboration with K. A. Cliffe. We will concentrate on giving an overview of the techniques, and details of the basic numerical methods can be found in Jepson and Spence (1984, 1985).

6.7.2 *The governing equations*

The two-dimensional nondimensionalized Navier–Stokes equations for an incompressible fluid with constant viscosity in Cartesian coordinates (x^*,y^*) are given by:

$$\left.\begin{aligned} &\frac{du}{dt} + u\frac{du}{dx} + v\frac{du}{dy} + \frac{dp}{dx} - \frac{1}{R}\left(\frac{d^2u}{dx^2} + \frac{d^2u}{dy^2}\right) = 0 \\ &\frac{dv}{dt} + u\frac{dv}{dx} + v\frac{dv}{dy} + \frac{dp}{dy} - \frac{1}{R}\left(\frac{d^2v}{dx^2} + \frac{d^2v}{dy^2}\right) = 0 \\ &\frac{du}{dx} + \frac{dv}{dy} = 0. \end{aligned}\right\} \qquad (6.7.1)$$

In the above equations, x,y are nondimensional coordinates, u and v are the nondimensionalized velocities in the x- and y-directions and p is the nondimensionalized pressure. The variables x, y, u, v and p are given by

$$x = x^*/d, \quad y = y^*/d, \quad u = u^*/U_0, \quad v = v^*/U_0, \quad p = p^*/U_0{}^2,$$

where U_0 is the maximum of the prescribed inlet flow and d is the half-height of the inflow channel. The finite element grid used in the calculations is shown in Fig. 6.29.

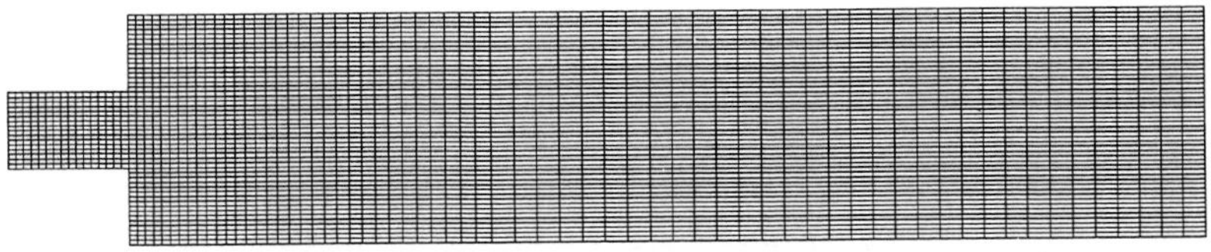

Fig. 6.29. The finite element grid used in the calculation of the flow through a sudden expansion.

We first consider the steady flow problem for which $d/dt = 0$ in Eqs (6.7.1). The origin of the coordinate system is defined to lie on the mid-line of the channel at the beginning of the expansion. Eqs (6.7.1) therefore hold on the region defined by:

$$D = (x,y) | -10 \le x \le 0, -1 \le y \le 1 \cup 0 \le x \le x_{max}, -3 \le y \le 3.$$

No-slip boundary conditions are applied on the walls of the channel such that $u = v = 0$ on all walls. The inlet flow conditions are prescribed to be a fully developed parabolic profile, such that $v = 0$ and $u = 1 - y^2$. Normal stress boundary conditions are imposed at the outlet such that $p = 0$, $du/dy = 0$ and $dv/dx = 0$ at $x = x_{max}$. The validity of these conditions was checked by extending the length of the calculation domain and showing that the solution is independent of the extension.

6.7.3 *Finite element calculations*

The steady two-dimensional Navier–Stokes equations, (6.7.1), are solved using a standard Galerkin discretization. They are converted into a nonlinear operator equation in an appropriate Hilbert space (see Rae and Sykes, 1976). When this Hilbert space is replaced by a finite-dimensional subspace, the result is a finite element approximation to Eqs (6.7.1). In this method, the domain D is covered by a mesh of nine-noded rectangular elements, symmetric about the line $y = 0$. The symmetry of the grid is necessary in order that the discrete problem retains the symmetry of the continuous one. At each node the components of velocity are interpolated by biquadratic polynomials, and the pressure is approximated at the central node of each element by linear functions which are not in general continuous across element boundaries.

As shown by Cliffe and Spence (1985), the finite element equations may be written in the form:

$$f(\boldsymbol{x},R) = 0, \tag{6.7.2}$$

where $\boldsymbol{x}$ is a vector containing all the velocity and pressure degrees of freedom. Solutions to Eq. (6.7.2) are found by use of an Euler–Newton continuation method. For calculations of symmetric flows, only half of the region D, corresponding to the region $y \geq 0$, need be discretized (Cliffe and Spence, 1984). The full solution for the whole domain D can then be obtained by reflecting the half-domain solution about the line $y = 0$. This technique of calculating the flow on half the grid is computationally more efficient but is only applicable to symmetric solutions.

Cliffe and Spence also show that the symmetry-breaking bifurcation point can be detected using the half-domain calculation in the following way. One of the features characterizing a bifurcation in a finite-dimensional system is that the determinant of the Jacobian matrix $f_{\boldsymbol{x}}$ is zero. Therefore, if there is a change in the sign of the determinant of the Jacobian matrix along a solution branch then a bifurcation has occurred on that branch. Numerical methods of following or continuing solution branches will be discussed below. Suffice it to say here that, if there is a change in the sign of the determinant of the Jacobian matrix between two steps n and $n + 1$ of the continuation process, it follows that at some intermediate value of the control parameter (R here) there is a bifurcation point.

Along a symmetric solution branch, it is possible to compute a symmetric and an antisymmetric Jacobian matrix, $f^{\mathrm{s}}_{\boldsymbol{x}}$ and $f^{\mathrm{a}}_{\boldsymbol{x}}$, on the half-grid, at each step of the continuation. The former is computed with symmetric boundary conditions on the line $y = 0$, and the latter with 'flipped' boundary conditions, given by $u = 0$, $dv/dy = 0$, along $y = 0$. A change in sign of the determinant of $f^{\mathrm{s}}_{\boldsymbol{x}}$ indicates the presence of a limit point, whereas a change in the sign of the determinant of $f^{\mathrm{a}}_{\boldsymbol{x}}$ indicates that a symmetry-breaking bifurcation has been stepped over in the continuation.

The stability of the steady state solutions on either side of the bifurcation point may be found by checking the stability of the calculated solutions against small time dependent perturbations. It can be shown that the linear stability of solutions $\boldsymbol{x}$ to Eq. (6.7.2) depends upon the generalized eigenvalues of the Jacobian $f_{\boldsymbol{x}}$ (see for example Jackson, 1987). In particular, a simple stability analysis shows that, for a stable solution, all generalized eigenvalues must have a positive real part. A necessary but not sufficient condition for stability is therefore that the product of the real parts of the generalized eigenvalues is positive. The sign of the determinant is, however, equal to the sign of the product of the eigenvalues. Thus, for any particular solution to be stable, the sign of the determinant of the Jacobian matrix must be positive (a necessary but not sufficient condition). In addition, a solution with a negative determinant of $f_{\boldsymbol{x}}$ is necessarily unstable. At a bifurcation point, therefore, because the determinant of the Jacobian matrix changes sign, an initially stable solution necessarily becomes unstable.

The solution obtained at a value of R immediately before the bifurcation point can be used as a first approximation, and the symmetry-breaking bifurcation point may then be accurately found from the solutions of the 'extended system' (Werner and Spence, 1984). It is necessary to use this extended version of the equations because the original system is singular at the bifurcation point. Here the extended system is:

$$F(\boldsymbol{y}) = \begin{matrix} f(\boldsymbol{x},R) \\ f_{\boldsymbol{x}}(\boldsymbol{x},R) \\ \boldsymbol{l}\boldsymbol{\varphi} - 1 \end{matrix} = 0, \qquad (6.7.3)$$

with $\boldsymbol{y} = (\boldsymbol{x},\boldsymbol{\varphi},R)$, $\boldsymbol{\varphi} \in \boldsymbol{X}$ and $\boldsymbol{l} \in \boldsymbol{X}'$ the dual of $\boldsymbol{X}$. The space $\boldsymbol{X}$ is the set of all possible $\boldsymbol{x}$'s and is equivalent to R_N, where N is the total number of degrees of freedom of the system. In Eq. (6.7.3) the basic solution $\boldsymbol{x}$ is symmetric and $\boldsymbol{\varphi}$ is an antisymmetric eigenvector.

Although the symmetric solutions and the location of the symmetry-breaking bifurcation point can be calculated using one half of the region D, the asymmetric solutions cannot be computed on this grid. Instead, the asymmetric solutions which branch from the symmetric solution at the bifurcation point are calculated on the full grid which covers the whole of the domain D. Each of the branches is followed by Euler–Newton continuation methods and using the solution $x_0 + \phi_0$ to Eq. (6.7.3) as an initial approximation. Along an asymmetric branch $f_{\boldsymbol{x}}$ cannot be decomposed into $f_{\boldsymbol{x}}^{\mathrm{s}}$ and $f_{\boldsymbol{x}}^{\mathrm{a}}$ and we must now compute the full Jacobian.

A second continuation method, known as the Keller arclength continuation technique (Keller, 1977), is often used for following a path of solutions with the variation of a parameter. This method introduces an arclength $\boldsymbol{s}$ to parametrize the solution. It has the advantage that, when there is a limit point on the solution branch, the Keller method will succeed in following the branch around the limit point, whereas ordinary continuation methods fail at such points.

In this way both symmetric and asymmetric steady flows and their stabilities can be predicted from the calculated solutions of the discretized steady, two-dimensional Navier–Stokes equations. The numerical results provide values of u, v and S (the streamfunction) at each node and p, p_x and p_y at the central nodes of each element, for selected values of Reynolds number.

When the above methods are applied to the Taylor–Couette flow problem we find that the bifurcation points now depend on two parameters. These are the Reynolds number, as in the present case, and the geometrical aspect ratio (nondimensionalized size of the system). The methods can then be adapted to investigate the dependence of the bifurcation points on the control parameters. By these means the complete steady solution bifurcation set in the control parameter plane can be uncovered. In addition, since the calculations are performed on the steady flow equations, both stable and unstable solutions can be found and bifurcations on both types of solution can also be computed. This gives a powerful new insight into the origins of more complicated dynamical behaviour, as discussed in the main text, since the interaction of different types of steady solutions can produce dynamical behaviour. Finally, these methods are particularly useful when used in combination with experimental investigations, as unstable flows cannot be observed in the laboratory. Nevertheless, they can play a crucial role in the nonlinear behaviour.

The onset of time dependence in the flow is investigated by searching for the occurrence of a Hopf bifurcation point along a solution branch. At a Hopf bifurcation point, a pair of eigenvalues of the Jacobian matrix $f_{\boldsymbol{x}}$, calculated from the discretized flow equations, crosses the imaginary axis at a finite imaginary value $\mathrm{i}\omega$. This bifurcation marks a transition to a periodic solution with angular frequency ω and period $2\pi/\omega$. The presence of a Hopf bifurcation is thus found by monitoring to see when a complex conjugate pair of eigenvalues crosses the imaginary axis in the complex plane with variation in R.

The calculation is performed using a subspace iteration technique on an extended system of the discretized steady state equations. This technique is described in Jackson (1987). The iteration process is used to compute the m eigenvalues with the smallest moduli, in the range of Reynolds number where the occurrence of a Hopf bifurcation is anticipated. If the particular value of R is close to a Hopf bifurcation, then one of the eigenvalues necessarily has a small real part. The value of m is determined to ensure that the size of the imaginary part ω of each eigenvalue is of the same order as the experimentally measured values of the oscillation frequency. This typically means that the calculation of the first ten eigenvalues will suffice since the calculation of the whole eigenvalue spectrum, while in principle possible, would be computationally very expensive.

This approach of identifying the onset of time dependence is computationally cheaper than a full time dependent calculation. It also allows the investigation of the dependence of the Hopf bifurcation on other parameters in multiple bifurcation problems such as the Taylor–Couette problem. In addition, it enables the calculation of Hopf bifurcation points on unstable branches, which is another very useful feature. However, the method does not allow the calculation of the time dependent flow field itself and only indicates the type of motion to which the steady flow loses stability. The calculation of the oscillatory flow field requires the direct integration of the time-dependent Navier–Stokes equations.

6.8 Appendix B: Reconstruction of attractors

6.8.1 *Introduction*

The aim of this appendix is to show how the information contained in the time series measurement of a physical experiment may be analysed using modern signal processing techniques, and to indicate how the results can be related to ideas from finite-dimensional dynamical systems. The ultimate objective is to see whether a direct connection can be made between the information contained in the time series obtained from a continuous, infinite degree of freedom system and the well-developed field of chaos in ordinary differential equations. We will introduce the concepts of phase portrait reconstruction using the straightforward 'method of delays'. An outline will be given of how the principles can be extended by using the singular value decomposition technique. This provides a systematic way of selecting a coordinate system in which the attractor can be constructed, and also deals with the universal problem of noise in experimental observations.

6.8.2 *Phase portrait reconstruction*

Method of delays

The technique which is central to modern nonlinear signal processing is phase portrait reconstruction, where the attractor is formed in a pseudo-phase space from a single time series measurement. Theoretical phase portraits usually consist of plots of variables versus their derivatives. Although it is in principle possible to carry out this procedure with experimental data, problems arise in practice due to the presence of noise on finitely sampled data. This gives artificially high derivatives, which will at best severely distort the reconstructed phase portrait. Therefore a simple practical alternative, called the method of delays, is used, which was first proposed by Packard, Crutchfield, Farmer and Shaw (1980). The connection between the phase portraits reconstructed by this technique can be shown to be topologically equivalent to the actual attractors, as shown by Takens (1981).

In the primitive method of delays, the first data sample from a time series is plotted against the second, the second against the third, and so on; i.e. $x(i)$ is plotted versus $x(i + 1)$, where i goes from 1 to N and N is the total number of samples. This method gives a two-dimensional projection of the phase portrait and thus a sine wave would give an ellipse as the attractor. The period of the oscillation is the time taken for one orbit of the ellipse. One immediate problem with this approach is that neighbouring samples are highly correlated so that the ellipse is strongly aligned along the (1,1)-direction, as shown in Fig. 6.30. This can be a major problem in practice, as a complicated time series needs to be highly sampled to retain as much information as possible. Therefore, there will be a high correlation over a range of samples and the resulting portrait would be highly distorted using this naive approach.

In order to overcome this practical limitation, a delay time $n\tau_s$ (where τ_s is the sample time) is chosen, and the results for the sine wave are shown in Fig. 6.31. In practice, there are a number of suggested techniques for choosing n, and these are reviewed in Buzug, Reimers and Pfister (1990). In fact, the choice of n will not alter the topology of the attractor, but this is of little consolation when dealing with some complicated geometrical form, as the compression and stretching of the structure by a poor choice may distort pertinent features to the extent that they are very difficult to analyse. In addition, if noise is present and the object is highly compressed in one or more directions by the reconstruction process, then the topology could become completely obscured.

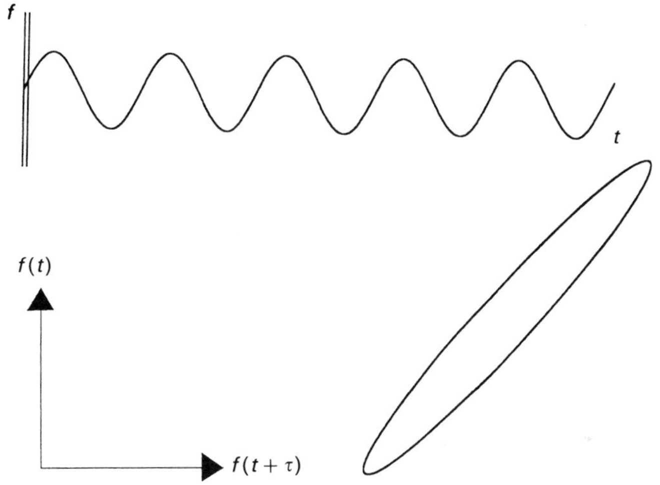

Fig. 6.30. The reconstruction of a limit cycle from a sine wave using the method of delays with a delay time of one sample.

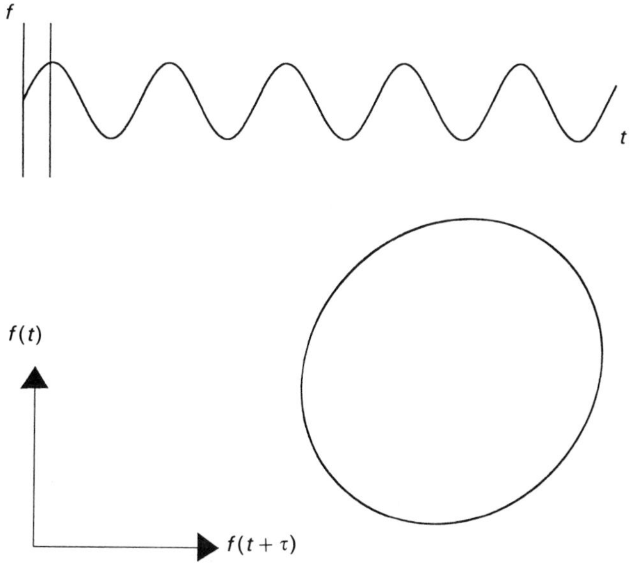

Fig. 6.31. A reconstructed attractor using a delay time corresponding to one quarter period of the cycle.

The above technique can be extended to three or higher dimensions by introducing second and third delayed coordinates etc. This method has been used to construct the torus of Fig. 6.32, which is the attractor formed from a quasiperiodic time series containing two incommensurate frequencies. It can be thought of as the combination of the motion on two ellipses, so that a trajectory will spiral around the torus or doughnut.

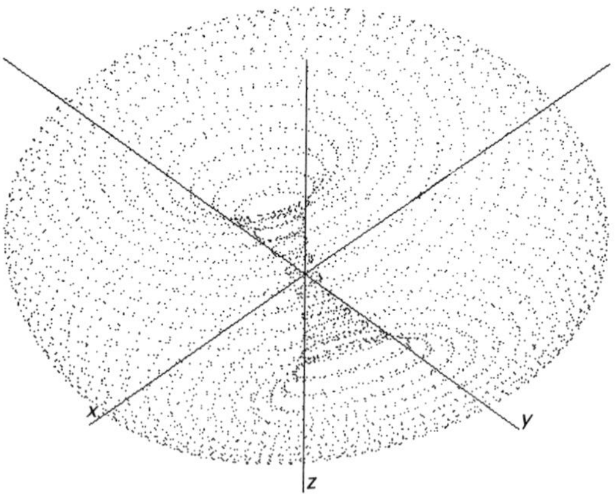

Fig. 6.32. A reconstructed torus from two sine waves. Here we have plotted the data points only to show the structure of the torus more clearly.

Singular value decomposition

The main practical problem with the standard method of delays is that the phase portrait reconstruction is carried out by projecting the time series onto an arbitrary basis. Thus no systematic account is taken of the information content of the signal in the reconstruction process, nor are the effects of finite sampling and experimental noise in the time series catered for. The technique of singular value decomposition was proposed as a solution to these difficulties by Broomhead and King (1986). It is used to calculate an optimal basis for the projection of the attractor which is reconstructed from the time series. In addition, the technique was developed to deal with noise and is, therefore, ideally suited to experimental data.

The procedure is carried out as follows. First of all we form a trajectory matrix **X**, whose rows contain the n-dimensional vectors used in the method of delays with the delay time set equal to τ_s. In practice, we set n large enough to capture the lowest frequency component in the signal, and n can be selected using the autocorrelation function. We refer to this as the window length of our trajectory matrix. Each successive row in the matrix is then given by sequential data windows.

The objective of the singular value technique is to find the unit vectors which are optimally aligned with the position vectors of the trajectory matrix **X**. These unit vectors will form the coordinate system onto which the time series will be projected in the phase portrait reconstruction. One may think of this technique as finding the principal moments of an 'object', which in this case is the attractor. Thus it is a method of extracting the optimum projection of the phase portrait from the data. The process involves diagonalization of the covariance matrix ($\mathbf{X}^T\mathbf{X}$) to obtain the set of eigenvectors which are the orthogonal singular vectors. The square roots of the corresponding eigenvalues give the singular values. Thus the singular vectors give the directions of the coordinate axes and the corresponding singular values give the weightings for each. The phase portrait is constructed on this coordinate system, and each point on the trajectory is a weighted version of the sampled point in the time series.

Noise will be present in all the singular values, and will appear in the singular value spectrum as a flat noise floor at its upper end. The significant singular values will appear above this floor and their number gives an upper limit on the embedding dimension of the attractor. Thus, this is a way of distinguishing the deterministic part of the signal from the noise. It should be noted that this process does not give the dimension of the attractor directly. However, the scaling of the singular values can be used to give an estimate of the local dimension of an attractor (see Broomhead and Jones, 1990, for details).

We now turn to a specific example to show some of the practical advantages of this technique when applied to experimental data. We show in Fig. 6.33(*a*) a time series obtained from the Taylor–Couette experiment, and below it in Fig. 6.33(*b*) the phase portrait reconstructed by the method of delays using the optimal delay time. The same data set was used to construct the phase portrait shown in Fig. 6.34(*a*) using the SVD

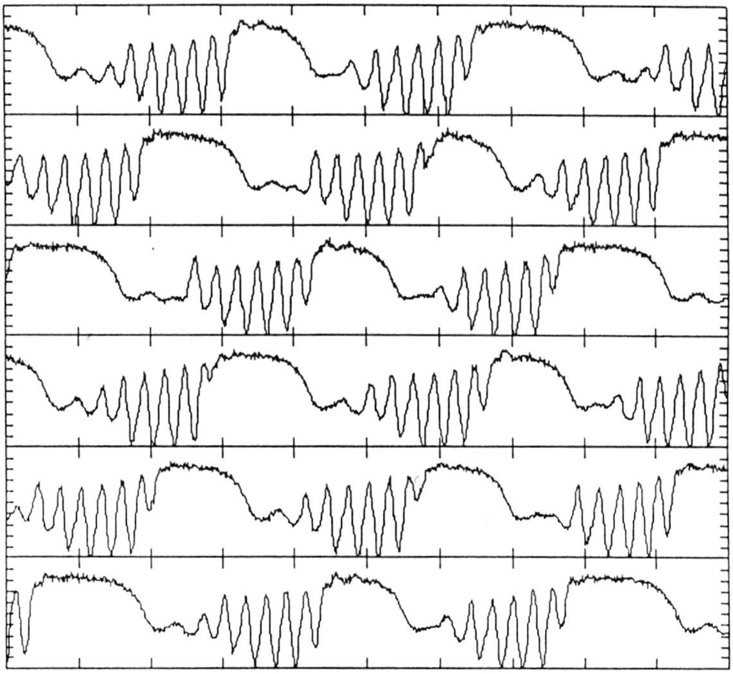

(*a*)

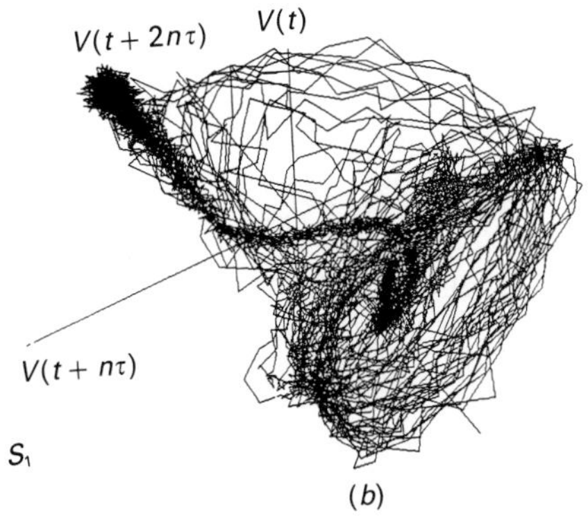

(*b*)

Fig. 6.33 (*a*) A velocity time series from the Taylor–Couette experiment. (*b*) The reconstructed attractor obtained using the method of delays with the optimum delay time.

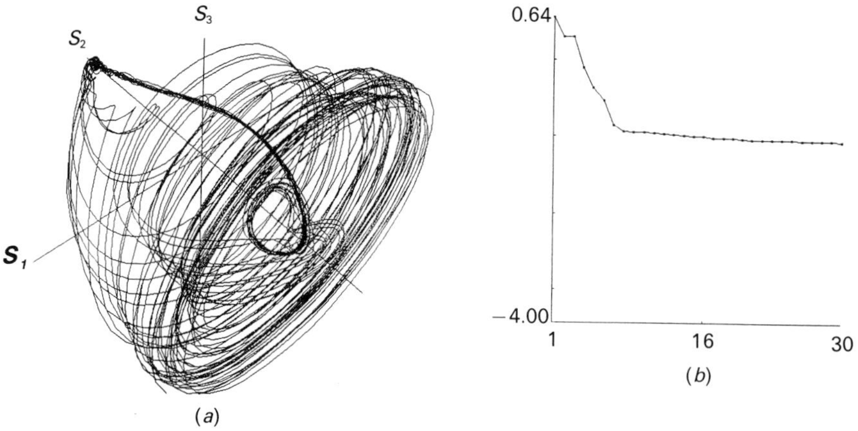

Fig. 6.34. The reconstructed attractor from the time series shown in Fig. 6.33. (*a*) Using the optimal coordinate basis obtained by SVD. (*b*) The singular spectrum showing three significant singular values and a noise floor.

technique. Its singular spectrum is shown plotted on a logarithmic scale in Fig. 6.34(*b*), where it can be seen that the presence of three significant singular values suggests that it is justifiable to plot the phase portrait in three dimensions. The difference between the reconstructed attractors in Figs 6.34(*a*) and 6.33(*b*) is evident, and is purely due to the nonoptimal projection of the trajectories in the reconstruction shown in Fig. 6.33(*b*). (NB This is not a filtering process.) In essence, the trajectories are attempting to explore dimensions which contain only noise, and their projection onto three dimensions can give an apparently poorer signal-to-noise content. The obvious improvement in the reconstruction process using SVD is not just cosmetic and may be of considerable importance when attempting to extract quantitative measures such as local dimensions from phase portraits reconstructed from experimental time series.

Acknowledgements

The main part of our research programe is supported through the SERC 'Nonlinear Initiative' with additional support on signal processing from RSRE, Malvern. I would also like to thank V. Heinrich, J. Kobine, F. Madden and T. Price for their help with the preparation of this manuscript and K. A. Cliffe for discussions of the numerical methods.

References

ANDERECK, C. D., LIU, S. S. and SWINNEY, H.. L. (1986). *J. Fluid Mech.*, **164**, 155–70.

BENJAMIN, T. B. (1978). *Proc. Roy. Soc. Lond.*, **359**, 1–43.

BROOMHEAD, D. S. and JONES, R. (1990). *J. Phys. A*, **20**, L563–9.

BROOMHEAD, D. S. and KING, G. P. (1986). *Physica D*, **20**, 217–36.

BUZUG, T., REIMERS, T. and PFISTER, G. (1990). *Euro. Phys. Lett.*, **13**, 605–10.

CLIFFE, K. A.(1988). *J. Fluid Mech.*, **197**, 52–79.

CLIFFE, K. A., KOBINE, J.J. and MULLIN, T. (1992). The role of anomalous modes in the Taylor flow problem. *Proc. Roy. Soc. London A*, **439**, 341–57.

CLIFFE, K. A. and GREENFIELD, A. C. (1982). Some comments on laminar flow in symmetric two-dimensional channels. AERE Report T.P. 939.

CLIFFE, K. A. and SPENCE, A. (1984). 'The calculation of high order singularities in

be taken to next order, leading to a velocity diffusion equation (Fokker–Planck) in place of Vlasov's equation (see, for example, Chapter 2). An estimate of τ_c is given by $\omega_p \tau_c \sim n\lambda_D^3$, where ω_p is the plasma frequency and n is the density. Both kinetic and Vlasov calculations are limited to relatively small systems because of constraints on the sizes of time step, Δt, mesh spacing, Δx, and simulation particle density ($N_D = n\lambda_D^3$); for example, explicit particle simulations (Hockney and Eastwood, 1988) require $\omega_p \Delta t < 2$, $\Delta x/\lambda_D \simeq 1$, and $N_D \gg 1$. Constraints similar to these arise for almost all explicit evolutionary calculations. Stability and accuracy of the numerical integration scheme usually require the time step to be small compared to the minimum characteristic period of the system even if the fast oscillations are not the ones of relevance to the situation being studied, although this can be relaxed by using implicit time integration.

Two-fluid model

To perform simulations of plasmas on longer time scales and for larger systems requires a less detailed plasma model. Both collisions and collisionless micro-instabilities in plasmas generally lead to Maxwell-like velocity distributions. Such velocity distributions are well described by their first two moments: the mean velocity and the thermal velocity.

The equations obtained by taking moments of the kinetic equations are themselves a hierarchy of equations, the first three of which describe mass, momentum and energy conservation; closure of the hierarchy is obtained by assuming some equation of state, $p = p(\rho)$. Although the moment equations eliminate the driving terms for velocity space micro-instabilities, they still describe a wide range of plasma waves. We know from Chapter 3 that the dispersion curve for the two-fluid component plasma has six branches: two differently polarized electromagnetic branches and an electrostatic branch at high frequencies, and three magnetoacoustic branches at low frequencies, which link into the cyclotron and acoustic branches.

The next stage of stretching to longer time and length scales is to eliminate the physical effects which lead to the fast, short wavelength dispersion branches. Some of the approximations used are listed below. If the time for a particular wave to cross the system is much shorter than the time scale of interest, then its velocity may be taken as infinite. In the case of the electromagnetic wave branches, $c \to \infty$ corresponds to dropping the displacement current from Maxwell's equation. Electrostatic plasma waves and electron cyclotron waves may similarly be eliminated by setting the electron mass to zero. An alternative to eliminating the fast waves is to compress time scales by using artificially reduced ion/electron mass ratios. In practice, both approaches have been used.

MHD (magnetohydrodynamics) model

In the limit of zero electron mass, the moment equations give the MHD plasma model. The electron equation of motion becomes an Ohm's law, the causal relationship between currents and fields is reversed and electrostatic forces are neglected. Dispersion in MHD plasmas comprises two magnetoacoustic modes and the pure Alfvén mode.

Despite the great simplifications used to obtain the MHD plasma model, the spread of scale lengths implicit in MHD may still be too great to lead to computations of manageable proportions. Consequently, further ordering is used.

• *Alfvén*: on fast time scales ($\tau_A = L/v_A$), changes of plasma field configurations due to finite electrical and thermal conductivities may be ignored.

• *Resistive*: slightly longer resistive time scales ($\tau_R = (\tau_A^2 \tau_\eta^3)^{1/5}$) require that changing field topologies be described.

• *Transport*: on a longer time scale still ($\tau_\eta = L^2/\eta$), inertial terms become unimportant, and the plasma is described by a sequence of quasi-equilibria; the evolution is determined by transport processes. Even transport processes, such as heat flow, are usually ordered, the flow of heat being much faster along than across field lines. The limit of infinite thermal conductivity parallel to magnetic field, $\kappa_\parallel \to \infty$, gives the '$1 - 1/2D$' transport models.

• *Equilibrium*: on the longest time scale of all ($\tau \to \infty$), all changes are taken to occur infinitely fast, giving the equilibrium force balance problem, $\nabla p = \mathbf{J} \times \mathbf{B}$.

7.3.1 *Scale length reduction*

The hierarchy of models listed above relies on reducing the range of scale lengths by using appropriate physical approximations. The need for restricting scale lengths becomes apparent when we consider some typical numbers associated with a plasma. Parameters for a typical hydrogen plasma are density $n = 10^{19}\,\mathrm{m}^{-3}$, temperature $T = 1\,\mathrm{keV}$, magnetic field $B = 3\,\mathrm{T}$, and dimension $L = 0.5\,\mathrm{m}$. These yield time scales of plasma period $\sim 3.5 \times 10^{-11}\,\mathrm{s}$, ion cyclotron period and Alfvén transit times $\sim 2 \times 10^{-8}\,\mathrm{s}$, resistive time $\sim 2\,\mathrm{ms}$ and Ohmic decay time $\sim 3\,\mathrm{s}$. Length scales also span several orders of magnitude, from $\lambda_D \sim 7 \times 10^{-5}\,\mathrm{m}$ to the plasma dimension L. It is clear that the time step $\Delta t \sim 10^{-11}\,\mathrm{s}$ and mesh size $\Delta x \sim 10^{-5}\,\mathrm{m}$ demanded by electrostatic waves are far too small to permit simulation of, say, resistive evolution ($\tau \sim 10^{-3}\,\mathrm{s}$) over device dimensions ($L \sim 0.5\,\mathrm{m}$).

Amongst the approximations used to reduce the range of time scales in the simulation model are the following:

(i) *The speed of light* is set to infinity to eliminate the electromagnetic wave branches. $c \to \infty$ corresponds to dropping the displacement current from Maxwell's equations. Alternatively, c is reduced to bring disparate frequencies closer together in order to speed up calculations.

(ii) *Mass ratios.* Electrostatic plasma waves and electron cyclotron waves may be eliminated by setting the electron mass to zero. An alternative to eliminating the fast waves is to compress time scales by using artificially reduced ion/electron mass ratios. In practice, both approaches have been used.

(iii) *Thermal conductivity* κ is set to zero in Alfvén time scale MHD, and $\kappa_\parallel$ is set to infinity in surface averaged transport models.

(iv) *Resistivity* is set to zero in 'ideal MHD', and increased to compress time scales in 'resistive MHD'.

(v) *Other approximations* include ignoring inertia in MHD transport models, and device-specific orderings, such as small aspect ratio ordering and ignoring kinetic pressure in tokamak computations.

Three crucial questions to be asked of any prospective plasma simulation are (i) what processes must be included in the model, (ii) what are their dimensionalities and (iii) what are their length and time scales? The dimensionality and the length and time scales of the relevant processes must be reflected in the simulation model. Slower processes than those of importance can be ignored, faster ones cannot. Fast processes must be either suppressed, by modifying the physical model or by using implicit integration schemes, or their constraints on time step be made less severe by compressing time scales.

7.4 Numerical methods

Considerations of the length and time scale constraints, and of the constraints imposed by finite computer resources, lead to the choice of mathematical model. This model is usually a coupled set of differential equations, although it may also involve integral equations and variational principles. The purpose of the numerical method is to approximate the differential equations relating continuous variables by a set of discrete algebraic equations for sets of values.

7.4.1 *Classification of equations*

It is important to classify the equations in order to determine the form of boundary conditions required and to help choose the appropriate numerical method. The first distinction, between ordinary differential equations (ODEs) and partial differential equations (PDEs), may arise from dimensionality or from the choice of formalism.

Ordinary differential equations are divided into initial value (time dependent on an open domain) and boundary value (time independent on a closed domain) problems. An example of an initial value ODE is the equation of motion of particles in a given field, and of a boundary value ODE is Poisson's equation in one dimension. Less obvious instances of ODEs are those obtained by representing hyperbolic PDEs by characteristics, using finite element spatial representations in initial value/boundary value problems, or transformation methods on eigenvalue problems.

Partial differential equations having time dependence are of two types, *hyperbolic* or *parabolic*. *Hyperbolic* equations have real characteristics, and so may be mapped forwards in time by particle methods. Instances of hyperbolic equations are the wave equation, the advection equation and conservation equations of ideal MHD. *Parabolic* equations are typified by the diffusion equation. The prototypical time-independent (*elliptic*) PDE is Poisson's equation; equations of this type arise in electrostatics, magnetostatics and long time (steady state) solutions of parabolic equations.

The different physical nature of the hyperbolic and parabolic PDEs is revealed by their dispersion relations; the hyperbolic equation has a nondecaying behaviour with time scale proportional to length scale, whilst the parabolic equation gives a decaying solution with time scale proportional to the square of the length scale.

An important result which emerges from the characteristic analysis of PDEs is the correct form of boundary condition (BC) for each type of equation. For second order PDEs, the appropriate BCs are: Cauchy (value and normal derivative given) on an open boundary for hyperbolic equations; Neumann (normal derivative given) or Dirichlet (value given) on an open boundary for parabolic equations; and Neumann or Dirichlet on a closed boundary for elliptic equations. The value or normal derivative boundary condition may be replaced by mixed or by periodic BCs.

7.4.2 *Discretization*

The continuum in space (space-time for time-dependent problems) is replaced by a finite set of values, and the differential equations are approximated by algebraic equations to obtain a system of equations which can be solved using digital computers. There are basically four methods of making the step from the differential continuum to the numerically computable discrete system:

(1) finite difference approximation (FDA);
(2) finite element method (FEM);
(3) spectral methods; and
(4) particle-in-cell method (PIC).

This distinction is to some extent arbitrary, in that combinations of methods are often used in obtaining simulation models, and that algorithms derived using one method may be identical to those derived using another.

FDA and FEM schemes are further classified into *Eulerian* and *Lagrangian* models depending on the motion of the grid (or elements). If a time-independent grid attached to some observer frame is used, then the scheme is Eulerian. In Lagrangian schemes, the grid moves with the fluid velocity. The advantage of the Lagrangian method is that the difficult advective terms (appearing as $\mathbf{v} \cdot \nabla \mathbf{v}$ in Eulerian formulations) can be treated accurately, but in more than one dimension this is accompanied by difficulties of mesh shearing. Some simulation programs use a grid velocity which is neither Eulerian nor Lagrangian, in an attempt to get the advantages of the Lagrangian method without its pathological meshes.

Finite difference
The FDA gets discrete equations on replacing

- the continuum by values on a lattice of points,
- the derivatives by value differences,
- the equations by difference equations on the lattice.

Advantages of the FDA are that the equations are simple to derive, there are few operations per lattice point, coding is easy in regular geometries, and there is an established body of methods that work. Historically, the FDA has been the favoured method for plasma simulations. In almost all time-dependent calculations, finite differences are used for the time coordinate.

Limitations of the FDA are that it is difficult to apply to irregular boundaries and anisotropic media, its numerical dispersion and diffusion may give a poor approximation to the physics of the plasma, and it is more susceptible to nonlinear instabilities than alternative methods.

Finite element
The FEM obtains discrete equations on replacing

- the continuum by local piecewise polynomials with nodal values,
- the derivatives by derivatives of the approximating polynomials,
- the equations by weighted residual or approximated variational equations.

Advantages of the FEM are that it overcomes the difficulties faced by the FDA; the FEM gives optimal schemes, allows straightforward generalization to high accuracies, and easily handles irregularly shaped regions. The local element description and assembly of the global structure give the high degree of modularity and flexibility which has made the FEM the dominant approach in engineering calculations.

Limitations are that the FEM equations are usually implicit, so there is a lot of work per node. Exploiting the flexibility of arbitrary element nets can lead to large computational overheads. The superiority of the FEM over the FDA for elliptic and parabolic equations is established. The situation is less clear for hyperbolic problems, where functions are in general not smooth, although finite element/particle methods offer the best prospect in these cases.

Spectral

Spectral and pseudospectral methods differ from the FEM in their choice of approximating functions. They replace

- the continuum by expansions using global orthogonal polynomial basis functions,
- the derivatives by derivatives of the approximating polynomials,
- the equations by projections onto basis functions (spectral) or onto meshes (pseudospectral).

In applications where simple boundary conditions allow fast transforms to be used, these methods generally give superior combinations of accuracy and computational speed.

Spectral methods correspond to Galerkin FEMs, with global orthogonal functions, such as Fourier harmonics and Chebychev polynomials, being used instead of local polynomials. Pseudospectral methods correspond to point collocation FEMs. For linear terms, spectral and pseudospectral methods give the same results. Quadratic terms differ in that the pseudospectral method has periodic conditions in wave number space. For poorly resolved spectra, this can lead to nonlinear instabilities (see Section 7.6.2). A 'de-aliased' pseudospectral scheme, where the top third of the spectrum is set to zero when using finite Fourier transforms, is mathematically identical to a spectral approximation.

Particle

Particle-in-cell (PIC) methods replace

- the continuum by a set of sample points (particles),
- the derivatives by the FDA, the FEM or spectral methods,
- the hyperbolic equations by ODEs for particle trajectories.

PIC methods provide an alternative route to achieving the advantages of both Lagrangian and Eulerian methods. To achieve this, they exploit the real characteristics of hyperbolic systems. The hyperbolic terms in the differential equations are treated in a Lagrangian fashion, whilst the parabolic and elliptic terms are dealt with using Eulerian meshes. To advance the hyperbolic terms, the continuum is replaced by a set of randomly located sample points (particles), each carrying attributes such as mass, position, entropy, charge. Ideally, these attributes are conserved quantities of the evolutionary equations obtained when the parabolic terms are set to zero; in this limit, the time advancement of the

PDEs consists simply of moving the particles. Parabolic terms are advanced using either finite difference or finite element schemes. The two parts of the calculation are linked together by interpolation and inverse interpolation: mesh values of continuum quantities are obtained by making weighted means of neighbouring particle values, and vice versa.

7.4.3 *Implicit methods*

The classifications by equation type and by discretization method may be complemented by a third, *implicit/explicit*. This third distinction refers to the time integration. If, in advancing quantities to a new time level, values at the new level are given only in terms of those at the old, then the scheme is said to be explicit, otherwise it is implicit. Explicit schemes make time advancement very simple, but often suffer severe restrictions on time steps. Implicit schemes generally have much better numerical properties, but entail the solution of matrix equations to perform the time integration. Although the implicit/explicit categorization of schemes refers strictly to equations with time dependence, it may be extended to include elliptic equations as extreme instances of implicit parabolic equations.

7.5 Computer models

The BBGKY hierarchy outlined in Section 7.3 provides a logically coherent framework into which the diversity of plasma simulation models can be ordered. A summary of the types of simulation models and their applications is given in the following subsections.

7.5.1 *Collisionless plasma*

The collisionless plasma model provides a detailed description of the motions of ions and electrons in self-consistent electromagnetic fields. The basic equations are the Vlasov equation for each plasma species plus Maxwell's equations for the electromagnetic fields. In the electrostatic approximation, Maxwell's equations are replaced by the Poisson equation. The PIC method is the pre-eminent method for collisionless plasmas, because it accurately models advection and avoids the need for a velocity space mesh.

The Vlasov equation

$$\frac{Df}{Dt} = \frac{\partial f}{\partial t} + v\frac{\partial f}{\partial x} + a\frac{\partial f}{\partial v} = 0 \tag{7.5.1}$$

is equivalent to saying that f is conserved along the trajectory of a particle moving according to the equations of motion

$$\frac{dx}{dt} = v; \frac{dv}{dt} = a. \tag{7.5.2}$$

That is,

$$f\left(x + \int_t^{t+\Delta t} v dt, v + \int_t^{t+\Delta t} \frac{F}{m} dt, t + \Delta t\right) = f(x, v, t). \tag{7.5.3}$$

Equations (7.5.2) are the characteristic equations of the hyperbolic Vlasov equation. The use of particles transforms the Vlasov equation into relatively straightforward ODEs. The self-consistent model is completed by a FDA or FEM calculation of the fields. Source terms for the field equation are obtained by assignment from particles, and accelerations at particles are interpolated from the mesh computed fields.

A physical interpretation of the particle model is to view each particle as a finite sized cloud of electrons (or ions), where internal degrees of freedom of the cloud are suppressed. The complete description of the cloud (slab in one dimension) is given by its velocity, position, density profile and charge/mass ratio. The clouds must have a finite width to soften short range interactions to the extent that they freely pass through each other, otherwise binary collisional effects will dominate collective effects. Each cloud will correspond to perhaps 10^5 electrons, so without the softening of short range forces, qualitatively incorrect behaviour would result. Generally, a cloud width of order the Debye length is used; this minimizes the collisional effects without seriously affecting collective (plasma wave) properties. The value of charge density at a given point is given by the weighted mean of the charges of neighbouring particles, and is computed by summing contributions from clouds which overlap that point, so the range of neighbouring particles for finding weighted mean values may be interpreted as the cloud width. To avoid statistical fluctuations in such mean values, we need a large number of clouds (say ~ 10) to overlap. This requirement implies that the number of particles per Debye slab (square or cube in higher dimensions) is large (say $\sim 10!$).

Applications:

- basic plasma processes,
- instability growth and saturation,
- resonant coupling,
- effect of turbulence,
- beam–plasma and wave–plasma interactions,
- microwave devices.

7.5.2 *Kinetic*

For plasma phenomena where collisions are not negligible, and yet the plasma is not near-Maxwellian, a kinetic model is required. The appropriate equation for the plasma species is the Fokker–Planck equation. This equation differs from the Vlasov equation in that the right-hand side is the sum of collision (drag and diffusion), source (S) and loss (L) terms:

$$\frac{\mathrm{D}f}{\mathrm{D}t} = \left(\frac{\partial f}{\partial t}\right)_{\mathrm{c}} + S + L. \tag{7.5.4}$$

The collision terms involve gradients in velocity, and therefore lead to the need for either Monte-Carlo methods for scattering particles, or for the use of a velocity space mesh to evaluate velocity gradients. Both methods have been used, but the favoured approach is to use either FDAs or FEMs to solve the parabolic PDE for the distribution function f.

Applications:
- beam–plasma heating,
- wave–plasma heating,
- current drive,
- loss cone processes.

7.5.3 *MHD equilibrium*

MHD equilibrium is the longest time scale plasma model. Dynamics are reduced to force balance $\nabla p = \boldsymbol{J} \times \boldsymbol{B}$, and fields are given by magnetostatics $\nabla \times \boldsymbol{B} = \boldsymbol{J}$. Assuming axisymmetry (in toroidal geometry) implies that magnetic field lines lie on nested flux surfaces, and in the equilibrium limit thermodynamic quantities equilibrate around the flux surfaces.

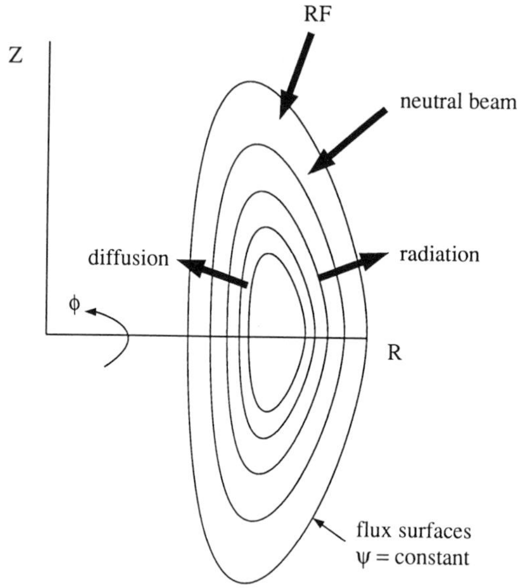

Fig. 7.2. MHD equilibrium solves for the location of axisymmetric flux surface at force balance, stability asks if these surfaces are stable to nonaxisymmetric perturbations, and transport computations follow the slow axisymmetric evolution of the surfaces due to diffusion and radiation processes.

Axisymmetry allows the field to be written in terms of poloidal and toroidal flux functions, ψ and f, $\boldsymbol{B} = f\hat{\phi}/R + \hat{\phi} \times \nabla\psi/R$. The toroidal component of Ampère's equation then becomes the Grad–Shafranov equation:

$$\nabla \cdot \frac{1}{R^2} \nabla\psi = -\frac{\partial p}{\partial\psi} - \frac{f}{R^2}\frac{\partial f}{\partial\psi}. \tag{7.5.5}$$

Given p and f as functions of ψ, Eq. (7.5.5) can be solved for the poloidal flux ψ. A variety of finite difference, finite element, and mixed FDA/spectral methods using both direct and iterative solvers have been used to solve Eq. (7.5.5).

Applications:
- engineering,
- data interpretation,
- stability,
- transport.

8

Tokamak experiments

M. R. O'BRIEN AND D. C. ROBINSON

8.1 Introduction

The aim of nuclear fusion research is to produce a plasma that will generate more energy through fusion reactions than is required to create and sustain that plasma. Such a plasma must satisfy the Lawson criterion: $n_i \tau_E \gtrsim 2 \times 10^{20}\,\mathrm{m}^{-3}\,\mathrm{s}$ for deuterium–tritium fusion. Here n_i is the number of ions per unit volume and τ_E is the energy confinement time, that is the ratio of total kinetic energy to input power in steady state. In addition, of course, the plasma must have a high enough ion temperature, for deuterium–tritium fusion: 10–30 keV. There are two main approaches to achieving such a plasma. In inertial confinement, small high density plasmas are created for times of order nanoseconds. In the second approach, magnetic confinement, ions and electrons of lower number density, $n \sim 10^{20}\,\mathrm{m}^{-3}$, are constrained to follow closed magnetic field lines for times of order a second, and thus not impinge on material surfaces which would rapidly cool them. Magnetic confinement is aided by the pinch phenomenon, in which a current-carrying plasma column tends to contract radially until the inward Lorentz force is balanced by the outward pressure gradient force: $\boldsymbol{J} \times \boldsymbol{B} = \nabla p$. The simplest closed system is the torus, and the most successful toroidal pinch, the tokamak (a Russian acronym for toroidal magnetic chamber), is the subject of this chapter.

Early research using toroidal magnetic confinement concentrated on stellarator and reverse field pinch experiments, but it soon became clear that instabilities seriously limited plasma performance. However, in 1969, Russian claims of stable operation, high temperatures and thermonuclear neutrons in tokamaks were confirmed by British scientists (Peacock *et al.*, 1969), and the tokamak has been the principal line of research in magnetic fusion ever since. In the 1970s and 1980s tokamak performance steadily improved as techniques for control, shaping and plasma exhaust developed, and heating additional to the inherent Ohmic heating of the tokamak gave large increases in the electron and ion temperatures. Progress towards ignition is summarized in Fig. 8.1. The biggest tokamaks now operating, JET (Europe, see Fig. 8.2), TFTR (USA) and JT-60 (Japan) have achieved densities, temperatures and confinement times sufficient for ignition, although these parameters have not been achieved simultaneously. Most tokamaks to date have used hydrogen, deuterium or helium plasmas, and the power produced in fusion reactions has been very small. However, in November 1991, tritium atoms were injected into a JET deuterium plasma generating $\lesssim 2$ MW of fusion power for ~ 2 s (JET Team, 1992). Further

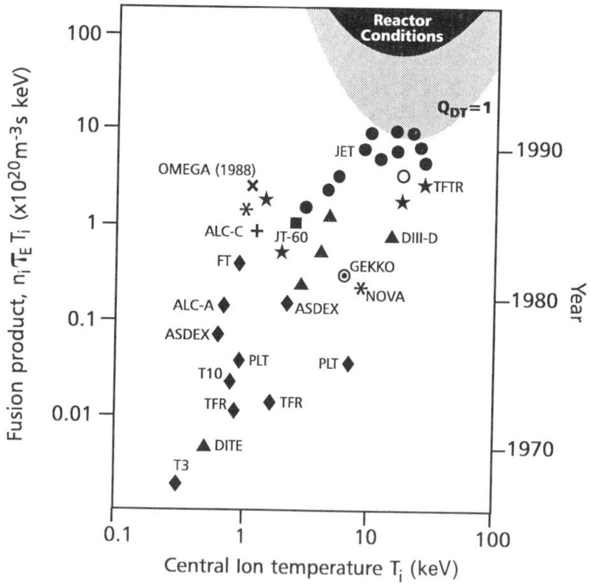

Fig. 8.1. Progress in nuclear fusion research showing the figure of merit $n_i \tau_E T_i$ as a function of central ion temperature for several tokamaks. Also shown are data from the inertial confinement experiments NOVA, GEKKO and OMEGA.

Fig. 8.2. The Joint European Torus (JET) at Culham in Oxfordshire, England. This photograph shows JET during assembly with one octant removed and before additional heating and diagnostic apparatus were added to the machine. Reproduced by permission of the JET Joint Undertaking, Abingdon, UK.

experiments with tritium on JET and the other large tokamaks are planned. The International Thermonuclear Experimental Reactor (ITER) project aims to build a tokamak in the early years of the 21st century, which will demonstrate thermonuclear ignition and be used to study the physics of burning plasmas.

In this chapter, the basic principles of the tokamak and the current status of tokamak research are outlined. It is impossible to discuss every aspect of tokamak physics here: the references cited give fuller discussions for the interested reader. A useful book describing tokamak physics in more detail is that by Wesson (1987).

8.2 Equilibrium

The geometry and magnetic field configuration of a tokamak are shown in Fig. 8.3. The major radius, R, varies from ~ 0.3 m in small tokamaks to 3 m in JET, and the aspect ratio (R/a with a the minor radius) is typically ~ 4, but is 2.5 for JET and as low as 1.3 for the British device START. A strong magnetic field around the torus (B_ϕ, usually $\sim$ 1–5 T, although tokamaks have operated with $B_\phi \sim 10$ T), varying inversely with the major radius R, is produced by cooled copper or superconducting coils external to the plasma. The plasma is either paramagnetic or diamagnetic depending on whether the kinetic pressure is less

Fig. 8.3. The magnetic field geometry of the tokamak.

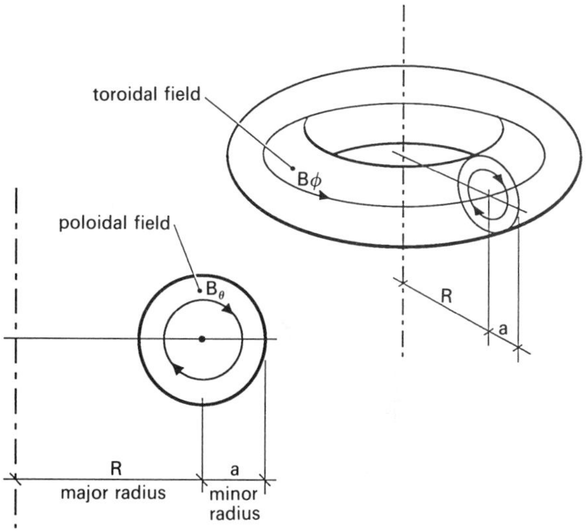

than or greater than the *poloidal* field pressure, and so makes a, usually small, contribution to B_ϕ. The toroidal and other equilibrium fields are usually assumed to be independent of toroidal angle ('axisymmetric') although the discrete number of coils results in a slightly 'rippled' B_ϕ, which, if the ripple is large enough, can affect plasma confinement. If the field were purely toroidal, ions and electrons would drift vertically in opposite directions due to the nonuniform $\boldsymbol{B}$. A vertical electric field would be established, resulting in an outward $\boldsymbol{E} \times \boldsymbol{B}$ motion of the plasma. Thus, an additional poloidal field B_θ is required to give orbits that short-circuit this electric field. This B_θ is smaller than B_ϕ in the tokamak and stellarator (at the edge $B_\theta \lesssim 2aB_\phi/R$), and comparable with B_ϕ in the reverse field pinch. In the stellarator, B_θ is generated by coils wound helically around the plasma, whereas in the tokamak B_θ is produced by an internal toroidal current I_P induced in the plasma by transformer action. The inductive loop voltage is high ($\lesssim 20$ V) in the initial breakdown phase of the discharge, dropping during the current ramp-up phase to ~ 1 V or less in the final, highly conducting, plasma. The pulse length of the discharge is limited by the flux-swing of the transformer, which may be air- or iron-cored, although methods for driving some or all of the current noninductively can be used to extend this, so it is important to minimize the volt-seconds required for breakdown and ramp-up of the current. Poloidal field coils surrounding the tokamak are required for plasma formation, control and shaping.

A vertical field, $B_Z \lessdot B_\theta$, is also required to balance the outward hoop force of the plasma current. This field is generated by external windings, or by using a highly conducting shell around the plasma in which image currents are induced by plasma motion. The usefulness of the shell is limited by the time constant for dissipation of the image currents. If a conducting shell is absent, as in many tokamaks and almost certainly in a reactor, the inhomogeneity of B_z will lead to instability in the vertical direction unless the quantity $\gamma \equiv (R/B_z)\partial B_z/\partial R$ is negative. This means that an additional quadrupole field producing a convex maintaining field is required. There is also an instability in the horizontal direction unless $\gamma < -3/2$, so that either careful design of the vertical field windings or active feedback must be used. Plasmas elongated in the vertical direction, which are often studied since they can operate with higher currents and consequently higher pressures than circular plasmas, require $\gamma > 0$ and so need active feedback or a shell. These and related issues are discussed in the review by Mukhovatov and Shafranov (1971).

The magnetic field lines define closed, nested toroidal surfaces. These are known as flux surfaces because the poloidal flux, ψ, and the plasma pressure are constant on them. The poloidal field is tangent to the flux surfaces and given by $B_\theta = |\nabla\psi|/R$. The centre of the nest of surfaces is called the *magnetic axis* and is where the poloidal field is zero and, usually, the plasma pressure and current density are maximum. The edge of the plasma is defined either by a material surface, a 'limiter', or by a magnetic separatrix, created by external 'divertor' coils.

The ions and electrons in the plasma follow the field lines and so are constrained to move on flux surfaces with only small departures due to $\boldsymbol{E} \times \boldsymbol{B}$, grad $\boldsymbol{B}$ and magnetic curvature drifts. Particles are either passing, in which case they explore the whole flux surface, or trapped, in which case they are restricted to the low-B region of the tokamak by the mirroring property of the nonuniform field ($B \sim 1/R$). The trapped particle orbits have finite width because of the vertical drifts, and have a characteristic 'banana' shape (see Fig. 1.5 in Chapter 1). A sizeable fraction of particles are trapped ($\sim [2\varepsilon/(1+\varepsilon)]^{1/2}$ for a flux surface of radius εR) and the division into passing and trapped particles can strongly influence properties such as the plasma conductivity, pressure-driven currents, instabilities, and transport across flux surfaces due to collisions and fluctuating electric and magnetic fields.

8.3 Stability

The safety factor of a flux surface, $q(\psi)$, is defined to be the number of times a field line on that surface goes around toroidally for one poloidal circuit. It is determined by the current density profile $\boldsymbol{J}(\psi)$ and the flux surface geometry, and usually increases monotonically from the magnetic axis to the edge of the plasma. MHD instabilities can form and grow on flux surfaces with rational safety factors

($q = m/n$) if the local current and pressure gradients are sufficiently destabilizing. The lowest order rational surfaces ($(m,n) = (1,1),(2,1),(3,2)$, etc.) tend to be least stable, and helical perturbations with these mode numbers are routinlely observed in tokamak plasmas. The best known of these is the sawtooth oscillation, illustrated in Fig. 8.4 and reviewed in Kuvshinov and Savrukhin (1990), in which a slow increase in the pressure at the centre of the plasma is followed by an abrupt fall. This is believed to be due to an $(m,n) = (1,1)$ instability forming as the current density grows and the central safety factor q_0 decreases, with the sawtooth crash occurring as the instability displaces the $q < 1$ region and the magnetic field lines reconnect. However, the details of the sawtooth are not fully understood, in particular the characteristic time scales and the absence of sawteeth for some operating conditions with $q_0 < 1$. Sawteeth can be removed by using radio-frequency heating of ions or electrons, particularly if it is localized near the $q = 1$ flux surface. So-called 'monster sawteeth', in which the central temperature rises for several seconds with no sawtooth collapses, have been produced on JET using ion cyclotron heating. This suppression may be due to the generation of a stabilizing population of fast trapped ions.

Other MHD modes often observed include tearing modes, which occur at low order rational surfaces ($m/n = 4/1$, $3/2$, etc.). The greatest tearing activity is usually at the $q = 2$ surface, and is mainly due to the $m/n = 2/1$ mode (here m and n are the poloidal and toroidal mode numbers). Fig. 8.5 shows a typical signal from a coil inside the vacuum vessel of the British tokamak COMPASS-D. The mode, initially rotating, slows and 'locks' at a particular position, due to the force between its field and currents in $(2,1)$ windings external to the plasma:

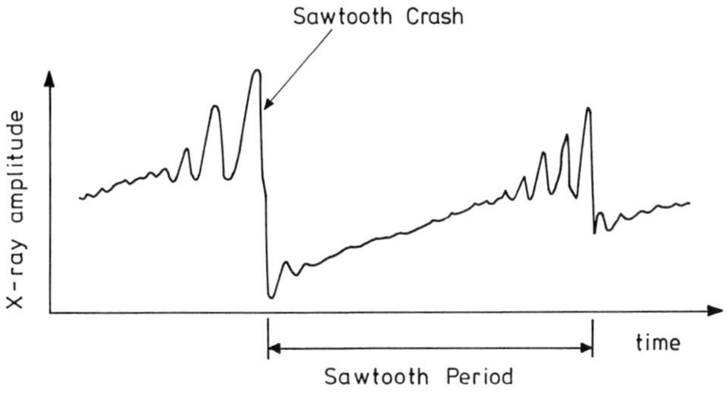

Fig. 8.4. Typical sawtooth oscillation in which the plasma pressure, measured for example by X-ray cameras, rises gradually and then falls abruptly. The sawtooth crash is often preceded by growing oscillations. The sawtooth period can range from milliseconds in small tokamaks to seconds in JET.

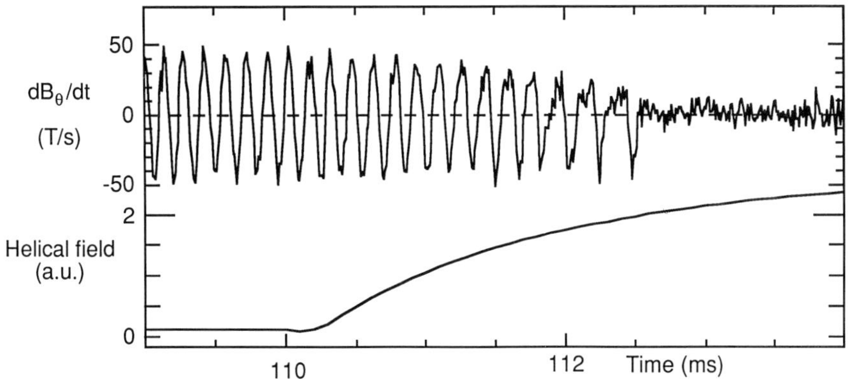

Fig. 8.5. Mode locking in the COMPASS-D tokamak. Interaction between the $m/n = 2/1$ mode and currents in external windings slows the mode rotation, eventually leading to locking at $t = 112.5$ ms. The magnitude of the field associated with the mode continues to grow, however.

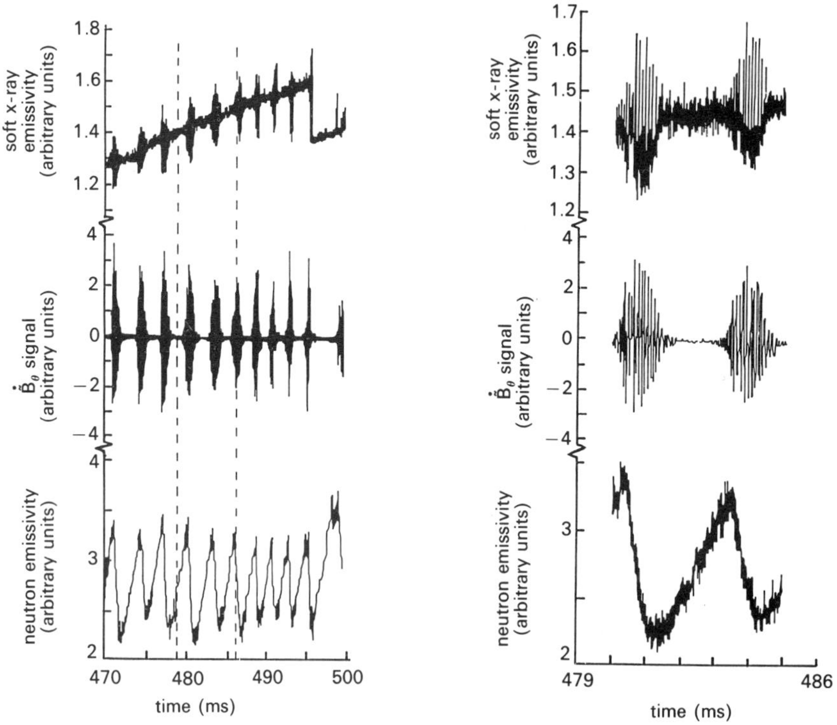

Fig. 8.6. The time evolution of central soft X-ray emission, rate of change of edge poloidal field, and the neutron emissivity during perpendicular neutral beam injection in the PDX tokamak. Expansion of the data near two 'fishbones' is shown on the right. From McGuire *et al.* (1983). Reproduced by permission of Princeton Plasma Physics Laboratory, Princeton, USA.

interaction with image currents in the resistive vessel can give similar effects. The mode does not disappear; in fact it generally grows faster when locked. Study of these effects is important because in large tokamaks locking and other effects of external helical fields, arising perhaps from very small errors in toroidal and poloidal field coils, can have deleterious effects (Morris *et al.*, 1992). MHD modes at different rational surfaces are coupled, particularly in low aspect ratio ($R/a \lesssim 3$) and noncircular tokamaks, and so the dominant modes are accompanied by sideband modes.

In general, experimental observations of global instabilities are well understood theoretically, though the sawtooth is a notable exception. An example of this is the 'fishbone' instability, which is sometimes observed when high energy neutral beams are injected into a tokamak (Fig. 8.6). A rapid burst of activity is observed on a range of diagnostics, but only if the beams are injected perpendicular to the magnetic field. An $(m, n) = (1, 1)$ mode is destabilized by a resonance between its toroidal wave velocity and the toroidal drift of fast trapped ions (Chen *et al.*, 1984). The bursts of activity are associated with the expulsion of trapped beam ions by the perturbed fields of the instability. There follows a quiescent phase until the trapped ion population has grown large enough to destabilize the mode again.

8.3.1 *Disruptions*

Low order, global instabilities are not thought to determine tokamak confinement unless the instabilities are large and overlap spatially giving large ergodic regions and consequently rapid particle loss. However, they do limit the operational regime of the tokamak to a region of $(q_a, \bar{n})$ space, where q_a denotes the safety factor at the edge of the plasma and $\bar{n}$ denotes the mean electron density. This is conveniently summarized by the Hugill diagram (Fig. 8.7) in which the operating regime is shown as a function of $1/q_a$, which is proportional to I_P, and of the Murakami parameter $\bar{n}R/B$. There are two phenomena which limit this region: 'low-q disruptions' and 'density limit disruptions'. These are similar dramatic events, in which the plasma current rapidly falls to zero, but are a result of two different mechanisms. Low-q disruptions occur when q_a drops to just above 2. At this stage, the large current density gradient just inside the surface where $q = 2$ is thought to destabilize the $(m, n) = (2, 1)$ tearing mode. This in turn interacts with the cold outer plasma, leading to a rapid fall in electron temperature and loss of the current. Interaction with the $m = 1$ sawtooth region can also occur. Operation with $q_a < 2$ is rare, although it is possible with a close conducting wall to stabilize the mode. Other minor disruptions, from which the plasma usually recovers, are observed as q_a passes through other integer values, in particular $q_a = 3$.

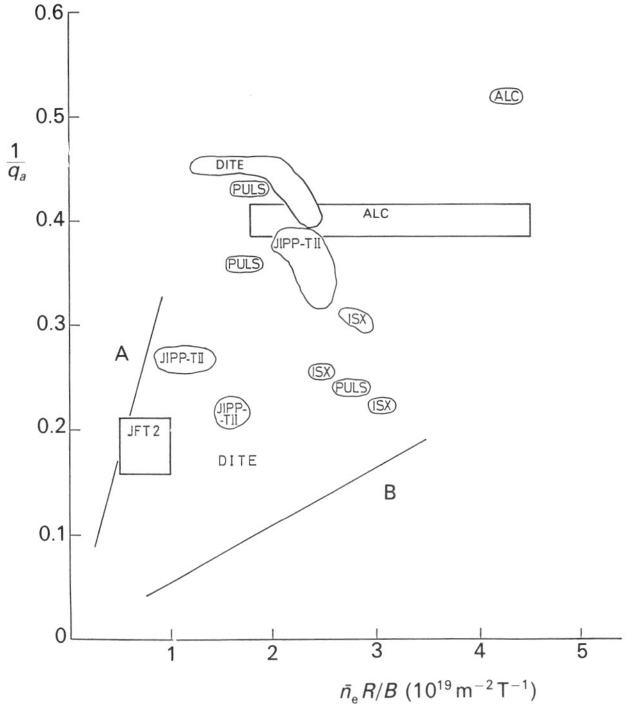

Fig. 8.7. An early (1978) Hugill diagram, showing operating conditions for a number of tokamaks. Plasmas are limited to $1/q_a \leqslant 0.5$, and to densities within lines A and B. To the left of line A, electrons 'run away' to very high energies since collisional drag is too weak to balance acceleration by the inductive loop voltage. Line B is the density limit boundary discussed in the text. More recent Hugill diagrams are more complicated as a number of techniques (additional heating, impurity control methods, etc.) have been found to distort and move these boundaries.

The second type of disruption, illustrated in Fig. 8.8, occurs at high density, and also involves $(m, n) = (2, 1)$ tearing activity. It is thought to occur as follows. As the density n rises, the power radiated at the edge increases as n^2, rapidly cooling the edge and leading to shrinking and peaking of the current density profile to maintain I_P. The increased current density gradient at the $q = 2$ surface destabilizes the $(m, n) = (2, 1)$ mode, leading to a disruption similar to that at low q_a. The Hugill diagram is perhaps a little misleading, as the disruption is triggered

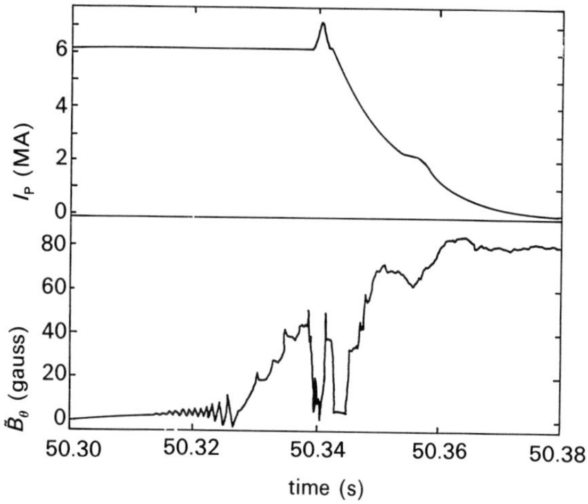

Fig. 8.8. A high density disruption on JET, showing the rapid termination of the plasma current. MHD activity, shown here as the poloidal magnetic field fluctuations at the edge, rises strongly before the disruption. Reproduced by permission of the JET Joint Undertaking, Abingdon, UK.

more by high edge density than by high average density $\bar{n}$. The density limit may be increased by a number of methods: in particular by heating additional to the basic Ohmic heating of the tokamak; by conditioning of the vessel to give plasmas with very low impurity content; and by control of MHD activity using localized electron heating or current in external helical windings. A comprehensive survey of disruptions in JET is given by Wesson *et al.* (1989).

Disruptions produce a very rapid decay of the plasma current, typically at a rate $\sim -10^9\,\mathrm{As}^{-1}$, giving rise to very large inductive forces on the vessel and coils. In future large tokamaks, such as ITER, these forces could cause great damage, and so techniques for preventing disruptions are being investigated. Since both types of disruption are usually, but not always, preceded by growing $(m, n) = (2, 1)$ precursors, shown in Fig. 8.8, one approach currently being explored is the suppression of these precursors. This might be by using currents in external helical windings or by using electron cyclotron current drive (ECCD) to flatten the current profile in the vicinity of the $q = 2$ surface, perhaps using fast feedback to follow the modes. Fast feedback control of $(m, n) = (2, 1)$ tearing modes using external helical windings was demonstrated on the British tokamak DITE, with an accompanying increase in the density limit. For more details concerning the control of tokamak instabilities, see the review by Robinson (1989).

8.3.2 *β limits*

Another limit to tokamak operation is the β limit, which usually results in a deterioration of confinement rather than a disruption. The average ratio of plasma pressure to magnetic field pressure, β, is a measure of how efficient the magnetic field is at confining plasma. Experimentally, the value of β is observed to be limited to $\beta < g_T I_P / a B_\phi$ (β in per cent, I_P in mega-amperes, a in metres, B_ϕ in tesla), where g_T is the 'Troyon factor'. Troyon factors as high as 6.5, associated with a peaked current profile, have been achieved on the San Diego tokamak DIII-D, but values of around 2.8 to 4 are more normal and agree with computer calculations of marginal stability of plasma profiles to global ideal modes, in particular to ballooning and kink modes. Note that the minor radius a is half the width of the plasma in the horizontal direction, so that β may be increased by

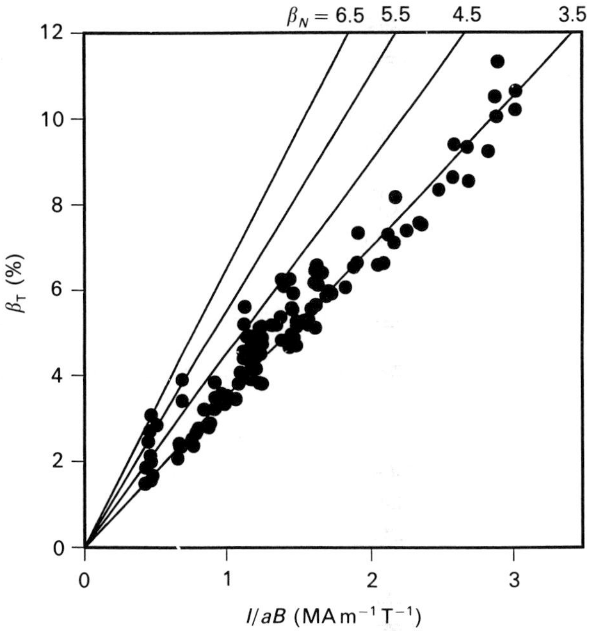

Fig. 8.9. A β limit diagram for the DIII-D tokamak, showing that the Troyon factor, here denoted β_N, has reached values as high as 6.5. Reproduced by permission of General Atomics, San Diego, USA.

elongating the plasma vertically, which allows higher I_P for a given q_a. Thus the highest values of β ($\lesssim 12\%$) have been achieved in elongated tokamaks such as DIII-D (elongation $\kappa \lesssim 2$), whose β limit diagram is shown in Fig. 8.9.

8.4 Heating and current drive

The Ohmic power that heats the tokamak is simply the Joule power $I_P^2\mathscr{R}$, where $\mathscr{R}$ denotes the plasma resistance. However, $\mathscr{R}$ is determined by the electron–electron collisions and has a similar dependence on the electron temperature, $T_e^{-3/2}$. Since the current is limited to $q_a \gtrsim 2$ for stable operation, the Ohmic power is insufficient to reach ignition temperatures unless the field B is very high allowing high I_P, or the aspect ratio R/a is very low giving a large fraction of trapped electrons which increase the resistance. Thus, in most tokamaks, which have $B \lesssim 5$ T and $R/a \gtrsim 2.5$, heating additional to the basic Ohmic heating is required to give temperatures greater than about 1 keV. There are two main classes of additional heating, neutral beam and radio-frequency heating, although other methods have been used with some success, for example adiabatic compression of the minor and major radii. Up to 30 MW of additional heating has been used on the largest tokamaks JET, TFTR and JT-60, dwarfing the Ohmic power (< 1 MW).

In neutral beam heating, fast atoms – for JET, typically 100 keV hydrogen or deuterium – are injected into the plasma. Since these are neutral, they are unaffected by the magnetic field and follow straight trajectories until they undergo collisions with ions involving charge exchange, or are ionized. In addition to heating the plasma, neutral beam injection refuels it, may add toroidal angular momentum to the plasma, and generates a high energy ion population which may influence plasma behaviour. Also it allows some flexibility in the location of the heating, whereas Ohmic heating is concentrated near the magnetic axis where the temperature is highest and there are few trapped electrons. The energy of the beam is important: too low, and the beams are stopped immediately at the edge of the plasma; too high, and there is 'shine-through' which can damage the tokamak vessel. Neutral beam heating has produced ion temperatures up to 30 keV in the JET and TFTR tokamaks.

In radio-frequency (RF) heating, high power microwaves are injected into the tokamak and impart energy by resonant interactions. Three main schemes are used: ion cyclotron and electron cyclotron resonance heating (ICRH, $\nu \sim 50$ MHz, and ECRH, $\nu \sim 100$ GHz), in which the wave frequency is chosen to match the cyclotron frequency or one of its harmonics; and lower-hybrid (LH) heating ($\nu \gtrsim 1$ GHz), in which the waves are Landau damped by either electrons or ions, depending on the density. Since cyclotron absorption occurs at or around one value of B determined by the wave frequency, ICRH and especially ECRH can be spatially localized. They therefore allow flexible control of the heating profile, and consequently some control of MHD activity. ECRH has produced electron temperatures up to 10 keV on the Russian tokamak T10, and ICRH has produced ion temperatures up to 14 keV on JET.

Current drive is a useful by-product of additional heating. If the beams or waves are injected preferentially in one toroidal direction, currents may be driven. The mechanism for current drive is subtle, and does not rely on momentum input: for further details the reader is referred to Chapter 15 and the review by Fisch (1987). Unlike the Ohmic current, which is determined by the transformer and is thus transient, noninductive current drive can give steady state currents. Also, the Ohmic current density J_Ω peaks sharply at the magnetic axis, since $J_\Omega \sim T_e^{3/2} g(r)$, where the trapping correction $g(r) \sim 1 - 2(r/R)^{1/2}$ with r the flux surface radius. The noninductive current appears where the heating occurs and so, particularly for ECRH current drive, might allow local control of the current profile, which may prove valuable for instability control. A useful figure of merit for current drive schemes is $\eta_{CD} = \bar{n}_{20} I(\mathrm{MA}) R(\mathrm{m}) / P(\mathrm{MW})$, where current I is driven by power P in a plasma of major radius R and average density $\bar{n}_{20} \times 10^{20}\,\mathrm{m}^{-3}$. Lower-hybrid current drive, in which electrons with energies ~ 100 keV are heated, has proved the most successful current drive scheme with $\eta_{CD} \lesssim 0.3$, and has driven the whole plasma current in several experiments (see Fig. 8.10). It has extended pulse lengths to tens of seconds in JET and the French tokamak TORE SUPRA, and to an hour in the Japanese device TRIAM-1M. This is in contrast to typical inductive pulse lengths which range from a few milliseconds in the smallest devices to a few seconds in JET; for ITER, discharges lasting up to 3000 s are planned. It has also been shown to influence current profiles, suppress sawteeth and affect other MHD activity. Current drive using neutral beams and ECRH has also been demonstrated. These and other methods are expected to give similar values for η_{CD} in reactor-grade plasmas, and a variety of non-inductive current drive schemes are being considered for next-generation tokamaks.

In general, the observed propagation, absorption and current drive of neutral beams and radio-frequency waves agree well with theoretical calculations. However, some of the details are not yet understood fully, particularly those of lower-hybrid heating and current drive. For an extended discussion of plasma heating, see Chapter 15.

In addition to the externally driven Ohmic, beam and RF currents, the plasma

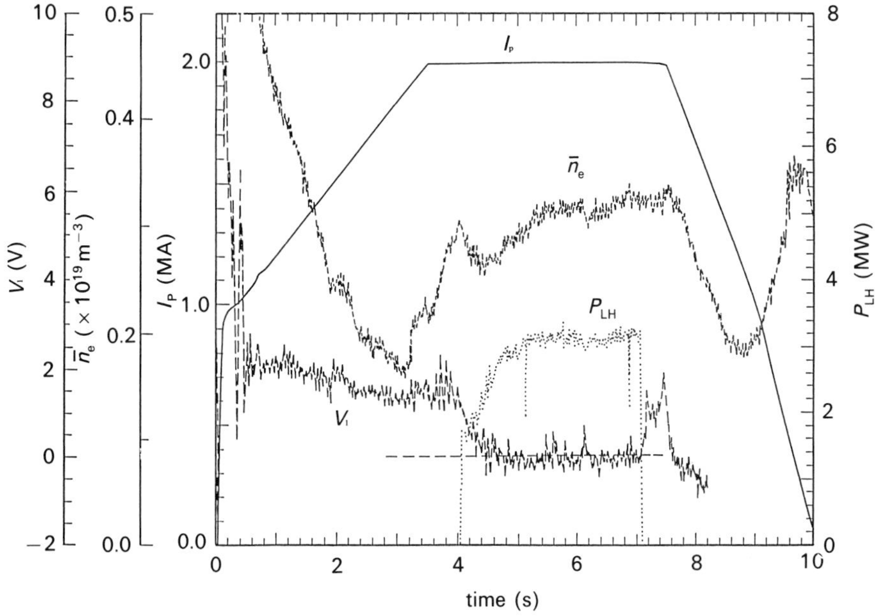

Fig. 8.10. Discharge from the JT-60 tokamak in which the loop voltage, V_l, drops to zero as lower-hybrid power, P_{LH}, is applied. In this discharge, 3 MW of power drove the whole plasma current, $I_P = 2$ MA. Also shown in the evolution of the average density, $\bar{n}_e$. (JT-60 Team, 1987.) Reproduced by permission of the Japanese Atomic Energy Research Institute.

has an internal current driven by its pressure gradient. This 'bootstrap current' may provide much of the plasma current in future large devices, perhaps allowing steady state operation. It peaks where the pressure gradient is greatest, however, and so an additional means of driving current near the magnetic axis, where $\nabla p = 0$, might also be necessary. The bootstrap current has been observed in a number of large tokamaks, notably the Japanese tokamak JT-60 in which it has contributed up to 80% of the plasma current.

8.5 Impurities and edge physics

The tokamak plasma consists of electrons, the majority ion species, which is usually hydrogen, deuterium or helium, and a small fraction of impurity ions. These impurities can greatly influence plasma performance, even when they are present in very small concentrations. In the bulk of the plasma they are fully ionized and therefore radiate strongly, since the power radiated is proportional to the square of the charge number Z_i. This power loss can inhibit plasma start-up, determine the density limit for tokamak operation, and contribute significantly to the tokamak power balance. Impurities also dilute the plasma: for example, if the density of fully ionized oxygen is 1% of the electron density n_e, then it follows from the quasineutrality condition $n_e = \Sigma_i n_i Z_i$ that the majority ion density (assuming hydrogen or deuterium) is only $0.84n_e$. Since the fusion power density in a deuterium–tritium reactor will scale as $n_D n_T$, it is clear that highly charged impurities must be minimized. The effective charge of the plasma, $Z_{eff} = \Sigma_i n_i Z_i^2 / n_e$, is the accepted measure of impurity content, and will have to be ~ 2 or less in a reactor.

The impurities come from plasma wall interactions, and their role is discussed in detail in Chapter 16. The presence of weakly bound molecules which are easily desorbed, such as water vapour and carbon monoxide, has to be minimized by regular baking of the vacuum vessel, and by glow discharge or rapid pulse discharge cleaning. Other techniques include 'gettering', the evaporation of materials such as titanium in the chamber between discharges, which gives a protective coating to the vessel and other surfaces. More recently, 'boronization' has been found to be very effective: a weak 'glow' discharge lasting several hours, into which a boron compound is introduced, is found to keep Z_{eff} low for many subsequent plasma shots. Even in well-conditioned tokamaks, surface material is introduced into the plasma by a variety of processes. These include sputtering, in which energetic ions or neutral atoms displace material, arcing between the plasma and the surface, and evaporation when the surface temperature approaches melting point. Operation with limiters made of low Z materials is clearly desirable, and graphite, and more recently on JET beryllium, have proved particularly successful limiter materials, keeping Z_{eff} close to unity at high densities in many tokamaks.

A flux of ions incident on a surface can give a flux of ions back into the plasma, either by direct backscatter or by other processes, with 'recycling' coefficients sometimes greater than unity. The influx of impurities and ions from the wall and limiter into the plasma, together with an inward particle pinch that is not fully understood, can maintain a peaked density profile against outward diffusion. Gas puffing at the edge has the same effect, with more peaked profiles obtained by central fuelling with small pellets or neutral beams.

Many problems associated with plasma wall interactions may be avoided or reduced by using a magnetic divertor. Limiters are not used to define the edge of the plasma: instead, external coils divert magnetic field lines to an external chamber where the plasma is neutralized and the products of sputtering and other plasma–surface interactions are pumped away to prevent contamination of the main plasma. In the most successful divertor configuration, coils produce an axisymmetric poloidal field null, shown for the British tokamak COMPASS-D in Fig. 8.11. This configuration is used on many tokamaks (for example, DIII-D, JET, JT-60 and the German machine ASDEX) and is proposed for ITER. The flux surface that passes through the null is a magnetic separatrix, and defines the plasma edge

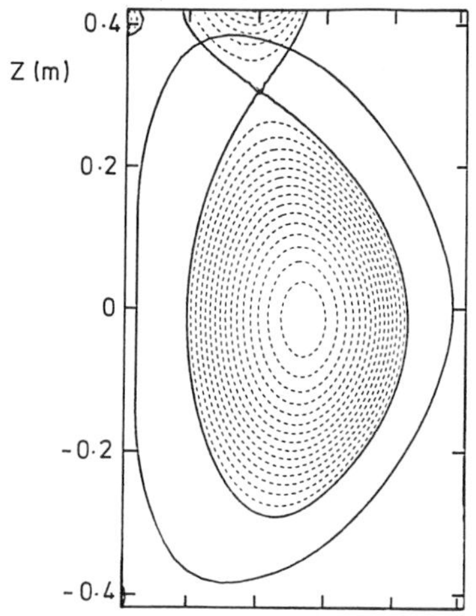

Fig. 8.11. Flux surface geometry for COMPASS-D showing a poloidal field null ('X-point') at $R = 0.50$ m, $Z = 0.30$ m. The magnetic axis, where the poloidal field is also zero, is at the centre of the nest of flux surfaces. The flux surface that passes through the X-point and defines the edge of the plasma is known as the separatrix, and in this example is well within the vacuum vessel. The magnetic geometry was calculated by an equilibrium code using coil and plasma parameters.

without acting as an impurity source. Nulls at different poloidal locations have been tried, with the most successful position being that shown in Fig. 8.11. Several tokamaks have operated with two poloidal field nulls, above and below the plane of the magnetic axis, giving an up–down symmetric plasma. Diverted plasmas have the disadvantages of smaller area and greater coil complexity than limited plasmas, but have shown superior impurity and confinement properties. It is unlikely that limiters would be able to bear the high heat fluxes in a reactor, and so diverted plasmas are a feature of designs for ITER and other future tokamaks.

The physics of the tokamak edge is in many ways more complicated and difficult to model than the rest of the plasma. Atomic processes are important, determining the balance between impurity charge states; radiation losses are a strong function of parameters and impurity content; both parallel and cross-field transport of particles, energy, and momentum are important; collision times are short compared with circulation times around the tokamak, unlike in the rest of the plasma; the plasma potential varies in the boundary, with a sheath forming close to material surfaces; there is a rich variety of possible plasma–surface interactions; and the edge plasma is generally highly turbulent, with electron density fluctuations $\delta n_e \sim n_e$, for example. All these effects contribute, and, though computer codes have been written to model this behaviour, progress has been slow, and understanding of the tokamak edge is still very limited. For further details of this complicated subject the reader is referred to Chapter 16 and to the review by Stangeby and McCracken (1990).

8.6 Confinement

In a tokamak reactor, the power liberated in fusion reactions would have to be much greater than that lost via radiation and transport across the magnetic field ('radial transport'). Thus a key tokamak parameter is the energy confinement time, τ_E, the ratio of the total kinetic energy to the power required to sustain the plasma in steady state. Whereas the radiative losses – principally Bremsstrahlung, but also synchrotron emission in a reactor – are well understood, this is not the case for the radial transport processes, which are normally the dominant energy loss channel. In this section, we describe the experimental measurements of confinement, and then discuss the present understanding of radial transport. Whilst we concentrate on energy confinement, we also briefly mention the related topics, particle and momentum confinement. The review paper by Callen (1992) gives further details of the experimental and theoretical understanding of radial transport.

8.6.1 *Scaling laws*

Because of the limited theoretical understanding of confinement, there have been many empirical studies of how τ_E varies with parameters, often using results from several tokamaks. The resulting scaling laws split into two groups: those for Ohmically heated tokamaks, and those for additionally heated tokamaks.

The most commonly quoted Ohmic scaling law, Neo–Alcator, is very simple: $\tau_E = 7 \times 10^{-22}\bar{n}_e a R^2 q_a^{1/2}$, where SI units are used and q_a is the edge safety factor. This has proved successful in fitting τ_E for a wide range of tokamaks, although at high densities τ_E is observed to saturate, perhaps because neoclassical ion transport (see below) becomes important. At low densities, when the Ohmic heating produces a large superthermal electron component, τ_E is often above the Neo–Alcator value.

In the early 1980s, Goldston proposed the first empirical scaling law for additionally heated tokamaks, which incorporated the observed degradation in τ_E as the input power is raised and a linear dependence on the current ($\tau_E \sim I_P P^{-0.5}$). At least a dozen such scaling laws have since been proposed. Some of these involve a large number of parameters. For example, one of the scalings used in studies for ITER (the ITER-89P scaling (Yushmanov *et al.*, 1990); see Fig. 8.12) is, in SI units,

$$\tau_E = 5.5 \times 10^{-5} A^{0.5} I_P^{0.75} B^{0.63} \bar{n}_e^{0.08} P^{-0.54} (R/a)^{-0.1} R^{1.84} \kappa^{0.85},$$

where κ denotes the vertical elongation, and A denotes the atomic number of the majority ions, though the dependence on A is disputed.

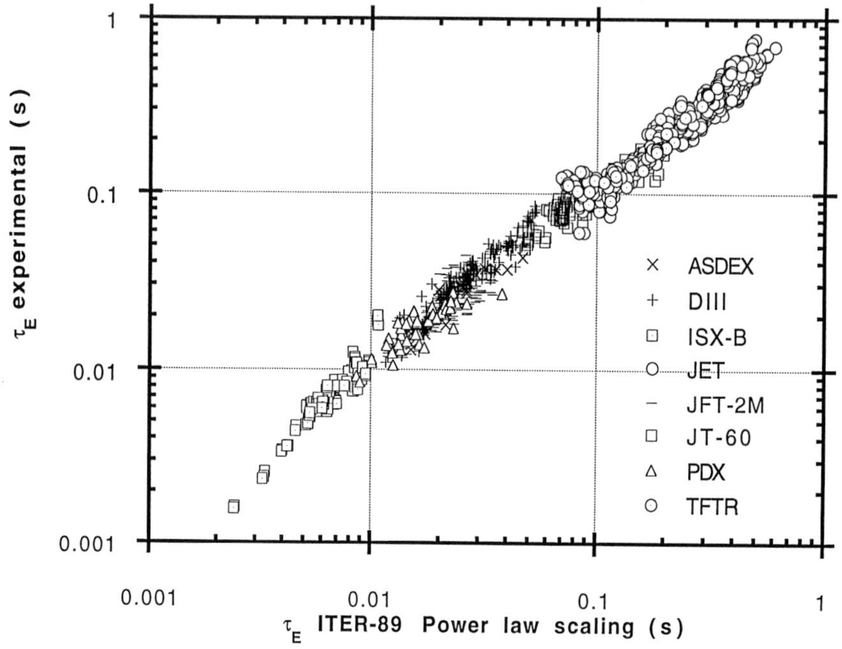

Fig. 8.12. Plot showing comparison of the energy confinement time from several tokamaks with the ITER-89P scaling law (Yushmanov *et al.*, 1990).

8.6.2 *H-modes*

The ITER scaling law given above applies to the normal additionally heated regime, the L-mode (*low confinement*). However, under some conditions the plasma confinement is observed to improve suddenly, though τ_E usually remains less than the Ohmic value. This *higher confinement* regime is called the H-mode. The scalings of τ_E with parameters in the L- and H-modes are similar, and usually the H-mode scaling is taken to be simply a multiple, often two, of the L-mode scaling. The first H-modes were observed using neutral beam heating in the ASDEX (Germany), PDX (Princeton, USA) and DIII-D (San Diego, USA)

9.2.3 *Lines of force*

Recall that a line of force – a concept much used in space plasma physics in representing magnetic fields – is a curve whose tangent gives the direction of the field at any point, and whose density is a measure of the field strength. In a mathematical sense, a line of force clearly has continuity. It may be considered to be continuous in a physical sense, too, under circumstances where the field is suitably stable and where it effectively guides particle motion, for example in the inner region of the Earth's radiation belt. The concept needs to be considered very carefully, though, in situations where the field is variable or where, for any reason, particles of different species or different energy within a single species follow significantly different paths. Continuity between inner and outer parts of a planetary magnetosphere may for such reasons be a somewhat questionable concept. In the present context we suggest that lines of force are considered to represent local conditions only.

Imagine that the magnetized planet has, at a given instant, fully laden zones of trapped radiation, i.e. having all possible trajectories being represented, but with low enough intensities not to distort the field unduly. Particles with magnetic moments too small for the particles to be magnetically mirrored will reach the atmosphere or surface and be lost by ionization. Secondary particles may be emitted as a result. A loss cone of directions will consequently be established in the trapped population. The half-angle of the loss cone at a point where the field strength is B is given by

$$\sin^2 \alpha = B/B_0,$$

where B_0 is the field strength at the top of the atmosphere, or other absorbing surface.

9.2.4 *Perturbation of motion*

Perturbations due to gyro-resonance with plasma waves can cause pitch-angle scattering resulting in diffusion into the loss cone and thence precipitation into the atmosphere (if the planet has one), or onto the planet's surface, to create, via ionization and excitation, one form of aurora, the diffuse aurora. The process is very effective in the Earth's radiation zones, though it is still not fully and quantitatively understood. Doppler-shifted gyro-resonance between electrons and whistler-mode electromagnetic waves is thought to play a key role in this process. Here, the waves, which propagate at frequencies below the electron gyro-frequency, are raised to the gyro-frequency in the frame of reference of an electron travelling in the opposite direction. Resonant interaction then leads to scattering in pitch angle and minor changes in energy.

Another form of perturbation, which may accompany pitch-angle scattering, is an acceleration directed parallel to B. This also lowers mirror points and causes aurorae. The consequent enhancement of energy flux results at the Earth in the bright, highly structured and often spectacular discrete aurora. The process appears to be one of the most fundamental process enacted within collisionless plasma. There are two main theories to account for it: (1) acceleration through static, or quasistatic, potential differences, which, as we shall see, is highly questionable on fundamental grounds, and (2) acceleration by resonance with lower-hybrid waves, which offers, we suggest, a promising explanation of the phenomenon. This issue, central to the understanding of space plasma physics, will be discussed in more detail in Section 9.4.3.

Resonant interaction with a time-dependent electric field can produce acceleration together with a drift towards the planet, or a retardation with an outward drift, depending on the phase relation. This radial diffusion is sometimes confusingly described in terms of Fermi acceleration due to approaching mirror points, or adiabatic compression due to the reduction in volume of magnetic flux tubes in which particles find themselves. But this can be misleading, since the mirror points, although seen to approach from a reference frame moving with the particle, are stationary in the frame in which energization is taking place.

9.2.5 *Force-free motion*

Angular rotation of the planet introduces complications. If the magnetic and rotational axes are aligned, and if the planet is not a conductor and has no conducting ionosphere, there is no effect. However, if currents can flow in the

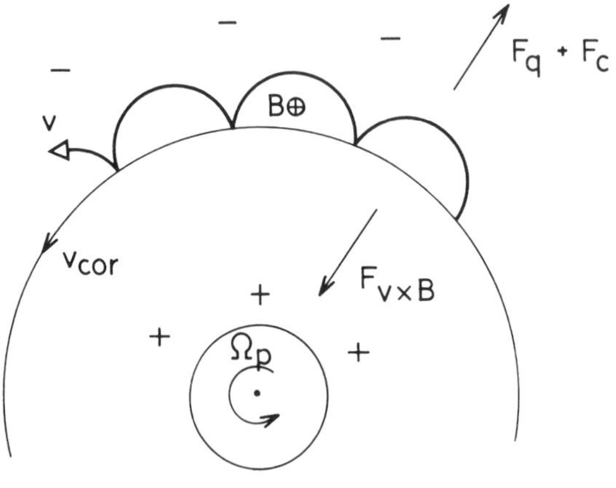

Fig. 9.3. Force-free motion, illustrated for co-rotation in the equatorial plane of a planet with an extensive neutral atmosphere rotating at an angular velocity Ω_p. Motion given to charged particles by friction with the atmosphere will lead to deflection by the magnetic field (via the force $F_{v\times B}$) – electrons outward, and positively charged particles inward. The charge separation this produces builds up until the consequent, opposing electrostatic force F_q, combined with the centripetal force F_c, balances $F_{v\times B}$ to allow force-free co-rotation. A freshly introduced charged particle, or one which for some other reason does not co-rotate initially, will experience at first just F_q and will be accelerated radially. The motion thus introduced brings $F_{v\times B}$ into play, resulting in the cycloidal motion illustrated. The average azimuthal velocity equates to the co-rotational azimuthal velocity $\boldsymbol{v}_{cor}$. Wave–particle interactions will serve to subsume such freshly introduced particles into the general co-rotation. Similar considerations apply to the propagation of the solar wind through interplanetary space, to the introduction of foreign material into the solar wind, and to convection of plasmas within magnetospheres.

planet or its frictionally driven co-rotating atmosphere or ionosphere, the $\boldsymbol{v} \times \boldsymbol{B}$ force arising from motion at velocity $\boldsymbol{v}$ in the magnetic field $\boldsymbol{B}$ will cause charge to separate until the electric field, $\boldsymbol{E}$, of the space charge balances the magnetic force (see Fig. 9.3).

At this stage

$$\boldsymbol{E} = -\boldsymbol{v} \times \boldsymbol{B}.$$

(In MHD, where there are no particles or charges, this same relation is considered to arise from motion of magnetic field lines.) Once this balance is established, particles of the conducting planet or its ionosphere will be able to rotate with the planet force-free. Such co-rotation will naturally be maintained out to distances where a collision-effected viscous drag provides the necessary torque. However, co-rotation can prevail beyond the atmosphere it currents flow along magnetic field lines to ensure that they are electrostatic equipotentials (this is one of the tenets of magnetohydrodynamics), but it is unclear just how well this holds in practice. To the extent that it does hold, it implies that the magnetic field fills the role of an extended ionosphere. Co-rotation is believed to be impossible at distances beyond that at which co-rotation exceeds the Alfvén speed.

9.2.6 *Limit of planetary control*

A distance is eventually reached where, due to diminishing magnetic field, the planet's surroundings take control. In the solar system it is the solar wind that sets the boundary conditions.

9.3 The solar wind

The solar wind is the expanding solar corona. It is a plasma of electrons and protons, with an admixture of helium and other solar constituents, streaming continuously and in a highly gusty fashion at speeds of 200–1000 km s^{-1} into the volume of space which includes the planets of the solar system, known as the heliosphere.

9.3.1 *Protons and other ions*

The protons and other ions form well defined radial beams in velocity space. To a first order, they may be considered as drifting Maxwellians, with the magnitude of the drift velocity being much greater than the thermal speed. The energy density of the solar wind is, beyond a few solar radii, considerably greater than that of the solar magnetic field, so the regime is one in which the particles control and shape the field. They are able to overwhelm the solar magnetic fields and, effectively, extend the solar fields into space. Due to the 27-day rotation of the sun, the radially flowing solar wind sets up a pattern of magnetic field, known as the garden-hose or Parker spiral (see Fig. 9.4). The reason for this may be seen very simply if we assume that plasma flow and the magnetic field direction at the sun are purely radial, and that, due to the high conductivity, currents are readily induced to counter any influences attempting to change magnetic flux in any element of the plasma. Simple geometric projection shows that the inclination, θ, of the magnetic vector at a radial distance r, much greater than the solar radius, is given by

$$\tan\theta = \frac{r\Omega_s}{v_{sw}},$$

where Ω_s is the solar angular rotation rate of 2.7×10^{-6} rad s^{-1}, and v_{sw}, the solar-wind speed. The pattern is (though this is rarely pointed out) very sensitive to these assumptions. Even slight ($\sim 1°$) departures from radial flow or field direction at the sun project into gross changes of direction at the orbit of Earth, where, as shown in Fig. 9.4, the nominal inclination to the Earth–sun line is 43°. At Mars, the nominal inclination is 55°, and at Jupiter it becomes 80°. Since the time taken for average speed (430 km s^{-1}) solar wind to reach the Earth at 1 astronomical unit, Au (1.5×10^{11} m) is approximately four days, or some 54° of the solar rotation period, energetic particles or solar cosmic rays released in solar flares are for this reason more readily able to reach the Earth from flares occurring towards the western (as seen from the Earth) limb of the sun. The solar

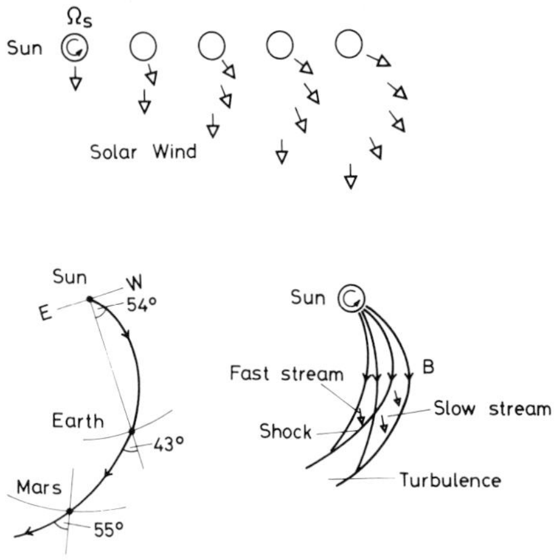

Fig. 9.4. The solar wind and the interplanetary magnetic field. The top part of the figure shows how the locus of elements of solar wind emanating from a given location on the sun, rotating with angular velocity Ω_s forms a spiral pattern in the ecliptic plane. Due to the high electrical conductivity of the plasma, a magnetic field that is radial at the point of emission is drawn out into a garden-hose, or Parker spiral, as illustrated in the lower parts of the figure. For a typical solar wind speed of 430 km/s, the interplanetary magnetic field is inclined to the radial direction by 43° at the Earth, and 55° at Mars. Under these conditions, both planets are magnetically conjugate to a region, seen from the planet to be towards the western limb of the sun. Variable solar-wind speeds lead (lower right) to gross departures from this picture, and to both shocks and turbulence.

wind continues to flow outward, with undiminished speed and geometrically reducing density, to meet, and eventually merge with, interstellar plasma and fields at the heliopause.

9.3.2 *Electrons*

The electron velocity distribution differs greatly from that of the ions. The solar wind is, therefore, not characterizable as a simple flowing plasma. This is because electron velocities are very much greater than the drift velocity. Electron drifts at the ion flow speed are almost insignificant. Solar-wind electrons have an approximately isotropic velocity distribution, occasionally showing a slight imbalance between opposite senses of flow parallel to the interplanetary field, about which direction the distribution is symmetric, or gyrotropic.

The electron velocity distribution, in common with the distribution of cosmic radiation, is very close to a power law. A power law, being self-similar on all scales, has no characteristic energy and, therefore, no meaningful temperature. For this reason concepts based on moments of the distribution function, including Debye lengths, must be treated with great care. This is an area of plasma physics requiring much more attention.

9.3.3 *Current sheets*

The necessity for the divergence of B to be zero ensures that there are regions of both polarities, one from the northern and one from the southern hemisphere of the sun. The current sheet dividing these lies generally in the region of the ecliptic, but has a wavy structure, often likened to a ballerina's skirt. Rotation of the sun introduces sector boundary crossings typically four times every 27 days. These crossings also introduce discontinuities in solar-wind properties.

The nature and behaviour of the solar wind out of the ecliptic plane can today only be inferred from remote sensing methods. However, the *Ulysses* space probe is now well on its journey of exploration which took it past Jupiter in February

1992 where it swung out of the ecliptic in order to pass over the southern solar pole in 1994, and the northern pole in 1995.

9.3.4 *Irregularities*

Irregularities in solar-wind density and speed give rise to interplanetary shocks, turbulence and 27-day (solar-rotation) effects in the interactions with the planets to be discussed below. Solar flares, with their production of solar cosmic rays and enhanced solar-wind density, also represent major perturbations.

9.3.5 *Ion pick-up*

Consider more closely the motion of ions and electrons forming the solar wind. Ions flow at an angle to B everywhere except, as assumed earlier, within the corona itself, so they experience a bending force. This is balanced, exactly as in the co-rotation effect discussed above, by a self-consistent space-charge field to allow the ions to flow effectively force-free at an angle to B. The space charge necessary to create the required density gradient is very weak, and the solar wind is still, to all intents and purposes, electrically neutral. It is a common misconception that the electrons need to flow with the ions to preserve this neutrality. Man-for-man marking is obviously unnecessary. Neither do the electrons need to flow at the same speed to avoid setting up currents. In fact the electron behaviour is totally different to that of the ions, being governed by almost (but not quite) isotropic velocity distributions.

The space-charge field manifests itself very clearly in a phenomenon known as ion pick-up. A low energy ion freshly formed in the solar wind as the result of photoionization of planetary or cometary material experiences the space-charge field created by the solar wind. It is therefore accelerated parallel to the electric field. This motion introduces a magnetic bending force. The combined effect is to cause the ions to follow cycloidal paths in the direction of the solar wind, their mean speed being equal to that of the solar wind but ranging each cycle from zero to as much as double v_{sw} if the magnetic field is normal to the flow. This is a form of collisionless friction. Instabilities and other causes of scattering serve gradually to subsume the injected ions into the solar wind. Naturally, the energy for the ion acceleration is drawn from the solar wind which is locally loaded and slowed down.

9.3.6 *Interaction with planets*

When elements of the solar wind confront the obstacles formed by planets with magnetic fields, currents are induced to resist a change in magnetic flux, thus neutralizing the planetary field within the solar wind, and enhancing it nearer to the planet. Eventually a point is reached, in the approach towards the planet, where the density and energy of the solar wind are no longer able to generate enough current to overcome the increasing magnetic field, and the solar wind is forced to divert and begin to flow around the obstacle. The process may be considered as a collisionless pressure balance, reached at the point where the combined pressures of the plasma and magnetic field are equal on the two sides of a boundary layer. The boundary layer which continues along the flanks of the magnetosphere is known as the magnetopause. The distance from the centre of the planet to the furthermost point upstream, or the 'nose', of the magnetosphere is known as the stand-off distance. If the planet does not have a magnetic field, interaction is directly with the gravitationally unrestrained exosphere, the atmosphere, or the surface itself. In practice, the situation is more complicated than this because the solar wind is supersonic and super-Alfvénic, so a standing bow shock forms upstream. The shock serves the function of slowing and randomizing the flow sufficiently to allow flow around the obstacle. In this process, both the ions and electrons gain thermal energy, as discussed in detail in Part II of this chapter.

A comet-shaped wake or magnetotail is generally created to form a downstream extension of the planet's magnetosphere. The exact nature of the interaction between the solar wind and the magnetosphere taking place across the magnetopause boundary layer serving to match one plasma to the other is still very much a mystery. It is clear that matter, momentum and energy are transmitted across the boundary layer in both directions – the planet loading the solar wind with planetary material, and the solar wind driving dynamical processes in the planet's magnetosphere. Plasma within the magnetotail is subject, through wave–particle interactions, to an effective viscosity applied across the magnetopause boundary layer. This sets up a circulation, or convection, of plasma within the magnetosphere, which may or may not be along closed trajectories. The situation is again analogous to that in the solar wind itself – a space charge is set up whose electric field balances the magnetic force to allow force-free drift. Particles not initially drifting with their fellows will, as in the solar wind and in the co-rotation phenomenon, be picked up and eventually subsumed into the general flow. Closer to the planet, co-rotation can become dominant.

A specific process that is considered by some workers to play a role in the transfer of matter and momentum across the boundary layer is that of magnetic reconnection, originally thought of as a steady state phenomenon, but increasingly seen as a transient, localized process known as a flux transfer event. This is an MHD concept in which interplanetary magnetic lines of force become joined to their planetary counterparts, allowing free flow of plasma particles from one regime to the other. The well established fact that the energy from the solar wind seems more readily transmitted to the Earth's magnetosphere when the interplanetary magnetic field has a southward component, and thus allows the joining to take place more readily, lends some support to this interpretation. Ion acceleration in the magnetopause is also often explained in these terms. However, there are problems with both the steady state and transient pictures. Unlike magnetic annihilation, in which magnetic energy is released and converted into particle kinetic energy, the steady state reconnection picture has a quasistatic magnetic field, which being unchanging can release no energy. The notion of the cutting and rejoining of magnetic lines of force is an issue whose physical foundations need much more attention, especially since the X-type neutral point, a major characteristic of reconnection can be produced very simply using two attracting magnetic dipoles. Some recent studies of the terrestrial boundary layer indicate, in any case, that the layer is one of gradual transition from magnetosheath to magnetosphere properties in all quantities, more suggestive of a diffusion-controlled intermixing than of boundary rupture.

With the above processes in mind, let us tour the solar system to observe them in action. The tour cannot be comprehensive – space does not permit that – but it will, hopefully, reveal that, though governed by a relatively small set of physical processes, the interactions at different planets have their own distinct character, and will perhaps whet the appetite for further, detailed study of the infinitely fascinating natural plasma laboratories of planetary magnetospheres and outer atmospheres, where comparison serves as a substitute for the control that can be exercised in a laboratory experiment.

9.4 Magnetospheres of the planets

9.4.1 *Mercury*

Magnetosphere

The proximity of Mercury to the sun and the relative weakness of its magnetic field lead to a magnetosphere which is largely filled by the planet itself, the stand-off distance being typically only $\sim 1.6R_{Me}$ (see Fig. 9.5). It is clear from this figure that the scope for stable radiation belts is rather limited. Nevertheless,

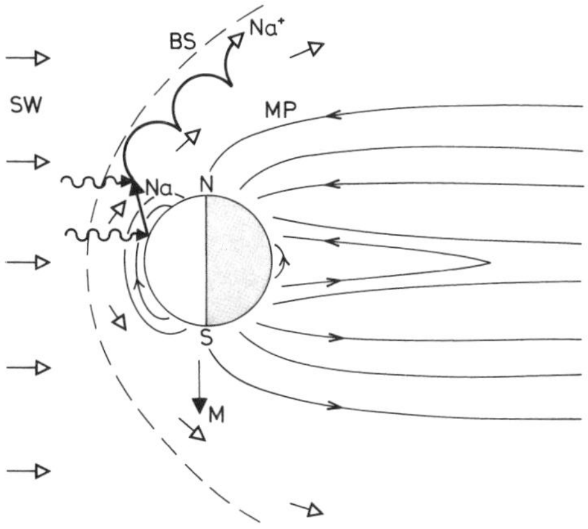

Fig. 9.5. The magnetosphere of Mercury, perpendicular to the ecliptic and in the noon–midnight plane. The planet itself occupies much of the magnetosphere, leaving little opportunity for the orderly motion depicted in Fig. 9.2. The obstacle presented by the magnetosphere and by the loading of the solar wind (SW) by sodium released from the surface generate a bow shock, BS. MP = magnetopause boundary layer; M = magnetic moment.

there are trapped particles (H, He, O, Na and K, and of course electrons) and a definite tail. Sodium atoms are sputtered from the dayside of the planet primarily by solar ultraviolet radiation and, to a smaller extent, by direct solar-wind impact. These atoms become ionized, again by solar UV, either within the planet's magnetic environment where they are the source of the planet's predominantly heavy-ion magnetosphere, or within the solar wind where they are 'picked up' and eventually subsumed. The slower atoms are drawn back to the surface where they may cause further sputtering. As depicted in Fig. 9.5, the nightside of the magnetosphere is considerably distended through the diamagnetic effect of the plasma it contains. There appear to be rapid changes in the particle fluxes and in the configuration of the magnetic field, similar in many respects to the substorms known at Earth (Section 9.4.3) even though Mercury has little or no atmosphere or ionosphere and its surface is a poor conductor. These facts seem to prove that an ionosphere is not essential to the substorm phenomenon as had previously been thought from studies at the Earth. Oxygen

Table 9.1. *Mercury, basic information*

Mean distance from sun	5.8×10^{10} m
Mean planetary radius, R_{Me}	2439 km
Magnetic moment	5.1×10^{12} Tm^3
Inclination of magnetic axis	10°
Upstream distance of bow shock	$\sim 2\,R_{Me}$
Stand-off distance	$\sim 1.6\,R_{Me}$
Visited by	*Mariner 10* (1974, 1975); *Orbiter* proposed

Note that the spacecraft listed in this and subsequent tables are restricted to those capable of plasma measurements.

and potassium from the planet, and protons and helium ions from the solar wind, are also found in the radiation zones. The exact manner of entry of the solar wind to the Hermean* magnetosphere is, as at Earth and the other planets, still very much a mystery.

In common with its counterparts at the other planets, the Hermean magnetosphere presents an obstacle to solar-wind flow, causing a standing bow shock to form upstream. The effect of the bow shock is to broaden the velocity spectrum of solar-wind ions, i.e. heating the ions and randomizing their directions, as required for them to find their way around the planet. The effect is greatest at the 'nose' of the magnetosphere, and gradually diminishes as the impact parameter increases. An augmentation of the magnetic field strength at the bow shock causes some ions to be reflected upstream a number of times before they penetrate the shock.

The piecing together of fragments from an incomplete survey, such as that resulting from a spacecraft encounter, is highly reminiscent of archaeology, another form of endeavour in which an understanding of the basic principles helps to bridge gaps in information. The perspective so gained may, however, be very limited, as has been realized from the restricted view of the terrestrial magnetosphere obtained from the 'fly-by' of *Galileo* in December 1990, en route to Jupiter via gravitationally assisted encounters with both Venus and Earth.

9.4.2 *Venus*

Solar-wind interaction

Venus has little or no intrinsic magnetic field, so the solar wind interacts directly with the ionosphere and atmosphere. A rich spectrum of neutral particles (C, N, O and their compounds) released by these interactions from the atmosphere become ionized by solar radiation, and are consequently picked up and swept downstream to form an irregular comet-like tail which contains also a corpuscular umbra and penumbra, threaded by the interplanetary (solar-wind) magnetic field, as shown in Fig. 9.6. The field, distorted by the need (in view of the high conductivity) to preserve continuity of magnetic flux linkage between plasma elements, appears to become 'draped' around the planet. This draping is

*A tour of the planets provides an excellent incentive for a revision of Greek as well as Roman mythology.

Table 9.2. *Venus, basic information*

Mean distance from sun	1.1×10^{11} m
Mean planetary radius, R_V	6161 km
Magnetic moment	$\to 0$
Upstream distance of bow shock	$\sim 1.4 R_V$
Stand-off distance	inapplicable
Visited by (fly-by)	*Mariner 2* (1962); *Galileo* (1990); and many others
Orbital missions	*Veneras 9* and *10* (1975); *Pioneer Venus* (1978–91)

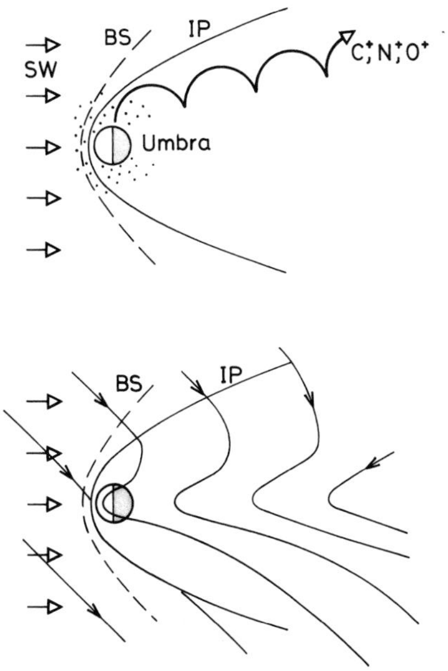

Fig. 9.6. Solar-wind interaction with Venus. The upper part of the figure, symmetric about the planet–sun line, shows the solar wind SW interacting directly with the atmosphere to create a bow shock BS and an ionopause IP. The lower part of the figure shows, in the ecliptic plane, the local distortion of the solar-wind magnetic field, sometimes described as 'draping', a phenomenon also associated with Mars and with comets.

also a characteristic of the solar-wind interaction with comets (Section 9.5) which, like Venus, have extensive exospheres.

Since Venus and the surrounding ions represent an obstacle to solar-wind flow, the planet gives rise to a bow shock and a magnetosheath or ionosheath of retarded and heated solar wind. Again, a similar situation prevails at comets where the solar wind is slowed by the loading caused by injected cometary material. A particularly revealing observation made by the *Galileo* spacecraft in 1990 was of the presence of very energetic ions outside the bow shock and streaming away from Venus. Since there are no radiation zones able to serve as a source for these particles, as is believed to be a contributory factor at other planets, it is clear that planetary bow shocks are effective accelerators of charged particles. The mechanism, though, still remains obscure. A mechanism termed shock-drift acceleration has been proposed; but, since it appears to draw on a static magnetic field and on the solar-wind electric field, which we have seen to be of space-charge origin, and therefore conservative, this process must be highly questionable. Wave–particle interactions with the electrostatic turbulence always associated with bow shocks seems a much more likely explanation (see Part II of this chapter).

9.4.3 *Earth*

Magnetosphere

Far more is known, for obvious reasons, about Earth than about the other planets. Its ionosphere and magnetosphere have been explored using radio-waves, high-altitude balloons, sounding rockets, and near-Earth, geosynchronous, and eccentric-orbit satellites. Photography and even unaided visual observations of aurorae have also played important rôles.

The stand-off distance (see Fig. 9.7) is large enough to allow the formation of a well developed and permanent radiation zone (named the Van Allen belts after

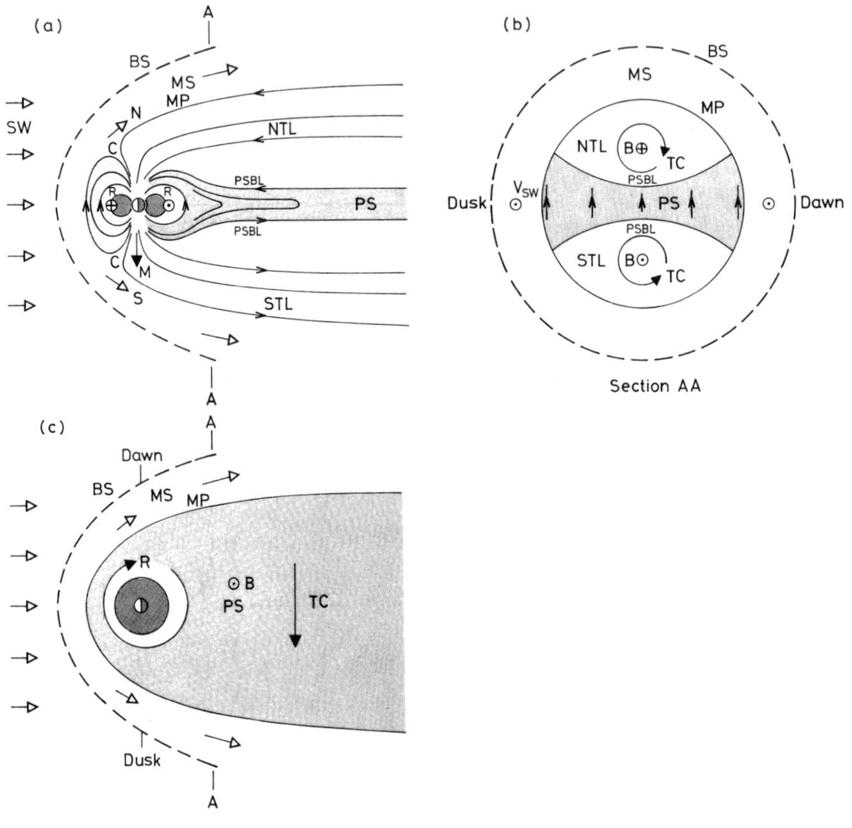

Fig. 9.7. The magnetosphere of Earth. (*a*) Noon–midnight plane perpendicular to the ecliptic; (*b*) dawn–dusk plane perpendicular to the ecliptic; (*c*) ecliptic plane. (*b*) is a cross-section of (*a*), and (*c*), along *AA*. In order to avoid complication, the magnetosphere is drawn with the dipole tilt out of the plane of the paper. BS is the bow shock and MS is the magnetosheath. MP is the magnetopause boundary layer, and NTL and STL are the northern and southern lobes of the magnetotail. PS is the plasma sheet (lightly shaded), and PSBL is its (outer) boundary layer. *R* represents the ring current resulting from drift of particles in the radiation zone (see Fig. 9.2), and TC are the northern and southern elements of the tail-lobe current. The heavily shaded area is the plasmasphere. C is the cusp, thought to permit ready access (as in Fig. 9.1) to solar wind (SW) from the magnetosheath.

their discoverer). The innermost part of the magnetosphere, effectively an extension of the ionosphere, and approximately coinciding with the inner part of the radiation zone, is the plasmasphere. The plasmasphere is composed of particles of ~ eV energies and is generally understood to be co-rotating with the Earth and its atmosphere. Next is the plasmasheet of ~ keV particles forming a

Table 9.3. *Earth, basic information*

Mean distance from sun	1.5×10^{11} m
Mean planetary radius, R_E	6370 km
Rotation period	24 h
Magnetic moment	7.9×10^{15} Tm^3
Inclination of magnetic axis	11°
Upstream distance of bow shock	$\sim 15\, R_E$
Stand-off distance	$\sim 10\, R_E$

giant, distorted plasma torus around the Earth. It is dynamic in nature, in response to sudden changes, known as substorms, in the ring current of higher energy particles making up the radiation belt which co-exists with the relatively low energy plasmasphere and plasmasheet. The exact nature of substorms is not fully understood, but it is clear that they cause the inflated magnetosphere to return at intervals (several times a day) to a more-nearly dipole configuration. It is still unknown whether these substorms are direct responses to changes in the solar wind, and in particular to the linking of solar and terrestrial magnetic fields through steady state reconnection and/or flux-transfer events, or whether there is an important element of spontaneous release of stored energy. The plasmasheet carries the solenoidal currents supporting the towards-Earth and away-from-Earth, northern and southern, lobes of the magnetotail magnetic field. It is widely believed that field-aligned currents into the ionosphere, and their closure through the ionosphere are essential elements of the substorm. However, the substorm-like changes at Mercury (Section 9.4.1), which are thought to be unable to support such currents, have raised serious doubts over this.

North and south of the plasmasheet lie the northern and southern tail lobes composed of low density plasma, mostly an outflow from the polar ionosphere. The magnetic field is stronger here than in the plasmasheet. The dawn–dusk flanks of the plasmasheet continue to the magnetopause boundary layer, separating it from the magnetosheath. This topology is not evident in a noon–midnight projection. Kelvin–Helmholtz instabilities between the magnetosheath and the plasmasheet are thought to be projected along the magnetic field as absolute instabilities to result ultimately in large scale folds in aurorae and in ionospheric characteristics. The plasmasheet boundary layer forms at the boundary between the hot and dense plasmasheet and the relatively cool, diffuse plasma of the tail lobes. This naturally active region appears to be the seat of more active and intense forms of aurorae such as the curtain-like auroral arcs.

There is much discussion at present over the nature of the magnetopause boundary layer, and in particular over whether it effects a smooth, diffusion-like transition, or whether it undergoes steady state or sporadic rupture. Some measurements have been interpreted in terms of sporadic reconnection in the equatorial plasmasheet at distances of $20R_E$, or so, with the formation of break-away clouds of plasma known as plasmoids. However, it should be noted that isolated measurements from a single spacecraft, or even from two or more, cannot give a unique confirmation of the overall configuration. The electric-current system supporting a plasmoid has still to be proposed. Temporal/spatial ambiguity is a real problem in such matters, and multipoint measurements are needed to resolve it. The European Space Agency's *Cluster* mission, due for launch in 1995, will employ four spacecraft in close formation in the latest attempt to cope with this major difficulty.

Bow shock

Moving further away from Earth we encounter the bow shock at typically $\sim 15R_E$. The shock which serves to slow and heat solar wind ions also has the effect of accelerating some ions and electrons. The shock is a region of intense wave activity, the free energy for which derives from the solar wind. Solar-wind electrons, having velocities generally much greater than the ions, are able to cross the shock in either direction. Wave–particle interactions produce a bulge in the velocity distribution which we interpret as a resonance with lower-hybrid electrostatic waves. There are also explanations in terms of the shock-drift acceleration, discussed in Section 9.4.2, and in terms of an electrostatic potential difference existing across the shock. We feel that the latter can be discounted on grounds of inconsistency with the existence of a low energy tail to the distribution, and indeed on the same grounds of the former, namely that a localized, static (relative to transit time), electrostatic field is conservative. We shall return to this crucial question of particle energization processes in several other connections. The rate and extent of change at the shock vary with position along the shock surface and on the angle between the solar-wind magnetic field and the shock normal. Effects are greater and sharper where these directions are more nearly perpendicular, i.e. at a quasiperpendicular *shock*. Quasiparallel shocks are generally more gradual and diffuse. The shock, like the magnetopause, is highly variable in position, over many Earth radii, in response to variations in the solar wind.

Upstream from the shock are found electrons which have penetrated from downstream and ions which have been reflected at the shock. These 'impurities' in the solar wind give rise to upstream wave activity. There is still no evidence, though, for the electron beams so confidently thought to be responsible for electron plasma oscillations in the upstream solar wind. There is further discussion of the bow shock in Part II of this chapter.

Sources of plasma for the Earth's magnetosphere are from the atmosphere/ionosphere in the polar regions (the ion fountain), cosmic-ray albedo, in which emitted neutrons decay into protons, electrons and neutrinos, and from the solar wind via the magnetopause and cusp regions. There is still much discussion over the relative importance of these different sources. The terrestrial magnetosphere contains a rich variety of components, including doubly ionized helium, almost certainly of solar origin, and oxygen ions of the full range of charge states, firmly indicative of a terrestrial source.

Precipitation and aurorae

Precipitation from the radiation belts arises from a number of causes. One is the straightforward result of a local weakness in the geomagnetic field near the south-east coast of South America known as the South Atlantic anomaly. Here, particles that would mirror and remained trapped at other longitudes penetrate into the atmosphere without encountering a field strong enough to effect mirroring. Another cause of precipitation is pitch-angle scattering from otherwise stable orbits by gyro-resonance with electromagnetic radiation, for example whistler-mode noise. It is interesting to note that, since whistler waves have frequencies, below the electron gyro-frequency, resonance is possible only through the upward shift in frequency produced by the Doppler effect between electrons and waves travelling in opposite directions. This leads to a diffuse, sometimes patchy and pulsating form of aurora, the underlying causes of which are still only partially understood. Diffuse aurorae frequently contain pulsating patches. It is clear from a velocity dispersion between electrons of different energies – the faster ones reaching the atmosphere earlier – that the pulsations arise close to the equatorial crossing point of the tubes of force exhibiting the pulsations. This is not unexpected since this is the region where only small deflections are required to drive electrons into the loss cone. The origin of the pulsations, typically of a period of several seconds, is still a complete mystery.

At the poleward edge of the diffuse aurora, which now seems well established as the 'magnetic footprint' of the plasmasheet boundary layer, are found the structured or discrete aurorae. It is clear that the electrons responsible for this are accelerated parallel to the magnetic field as they precipitate into the atmosphere. The acceleration occurs between the altitudes of 10000 and 1000 km, and commonly produces an electron velocity distribution with a peak in the region 5–10 keV. The cause will be found in most articles on the subject to be a static (relative to the electron transit time of ≤ 1 s) electric field, possibly in the form of many (tens of thousands) electrostatic double layers. However, this 'explanation' unaccountably overlooks the fact that electric fields due to space charge are conservative, and therefore incapable of producing any net effect. Note the similarity with acceleration at the bow shock. It will be found that models based on this interpretation all contain a major flaw, such as electrostatic equipotentials that are not closed surfaces. We have advanced and developed over recent years an explanation in which the acceleration is caused by electrostatic waves (of the lower-hybrid mode) whose phase velocity is comparable with that of the precipitating electrons. The electrons are able to 'surf-ride' on these waves, drawing energy from them in a process known as Landau damping. The process envisaged is an electrostatic equivalent of Fermi's model of magnetic reflection from moving gas clouds put forward to account for the acceleration of cosmic rays. The motion of the electrostatic barrier in the former case, and the magnetic barrier in the latter, is essential for energy transfer. Consider the kinematic of a simple two-body interaction. It is readily shown that one of the bodies can gain energy from a second body only in a reference frame in which the second has kinetic energy, i.e. in a frame in which the donor is moving. It can readily be shown, moreover, that the momentum of the donor has to be greater and oppositely directed to that of the recipient. In short, energy may be gained by a body only in a frame of reference in which, along the net direction of interaction, the centre of mass of the system moves towards the body.

The terrestrial auroral zone is an exceedingly complex plasma physics laboratory, exhibiting a wide range of processes whose relationships and cause-and-effect hierarchy have yet to be established. In addition to the precipitation of electrons and ions, with and without acceleration, there is also an upward streaming of both types of particle (sometimes simultaneously), and variously magnetic-field-aligned and perpendicular acceleration. There is also a rich spectrum of electromagnetic and electrostatic wave activity.

9.4.4 *Mars*

Solar-wind interaction

Mars has no measurable intrinsic magnetic field. Interaction with the solar wind is, therefore, basically similar to that at Venus. Ionized hydrogen, oxygen and carbon dioxide from the atmosphere are picked up by the solar wind to create comet-like tails (Fig. 9.8). Near the centre of the tail there is a plasma regime characterized by anti-sunward streaming oxygen ions reminiscent of terrestrial auroral ion beams. This removal of material represents, on cosmic time scales, a significant erosion of the Martian atmosphere.

Table 9.4. *Mars, basic information*

Mean distance from sun	2.3×10^{11} m
Mean planetary radius, R_{Ma}	3332 km
Magnetic moment	$\rightarrow 0$
Upstream distance of bow shock	$\sim 1.5\, R_{Ma}$
Stand-off distance	inapplicable
Visited by	*Mariner 4* (1965); *Phobos* (1989)

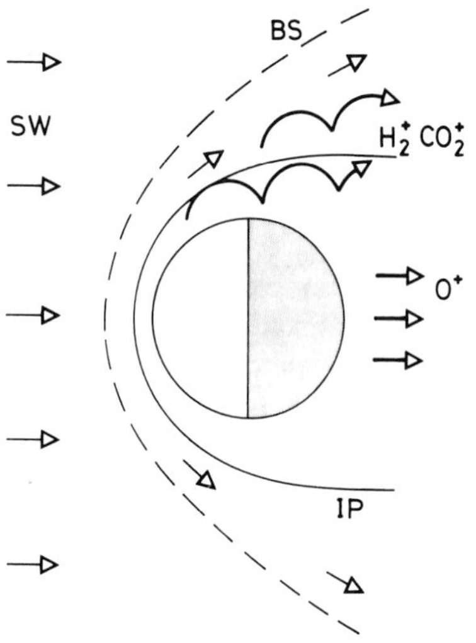

Fig. 9.8. Solar-wind interaction with Mars showing, in a figure symmetric about the planet–sun line, the pick-up and erosion of ionospheric material.

The loading of the solar wind by the implanted ions introduces a bow shock with the now familiar retardation and thermalization of solar-wind protons, strong wave activity and resonant acceleration of electrons, again possibly attributable to lower-hybrid wave turbulence.

9.4.5 *Jupiter*

Magnetosphere

Jupiter's huge magnetic moment, combined with the reduction in solar-wind pressure at the planet's considerable distance from the sun, allow Jupiter, most

Table 9.5. *Jupiter, basic information*

Mean distance from sun	7.8×10^{11} m
Mean planetary radius, R_J	70 000 km
Rotation period	10 h
Magnetic moment	1.4×10^{20} Tm^3
Inclination of magnetic axis	10°
Upstream distance of bow shock	$\geq 100 R_J$
Stand-off distance	60–90 R_J
Visited by	*Pioneer 10* (1973); *Pioneer 11* (1974); *Voyagers 1* and *2* (1979); *Ulysses* (1992); *Galileo* (due 1996)

fittingly, to dominate its environment out to distances of $100R_J$ or more. The ordered, dipolar magnetic field close to the planet gives rise to well defined, high intensity radiation zones within $\sim 20R_J$. Protons acquire energies in excess of 10 MeV, and bidirectional streams of electrons are found up to several tens of kilo-electron-volts. Currents flowing within the magnetosphere exceed 10^9 A. Jupiter is a strong source of radio emission at all frequencies up to several megahertz.

Jupiter's many moons serve as both sources and sinks for the radiation belt particles. In fact, a new moon was actually discovered from the channel it carved in the radiation. Precipitation, as at the Earth, produces aurorae in Jupiter's atmosphere, and there is synchrotron emission from gyrating trapped electrons. The moon Io is of special interest since it is highly volatile through volcanic action, and generates its own plasma torus of sulphur and oxygen ions around the planet. There is also a dynamo action resulting from its motion through the magnetic field, leading to voltages of ~ 400 kV and a power of $\sim 10^{13}$ W. Note here that the constraints set by the moon's physical structure and by the dynamics of its orbital motion serve the role of equivalent constraints in an engineered dynamo. This is unlike the situation sometimes envisaged for MHD dynamos, where the constraints vital to the functioning are often unspecified.

Much of the magnetosphere consists (see Fig. 9.9) of a co-rotating plasmadisc

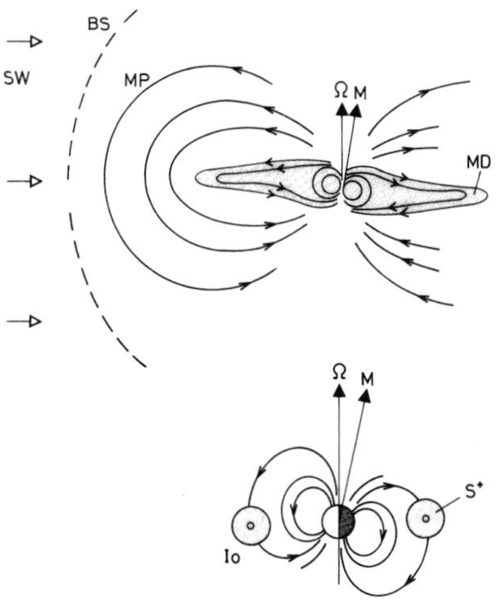

Fig. 9.9. The magnetosphere of Jupiter, showing (shaded in upper figure) the co-rotating magnetodisc, MD, and the Io plasma torus (shaded in lower figure). The increased spacing of solar-wind arrows in this and subsequent figures reflects the reducing density of the solar wind with increasing distance from the sun.

or magnetodisc into which fresh ions from planetary and lunar sources are picked up in just the same way as ions are subsumed into the solar wind. Close to the planet the disc lies perpendicular to the magnetic axis, whereas at great distances centripetal forces tend to align it perpendicular to the rotation axis. Co-rotation continues out to distances $\sim 50\,R_J$, where the velocity reaches the Alfvén speed, when the coupling necessary to the process can no longer be maintained. This giant obstacle generates, of course, a well developed bow shock. Particles accelerated within the magnetosphere escape from the polar regions into the solar wind, electrons being detectable even near the Earth, making Jupiter something of a rival to the sun itself. Precipitation from the ring current and other plasma regions produces aurorae.

9.4.6 *Saturn*

Magnetosphere

Saturn, in complete contrast to Jupiter's majestic dynamism,* has an extremely well ordered magnetosphere (see Fig. 9.10). The orderliness is due to the magnetic moment being almost precisely aligned with the axis of rotation, to the many moons lying in the equatorial plane, and to the strong and extensive fields

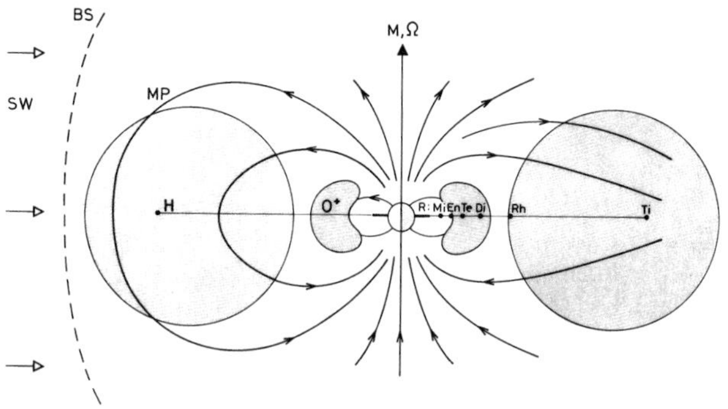

Fig. 9.10. The magnetosphere of Saturn, showing the hydrogen torus surrounding the orbit of Titan, Ti, and the oxygen ion torus associated with Dione, Di, and Tethys, Te. Positions of the rings, Ri, Mimas, Mi, Enceladus, En, and Rhea, Rh, are also shown.

being more than a match for the weakening solar wind at Saturn's great distance from the sun. Cosmic rays are almost totally excluded. Plasma co-rotates with the planet out to some $15R_S$, and there is a ring current composed of H^+ and He^+ from the ionosphere, O^+ from the icy satellites and rings, and N^+ from Titan through the pick-up process. Hydrogen atoms continuously emitted from Titan, but unable to escape from Saturn's gravity, form a giant neutral torus centring on Titan's orbit. The inner magnetosphere, within $3R_S$ is devoid of radiation due to absorption by the rings. The many moons carve neat channels in the radiation zones, and in some instances (Dione and Tethys, for example) sputtered O^+ creates a rich plasma torus. Precipitation from the radiation zones produces visible aurorae. As with the other planets, there is a plasmasheet, extended on the night-side of the planet, composed of ions of both planetary and solar-wind origin.

The magnetosphere contains an important and significant new plasma

*Holst's orchestral suite *The Planets* captures the contrast in character perfectly.

dence now exists (Cowley, 1982) to support the occurrence of both types of processes.

During magnetic field line reconnection at the day-side magnetopause, solar-wind plasmas enter along flux tubes leading to the day-side auroral zone. On the tailward-side, plasma which drifts into the X-type magnetic neutral point, where reconnection is taking place, can be accelerated earthwards or tailwards. Very rarely does the magnetosphere exist in a steady manner for any extended period. The connection patterns occur in bursts, and are associated with a variety of other phenomena on a global scale, particularly the enhancement of the ring current plasma and break-up of the aurorae. The notion of magnetospheric substorms or global perturbations of the magnetosphere is now well established. The southward turning of the interplanetary magnetic field initiates increased geomagnetic activity and magnetospheric substorms. These substorms lead to reconnection of magnetic field lines in the tail producing particle flows towards the Earth, enhancing the ring currents and increasing auroral particle precipitation. Whether reconnection or diffusion plays the major role has still to be determined, although it appears that reconnection is favoured at the moment.

The use of satellites in the exploration of the planetary magnetospheres has yielded a great deal of plasma data, from distribution functions for electrons and ions to wave spectra and magnetic field topology. It seems that we can now measure almost all the parameters needed by theorists to understand the physical processes going on. Most data sets, however, only contain single point measurements; what is really needed is a cluster of data points taken in the same locality at the same time – this will be the aim of *Cluster*, a four-satellite mission to be launched in the mid-1990s. Measurements made in the Earth's magnetosphere are now so precise that they exceed, in some cases, laboratory measurements in resolution and detail, especially in measurements of particle distribution functions. This makes the magnetosphere a unique laboratory for understanding the physics of collisionless plasmas. For a detailed discussion of the Earth's and other planetary magnetospheres see Part I of this chapter. In Part II we will concentrate on some of the processes that operate in the magnetosphere.

9.8 The Earth's bow shock

When the supersonic solar-wind flow meets the Earth's magnetosphere, a shock is launched upstream of the magnetosphere. A shock represents a discontinuous jump in the bulk parameters of the flow. At the shock the solar-wind plasma is slowed down and heated as it passes through the shock layer, transforming a supersonic flow to sub- or trans-sonic flow. The subsonic flow is then deflected around the obstacle, in this case the magnetosphere. The bow shock represents a collisionless shock (the Coulomb mean free path is of the order of 1 AU, i.e. the sun–Earth distance). Collisionless shocks have been proved to exist both in laboratory and in space plasma. The first clear observational evidence for collisionless shocks came from the study of solar flares. In particular, an apparent paradox arose when one considered the sudden onset of magnetic storms caused by the solar flare. The rise time was extremely short (of the order of minutes), which meant that the particles from the flare had to be extremely monoenergetic. Such monoenergetic particle distributions could hardly be 'produced by plasma acceleration process'. T. Gold suggested in 1950 that there would be no paradox if the interplanetary medium could propagate shocks.

The Earth's bow shock falls into three main categories: perpendicular, oblique and parallel. (Perpendicular shocks are those where the shock normal is perpendicular to $\boldsymbol{B}$; parallel shocks occur when the shock normal is parallel to $\boldsymbol{B}$, and oblique shocks form the intermediate state.) All shocks must conserve mass, momentum and energy. These conservation laws fix the shock jump conditions known as the Rankine–Hugoniot relations. Shocks must be dissipative, that is entropy must increase as the fluid traverses the shock. In collisionless shock physics the main problem is in understanding the dissipation mechanism, which must of course be caused by microturbulence. For the Earth's bow shock the relevant mode is the fast magnetosonic wave rather than ordinary sound waves. These waves propagate at a speed v_m given by

$$v_m^2 = \frac{1}{2}(v_A^2 + c_s^2)\left\{1 + \left[1 - \frac{4v_A^2 c_s^2 \cos\theta}{(c_s^2 + v_A^2)^2}\right]^{1/2}\right\} \tag{9.8.1}$$

for perpendicular shocks $\theta = 90°$ and $v_m^2 = (v_A^2 + c_s^2)$ and the Mach number corresponds to the magnetosonic Mach number $M_m = v_{sw}/v_m$, where v_{sw} is the solar-wind speed.

The general structure of a fast collisionless planar shock is shown in Fig. 9.13. In the absence of collisions to provide a deceleration of the upstream solar-wind ions, electric fields must be responsible for their slowing down and deflection. Some ions suffer reflection from the sharp gradient and gyrate back into the upstream region before passing downstream. These reflected ions sometimes cause a broad 'foot' in the magnetic profile about an ion gyro-radius in width.

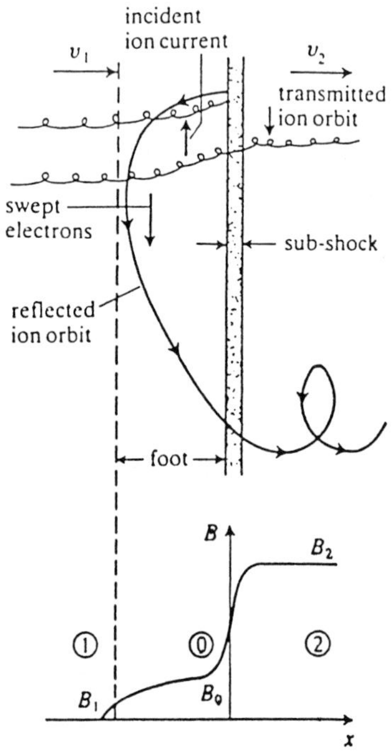

Fig. 9.13. Magnetic field and structure of a planar perpendicular shock.

The partial reflection provides the shock with two characteristic length scales for the fast magnetosonic shock: the ion Larmor radius $r_i = v_{sw}/\omega_{ci}$ associated with the solar-wind speed (not thermal speed), and the ion inertial length c/ω_{pi} corresponding to the thickness of the current layer; ω_{ci} and ω_{pi} are the ion cyclotron and plasma frequencies, respectively. The first length is associated with the foot and the second with the ramp or current layer. These have values of about 1000 km and 100 km for the case of the Earth's bow shock. In the derivation of the shock profile and electric fields responsible for subsequent ion and electron motion in planar perpendicular shocks, it is sufficient to use a two-fluid model. The fluid description will be adequate to describe the gross features, but will be unable to provide the necessary dissipation, which we will treat by adding an effective damping to the final equations. Following Tidman and Krall (1971) we assume that the shock is represented by a step function in magnetic field and consider a planar perpendicular shock with B in the z-direction and the upstream velocity v_1 in the x-direction. We seek constant profile waves in the wave frame and thus set the time derivatives to zero and assume variations in the x-direction only. The fluid equations for each species are then

$$\frac{\partial}{\partial x}(n_\alpha \boldsymbol{v}_\alpha) = 0 \tag{9.8.2}$$

$$v_\alpha \frac{\partial}{\partial x} \boldsymbol{v}_\alpha = \frac{q_\alpha}{m_\alpha}(\boldsymbol{E} + \boldsymbol{v}_\alpha \times \boldsymbol{B}); \tag{9.8.3}$$

α is the species index 'i' or 'e'.

From Maxwell's equations we have

$$\boldsymbol{\nabla} \times \boldsymbol{B} = \mu_0 \boldsymbol{J}, \tag{9.8.4}$$

$$\boldsymbol{\nabla} \times \boldsymbol{E} + \frac{\partial \boldsymbol{B}}{\partial t} = 0, \tag{9.8.5}$$

$$\boldsymbol{\nabla} \cdot \boldsymbol{B} = 0. \tag{9.8.6}$$

Note that we have neglected the displacement current in Eq. (9.8.4); this is possible since we are dealing with low frequency phenomena. From Eq. (9.8.2) together with the quasineutrality condition $n_e \simeq n_i$ we get the results

$$v_{ix} = v_{ex} = v_x \tag{9.8.7}$$

and

$$n = \frac{n_1 v_1}{v_x}, \tag{9.8.8}$$

where v_1, n_1 are the upstream conditions. Taking the y-component of the momentum equation for both species and using result Eq. (9.8.7) we get

$$v_x \frac{\mathrm{d}v_{ey}}{\mathrm{d}x} = -\frac{e}{m_e} E_y + \frac{e}{m_e} v_x B_z, \tag{9.8.9}$$

$$v_x \frac{\mathrm{d}v_{iy}}{\mathrm{d}x} = \frac{e}{m_i} E_y - \frac{e}{m_i} v_x B_z. \tag{9.8.10}$$

Adding Eqs (9.8.9) and (9.8.10) we get the result

$$v_{iy} = -\frac{m_e}{m_i} v_{ey}. \tag{9.8.11}$$

From this equation we can see that the ion velocity in the y-direction is much smaller than the electron velocity. This velocity difference in the y-direction results in a large transverse current J_y, which is responsible for supporting the increase in B_z.

Subtracting Eq. (9.8.10) from Eq. (9.8.9) results in

$$E_y = v_x B_z, \tag{9.8.12}$$

which is the convective electric field in the rest frame of the shock. Next, using the x-component of Eq. (9.8.5), i.e.

$$v_x \frac{dv_x}{dx} = -\frac{e}{m_e} E_x - \frac{e}{m_e} v_{ey} B_z, \tag{9.8.13}$$

$$v_x \frac{dv_x}{dx} = \frac{e}{m_i} E_x + \frac{e}{m_i} v_{iy} B_z, \tag{9.8.14}$$

and subtracting Eq. (9.8.13) from Eq. (9.8.14) results in

$$E_x = -v_{ey} B_z. \tag{9.8.15}$$

This states that the electrons are in force balance in the x-direction. Substituting E_x from Eq. (9.8.15) in Eq. (9.8.14) and using Eq. (9.8.11) results in

$$v_x \frac{dv_x}{dx} = -\frac{eB_z}{m_i} v_{ey}. \tag{9.8.16}$$

This shows that the Lorentz force on the electrons $(-ev_{ey}B_z)$ is transferred to the ions through the electric field. The y-component of Ampère's law gives us an equation for v_{ey}:

$$v_{ey} = \frac{1}{\mu_0 en} \frac{dB_z}{dx}, \tag{9.8.17}$$

and from Eqs (9.8.17) and (9.8.16) it gives

$$nv_x \frac{dv_x}{dx} = -\frac{B_z}{\mu_0 m_i} \frac{dB_z}{dx}. \tag{9.8.18}$$

Since the density n also changes appreciably across the shock region, it cannot be considered as a constant, and its variation across the shock region is important in obtaining the shock jump conditions. From the continuity equation we find that nv_x is a constant, i.e. the flux is the same in the shock region as in the solar wind:

$$nv_x = [nv_x]_{sw} = [nv_x]_s = \text{const}. \tag{9.8.19}$$

Using this condition in Eq. (9.8.18) finally gives us a jump condition for the magnetic field in the shock as

$$[m_i nv_x^2]_s = -\frac{1}{2\mu_0}([B_z]_s^2 - [B_z]_{sw}^2) + [m_i nv_x^2]_{sw}. \tag{9.8.20}$$

This equation makes sense since, in the shock region, the solar wind slows down whilst the magnetic pressure increases. We now use Eqs (9.8.15) and (9.8.17) to derive the electric field responsible for slowing the solar wind, resulting in

$$E_x = -\frac{1}{2\mu_0 en}\frac{\mathrm{d}B_z^2}{\mathrm{d}x}. \tag{9.8.21}$$

The electric field along the x-direction points in a direction opposite to the gradient of the magnetic pressure, i.e. this component of the field, which we refer to as the ambipolar field, is proportional to the Maxwell stress tensor. This field E_x is responsible for slowing the ions in the solar wind. It can easily be demonstrated that if the shock reflects solar-wind ions then

$$eE_x \approx \frac{1}{2}\frac{m_i v_{sw}^2}{L}, \tag{9.8.22}$$

where L is the shock thickness.

Substituting Eq. (9.8.17) into Eq. (9.8.16) and integrating we obtain

$$v_x = -\frac{1}{\lambda}(B^2 - B_{sw}^2) + v_{sw}, \tag{9.8.23}$$

where $\lambda = 2n_1 v_1 \mu_0 m_i$; note $v_1 = v_{sw}$. Finally, to obtain an equation for the shock profile we have to obtain an equation involving B_z only. Multiplying Eq. (9.8.17) by $v_x(\mathrm{d}/\mathrm{d}x)$ gives us

$$v_x \frac{\mathrm{d}}{\mathrm{d}x}\left(v_x \frac{\mathrm{d}B_z}{\mathrm{d}x}\right) = \mu_0 n_1 v_1 e v_x \frac{\mathrm{d}v_{ey}}{\mathrm{d}x}, \tag{9.8.24}$$

and using Eq. (9.8.8) this becomes

$$v_x \frac{\mathrm{d}}{\mathrm{d}x}\left(v_x \frac{\mathrm{d}B_z}{\mathrm{d}x}\right) = \frac{\mu_0 n_1 v_1 e^2}{m_e}(B_z v_x - E_y), \tag{9.8.25}$$

but we find that from Eq. (9.8.23) v_x is a function of B_z only, and using $E_y = v_1 B_1$ results in an equation for B_z only, namely

$$f(B_z)\frac{\mathrm{d}}{\mathrm{d}x}\left(f(B_z)\frac{\mathrm{d}B_z}{\mathrm{d}z}\right) = g(B), \tag{9.8.26}$$

where $f(B_z) = 1 - (B_z^2 - B_1^2)/\lambda v_1$ and

$$g(B_z) = \frac{\omega_{pe}^2}{c^2}(B_z - B_1)\left(1 - \frac{B_z(B_z + B_1)}{\lambda v_1}\right).$$

Eq. (9.8.26) can be integrated once by multiplying it by $\mathrm{d}B_z/\mathrm{d}x$, resulting in

$$\frac{1}{2}\left(\frac{\mathrm{d}B_z}{\mathrm{d}x}\right)^2 + \Phi(B_z) = 0, \tag{9.8.27}$$

where

$$\Phi(B_z) = \frac{-\left(\frac{\mathrm{d}B_z}{\mathrm{d}x}\right)_1^2 - \frac{\omega_{pe}^2}{c^2}(B_z - B_1)^2\left(1 - \frac{(B_z + B_1)^2}{2v_1\lambda}\right)}{2\left(1 - \frac{(B_z^2 - B_1^2)}{\lambda v_1}\right)^2}$$

and $(\mathrm{d}B_z/\mathrm{d}x)_1$ is evaluated upstream where $B_z = B_1$.

It is instructive to note that Eq. (9.8.27) has the form for particle motion in a potential well Φ (B_z plays the role of particle coordinate and x replaces time). It can be formally integrated as

$$x = \pm \int \frac{dB_z}{[-2\Phi(B_z)]^{1/2}} + \text{const.} \tag{9.8.28}$$

Part of the function Φ is sketched in Fig. 9.14.

By analogy with the particle in the well problem, it is evident that periodic solutions exist corresponding to infinite nonlinear wavetrain solutions for B_z,

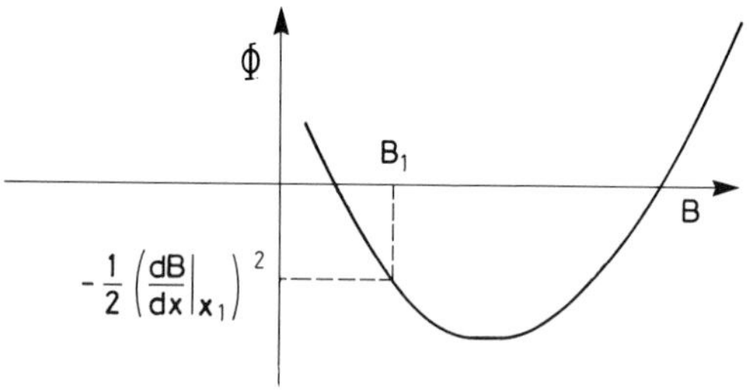

Fig. 9.14. Potential function $\Phi(B)$ for a magnetosonic wave. $\Phi(B)$ is sometimes referred to as the Sagdeev potential.

these solutions correspond to 'reflections' at the 'walls' of the well.

The case of interest is when $(dB_z/dx)_1 = 0$ so that $\Phi(B_z)$ has the form shown in Fig. 9.15. By considering a particle in the classical well incident from C we see it will be reflected at point D but will never be reflected again. This corresponds to a solitary pulse since it undergoes one 'infinite period' (see Fig. 9.16), oscillating from B_1 to B_m and back to B_1. It is clear that shock-like solutions do not exist in Eq. (9.8.27). One important ingredient missing in Eq. (9.8.27) is a dissipative term. We can go back to the original fluid momentum equations and include an anomalous collision term $\nu_e v_e$ due to microturbulence, where ν_e is the 'collision' frequency obtained from plasma kinetic theory.

Following a similar procedure as before, Eq. (9.8.25) can be written as

$$v_x \frac{d}{dx}\left(v_x \frac{dB_z}{dx}\right) = \mu_0 e n_1 v_1 \left(\frac{-eE_y}{m_e} + \frac{ev_x B}{m_e} - \nu_e v_{ey}\right) \tag{9.8.29}$$

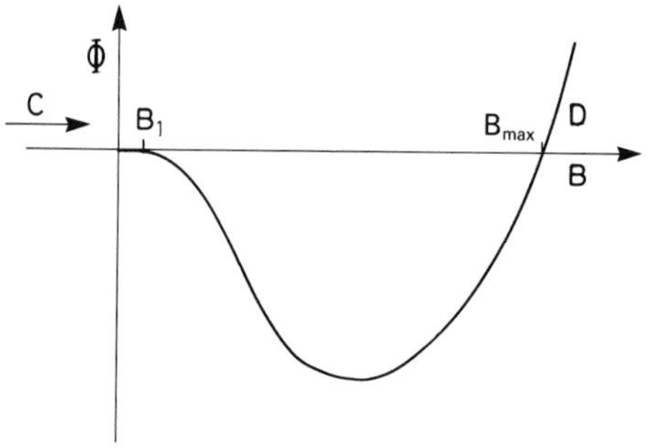

Fig. 9.15. Potential function $\Phi(B)$ for a magnetosonic soliton.

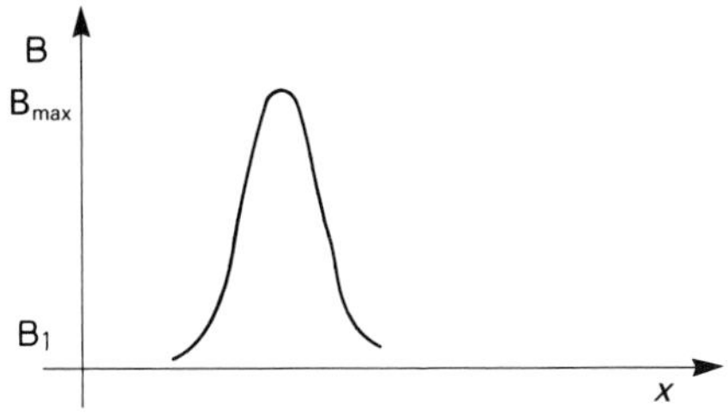

Fig. 9.16. Solitary pulse or soliton.

and the equation for B_z reduces to

$$\frac{v_x}{v_1}\frac{\mathrm{d}}{\mathrm{d}x}\left(\frac{v_x}{v_1}\frac{\mathrm{d}B_z}{\mathrm{d}x}\right) = \frac{\omega_{\mathrm{pe}}^2}{c^2}(B_z - B_1)\left(1 - \frac{B_z(B_z + B_1)}{\lambda v_1}\right) - \frac{\nu_{\mathrm{e}}}{v_1^2}v_x\frac{\mathrm{d}B_z}{\mathrm{d}x}. \tag{9.8.30}$$

Defining a new variable τ by $\mathrm{d}x/\mathrm{d}\tau = v_x/v_1$, Eq. (9.8.30) becomes

$$\frac{\mathrm{d}^2B_z}{\mathrm{d}\tau^2} = -\frac{\partial\Phi}{\partial B_z} - \frac{\nu_{\mathrm{e}}}{v_1}\frac{\mathrm{d}B_z}{\mathrm{d}\tau}, \tag{9.8.31}$$

where the potential Φ has the form

$$\Phi(B_z) = \frac{\omega_{\mathrm{pe}}^2}{c^2}\frac{(B_z - B_1)^2}{2}\left(\frac{(B_z + B_1)^2}{2\lambda v_1} - 1\right). \tag{9.8.32}$$

Eq. (9.8.31) is in the form of a standard potential well problem with damping present. The maximum value of B_z that can occur in the well is denoted by $B_z = B_{\mathrm{max}}$ and occurs when $\Phi = 0$, i.e.

$$B_{\mathrm{max}} = -B_1 + 2v_1(n_1 m_{\mathrm{i}}\mu_0). \tag{9.8.33}$$

The motion of a particle in a potential well with damping is indicated schematically by the dashed line in Fig. 9.17, and Fig. 9.18 shows the shape of B as a function of τ, and a shock-like solution for B_z exists for a finite value of ν_{e}. The minimum value of Φ occurs when $B_z = B_2$, where

$$B_2 = (\lambda v_1 + B_1^2/4) - B_1/2. \tag{9.8.34}$$

As dissipation increases, i.e. $\nu_{\mathrm{e}} \to \infty$, B_{m} approaches B_2; there is also no upper bound for M_{m}, the mach number for shocks to exist. We have shown that the inclusion of dissipation turns a solitary pulse solution into shock-like solutions. It is important to note that we neglected the pressure term in the fluid equations and did not allow the plasma to heat up. To treat the problem correctly we would need to solve the exact Vlasov–Maxwell equations. It should be noted that the inclusion of plasma pressure results in the following expression for E_x:

$$E_x = -\frac{1}{ne}\frac{\mathrm{d}P_{\mathrm{e}}}{\mathrm{d}x} - \frac{1}{2\mu_0 en}\frac{\mathrm{d}|B_z|^2}{\mathrm{d}x}, \tag{9.8.35}$$

and therefore the potential change $\Delta\Phi$ acting on the solar-wind ions is given by

$$\Delta\Phi = \frac{1}{en}\Delta nkT_{\mathrm{e}} + \frac{1}{2\mu_0 en}\Delta|B|^2. \tag{9.8.36}$$

For typical bow shock conditions the potential change is of the order of 1 kV.

The particle dynamics at collisionless shocks is extremely complicated; some electrons can be accelerated out of the shock and travel upstream away from the shock, forming beams or high energy tails in the solar wind. The acceleration is thought to be due to electrostatic turbulence generated by cross-field streaming instabilities, such as the modified two-stream instability or anisotropic ion velocity distributions, i.e. ring distributions. These backstreaming electrons form the electron foreshock in the solar wind (Fig. 9.19). The backstreaming ions form the ion foreshock region, where low frequency ion–acoustic-like waves are observed. The ion distribution functions formed in the solar wind by backstream-

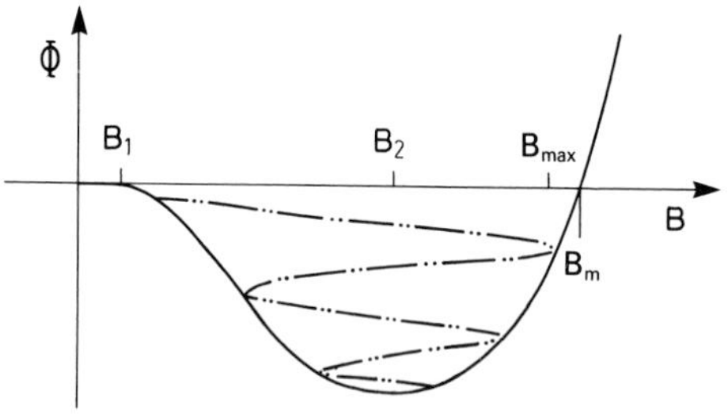

Fig. 9.17. The damped motion of a 'particle' in a potential well $\Phi(B)$.

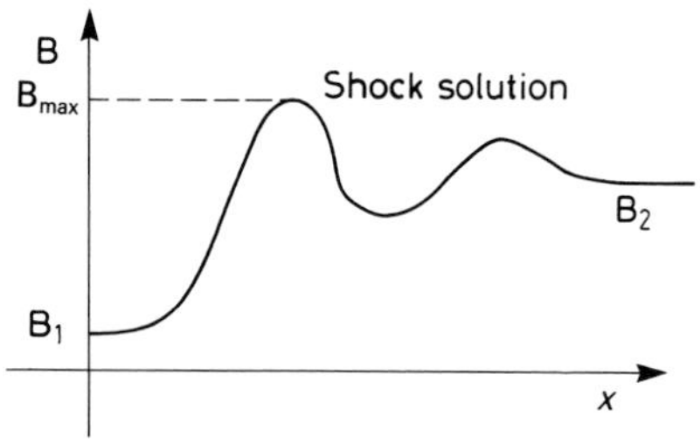

Fig. 9.18. Variation of the magnetic field through a magnetosonic shock.

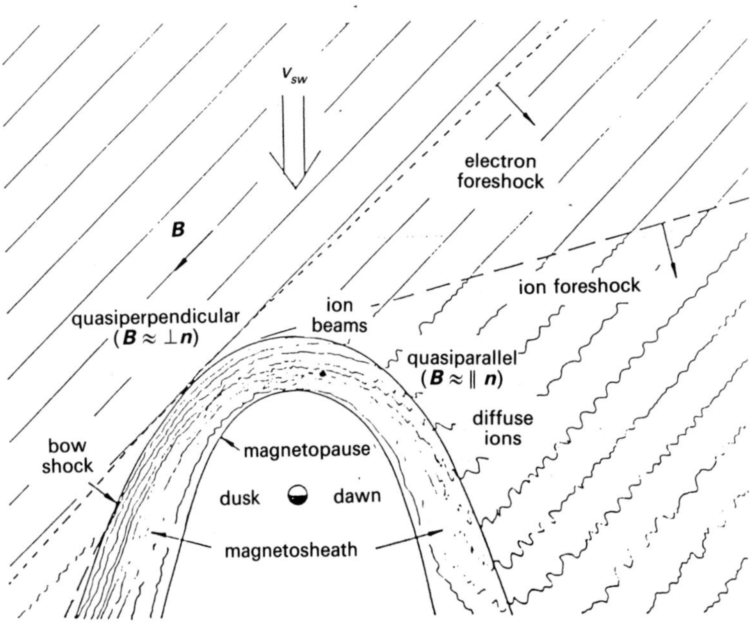

Fig. 9.19. The electron and ion foreshock regions upstream of the bow shock.

ing ions from perpendicular shocks form ring distributions in velocity space centred on the magnetic field.

9.9 Upstream waves

There is a great deal of interest in the Earth's foreshock region, where there is a profusion of particle and wave phenomena, which provide ideal tests for a great number of wave–particle interaction processes. Similar waves and particle distributions are seen upstream of other planets and also in association with interplanetary shocks. Astrophysicists are particularly interested in the results of research into acceleration processes in the upstream region, as it may provide insight into the physics of similar processes upstream of interplanetary and interstellar shocks.

Observations of backstreaming waves and electrons obtained by satellite are shown in Fig. 9.20. There are four main periods of strong wave activity in the frequency range 40–70 kHz, lasting for about 1 h each, and there is a strong line which runs through each of these periods which has been assumed to correspond to emission at the local electron plasma frequency. This line indicates that the average value of the plasma frequency is about 50 kHz, corresponding to an electron density of about 30 cm^{-3}. This is rather higher than the typical electron density in this region, which is about 10 cm^{-3} giving a plasma frequency of $f_{pe} \simeq 30$ kHz.

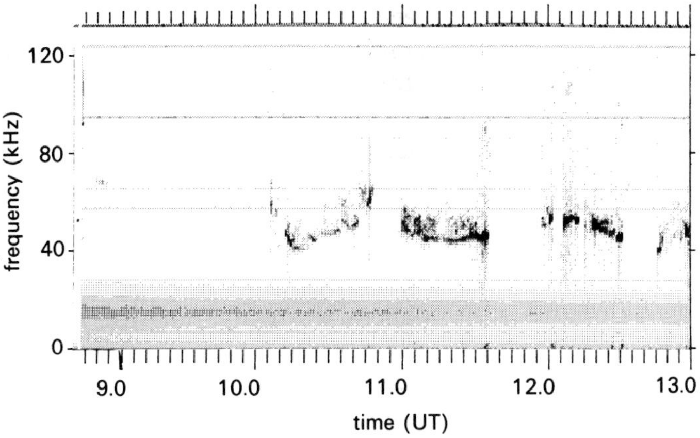

Fig. 9.20. Grey-scale plot against time of the intensity of the electric component of wave activity at different frequencies, measured by the UKS wave experiment. The data have been averaged over 60 s, and certain horizontal lines, such as the one at 58 kHz, are the result of residual interference, as is the band of noise a few kilohertz wide at 15 kHz. (Courtesy of A. Derbyshire.)

At the times when the high frequency electrostatic waves were observed, enhanced electron fluxes were simultaneously detected moving parallel to the magnetic field in the direction away from the bow shock and towards the sun. By comparing electron distributions corresponding to times when there are waves and times when there are no waves observed, it can be seen that the only electron component which correlates positively with the enhanced wave activity is the one corresponding to electrons backstreaming from the bow shock parallel to the magnetic field direction (this component will be referred to as the parallel component from now on).

Normally in the solar wind the electron distribution function is anisotropic with $T_{\parallel} > T_{\perp}$. Within the foreshock, when the waves are observed, the electron distribution function becomes isotropic except for a small bump or plateau, which appears frequently in the tail of the parallel component. The small bump or plateau in the parallel component corresponds to an increased flux of electrons with energies in the range 200–1000 eV and a temperature of about 50 eV, backstreaming from the shock.

Measurements of the heat flux show that it is predominantly directed away from the sun, but that the enhanced flux of backstreaming electrons can reduce the net heat flux to zero or give rise to a flux in the direction towards the sun. It is generally accepted that the energy flux in the interplanetary medium flows along the magnetic field, directed away from the sun; however, when the local magnetic field connects to the shock, the net flux is towards the sun, the change being brought about by the backstreaming particles.

In summary, the data indicate that the observed increase in intensity of electrons moving away from the shock in the magnetic field direction is responsible for the generation of the observed electrostatic plasma oscillations, and in particular there is a correlation with the low energy electrons with energies in the range 400–500 eV. The electron intensity is highest at the foreshock boundary, as would be predicted by the standard model of the foreshock, and it is here that the plasma line is at its most thin and intense.

In this section we use a Vlasov stability analysis to show that the streaming interaction between the backstreaming electrons and the solar wind could generate waves in the observed frequency range. The observed solar wind consists of a slow-moving cold Maxwellian component with a temperature of 10 eV and a fast hot Maxwellian component with a temperature of 200 eV both streaming towards the shock with a solar-wind velocity of about 445 km s^{-1} and a total density of 30 cm^{-3}, the hot population being 10% of the total.

We include an electron beam population backstreaming from the shock along the magnetic field direction with a beam energy of 400 eV ($v_{\mathrm{db}} \simeq 12\,000$ km s^{-1}). The reduced particle distribution is sketched in Fig. 9.21. The dispersion relation of such a plasma describing electrostatic waves propagating parallel to the magnetic field is given by Eq. (9.14.23) (see Appendix 9.14; Eq. (9.14.23)):

$$1 - \sum_{\alpha} \frac{1}{2k_{\parallel}^2 \lambda_{\mathrm{D}\alpha}^2} Z'\left(\frac{\omega}{2^{1/2} k_{\parallel} v_{\mathrm{th}\alpha}} - \frac{v_{\mathrm{d}\alpha\parallel}}{2^{1/2} v_{\mathrm{th}\alpha}}\right) = 0, \qquad (9.14.23)$$

where α denotes the species (as before, we shall denote the beam component by the subscript b, the hot solar wind electrons by subscript h and the cold solar wind electrons by subscript c'), and $v_{\mathrm{d}\alpha\parallel}$ is the drift velocity parallel to the magnetic field.

Substituting the plasma parameters into this dispersion relation and solving it numerically, we obtain solutions corresponding to growing electrostatic waves. Fig. 9.22 shows plots of the normalized frequency $\omega_{\mathrm{r}}/\omega_{\mathrm{pe}}$ and growth rate $\gamma/\omega_{\mathrm{pe}}$, against $k\lambda_{\mathrm{Dc}}$, for three values of n_{b}/n_0, where n_{b} is the beam density, n_0 is the total density of all the electrons, i.e. $n_0 = n_{\mathrm{h}} + n_{\mathrm{c}} + n_{\mathrm{b}}$; $\omega_{\mathrm{pe}} = (n_0 e^2/m\varepsilon_0)$, and $k\lambda_{\mathrm{Dc}} = (\varepsilon_0 k T_{\mathrm{c}}/n_{\mathrm{c}} e^2)$. The growth rate has a positive value for $n_{\mathrm{b}}/n_0 \gtrsim 0.02$. The waves grow over a range of frequencies lying just below the plasma frequency ω_{pe}. As the beam density is increased, the bandwidth of the growing waves increases, which is in agreement with the observations, the frequency of the fastest growing

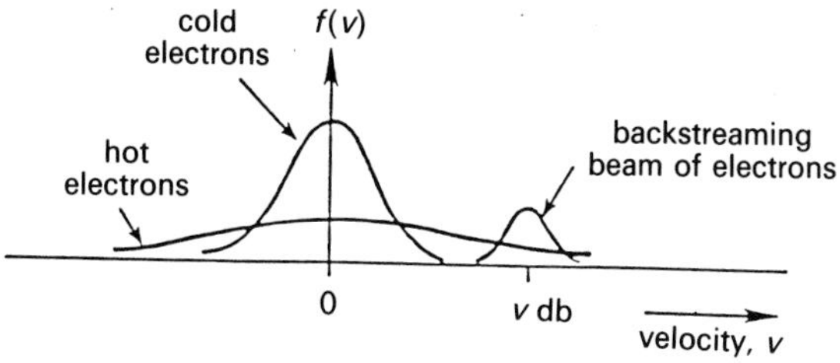

Fig. 9.21. Sketch of the one-dimensional electron distribution function used to model the observed distribution.

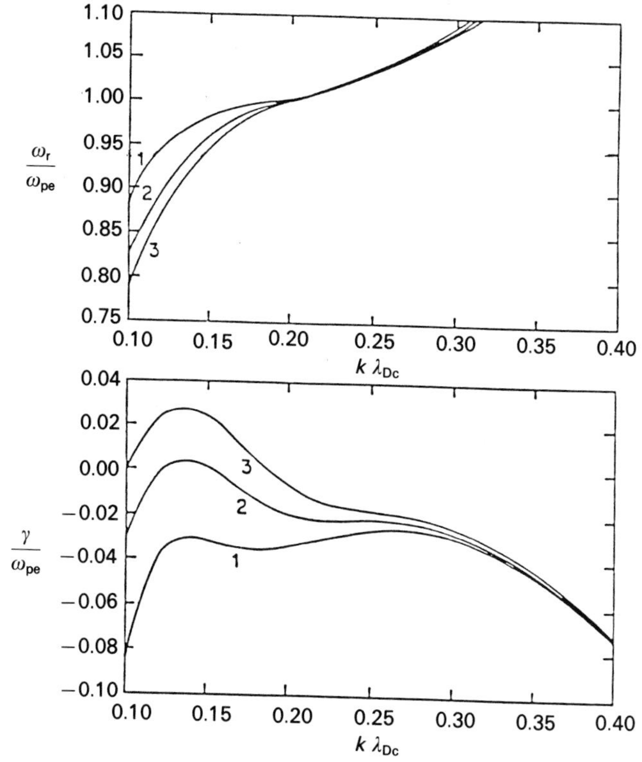

Fig. 9.22. The complex frequency as a function of wave number for three values of the beam density. The labels 1, 2 and 3 correspond to $n_b/n_0 = 0.01$, 0.02 and 0.03, respectively. $n_0 = 30\,\mathrm{cm}^{-3}$, $n_h = 3\,\mathrm{cm}^{-3}$, $T_e = 10\,\mathrm{eV}$, $T_h = 200\,\mathrm{eV}$, $T_b = 50\,\mathrm{eV}$, $v_{db} = 12\,000\,\mathrm{km\,s}^{-1}$.

wave decreases, and the value of the maximum growth rate also increases. The Doppler shift in frequency if we transform back into the spacecraft frame is about 1%, which is small enough that we are justified in comparing the results of the analysis with the experimental results.

Fig. 9.23 shows the normalized growth rate γ plotted as a function of the normalized frequency for $n_b/n_0 = 0.02$ and 0.03, which again illustrates the increase in unstable frequencies and the increase in growth rate with increased beam density. For $n_b = 0.02$, the maximum value of the growth rate, $\gamma_{max} \approx 0.004\omega_{pe}$ and the bandwidth of the growing waves is about $0.06\omega_{pe}$, while $n_b/n_0 = 0.03$ corresponds to $\gamma_{max} \approx 0.027\omega_{pe}$ and a bandwidth of about $0.2\omega_{pe}$. Using $n_0 = 30\,\mathrm{cm}^{-3}$, $\omega_{pe} = 49\,\mathrm{kHz}$, and these growth rates and bandwidths correspond to the values shown in Table 9.10. The observed bandwidth varies, and as can be seen from Table 9.10 this would correspond to a beam density of $0.03n_0$. Fig. 9.24 is a plot of the real and imaginary frequencies as a function of $k\lambda_{Dc}$. The solid line is the electron beam mode and the dashed curve is the

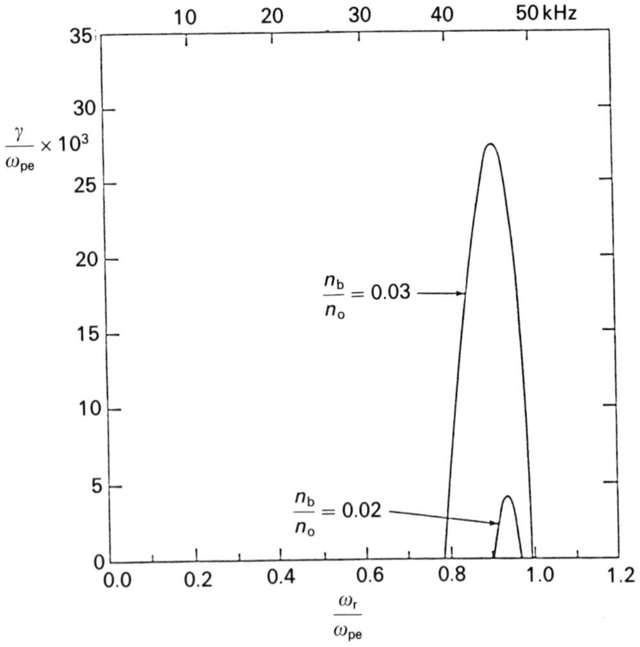

Fig. 9.23. Growth rate as a function of the real part of the frequency for two values of the beam density. $n_0 = 30\,\mathrm{cm}^{-3}$, $n_h = 3\,\mathrm{cm}^{-3}$, $T_e = 10\,\mathrm{eV}$, $T_h = 200\,\mathrm{eV}$, $T_b = 50\,\mathrm{eV}$, $v_{db} = 12\,000\,\mathrm{km\,s}^{-1}$.

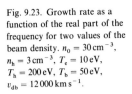

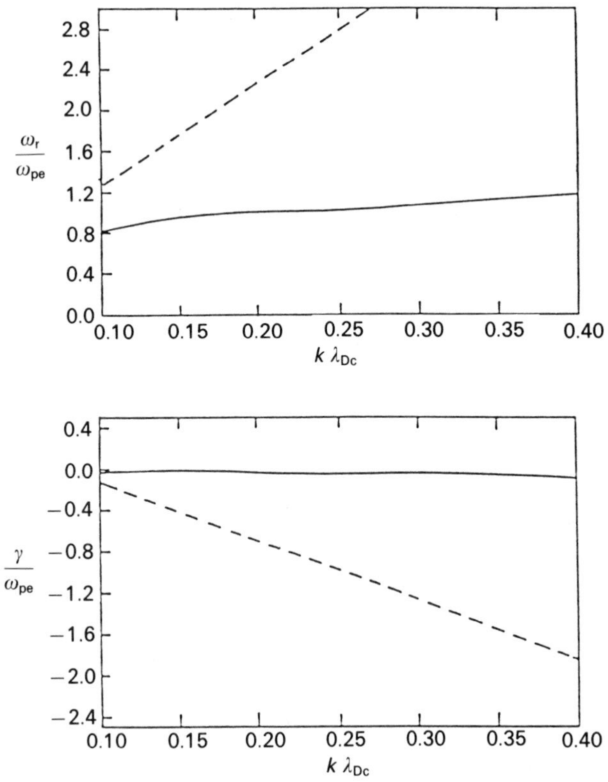

Fig. 9.24. The complex frequency as a function of wave number for the $n_b = 0.02n_0$ case. The solid curves show solutions to the dispersion relation, and the dashed curves represent another solution which is heavily damped. $n_0 = 30\,\mathrm{cm}^{-3}$, $n_h = 3\,\mathrm{cm}^{-3}$, $T_e = 10\,\mathrm{eV}$, $T_h = 200\,\mathrm{eV}$, $T_b = 50\,\mathrm{eV}$, $v_{db} = 12\,000\,\mathrm{km\,s}^{-1}$.

Table 9.10.

n_b/n_0	γ_{max}, kHz	Bandwidth, kHz
0.02	0.2	3
0.03	1.3	10

Langmuir-like wave generated by the Langmuir beam instability. The analysis shows that the electron population backstreaming from the bow shock can only generate the waves if the density of beam electrons is $\geq 0.02n_0$, i.e. 2% of background, and to explain the observed bandwidth it would have to be around 3%. The observed distributions frequently, but not always, show evidence of a bump at the times when the waves are seen. The growing waves belong to the electron beam branch, whereas the Langmuir mode is heavily damped due to enhanced Landau damping, which is the result of the electron tail.

9.10 Particle acceleration

The study of acceleration processes remains one of the most important areas of research in both laboratory and space plasmas. The subject is also vast, making it almost impossible to cover all aspects. We will concentrate here on particle acceleration which occurs in both space and laboratory plasmas.

We begin by defining what we mean by acceleration in plasmas, which can be summed up by the simple equation

$$\Delta T = q \int_0^{\Delta t} \boldsymbol{v}_0(t) \cdot \boldsymbol{E}(\boldsymbol{r}_0(t), t) \mathrm{d}t, \tag{9.10.1}$$

where $\boldsymbol{E}$ is the electric field determined by currents and charges in the plasma, ΔT is the change in the particle kinetic energy, $\boldsymbol{v}_0$ is the particle velocity, and q is the particle charge. A deterministic system is defined as one where $\boldsymbol{E}(\boldsymbol{r}, t)$ does not change significantly during the acceleration phase, whereas for a stochastic system $\boldsymbol{E}(\boldsymbol{r}, t)$ changes in a random fashion. We will only be concerned in this chapter with those acceleration processes which result in energetic particles which have been appreciably accelerated rather than processes that lead to bulk plasma heating. The processes considered are distinguished by the formation of an energetic tail or 'runaways' in the particle distribution function. Such energetic particle distributions are prevalent in all plasmas, examples being:

- cosmic rays, e^-, p^+, He^+, heavy ions, 10^7–10^{22} eV
- solar flares, e^-, p^+, 10^5–10^9 eV
- aurorae in planetary magnetospheres, e^-, p^+, 10^4–10^6 eV
- runaways in tokamaks, e^-, 10^5–10^6 eV
- laser fusion, e^-, 10^5–10^6 eV.

Although the above plasma environments and scales are vastly different, a general trend appears, namely a sizeable fraction of the free energy goes into a small fraction of the available particles. The result is the formation of nonthermal tails in the distribution function. The fundamental question therefore arises of how the particles are raised from the thermal level to higher energies. In space plasmas they are usually preceded by violent events, with the energy coming from magnetic fields in, for example, magnetospheric substorms, solar flares, supernovae, etc. In laboratory plasmas the accelerated particles can be produced by injecting high power electromagnetic radiation into the plasmas as, for example, in radio-frequency current drive in tokamaks or laser fusion.

We will concentrate on stochastic acceleration processes that occur in magnetospheric plasmas. Although they are minute in energy content relative to astrophysical plasmas, the magnetospheres of the planets are essential links in the chain of studies that connect plasmas in the laboratory and cosmic plasmas. They and the solar wind form the largest plasmas that can be studied *in situ*.

Although coherent wave acceleration processes are important in laboratory plasmas, where large coherent fields such as lasers or microwaves can easily be generated, it is rather doubtful that they play a major role in space plasmas, where the majority of waves are observed to be random turbulent fluctuations. This, however, may not be the case in pulsars where very large amplitude coherent fluctuations have been predicted (Gunn and Ostriker, 1969). There are exceptions also in space plasmas where coherent waves have been detected, an example being coherent electrostatic ion cyclotron waves on auroral field lines observed by the s3-3 polar orbiting satellite. These waves may be responsible for perpendicularly heating the ions by wave trapping. Some of the perpendicular energy can then be converted into parallel energy by the dipole mirror forces of the Earth.

Stochastic acceleration by turbulent wave fields is an efficient mechanism for transferring plasma turbulence to fast particles, increasing their energy even further. Charged particles interacting with the turbulent fields gain a small amount of energy during each collision.' In astrophysics the best known stochastic acceleration mechanism is Fermi acceleration. Fermi suggested that cosmic rays could collide' with moving magnetized clouds. The collision is effective via the magnetic mirror effect and the particle suffers an energy gain or loss proportional to the velocity of the cloud and the energy of the particle, as illustrated in Fig. 9.25.

For relativistic particles, the average gain is $\Delta T = \pm vT/c$, where T is the kinetic energy of the particle and v is the speed of the magnetic cloud; the sign of the increment depends on the relative direction of motion between the cloud and particle, being positive if the cloud and particle are moving towards each other, and negative if not . Statistically head-on collisions predominate with particles gaining energy on average. In fact $dT/dt = n(v/c)^2 T$, where n is the number of

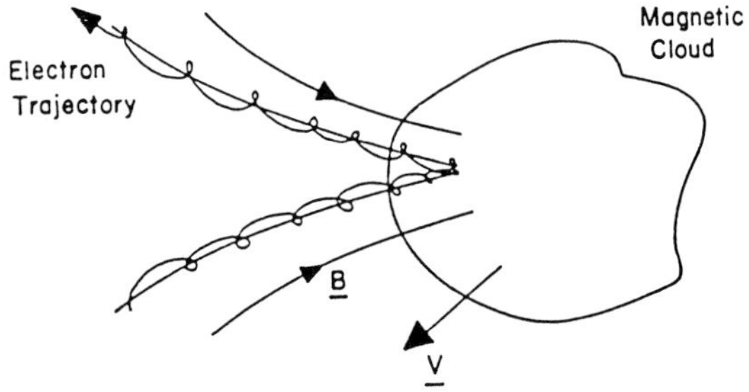

Fig. 9.25. Fermi acceleration of charged particle colliding with a moving magnetic cloud.

collisions per unit time. If the particles are lost at a rate $1/\tau$, the particles then take on a power law distribution. If f is the number of particles in range dT then

$$\frac{df}{dt} = -\frac{\partial}{\partial T}(\alpha T f) - \frac{f}{\tau}, \quad \alpha = \left(\frac{v}{c}\right)^2 n, \tag{9.10.2}$$

and the steady state solution is

$$f = \text{const.} \times T^{-(1+/\alpha\tau)}.$$

This expression fits the data extremely well, with a coefficient between 2 and 3. A number of things about Fermi acceleration are worth noting since they are common to other stochastic acceleration processes:

(a) The acceleration takes place in very small steps with many collisions off different clouds to reach the final energy.

(b) The particles are accelerated in a nonisotropic fashion – energy being put into the parallel direction.

(c) Energy distributed among a large number of particles is redistributed among a few of very high energy. Fermi acceleration illustrates the mechanism of stochastic acceleration in general.

In laboratory plasmas wave–particle interactions play a fundamental role in transferring energy and momentum, accelerating particles through stochastic processes. In magnetic fusion devices radio-frequency heating is regarded as a major area of research.

In the sun, solar-wind and planetary magnetospheres, acceleration and heating by plasma turbulence is of fundamental importance. Phenomena such as aurorae, X-ray bursts in solar flares, and radio-wave generation, producing for example auroral kilometric radiation in the magnetosphere, rely on wave–particle interaction processes where particles are stochastically accelerated by the broad spectrum of turbulent waves. These wave–particle interactions can take place through two main mechanisms: Landau resonance or gyro-resonance. In Landau resonance, sometimes called a Cherenkov process, the longitudinal electric field component of the wave interacts with the particle and changes its parallel energy. The acceleration through absorption of longitudinal plasma waves is by no means the only mechanism by which particles can accumulate energy. Absorption through the cyclotron mechanism or gyro-resonance increases the velocity component at right angles to the magnetic field. We will concentrate on the acceleration by longitudinal fields since we believe it is the one responsible for beam production on auroral field lines. In particular we consider the stochastic acceleration of electrons parallel to the magnetic field by lower-hybrid turbulence.

In the laboratory, waves are created by mode converting radio-frequency power in the plasma. However, in the magnetospheric plasmas, the waves must be generated by an instability. The nature of the instability depends on the origin of the free energy (in a nonequilibrium plasma). In the magnetosphere, nonequilibrium plasma states are created during magnetospheric substorms resulting in bulk plasma motion and the creation of shocks and steep gradients; such conditions exist on auroral field lines and can trigger either the modified two-stream instability or the lower-hybrid drift instability. In either case, the fluctuating fields are primarily electrostatic with $k_\perp \gg k_\parallel$, where $k_{\perp,\parallel}$ are the perpendicular and parallel wave-number components, respectively. These waves are lower-hybrid waves with frequencies $\omega_{\rm lh}^2 = \omega_{\rm pi}^2/(1 + \omega_{\rm pe}^2/\omega_{\rm ce}^2)$. Lower-hybrid waves are seen with large amplitudes in auroral regions of the magnetosphere, the amplitudes being large enough to produce strong turbulence effects. The nature of the waves are such that they have velocities along the field lines ranging from a cut-off above the electron thermal, $v_{\rm Te}$, to greater than the speed of light. Thus lower-hybrid waves have the range of phase velocities necessary to accelerate electrons parallel to the magnetic field producing high energy tails in the distribution function.

When a microinstability develops, fluctuating fields generated by the instability scatter plasma particles and cause diffusion of the velocity distribution function. If the spectrum of the fluctuating field is characterized by sufficiently small values of the parallel wave number $k_\parallel$, such that the phase velocity of the wave parallel to B is greater than the electron thermal speed, then the waves resonate with and accelerate only electrons moving parallel to B with velocities greater than $v_{\rm Te}$.

The statistical acceleration of electrons in the tail of the distribution function due to the resonant interaction can be demonstrated by solving the one-dimensional quasilinear diffusion equation (Davidson, 1972; see also Chapter 15 of this book)

$$\frac{\partial f_{\rm e}}{\partial t} = \frac{\partial}{\partial v_\parallel} D_\parallel \frac{\partial f_{\rm e}}{\partial v_\parallel}, \tag{9.10.3}$$

where $f_{\rm e}$ is the electron distribution function, and $D_\parallel$ is the quasilinear diffusion operator resulting from the Landau wave–particle interaction, $D_\parallel = \Delta v^2/\tau$ where Δv is the change in velocity in time τ and $D_\parallel$ is given by

$$D_\parallel = \frac{4\pi}{\varepsilon_0} \frac{e^2}{m_{\rm e}^2} \frac{k_\parallel}{\omega} \varepsilon_{k_\parallel}, \tag{9.10.4}$$

where $\varepsilon_{k_\parallel}$ is the wave energy density per unit wave number,

$$\frac{1}{2}\varepsilon_0 E_{\parallel\rm rms}^2 = \int_0^\infty \varepsilon_{k_\parallel} dk_\parallel, \tag{9.10.5}$$

and $E_{\parallel\rm rms}$ is the rms value of the lower-hybrid field. For an acceleration region of finite length, the effective range of $k_\parallel$ is

$$k_{\rm m} \le k_\parallel \le k_0,$$

where $k_{\rm m}$ is determined by the scale size of the plasma, and k_0 is determined by the strong Landau damping that takes place for $k_\parallel = \omega_{\rm lh}/2v_{\rm Te}$. The total wave energy density is thus

$$W_{\rm lh} = (k_0 - k_{\rm m})\varepsilon_k \simeq k_0 \varepsilon_k. \tag{9.10.6}$$

By substituting Eq. (9.10.4) into Eq. (9.10.3) and integrating, it can be shown that the asymptotic solution is given by

$$f_{\rm e} = \frac{n_0}{(4\pi D_\parallel t_0)^{1/2}} \exp(-v_\parallel^2/4D_\parallel t_0), \tag{9.10.7}$$

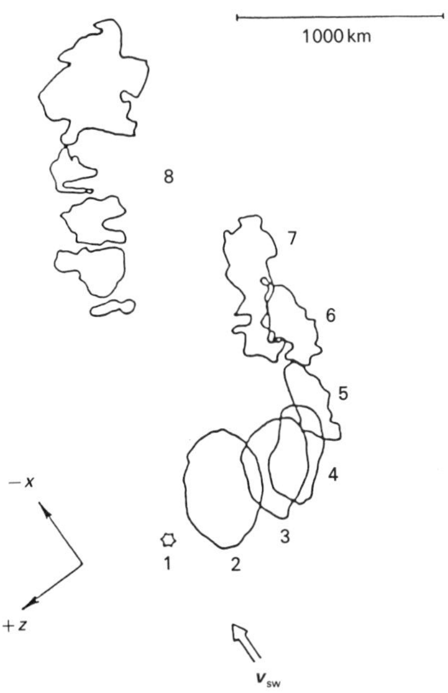

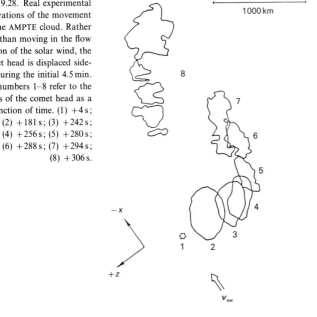

Fig. 9.28. Real experimental observations of the movement of the AMPTE cloud. Rather than moving in the flow direction of the solar wind, the comet head is displaced sideways during the initial 4.5 min. The numbers 1–8 refer to the outlines of the comet head as a function of time. (1) +4 s; (2) +181 s; (3) +242 s; (4) +256 s; (5) +280 s; (6) +288 s; (7) +294 s; (8) +306 s.

Similar observations of the comet taken from the South Pacific and from Boulder, Colorado, reveal an asymmetry in the density structure of the comet. The northern flank has very definite protrusions pointing in the direction of the convective electric field, whereas the southern flank appears smooth with a more tightly confined boundary and steeper density gradients. Using the ion energy per charge spectrometer on board the spacecraft, experimentalists were able to observe the momentum flow of the observed barium ions and also the solar-wind protons as they traversed the cavity. It is clear from the measurement made by the ion analyser that the effect of the release was to slow and divert the solar wind; however, owing to its initial high speed, the net effect was a deflection in the southward direction, i.e. in a direction opposite to the convective electric field direction. Further studies have shown that the solar-wind protons are deflected southwards, whereas there was a component of barium ions accelerated northwards, i.e. in the direction of the convection electric field.

The magnetic field showed a very sharp increase from zero in the cavity to a maximum of around 120 nT, which is a 12-fold increase over the ambient field strength at the front, creating a pile-up region upstream of the diamagnetic cavity characterized by an enhanced level of wave activity. This region is followed by a series of magnetic pulses with a period of about 10 s; these pulses have been identified as belonging to the magnetoacoustic branch.

A simple theoretical understanding describing the interaction of the solar-wind particles and the cavity can be achieved using a two-fluid model for the plasma and a constant profile structure for the magnetic field which only varies in the x-direction. The analysis treats the ions as unmagnetized, therefore the solar-wind ions do not respond directly to the change in the magnetic field created by the cavity; only the electrons (because of their small Larmor orbit) respond to this field change, which slows them down, the ions continue to penetrate, causing a charge separation resulting in an ambipolar field. This ambipolar field is responsible for retarding the ion flow. Within the cloud, the ambipolar field, however, has the opposite direction to that at the front region; here it opposes the expansion of the barium ions.

Using the steady state two-fluid model described by Eqs (9.8.2) and (9.8.3), assuming a constant profile for the magnetic field structure varying in the x-direction, adding the y-components of Eq. (9.8.3) for electrons and ions and using the boundary condition $v_{ix} = v_{ex} = v_x$ at $x = -\infty$ (i.e. in the solar wind, electron and ion fluids have the same v_x velocities) we obtain Eq. (9.8.11), $v_{iy} = -(m_e/m_i)v_{ey}$. Clearly $v_{iy} \leq v_{ey}$, therefore the electron velocity produces most of the J_y current which is responsible for the perturbed B_z field; the ambipolar field E_x from Eq. (9.8.15) is $E_x = -v_{ey}B_z$, which simply means that the electrons are in force balance in the x-direction. From the y-component of Ampere's law Eq. (9.8.4) together with Eq. (9.8.11) the ambipolar field can be written in the form of Eq. (9.8.21), namely $E_x = -(1/2\mu_0 en)(\mathrm{d}B_z^2/\mathrm{d}x)$. This electric field, pointing along the solar-wind direction, points in the direction opposite to the gradient of the magnetic pressure, i.e. this component of the field, which is the ambipolar field, is proportional to the Maxwell stress tensor. This supports our earlier assertion that the electric field points one way in the compressed field region, so as to slow down the incoming solar-wind particles, and points the opposite way in the cloud region, so as to push back on the cloud, confining the released plasma in the compressed region.

From the y-components of Eq. (9.8.14) and using $v_{iy} = -(m_e/m_i)v_{ey}$ we get $E_y = v_x B_z$. This simply means that, in the rest frame of the cloud, there exists a convective electric field given by $\boldsymbol{E} = -\boldsymbol{v}_{\mathrm{sw}} \times \boldsymbol{B}$.

The force on an incoming solar-wind ion due to the electric field E_x is given by

$$F = eE_x = -\frac{1}{2\mu_0 n}\frac{\mathrm{d}B_z^2}{\mathrm{d}x}. \tag{9.11.1}$$

The force required to reduce the velocity of such an ion from u_x to v_x in a distance Δx can be written as

$$F = \frac{m_i(v^2 - u^2)}{2\Delta x}. \tag{9.11.2}$$

Balancing these forces, we obtain the expression

$$u^2 - v^2 \simeq \frac{1}{2\mu_0 m_i n}\Delta B_z^2, \tag{9.11.3}$$

where ΔB_z^2 is the change in B_z^2 in a distance Δx. The force exerted by E_x is sufficient to account for the slowing down of the solar wind. Including a y-variation for the field quantities allows us to produce fields which deflect the incoming solar-wind ions.

The convective electric field is seen by the barium ions, which are effectively at rest with respect to the solar-wind ions, and as a result are ejected out of the cloud, producing a northward-flowing jet of barium ions. These cause the main part of the barium plasma cloud to recoil and move in the southward direction. This is the main reason for the mysterious southward deflection seen by the observers.

The dynamics of the interaction, including the generation of energetic release ions, low frequency wave activity and overall dynamics, can only fully be appreciated using a kinetic approach which allows for finite Larmor radius effects. Such effects have been successfully studied using hybrid simulation codes which have kinetic ions and fluid electrons, these codes being capable of treating finite Larmor radius effects. A hybrid code is intermediate between a full particle and an MHD code. It can resolve disturbances on the scale of the ion gyro-radius, its basic time scale being the ion gyro-period.

9.12 Conclusions to Part II

Some aspects of plasma processes occurring in space plasmas have been presented; these include perpendicular shocks, beam-plasma instabilities, particle acceleration and active experiments. The plasma physics is extremely rich in collisionless wave–particle processes and is ideal for studying anomalous transport processes. The role of precise *in-situ* plasma measurements makes space plasmas a unique laboratory for exploring such processes which can lead to the understanding of a number of astrophysical plasma phenomena.

9.13 Appendix A: Derivation of the general dispersion relation

In this section the general dispersion relation for small amplitude waves in a uniformly magnetized plasma is derived using the Vlasov theory of plasma waves. The derivation starts from the Vlasov equation in terms of the distribution function $f_\alpha(\boldsymbol{x}, \boldsymbol{v}, t)$:

$$\frac{\partial f_\alpha}{\partial t} + \boldsymbol{v} \cdot \nabla f_\alpha + \frac{q_\alpha}{m_\alpha}(\boldsymbol{E} + \boldsymbol{v} \times \boldsymbol{B}) \cdot \nabla_v f_\alpha = 0. \tag{9.13.1}$$

The electric and magnetic fields $\boldsymbol{E}$ and $\boldsymbol{B}$ depend on the particle distribution f_α through Maxwell's equations, thus Eq. (9.13.1) is a nonlinear equation in f_α and difficult to solve. The problem is simplified by examining the behaviour of a small departure $f_{\alpha 1}$ from some plasma equilibrium state $f_{\alpha 0}$ using the linearized Vlasov–Maxwell equations.

The plasma is assumed to be uniformly magnetized, with $\boldsymbol{B}_0 = B_0\hat{\boldsymbol{z}}$. Linearizing about $f_{\alpha 0}$,

$$f_\alpha = f_{\alpha 0} + f_{\alpha 1},$$
$$\boldsymbol{B} = B_0\hat{\boldsymbol{z}} + \boldsymbol{B}_1,$$
$$\boldsymbol{E} = \boldsymbol{E}_1,$$

the Vlasov equation becomes

$$\left[\frac{\partial}{\partial t} + \boldsymbol{v} \cdot \nabla + \frac{q_\alpha}{m_\alpha}\boldsymbol{v} \times \boldsymbol{B}_0 \cdot \nabla_v\right] f_{\alpha 1} = -\frac{q_\alpha}{m_\alpha}[\boldsymbol{E}_1 + \boldsymbol{v} \times \boldsymbol{B}_1] \cdot \nabla_v f_{\alpha 0}, \tag{9.13.2}$$

and the equilibrium distribution satisfies

$$\left[\boldsymbol{v} \cdot \nabla + \frac{q_\alpha}{m_\alpha}\boldsymbol{v} \times \boldsymbol{B}_0 \cdot \nabla_v\right] f_{\alpha 0} = 0. \tag{9.13.3}$$

In the equilibrium state the plasma is quasineutral and contains zero current. It is assumed that the waves investigated propagate in a plasma with an equilibrium distribution of the form $f_{\alpha 0} = f_{\alpha 0}(v_\perp^2, v_z)$, where $v_\perp$ is the component of the velocity perpendicular to the magnetic field.

The Vlasov equation is solved by integrating along the orbits of the particles in the unperturbed magnetic field (the method of characteristics). We define

$$\frac{\mathrm{d}\boldsymbol{x}'}{\mathrm{d}t'} = \boldsymbol{v}', \quad \frac{\mathrm{d}\boldsymbol{v}'}{\mathrm{d}t'} = \frac{q_\alpha}{m_\alpha}[\boldsymbol{v}' \times \boldsymbol{B}_0(\boldsymbol{x}', t')], \tag{9.13.4}$$

with boundary conditions

$$x'(t'=t)=x, \quad v'(t'=t)=v. \tag{9.13.5}$$

From the linearized Vlasov equation the perturbed distribution $f_{\alpha 1}(\mathbf{x}, v, t)$ is obtained by considering the function $f_{\alpha 1}(x(t'), v(t'), t')$, which is a function of t', and satisfies the linearized Vlasov equation:

$$\begin{aligned}\frac{\mathrm{d}f_{\alpha 1}}{\mathrm{d}t'} &\equiv \frac{\partial f_{\alpha 1}}{\partial t'}(x', v', t') + \frac{\mathrm{d}x'}{\mathrm{d}t'}\cdot\nabla_{x'}f_{\alpha 1}(x', v', t') + \frac{\mathrm{d}v'}{\mathrm{d}t'}\cdot\nabla_{v'}f_{\alpha 1}(x', v', t') \\ &= -\frac{q_\alpha}{m_\alpha}[E_1(x', t') + v'\times B_1(x', t')]\cdot\nabla_{v'}f_{\alpha 0}(x', v').\end{aligned} \tag{9.13.6}$$

B_0 is still the field that determines the equilibrium plasma distribution $f_{\alpha 0}$. Because of the boundary conditions given by Eq. (9.13.5) it follows that $f_{\alpha 1}(x', v', t') = f_{\alpha 1}(x, v, t)$ at time $t' = t$. So the solution to Eq. (9.13.6) at $t' = t$ is also the solution to the Vlasov equation. Integrating Eq. (9.13.6) from $t' = -\infty$ to $t' = t$ can be thought of as integrating the Vlasov equation from $-\infty$ to t along a path in (x, v)-space which coincides with the orbit of a charged particle in the equilibrium field B_0, and gives

$$\begin{aligned}f_{\alpha 1}(x, v, t) = &-\frac{q_\alpha}{m_\alpha}\int_{-\infty}^{t}\mathrm{d}t'\,[E_1(x', t') + v'\times B_1(x', t')]\cdot\nabla_{v'}f_{\alpha 0}(x', v') \\ &+ f_{\alpha 1}(x(-\infty), v(-\infty), t' = -\infty);\end{aligned} \tag{9.13.7}$$

x' and v' are generally functions of x and v as well as time.

The time-asymptotic solutions are obtained by assuming a solution of the form $E_1(x, t) = \bar{E}_1\exp[\mathrm{i}(k\cdot x - \omega t)]$ and that $f_{\alpha 1}(x', v', t' \to -\infty) = 0$. The last term in Eq. (9.13.7) disappears, and we are left with

$$f_{\alpha 1} = -\frac{q_\alpha}{m_\alpha}\int_{-\infty}^{0}[\bar{E}_1 + v'\times\bar{B}_1]\cdot\nabla_{v'}f_{\alpha 0}(v')\exp[\mathrm{i}(k\cdot X - \omega\tau)]\mathrm{d}\tau, \tag{9.13.8}$$

with $\mathrm{Im}(\omega) > 0$, where $\tau = (t' - t), X = (x' - x)$, and $\bar{E}_1$ and $\bar{B}_1$ are vector amplitudes that are constant in space and time. The solution $f_{\alpha 1}$ for $\mathrm{Im}(\omega) < 0$ is obtained by analytic continuation of Eq. (9.13.8). The field $\bar{B}_1$ can be expressed in terms of $\bar{E}_1$ using Maxwell's equations. From

$$\nabla\times E = -\frac{\partial B}{\partial t} \tag{9.13.9}$$

we obtain

$$\mathrm{i}k\times\bar{E}_1 = \mathrm{i}\omega\bar{B}_1. \tag{9.13.10}$$

By substituting $f_{\alpha 1}$ into Maxwell's equation,

$$\nabla\times B = \mu_0 J + \varepsilon_0\frac{\partial E}{\partial t}, \tag{9.13.11}$$

and using the following relation to calculate the perturbed currents:

$$J_{\alpha 1} = \sum_\alpha n_\alpha q_\alpha\int_L v f_{\alpha 1}\mathrm{d}v, \tag{9.13.12}$$

we obtain

$$-k\times k\times\bar{E}_1 = \omega^2\bar{E}_1 + \mathrm{i}\omega\mu_0\sum_\alpha n_\alpha q_\alpha\int_L v f_{\alpha 1}\mathrm{d}v, \tag{9.13.13}$$

where L is the Landau contour. By considering the $\hat{\imath}$-, $\hat{\jmath}$- and $\hat{k}$-components of (9.13.13) separately, we obtain three equations of the form

$$D_{xx}E_x + D_{xy}E_y + D_{xz}E_z = 0,$$
$$D_{yx}E_x + D_{yy}E_y + D_{yz}E_z = 0,$$
$$D_{zx}E_x + D_{zy}E_y + D_{zz}E_z = 0,$$

which can be written

$$D \cdot \boldsymbol{E} = 0, \tag{9.13.14}$$

where

$$D = \begin{bmatrix} D_{xx} & D_{xy} & D_{xz} \\ D_{yx} & D_{yy} & D_{yz} \\ D_{zx} & D_{zy} & D_{zz} \end{bmatrix}, \tag{9.13.15}$$

and the elements of the determinant are given by

$$D_{xx} = 1 - \frac{k_\parallel^2 c^2}{\omega^2} - \frac{2\pi}{\omega}\sum_\alpha\sum_n\left(\frac{\omega_p^2}{\omega_c}\right)_\alpha \mathscr{I}\left[\frac{n^2\omega_{c\alpha}^3}{k_\perp^2}J_n^2\chi_\alpha\right], \tag{9.13.16}$$

$$D_{xy} = -\frac{2\pi i}{\omega}\sum_\alpha\sum_n\left(\frac{\omega_p^2}{\omega_c}\right)_\alpha \mathscr{I}\left[\frac{n\omega_{c\alpha}^2 v_\perp}{k_\perp}J_n\frac{dJ_n}{d(k_\perp v_\perp/\omega_{c\alpha})}\chi_\alpha\right], \tag{9.13.17}$$

$$D_{xz} = \frac{k_\parallel k_\perp c^2}{\omega^2} - \frac{2\pi}{\omega}\sum_\alpha\sum_n\left(\frac{\omega_p^2}{\omega_c}\right)_\alpha \mathscr{I}\left[\frac{n\omega_{c\alpha}^2 v_\parallel J_n^2}{k_\perp}\Lambda_\alpha\right], \tag{9.13.18}$$

$$D_{yx} = -D_{xy}, \tag{9.13.19}$$

$$D_{yy} = 1 - \frac{(k_\perp^2 + k_\parallel^2)c^2}{\omega^2} - \frac{2\pi}{\omega}\sum_\alpha\sum_n\left(\frac{\omega_p^2}{\omega_c}\right)_\alpha \mathscr{I}\left[\omega_{c\alpha}\left(\frac{dJ_n}{d(k_\perp v_\perp/\omega_{c\alpha})}\right)^2 v_\perp^2\chi_\alpha\right], \tag{9.13.20}$$

$$D_{yz} = \frac{2\pi i}{\omega}\sum_\alpha\sum_n\left(\frac{\omega_p^2}{\omega_c}\right)_\alpha \mathscr{I}\left[\omega_{c\alpha}v_\perp v_\parallel J_n\frac{dJ_n}{d(k_\perp v_\perp/\omega_{c\alpha})}\Lambda_\alpha\right], \tag{9.13.21}$$

$$D_{zx} = \frac{k_\parallel k_\perp c^2}{\omega^2} - \frac{2\pi}{\omega}\sum_\alpha\sum_n\left(\frac{\omega_p^2}{\omega_c}\right)_\alpha \mathscr{I}\left[v_\parallel\frac{n\omega_{c\alpha}^2}{k_\perp}J_n^2\chi_\alpha\right], \tag{9.13.22}$$

$$D_{zy} = -\frac{2\pi i}{\omega}\sum_\alpha\sum_n\left(\frac{\omega_p^2}{\omega_c}\right)_\alpha \mathscr{I}\left[\omega_{c\alpha}v_\perp v_\parallel J_n\frac{dJ_n}{d(k_\perp v_\perp/\omega_{c\alpha})}\chi_\alpha\right], \tag{9.13.23}$$

$$D_{zz} = 1 - \frac{k_\perp^2 c^2}{\omega^2} - \frac{2\pi}{\omega}\sum_\alpha\sum_n\left(\frac{\omega_p^2}{\omega_c}\right)_\alpha \mathscr{I}[v_\parallel^2\omega_{c\alpha}J_n^2\Lambda_\alpha], \tag{9.13.24}$$

where the summation over n goes from $-\infty$ to $+\infty$. The plasma frequency is given by $\omega_{p\alpha} = (n_\alpha e^2/m_\alpha\varepsilon_0)$ and the cyclotron frequency by $\omega_{c\alpha} = q_\alpha B/m_\alpha$. The integral operator $\mathscr{I}$ is defined by

$$\mathscr{I}[F(v)] \equiv \int_{-\infty}^{\infty} dv_\parallel \int_0^\infty \frac{2v_\perp F(v_\perp, v_\parallel)}{k_\parallel v_\parallel + n\omega_{c\alpha} - \omega} dv_\perp,$$

and χ_α and Λ_α have been used to represent frequently appearing combinations of velocity derivatives: $\chi_\alpha \equiv \frac{\partial f_{\alpha 0}}{\partial v_\perp^2}\left(1 - \frac{k_\parallel v_\parallel}{\omega}\right) + \frac{k_\parallel v_\parallel}{\omega}\frac{\partial f_{\alpha 0}}{\partial v_\parallel^2}$

and

$$\Lambda_\alpha \equiv \frac{\partial f_{\alpha 0}}{\partial v_\parallel^2} - \frac{n\omega_{c\alpha}}{\omega}\left(\frac{\partial}{\partial v_\parallel^2} - \frac{\partial}{\partial v_\perp^2}\right)f_{\alpha 0}.$$

The argument of all the Bessel functions, J_n, is $k_\perp v_\perp/\omega_{c\alpha}$, and the integrals $\int_{-\infty}^{\infty} dv$ are to be taken along the Landau contour. The equations have a nontrivial solution provided

$$|D| = 0. \tag{9.13.25}$$

The quantity D is referred to as the dispersion tensor, and equation (9.13.25) as the dispersion equation. In solving dispersion relations we shall assumed that $\boldsymbol{k}$ is real and that the complex frequency ω is of the form $\omega = \omega_r + i\gamma$, where ω_r is the real frequency and γ is the growth rate of the wave with frequency ω_r.

9.14 Appendix B: Dispersion relations for Maxwellian distributions

We shall start by taking the velocity distribution function of species α to be a nondrifting Maxwellian of the form

$$f_{\alpha 0} = \frac{1}{(2\pi v_{\text{th}\alpha}^2)^{3/2}} \exp\left\{ -\frac{v^2}{2v_{\text{th}\alpha}^2} \right\}, \tag{9.14.1}$$

where $v_{\text{th}\alpha} = (k_B T_\alpha/m_\alpha)^{1/2}$ is the thermal velocity; drifting Maxwellians will be introduced later.

If we use a long wavelength approximation, then the argument of the Bessel functions is very small, i.e. $k_\perp v_\perp/\omega_{c\alpha} \ll 1$, so that we only need take into account functions for a small number of values of n. The nth order Bessel function is given by

$$J_n(z) = \left(\frac{z}{2}\right)^n \sum_{k=0}^{\infty} \frac{(-\frac{1}{4}z^2)^k}{k!\Gamma(n+k+1)}. \tag{9.14.2}$$

Terms of second order and higher in $k_\perp v_\perp/\omega_{c\alpha}$ will be neglected when we substitute into the elements of the dielectric tensor, so the only Bessel functions that we shall need are those corresponding to $n = 0$, $n = 1$ and $n = -1$. For $n = 0$ and $n = 1$ we have

$$J_0(z) = 1 - \frac{z^2}{4} + \cdots,$$

$$J_1(z) = \frac{z}{2} + \cdots.$$

For $n = -1$,

$$J_{-1}(z) = \left(\frac{z}{2}\right)^{-1} \sum_{k=0}^{\infty} \frac{(-\frac{1}{4}z^2)^k}{k!\Gamma(k)}. \tag{9.14.3}$$

$\Gamma(k)$ is only defined for $k > 0$; as $k \to 0, \Gamma(k) \to \infty$. So the first term obtained from the summation in Eq. (9.14.3) corresponds to $k = 1$:

$$J_{-1}(z) = -\frac{z}{2} + \cdots.$$

Eqs (9.13.16)–(9.13.24) can now be greatly simplified. Substituting Eq. (9.14.1) into the expressions for χ_α and Λ_α, we obtain

$$\chi_\alpha = \Lambda_\alpha = -\frac{1}{2v_{\text{th}\alpha}^2} f_{\alpha 0}, \tag{9.14.4}$$

and Eqs (9.13.16)–(9.13.24) become

$$D_{xx} = 1 - \frac{k_\parallel^2 c^2}{\omega^2} + \sum_\alpha \frac{\omega_{p\alpha}^2}{2\omega} \frac{1}{2^{1/2} k_\parallel v_{\text{th}\alpha}} \left\{ Z\left(\frac{\omega - \omega_{c\alpha}}{2^{1/2} k_\parallel v_{\text{th}\alpha}}\right) + Z\left(\frac{\omega + \omega_{c\alpha}}{2^{1/2} k_\parallel v_{\text{th}\alpha}}\right) \right\}, \tag{9.14.5}$$

$$D_{xy} = \sum_\alpha \frac{i\omega_{p\alpha}^2}{2\omega} \frac{1}{2^{1/2} k_\parallel v_{\text{th}\alpha}} \left\{ Z\left(\frac{\omega - \omega_{c\alpha}}{2^{1/2} k_\parallel v_{\text{th}\alpha}}\right) - Z\left(\frac{\omega + \omega_{c\alpha}}{2^{1/2} k_\parallel v_{\text{th}\alpha}}\right) \right\}, \tag{9.14.6}$$

$$D_{xz} = \frac{k_\parallel k_\perp c^2}{\omega^2} - \sum_\alpha \frac{\omega_{p\alpha}^2}{4\omega\omega_{c\alpha}} \frac{k_\perp}{k_\parallel} \left\{ Z'\left(\frac{\omega - \omega_{c\alpha}}{2^{1/2} k_\parallel v_{\text{th}\alpha}}\right) - Z'\left(\frac{\omega + \omega_{c\alpha}}{2^{1/2} k_\parallel v_{\text{th}\alpha}}\right) \right\}, \tag{9.14.7}$$

$$D_{yx} = -D_{xy}, \tag{9.14.8}$$

$$D_{yy} = D_{xx}, \tag{9.14.9}$$

$$D_{yz} = \sum_\alpha \frac{i\omega_{p\alpha}^2}{\omega}\frac{k_\perp}{2k_\parallel\omega_{c\alpha}}\left\{Z'\left(\frac{\omega}{2^{1/2}k_\parallel v_{th\alpha}}\right) + \frac{1}{2}\left[Z'\left(\frac{\omega - \omega_{c\alpha}}{2^{1/2}k_\parallel v_{th\alpha}}\right) + Z'\left(\frac{\omega + \omega_{c\alpha}}{2^{1/2}k_\parallel v_{th\alpha}}\right)\right]\right\}, \tag{9.14.10}$$

$$D_{zx} = D_{xz}, \tag{9.14.11}$$

$$D_{zy} = -D_{yz}, \tag{9.14.12}$$

$$D_{zz} = 1 + \sum_\alpha \frac{\omega_{p\alpha}^2}{\omega}\frac{1}{(2\pi)^{1/2}}\frac{1}{v_{th\alpha}^3}\int_{-\infty}^{\infty}\frac{v_\parallel^2\exp(-v_\parallel^2/2v_{th\alpha}^2)}{(k_\parallel v_\parallel - \omega)}dv_\parallel, \tag{9.14.13}$$

where $Z(\xi)$ is the Fried–Conte function or plasma dispersion function, which is given by

$$Z(\xi) = \frac{1}{\pi^{1/2}}\int_{-\infty}^{\infty}\frac{\exp(-t^2)}{t-\xi}dt \tag{9.14.14}$$

and tabulated by Fried and Conte (1961). $Z'(\xi)$ is the derivative of the dispersion function with respect to its argument, and it is related to the dispersion function by the relation $Z'(\xi) = -2(1 + \xi Z(\xi))$.

For waves propagating exactly parallel to the magnetic field, $k_\perp = 0, D_{xz} = D_{yz} = D_{zx} = D_{zy} = 0, D_{xx}, D_{xy}, D_{yy}$ and D_{yx} are unchanged, and D_{zz} takes the form

$$D_{zz} = 1 - \sum_\alpha \frac{\omega_{p\alpha}^2}{2k_\parallel^2 v_{th\alpha}^2} Z'\left(\frac{\omega}{2^{1/2}k_\parallel v_{th\alpha}}\right)$$

$$= 1 - \sum_\alpha \frac{1}{2k_\parallel^2\lambda_{D\alpha}^2} Z'\left(\frac{\omega}{2^{1/2}k_\parallel v_{th\alpha}}\right), \tag{9.14.15}$$

where $\lambda_{D\alpha}$, the Debye length, is given by $\lambda_{D\alpha} = (\varepsilon_0 kT_\alpha/n_\alpha e^2)^{1/2}$. The dispersion equation is now of the form

$$D_{zz}(D_{xx}D_{yy} - D_{xy}D_{yx}) = 0. \tag{9.14.16}$$

We can split the solutions to Eqs (9.14.16) and (9.13.14) into two types. First, those corresponding to longitudinal electrostatic perturbations ($\boldsymbol{E} \parallel \boldsymbol{k}$), with perturbed electric fields parallel to the magnetic field:

$$D_{zz} = 0, E_z \neq 0, E_x = E_y = 0. \tag{9.14.17}$$

It follows from Eq. (9.14.15) that the dispersion relation for electrostatic waves propagating parallel to the magnetic field in a homogeneous hot magnetized plasma is

$$1 - \sum_\alpha \frac{1}{2k_\parallel^2\lambda_{D\alpha}^2} Z'\left(\frac{\omega}{2^{1/2}k_\parallel v_{th\alpha}}\right) = 0. \tag{9.14.18}$$

The second class of solutions are those corresponding to transverse electromagnetic perturbations with perturbed electric fields perpendicular to the magnetic field:

$$(D_{xx}D_{yy} - D_{xy}D_{yx}) = 0, E_z = 0. \tag{9.14.19}$$

The dispersion relation for the electromagnetic case can be simplified to

$$D_{xx}^2 + D_{xy}^2 = 0, \tag{9.14.20}$$

which can be written

$$(D_{xx} + iD_{xy})(D_{xx} - iD_{xy}) = 0$$

and reduces to

$$D^\pm = 1 - \frac{k_\parallel^2 c^2}{\omega^2} + \sum_\alpha \frac{\omega_{p\alpha}^2}{\omega}\frac{1}{2^{1/2}k_\parallel v_{th\alpha}} Z\left(\frac{\omega \pm \omega_{c\alpha}}{2^{1/2}k_\parallel v_{th\alpha}}\right) = 0, \tag{9.14.21}$$

where $D^+ = D_{xx} + iD_{xy} = 0$ and $D^- = D_{xx} - iD_{xy} = 0$ correspond to circularly polarized electromagnetic waves with right-hand ($E_x = -E_y$) and left-hand ($E_x = E_y$) polarizations, respectively. The general dispersion relation can also be simplified if we look for waves propagating perpendicular to the magnetic field, by setting $k_\parallel = 0$.

For a Maxwellian drifting parallel to the magnetic field, of the form

$$f_{\alpha 0} = \frac{1}{(2\pi v_{\mathrm{th}\alpha}^2)^{3/2}} \exp\left\{ -\frac{(v_\parallel - v_{\mathrm{d}\alpha\parallel})^2 + v_\perp^2}{2v_{\mathrm{th}\alpha}^2} \right\}, \tag{9.14.22}$$

where $v_{\mathrm{d}\alpha\parallel}$ is the drift velocity parallel to the magnetic field, the dispersion relation (9.14.18) becomes

$$1 - \sum_\alpha \frac{1}{2k_\parallel^2 \lambda_{\mathrm{D}\alpha}^2} Z'\left(\frac{\omega}{2^{1/2} k_\parallel v_{\mathrm{th}\alpha}} - \frac{v_{\mathrm{d}\alpha\parallel}}{2^{1/2} v_{\mathrm{th}\alpha}} \right) = 0. \tag{9.14.23}$$

For a nonmagnetized plasma the dispersion relation for electrostatic waves is of a similar form to Eq. (9.14.23) and is given by

$$1 - \sum_\alpha \frac{1}{2k^2 \lambda_{\mathrm{D}\alpha}^2} Z'\left(\frac{\omega}{2^{1/2} k v_{\mathrm{th}\alpha}} - \frac{v_{\mathrm{d}\alpha}}{2^{1/2} v_{\mathrm{th}\alpha}} \right) = 0. \tag{9.14.24}$$

It should be noted that electrostatic wave modes in magnetized plasmas are not affected by the magnetic field when the perturbation frequency is much greater than the cyclotron frequencies of the plasma components and the perturbation wavelength is much shorter than the thermal gyro-radii. For such plasmas the dispersion relation given by Eq. (9.14.24) can be used to determine the dispersion properties of electrostatic waves.

9.15 Acknowledgements

R. Bingham would like to thank C. M. C. Nairn for the use of material from her thesis.

Bibliography to Part I

For the basics of the subject I recommend

ROEDERER, J. G. (1970). *Dynamics of Geomagnetically Trapped Radiation*. Springer, New York.

and

JURSA, A. S., ed. (1985). *Handbook of Geophysics and the Space Environment*. Air Force Geophysics Laboratory, Boston, Mass.

For the latest information and for tracing the stages of development, papers within and cited in the following journals will be found helpful: *Journal of Geophysical Research A (Space Science), Planetary and Space Science, Annales Geophysicae, Space Science Reviews, The Space Science Newsletter of the European Space Agency, EOS, Transactions of the American Geophysical Union*, and *Geophysical Research Letters*.

A useful compendium of references appears in

Contributions in Solar–Planetary Relationships, US National Report 1987–90, American Geophysical Union (1991).

In perusing the wealth of literature on space plasma science, the reader is urged to exercise his or her critical facilities to the full, questioning in particular each adopted premise, assumption and simplification.

References to Part II

AXFORD, W. I. and HINES, C. O. (1961). *Can. J. Phys.*, **39**, 1433.
COWLEY, S. W. H. (1982). *Rev. Geophys. Space Phys.*, **20**, 531.
DAVIDSON, R. C. (1972). *Methods in Nonlinear Plasma Theory*. Academic Press, New York.
DUNGEY, J. W. (1961). *Phys. Rev. Lett.*, **6**, 47.
DUNCKEL, N. *et al.* (1970). *J. Geophys. Res.*, **75**, 1854.
FRIED, B. D. and CONTE, S. (1961). *The Plasma Dispersion Function*. Academic Press, New York.
GOLD, T. (1959). *J. Geophys. Res.*, **64**, 1219.
GUNN, J. and OSTRIKER, J. P. (1969). *Phys. Rev. Lett.*, **22**, 728.
GURNETT, D. A. (1974). *J. Geophys. Res.*, **79**, 4227.
HAERENDEL, G., PASCHMANN, G., BAUMJOHANN, W. and CARLSON, C. W. (1986). *Nature*, **320**, 21.
MENYUK, C. R., DROBOT, A. T., PAPADOPOULOS, K. and KARIMABADI, H. (1987). *Phys. Rev. Lett.*, **58**, 2071.
TANAKA, M. and PAPADOPOULOS, K. (1983). *Phys. Fluids*, **26**, 1697.
TIDMAN, D. and KRALL, N. A. (1971). In *Shock Waves in Collisionless Plasmas* (S. C. Brown, ed.). Wiley and Sons.

10
Solar plasmas

A. W. HOOD

10.1 Introduction

This chapter aims to describe some of the extremely interesting observations and theories of plasma phenomena occurring on the sun. Observations are obtained from a variety of different observatories, including ground-based and space telescopes, in a variety of different wavelengths, for example white light, Hα, radio, X-rays, ultraviolet and even gamma rays. In a chapter of this length it is only possible to introduce a few of the exciting features observed and the simple models that have been put forward to explain them. In addition, the topics presented reflect the author's interests, and no attempt has been made to make them exhaustive.

Theories describing large scale phenomena are frequently based on magnetohydrodynamic (MHD) models and more detailed discussions are referenced. Before proceeding with a description of the sun, we therefore give the basic MHD equations:

$$\rho \frac{\mathrm{D}\boldsymbol{v}}{\mathrm{D}t} = -\nabla p + \boldsymbol{J} \times \boldsymbol{B} + \rho \boldsymbol{g}, \tag{10.1.1}$$

$$\frac{\partial \rho}{\partial t} + \nabla \cdot \rho \boldsymbol{v} = 0, \tag{10.1.2}$$

$$\frac{\partial \boldsymbol{B}}{\partial t} = \nabla \times (\boldsymbol{v} \times \boldsymbol{B}) + \eta \nabla^2 \boldsymbol{B}, \tag{10.1.3}$$

$$\frac{\rho^\gamma}{\gamma - 1} \frac{\mathrm{D}}{\mathrm{D}t}\left(\frac{p}{\rho^\gamma}\right) = \nabla \cdot (\kappa \nabla T) - \mathscr{L}, \tag{10.1.4}$$

$$p = \frac{\rho \mathscr{R} T}{\tilde{\mu}}. \tag{10.1.5}$$

Here ρ is the density, n the number density ($\rho \approx nm_\mathrm{p}$, m_p is the proton mass), $\boldsymbol{v}$ the velocity, p the gas pressure, $\boldsymbol{B}$ the magnetic induction (but normally referred to as the magnetic field), $\boldsymbol{J}$ the current density, μ the magnetic permeability ($\mu = 4\pi \times 10^{-7}\,\mathrm{H\,m^{-1}}$), $\boldsymbol{g}$ the gravitational acceleration (the surface gravity is $g = 274\,\mathrm{m\,s^{-2}}$), γ the ratio of specific heats ($\gamma = 5/3$ for a fully ionized plasma), $\mathscr{R} = k_\mathrm{B}/m_\mathrm{p} = 8.25 \times 10^3\,\mathrm{m^2\,s^{-2}\,deg^{-1}}$), k_B Boltzmann's constant, $\tilde{\mu}$ the mean atomic weight ($\tilde{\mu} = 0.5$ for a hydrogen plasma and approximately 0.6 for the solar corona), η the magnetic diffusivity ($\eta = 10^9 T^{-3/2}\,\mathrm{m^2\,s^{-1}}$, with T the temperature measured in kelvin), σ the electrical conductivity ($\eta = 1/(\mu\sigma)$). The total loss function $\mathscr{L}$ can be written as

$$\mathscr{L} = n^2 Q(T) - h = \rho^2 Q_1(T) - h. \tag{10.1.6}$$

The unknown coronal heating function, h, may depend on, for example, T, ρ and B. It is taken as constant in this review. Optically thin radiation is represented by $n^2 Q(T)$. This radiative loss function has been estimated by several authors (see, for example, Hildner, 1974). Normally it is approximated by a piecewise continuous function of the form $Q(T) = \chi T^\alpha$, where χ and α are given in Table 10.1 (from Hildner, 1974). In addition, Ohmic dissipation in Eq. (10.1.6) is neglected.

Table 10.1. *The constants χ and α used to approximate the radiative loss function $Q(T) = \chi T^\alpha$ (Hildner, 1974).*

T	χ	α
$8 \times 10^5\,\mathrm{K} \leqslant T \leqslant 10^7\,\mathrm{K}$	5.51×10^{-30}	-1.0
$3 \times 10^5\,\mathrm{K} \leqslant T \leqslant 8 \times 10^5\,\mathrm{K}$	3.94×10^{-21}	-2.5
$8 \times 10^4\,\mathrm{K} \leqslant T \leqslant 3 \times 10^5\,\mathrm{K}$	8.00×10^{-35}	0.0
$1.5 \times 10^4\,\mathrm{K} \leqslant T \leqslant 8 \times 10^4\,\mathrm{K}$	1.20×10^{-43}	1.8
$\leqslant 1.5 \times 10^4\,\mathrm{K}$	4.92×10^{-67}	7.4

Finally, in Eq. (10.1.4), κ is the anisotropic thermal conductivity tensor; it can be expressed in terms of conduction parallel ($\kappa_\parallel$) and perpendicular ($\kappa_\perp$) to the magnetic field. Thus, heat conduction can be expressed as

$$\nabla \cdot (\kappa \nabla T) = \boldsymbol{B} \cdot \nabla\left(\frac{\kappa_\parallel}{B^2} \boldsymbol{B} \cdot \nabla T\right) - \nabla \cdot \left(\frac{\kappa_\perp}{B^2} \boldsymbol{B} \times (\boldsymbol{B} \times \nabla T)\right), \tag{10.1.7}$$

where (Braginskii, 1965)

$$\kappa_\parallel = 10^{-11} T^{5/2}\,\mathrm{W\,m^{-1}\,K^{-1}}, \tag{10.1.8}$$

and

$$\kappa_\perp/\kappa_\parallel = 2 \times 10^{-31} n^2/T^3 B^2, \tag{10.1.9}$$

with B in tesla, n in $\mathrm{m^{-3}}$ and T in kelvin.

10.2 The internal structure of the sun

For theoretical convenience, the solar interior can be divided into three regions: the *nuclear core*, the *radiative zone* and the outer *convection zone*. The nuclear core extends from the centre out to about one-quarter of the solar radius (the solar radius, $R_\odot$, is 6.96×10^8 m, making the nuclear core very much larger than a laboratory fusion machine). It is in this core, where the plasma temperature is 1.5×10^7 K and the density is $1.6 \times 10^5\,\mathrm{kg\,m^{-3}}$, that the sun converts hydrogen into helium. The energy is released mainly in the form of gamma rays. However, at such high densities, these short wavelength photons cannot travel far without being absorbed. It can take a particular packet of photons the order of 10^7 y to make its way to the surface via absorption and emission in the radiative zone. The radiative zone extends from $0.25R_\odot$ to $0.7R_\odot$. During the various absorptions and emissions the energy is continually downgraded, until at the photosphere – the 'surface' of the sun – the photons emerge as visible light.

From $0.7R_\odot$ to the surface is the convection zone. The temperature gradient is now so large that convection sets in. In convection cells, the hot plasma rises up in a central plume, then spreads out and cools at the photosphere sinking at the edge of the convection cell. In ideal convection, the cells have an approximately hexagonal shape. At the surface, convection cells are observed as granules. Fig. 10.1 shows, besides the large sunspot, that the solar surface is covered with many small granules. The diameter of these granules is approximately 10^3 km and the lifetime is the order of 8 min. The velocity of the plasma, as it spreads out across the surface before sinking back down into the convection zone, is between 1 and 3 km s^{-1}. To put this into perspective with earthly scales, a velocity of 1 km s^{-1} is about ten times greater than the largest wind speed ever recorded on the Earth. In addition, we could lose London and much of England inside a granule! Convection on the sun occurs on several different scales. Also observed are supergranules with a diameter of 3×10^4 km, lifetimes of 12–24 h and velocities

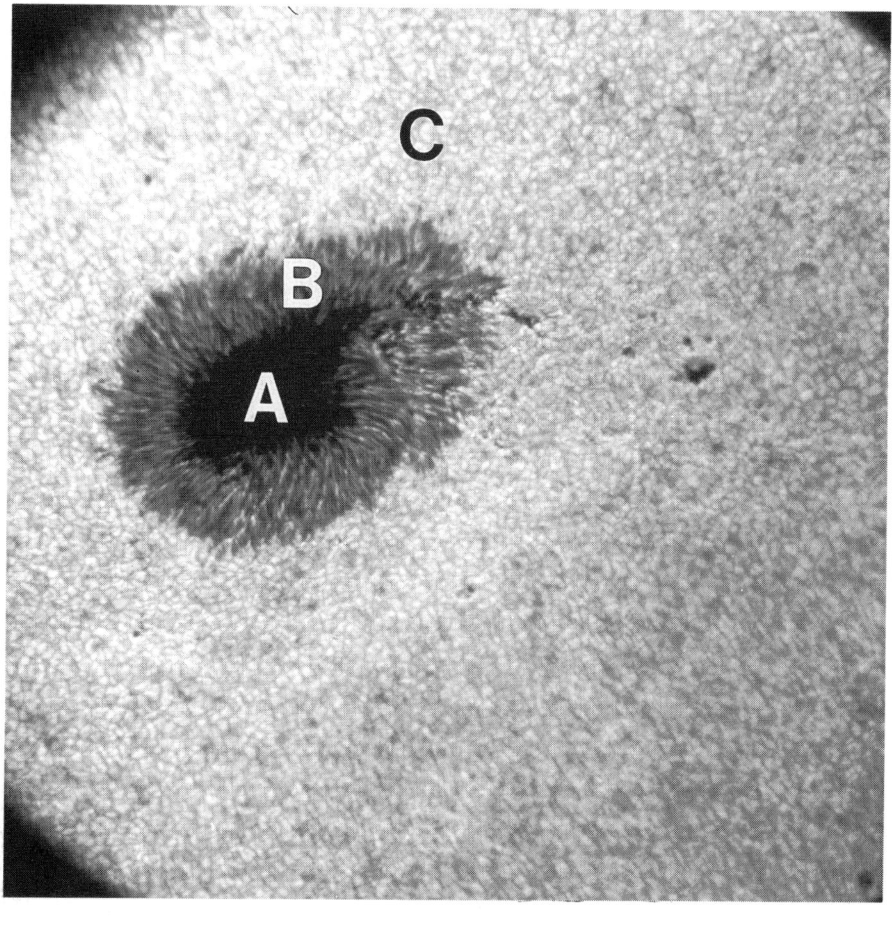

Fig. 10.1. A single medium sized sunspot surrounded by the convection pattern of the photosphere. Region A denotes the cool, dark umbra; region B shows the radial structure of the penumbra; and region C shows the photospheric granulation that outlines the top of the convection cells. (Courtesy Big Bear Solar Observatory, BBSO.)

of 0.1–0.2 km s^{-1} so that the Earth could easily fit into one. Because of the horizontal scales, it is thought that the plasma in a supergranule penetrates deep into the convection zone.

There are indications that even larger scales of convection occur. Giant cells have been predicted with a diameter of 3×10^5 km, so that the plasma would penetrate right to the bottom of the convection zone. Mesogranules have a scale intermediate between granules and supergranules, but their properties are less well known.

The properties of the solar interior are determined from standard stellar evolution models, but there are two possible ways to 'observe' the solar interior, using either the neutrinos emitted from the nuclear core or the oscillations of the solar surface.

In the nuclear core, hydrogen is converted into helium by a variety of routes in addition to the main three-step method (see Nicholson, 1982, p. 82). In the rarer, alternative routes, neutrinos of different energies are released, and it is the higher energy neutrinos that can be detected by a neutrino 'telescope'. One such 'telescope', situated 1500 m underground in an old gold mine, consists of a 4×10^5 l tank of dry cleaning fluid (basically chlorine-37). Placed so far underground, any spurious effects – for example due to cosmic rays – are eliminated and the solar neutrinos interact, albeit very weakly, with chlorine-37 to release atoms of argon. The argon atoms are then counted. Using a large tank and leaving it for a long time, theory estimates that between 6 and 7 SNU should be observed, where 1 SNU (solar neutrino unit) is 10^{-36} neutrino captures per second per target atom. This experiment has been running for many years but, surprisingly, the detection rate is only 2 SNU, which is a factor of three smaller than that predicted by the theoretical solar interior models. However, the number of these neutrinos emitted from the nuclear core is highly dependent on the central temperature, so one possible explanation is that the central temperature is 10% lower than previously believed. However, this would tend to reduce the luminosity of the sun. The neutrino problem has not been resolved, and there are almost as many proposed solutions as theorists! An excellent discussion of the neutrino problem is given in the review lecture by Taylor (1989).

An alternative way of probing the solar interior is through helioseismology. Careful observations of the solar disc reveal that the sun is oscillating on a global scale. The amplitudes of the various oscillations are very small, only 5 km in comparison to the solar diameter of 1.4×10^6 km. However, despite the extremely small signal-to-noise ratio, long-time observations allow a power spectrum of the modes to be constructed and the dominant periods of oscillations identified, with periods between 5 min and 1 h. These oscillations are like seismic waves, and by carefully analysing their behaviour it is possible to probe the solar interior.

Consider the following simple model hydrostatic atmosphere. The equilibrium satisfies

$$\frac{dp}{dz} = -\rho g, \tag{10.2.1}$$

where g is the gravitational acceleraton, p the plasma pressure, ρ the density and z the vertical coordinate. Now it is assumed that the plasma is slightly disturbed from its equilibrium state. Thus, for example, the density is expressed as

$$\rho(\boldsymbol{r},t) = \rho_0(z) + \rho_1(\boldsymbol{r},t),$$

where ρ_1 is much smaller in magnitude than ρ_0. Substituting into the MHD equations, using Eq. (10.2.1) and neglecting products of perturbed quantities – a process called linearization – gives:

$$-\,\mathrm{i}\omega\rho\boldsymbol{v} = -\,\nabla p_1 + \rho_1\boldsymbol{g}, \tag{10.2.2}$$

$$-\,\mathrm{i}\omega\rho_1 + \nabla\cdot(\rho\boldsymbol{v}) = 0, \tag{10.2.3}$$

$$-\,\mathrm{i}\omega p_1 + \boldsymbol{v}\cdot\nabla p = -\,\gamma p\nabla\cdot\boldsymbol{v}, \tag{10.2.4}$$

where p_1, ρ_1 and $\boldsymbol{v}$ are the pressure, density and velocity perturbations, and subscripts '0' on equilibrium quantities have been dropped. In addition, normal modes of the form

$$f(\boldsymbol{r},t) = f(z)\exp(\mathrm{i}kx - \mathrm{i}\omega t)$$

have been assumed, and variations in the y-direction are neglected. Eqs (10.2.2)–(10.2.4) can be manipulated, using Eq. (10.2.1), to obtain a single, second order differential equation for the divergence of the velocity, $\Delta \equiv \nabla\cdot\boldsymbol{v}$:

$$\Delta'' + \left[\frac{(c_s^2)'}{c_s^2} - \frac{\gamma g}{c_s^2}\right]\Delta' + \left\{\frac{\omega^2 - k^2c_s^2}{c_s^2} + \frac{gk^2}{\omega^2}\left(\frac{(c_s^2)'}{c_s^2} + \frac{(\gamma-1)g}{c_s^2}\right)\right\}\Delta = 0, \tag{10.2.5}$$

where $c_s^2 = \gamma p/\rho$ is the square of the sound speed. This may be rewritten in standard form by defining $Q = \rho^{1/2}c_s^2\Delta$, so that

$$\frac{\mathrm{d}^2Q}{\mathrm{d}z^2} + K^2(z)Q = 0. \tag{10.2.6}$$

In Eq. (10.2.6)

$$K^2(z) = \frac{\omega^2 - \omega_a^2}{c_s^2} + k^2\left(\frac{\omega_g^2}{\omega^2} - 1\right), \tag{10.2.7}$$

$\omega_a^2 = (1 - 2H')c_s^2/4H^2$ is the acoustic cut-off frequency squared, $\omega_g^2 = -\,(g/H + g^2/c_s^2)$ is the Brunt–Väisälä frequency squared, and the density scale height is defined by $H = -\,\rho/\rho'$. If $K^2(z) < 0$ then the solutions for Q are exponentially decaying (towards the boundaries), and if $K^2(z) > 0$ they are oscillatory. Thus, the solar oscillations can be described as *cavity* modes, with the cavity corresponding to the region with $K^2 > 0$. Solving the relevant boundary value problem gives the oscillation frequency, with the computed value of ω depending on the equilibrium properties. Thus, comparing predicted values of ω with the observed values gives some information about the density and temperature in the solar interior. It is seen that there are two distinct classes of mode, namely the *p-modes* (basically sound waves) with periods of the order of five minutes, and *g-modes* (gravity modes) with periods of the order of an hour or more. The p-modes propagate near the surface, are 'easily' observed, and provide information about the convection zone. Solving the boundary value problem for Eq. (10.2.4) gives a dispersion relation of the form

$$\omega^2 \approx gk, \tag{10.2.8}$$

which is in extremely good agreement with the observed frequencies. The g-modes have their largest amplitude near the centre of the sun and so should give information about the plasma conditions of the nuclear core. However, g-modes have not yet been unambiguously observed. For more details about solar oscillations and other references, see the review by Roberts (1989).

10.3 The atmosphere of the sun

The lowest level of the solar atmosphere, the photosphere, is visible in white light and defines the solar 'surface'. The photosphere is a thin region, some 500 km thick, through which the temperature falls from the surface value of 6500 K towards the temperature minimum value of 4300 K. The density of the photosphere is about 10^{-4} kg m^{-3}. The next level is the chromosphere, where the temperature passes through a minimum before beginning to rise again. At this level, a tremendous amount of structure is observed in Hα and this outlines the shape of the chromospheric magnetic field. As we continue to move up through the chromosphere, the temperature suddenly increases rapidly through a thin transition region until it reaches a value of about 2×10^6 K. This is the start of the corona. Meanwhile the density continues to fall from its photospheric value to a coronal value of only 10^{-12} kg m^{-3}. Thus, the photospheric density is some 10^8 times the coronal value.

The corona is so hot that it is now observable in soft X-rays, and indeed the kinetic temperature remains high even out to the radius of the Earth (about 10^6 K). In fact, the *Pioneer* and *Voyager* spacecraft have observed evidence of the corona out to Saturn and beyond. One of the major problems in solar physics is to explain why the corona is so hot and to determine the mechanism reponsible for heating the corona. The corona is a low-β plasma ($\beta = 2\mu p/B^2 \approx 10^{-2}$), so it is dominated dynamically by the coronal magnetic field. Thus it is generally agreed that the heating mechanism must be magnetic in origin – but how the magnetic energy is dissipated is not yet known.

If the sun were observed from stellar distances, it would appear as an ordinary star. Yet when we observe it in detail, we realize that the coronal magnetic field plays an important role in controlling the dynamics and structure of its atmosphere. Before applying theories to the stars, it is essential that we understand what makes the sun tick. Observations of the solar atmosphere reveal an amazing variety of different phenomena, ranging from quiescent (or slowly evolving) structures like sunspots, prominences and coronal holes, to dynamic (or rapidly evolving) phenomena like solar flares, spicules and the solar wind. Some of these are dealt with in the next few sections. However, it is worth noting that stellar observations have shown the existence of the stellar equivalents of sunspots and solar flares. Starspots are gigantic dark spots that cover a substantial fraction of the stellar surface, and stellar flares have similar characteristics to solar flares but on a much larger scale.

10.4 Sunspots

When the sun is observed in white light, the main features seen are sunspots. Sunspots are regions of extremely strong magnetic field with a field strength of about 3000 G (0.3 T). The sunspot can be divided into two regions: the umbra, where the magnetic field is essentially vertical, and the penumbra, in which the field is almost horizontal. These regions are clearly identified in Fig. 10.1, with the cool, dark umbra at the centre (region A) and the surrounding radial structure of the penumbra (region B). As pointed out earlier, the gas pressure drops sharply from the convection zone, through the photosphere, and out to the chromosphere. Since horizontal pressure balance must be maintained, the magnetic pressure must also drop sharply, and conservation of magnetic flux implies that the sunspot must expand rapidly in a radial direction, generating the penumbra. To put this into an earthly context, a typical sunspot's magnetic field is approximately 10^4 times as strong as the Earth's field. However, sunspots do not always occur in such a simple configuration. Fig. 10.2 shows a typical complex sunspot group. Here the magnetic field will be in a very complex configuration, providing a real challenge for the theorist!

The daily number of sunspots visible on the disc of the sun varies with an approximately 11-year cycle, the solar cycle. At the start of a new cycle, sunspots appear at the approximate latitudes $\pm 30°$, with one polarity in the northern hemisphere and the opposite polarity in the southern hemisphere. As the cycle progresses, the sunspots appear closer and closer to the equator. When the

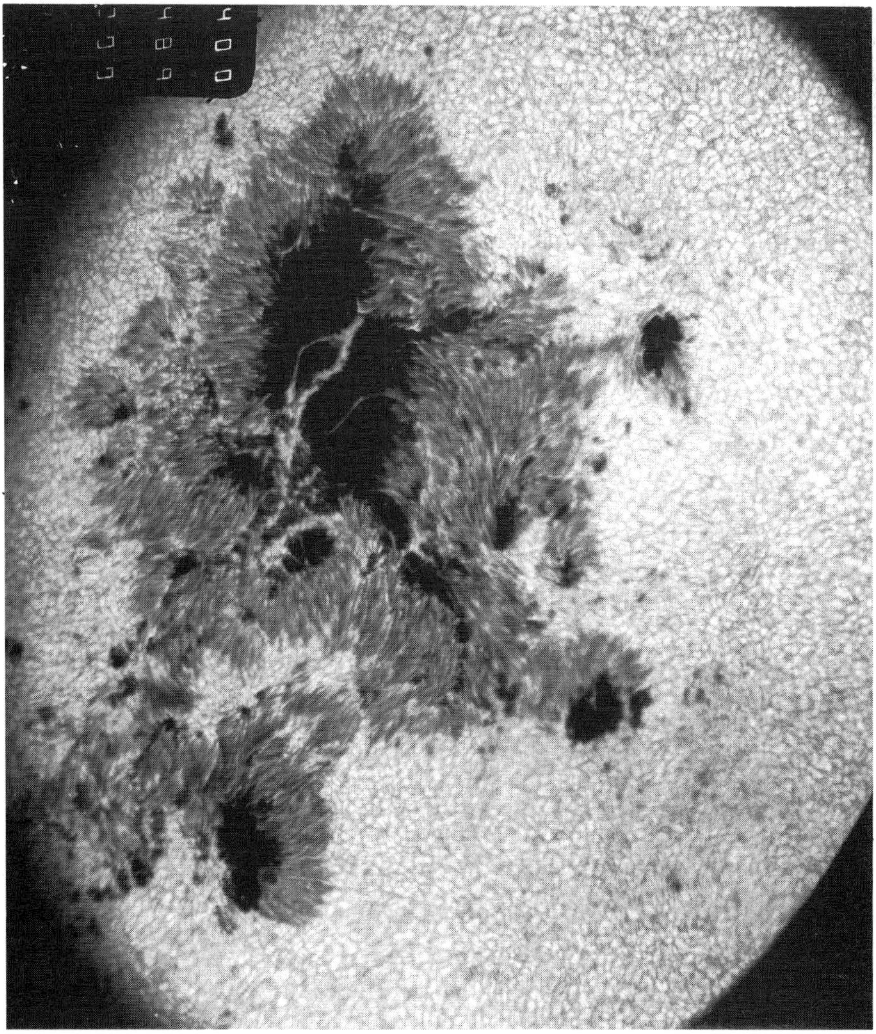

Fig. 10.2. A complex sunspot group of a large active region. (Courtesy BBSO.)

number of sunspots reaches the maximum value for that cycle, they appear with a mean latitude of $\leq 15°$. At the end of the cycle, sunspots from the old cycle are seen near the equator, whilst sunspots from the next cycle begin to appear at high latitudes. During the next cycle, the polarity of the sunspots is reversed so that the actual period of the sunspot cycle is approximately 22 years. A common way of illustrating the sunspot cycle is to plot the latitude of the spots versus time. The resulting diagram resembles the wings of a butterfly, and is called the butterfly diagram (see, for example, Fig. 1.30 in Priest, 1982).

The sunspot cycle is only approximately periodic. Plotting the daily number of sunspots observed on the solar disc as a function of time shows three other effects. First, the cycle is not symmetric, with the rise in sunspot number to maximum being more rapid than the decline to minimum. Secondly, the maximum sunspot number varies from cycle to cycle, and there are indications of an oscillation with longer period. Thirdly, sunspot records show that there have been times when the cycle seems to switch off and there are very few sunspots on the solar disc. The last occasion that the solar cycle stopped was during the years 1645 to 1720, called the *Maunder minimum*; this coincided with a mini ice age in northern Europe, when regular ice fairs were held on the Thames in London. Thus, the solar cycle is not as regular as we might initially have believed. Indeed, this irregular behaviour has been modelled by a system of differential equations that exhibit chaotic behaviour.

The lifetime of individual sunspots varies greatly depending on their size. Most spots disappear after only a few days, but large spots may exist for a few months. It was noted that some sunspot groups live for several solar rotations, and by timing the period of rotation of sunspots (and other transient phenomena) it was discovered that the sun does not rotate as a solid body. Because of its differential rotation the photosphere, at the equator, rotates every 26 days, whereas at the poles it rotates every 37 days. Sunspots, on the other hand, tend to rotate faster than the photosphere since they originate from – and are anchored near – the base of the convection zone.

The sunspot cycle points to the existence of a solar dynamo that is similar to the one responsible for the Earth's magnetic field. Dynamo theory involves the interplay between the differential rotation of the sun and the solar magnetic field. For example, because of the differential rotation, a purely poloidal magnetic field will be stretched out, near the equator, to generate a toroidal field component. Details of dynamo theory can be found in, for example, Moffatt (1978) and Priest (1982).

It is necessary to show how the magnetic field moves from the region where it is generated by the dynamo, through the convection zone and out to the photosphere, where it appears in the form of a spot or pair of spots. This was done by a simple mechanism proposed by Parker (1955). The strong magnetic field generated at the base of the convection zone is subject to an instability called magnetic buoyancy. To illustrate the concept of magnetic buoyancy, consider a horizontal, uniform magnetic flux tube of field strength B_i, pressure p_i, density ρ_i and temperature T_i surrounded by a field-free region of pressure p_e, density ρ_e and temperature T_e. From pressure balance we have

$$p_e = p_i + \frac{B_i^2}{2\mu}. \tag{10.4.1}$$

This implies that $p_e > p_i$ and, if the temperatures are equal so that $T_e = T_i$, it follows that $\rho_i < \rho_e$. Thus, the gas inside the magnetic flux tube is lighter than the external gas. Since the flux tube is lighter than its environment, it is buoyant and begins to rise. If the tube is distorted, as shown in Fig. 10.3, the small kink will rise and break through the photosphere as a pair of sunspots. As the tube continues to rise, the photospheric footpoints of the field will move apart: this is accordingly a signature of emerging magnetic fields. Such a simple model for magnetic buoyancy gives a risetime that is much shorter than the 11 y necessary for the solar cycle. Various suggestions have been put forward to delay the onset of the magnetic buoyancy instability, and recent theories predict that the field is held down, while the dynamo builds up the field strength, in the overshoot region at the base of the convection zone.

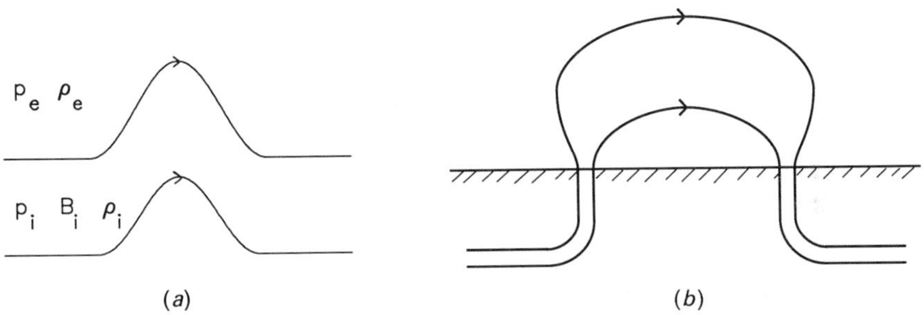

Fig. 10.3. (*a*) An isolated flux tube, surrounded by a field-free region, is perturbed in the manner shown. The perturbation continues to grow when $\rho_i < \rho_e$. (*b*) The rising part of the flux tube will break through the photosphere to produce a bipolar region of the form shown.

Apart from sunspots, there is strong evidence of magnetic fields when observing in the light of Hα. Hα is formed in the chromosphere and provides information about temperature, velocity and the magnetic field from its intensity, Doppler shifts and Zeeman splitting. Fig. 10.4 shows a typical Hα photograph of a small active region with a sunspot in the centre. The Hα structure appears to outline the shape of the magnetic field.

10.5 Solar wind

Parker (1958) realized that, since the corona is so hot and interstellar space is so cold, the corona cannot be in static equilibrium. He imagined that the corona is continually expanding outwards, with an outflow of plasma from the sun. This model was a triumph for theory, since it predicted some of the properties of the solar wind that were subsequently verified by satellites. Assuming a steady, spherically symmetric flow, the MHD equations can be written in the following form. Mass continuity becomes

$$\frac{d}{dr}(r^2 \rho v) = 0, \tag{10.5.1}$$

the momentum equation is

$$\rho v \frac{dv}{dr} = -\frac{dp}{dr} - \frac{\rho g R_\odot^2}{r^2}, \tag{10.5.2}$$

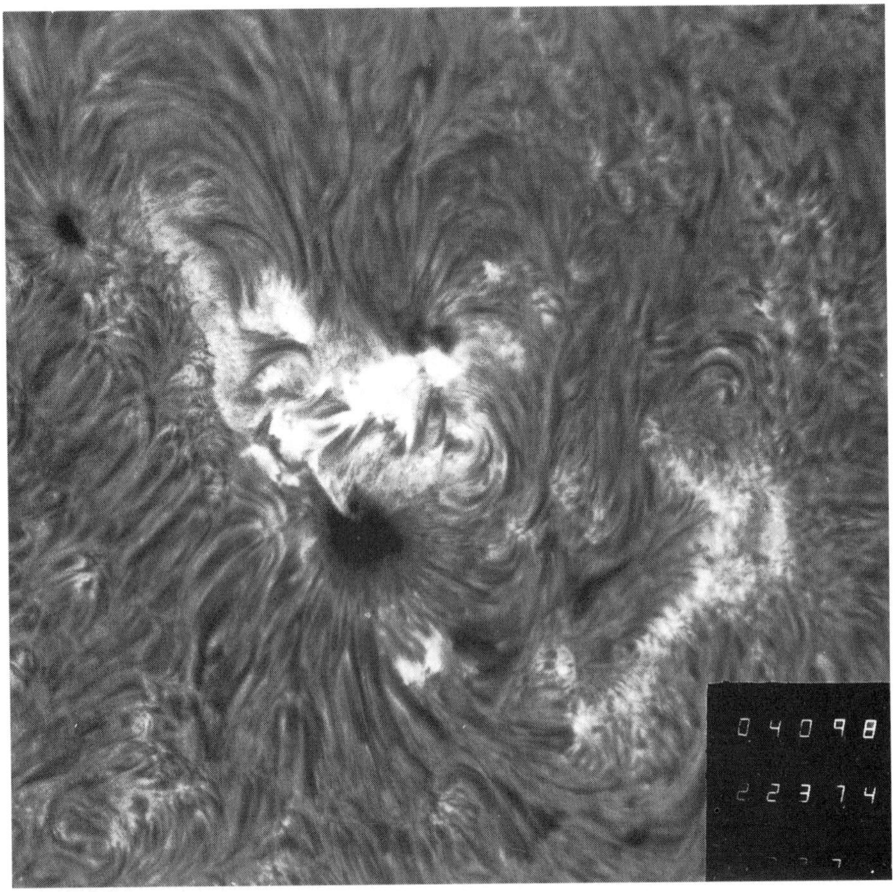

Fig. 10.4. A small active region in the light of Hα. The structures outline the local magnetic field in a manner similar to the way iron filings show up the field lines near a bar magnet. A sunspot is clearly visible in the lower centre of the photograph. (Courtesy BBSO.)

the ideal gas law is

$$p = \rho \mathscr{R} T, \tag{10.5.3}$$

and the energy equation is

$$T = \text{const.}, \tag{10.5.4}$$

where v is the radial velocity, ρ the density, p the pressure, g the surface gravitational acceleration, T the temperature, r the radius and $R_\odot$ the solar radius. Eqs (10.5.1)–(10.5.4) can be rearranged in terms of the radial velocity as

$$\left(v - \frac{v_c^2}{v}\right)\frac{dv}{dr} = 2\frac{v_c^2}{r} - g\frac{R_\odot^2}{r^2}, \tag{10.5.5}$$

where $v_c^2 = \mathscr{R}T$ is the square of the isothermal sound speed. Eq. (10.5.5) has a singularity at the sonic point $v = v_c$ and $r = r_c$, where the critical radius is $r_c = gR_\odot^2/2v_c^2$. Solving Eq. (10.5.5) gives the classic Parker solar-wind solution, namely

$$\left(\frac{v}{v_c}\right)^2 - \log\left(\frac{v}{v_c}\right)^2 = 4\log\left(\frac{r}{r_c}\right) + 4\frac{r_c}{r} + \text{const.} \tag{10.5.6}$$

Choosing different values for the constant gives the family of solutions shown in Fig. 10.5. Two physically acceptable solutions are possible, namely the solar breeze and the solar wind corresponding to subsonic and supersonic flows at large distances, respectively. The solar wind is the correct solution, since the pressure falls to zero at large distance whereas the solar breeze solution predicts an unphysical finite pressure. This simple model correctly predicts that the solar wind becomes supersonic, with $v > v_c$ beyond the critical radius r_c, and has been confirmed by satellite missions. More details of solar-wind models can be found in Hundhausen (1972).

Fig. 10.5. Solutions of the solar-wind equation (10.5.6). The physically relevant solution starts at a low velocity near the solar surface, passes through the critical point at $r/r_c = 1$ and $v/v_c = 1$, and becomes supersonic at large distances.

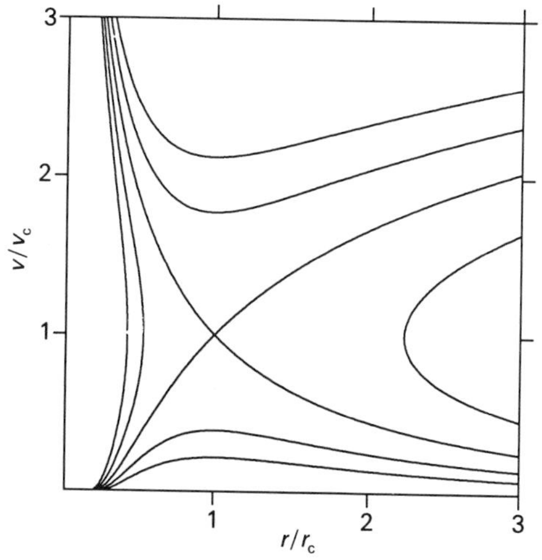

10.6 Prominences

Prominences, or synonymously filaments, consist of cool dense plasma supported high in the corona by the local magnetic field. On the disc, they appear as dark ribbon-like structures, as shown in Fig. 10.6, whereas on the limb they are bright since the corona is so tenuous. Quiescent prominences have a lifetime of approximately 200 days and tend to be situated away from active regions and sunspots, whereas active-region filaments have a much shorter lifetime and are located in active regions. The two types of prominence have different characteristics. Quiescent prominences have a temperature between 5000 and 10 000 K, and a number density of 10^{16}–$10^{17}\,\mathrm{m}^{-3}$. The typical dimensions are: length = 2×10^8 m, width = 6×10^6 m, and height = 5×10^7 m. From this we see that prominences are much longer than either their height and width and, as a first step in modelling, are usually assumed to be two-dimensional. Active-region prominences are denser ($> 10^{17}\,\mathrm{m}^{-3}$) and smaller by a factor of three or four. The line-of-sight magnetic field has been measured in quiescent prominences lying at the solar limb with a value between unobservable and 4×10^{-3} T (or 40 G). A typical value would appear to be about 10^{-3} T (or 10 G).

Since prominences exist for many Alfvén time scales, they are globally stable equilibria. However, instabilities can be triggered when critical conditions are reached, and the prominence erupts in a matter of an hour or so. An erupting

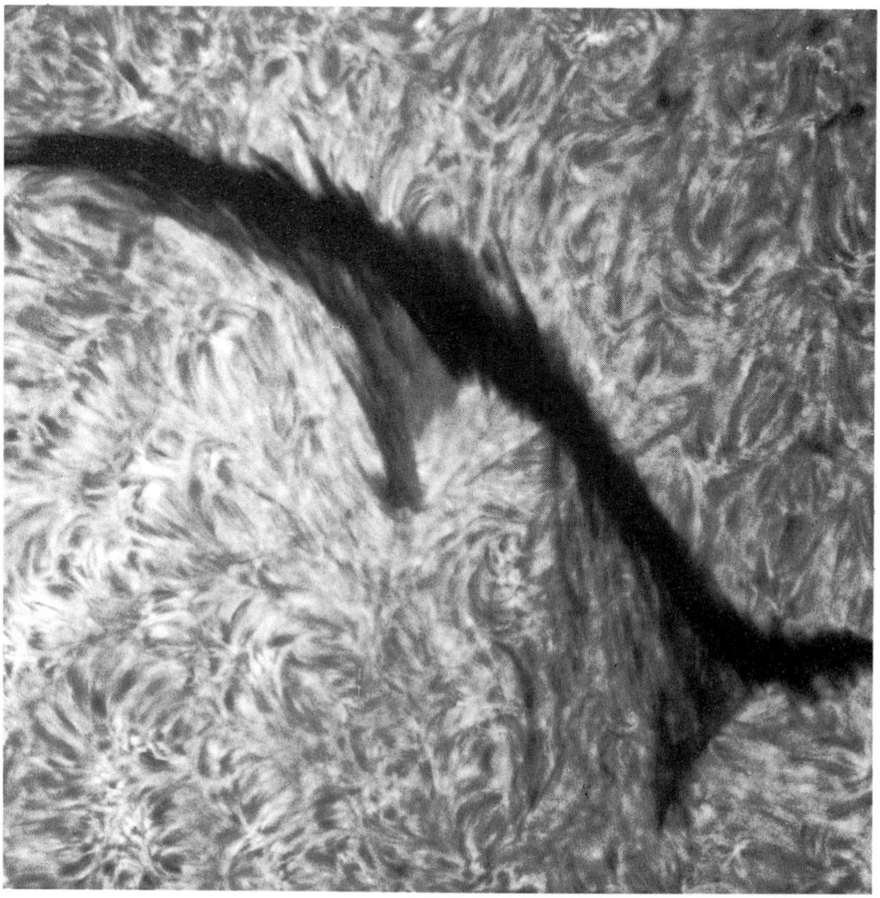

Fig. 10.6. A quiescent prominence observed on the solar disc. (Courtesy BBSO.)

prominence on the solar limb is seen in Fig. 10.7.

The number of prominences observed at any one time on the solar disc varies with the solar cycle. Fig. 10.7 shows the solar surface near sunspot maximum, and there are many quiescent prominences visible, as well as an erupting prominence. Quiescent prominences tend to form close to an active region, but then drift polewards. As they do so, differential rotation tends to stretch them until they lie almost in the toroidal direction, forming the polar crown. This can be seen in the prominence in Fig. 10.7 that is near the pole.

Prominences are always observed to lie above the polarity inversion line. This line separates regions of positive magnetic polarity from negative polarity, and is clearly shown in solar magnetographs that map out the light-of-sight photospheric magnetic field. Two classes of prominence have been observed. These are *normal polarity* prominences, in which the magnetic field passes through the prominence in the same direction as would be expected from the photospheric field, and *inverse polarity* prominences, in which the field passes through the prominence in the opposite direction to the photospheric field.

Since prominences are about 100 times denser than the surrounding coronal plasma, they must be supported against the force of gravity by the local magnetic field. Thus the Lorentz force must balance gravity. One of the first simple models

Fig. 10.7. Several quiescent prominences are seen on the solar disc. Near the solar pole, a quiescent prominence has been stretched out by differential rotation to form the polar crown. Also visible is an erupting prominence at the solar limb. (Courtesy BBSO.)

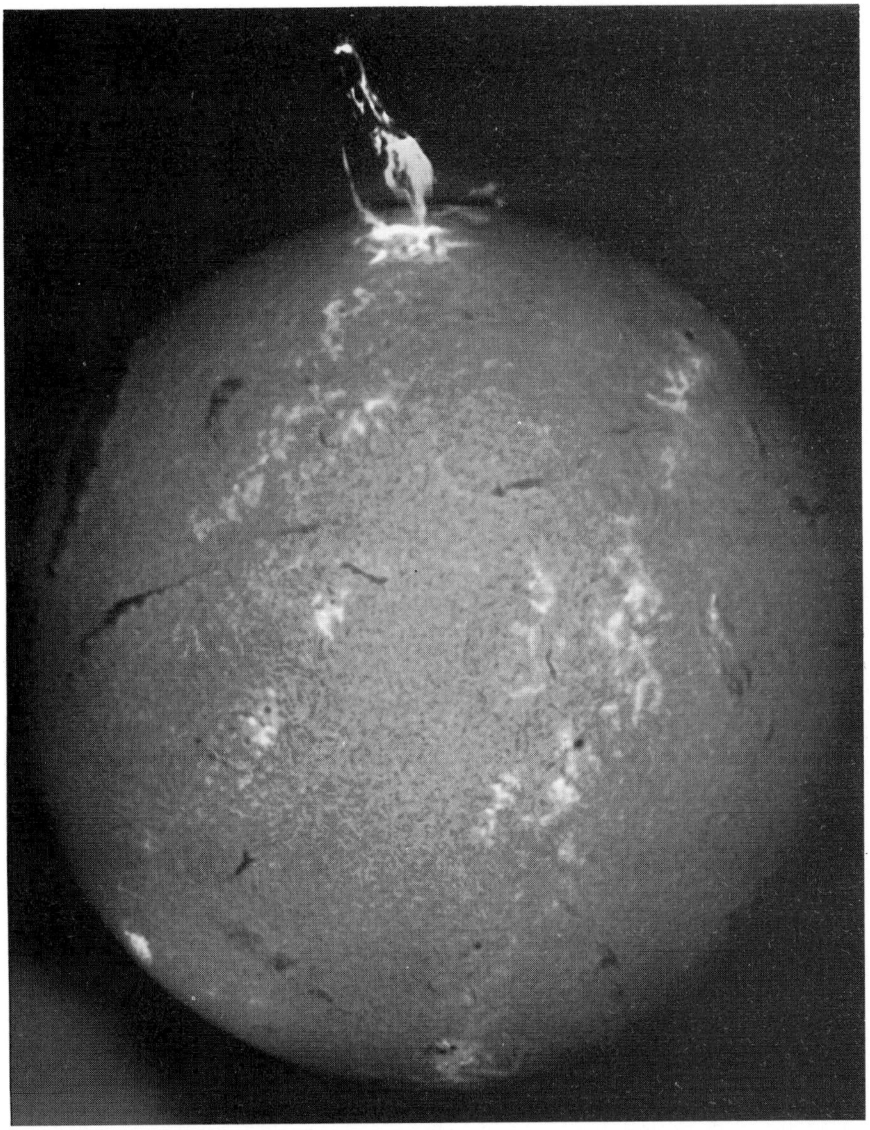

to explain the magnetic support of normal polarity prominences is the classic Kippenhahn–Schlüter model (Kippenhahn and Schlüter, 1957), which we now describe.

Let us assume that the prominence is in static, isothermal equilibrium, and that the plasma only varies in the horizontal x-direction (see Fig. 10.8) so that

$$\boldsymbol{B} = (B_{x0}, B_{y0}, B_z(x)), \tag{10.6.1}$$

$$p = p(x), \quad \rho = \rho(x), \quad T = \text{const.} \tag{10.6.2}$$

Then the MHD equations reduce to

$$\frac{\mathrm{d}p}{\mathrm{d}x} = -\frac{1}{2\mu}\frac{\mathrm{d}}{\mathrm{d}x}(B_z^2), \tag{10.6.3}$$

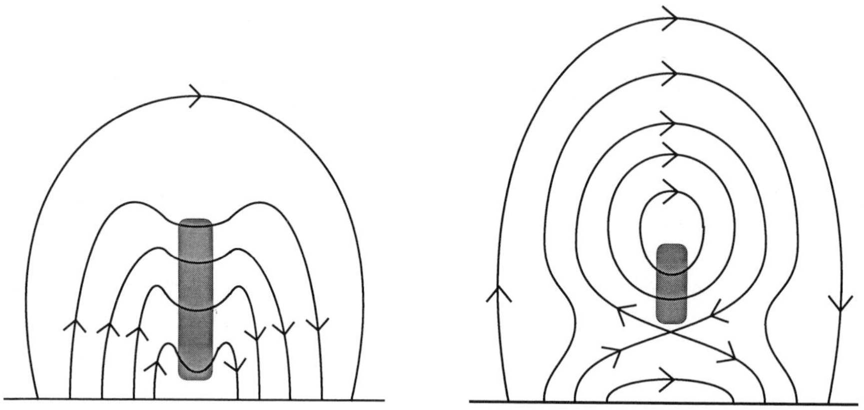

Fig. 10.8. (*a*) The Kippenhahn–Schlüter (1957) normal polarity prominence mode. (*b*) The inverse polarity model of Kuperus and Raadu (1974).

$$\frac{B_{x0}}{\mu}\frac{dB_z}{dx} = \rho g, \tag{10.6.4}$$

$$p = \rho \mathcal{R} T. \tag{10.6.5}$$

Integrating Eq. (10.6.3) and using the boundary condition that $p \to 0$ and $B_z \to B_{z\infty}$ as $x \to \infty$ gives

$$2\mu p + B_z^2 = B_{z\infty}^2, \tag{10.6.6}$$

and, using Eqs (10.6.5) and (10.6.6) to eliminate the pressure and density, Eq. (10.6.4) reduces to a simple equation for B_z. The final solutions are

$$\mathbf{B} = (B_{x0}, B_{y0}, B_{z\infty} \tanh(x/l)), \tag{10.6.7}$$

$$p = \frac{B_{z\infty}^2}{2\mu} \operatorname{sech}^2(x/l), \tag{10.6.8}$$

where the typical width of the prominence is $l = 2\mathcal{R} T B_{x0}/g B_{z\infty}$. This model exhibits two effects of the magnetic field. First, the horizontal magnetic pressure compresses the plasma and enhances the pressure and density in the prominence. Secondly, the vertical magnetic tension force supports the dense plasma against gravity. However, the model has several weaknesses: for example, the isothermal assumption and the one-dimensional nature of the variations mean that the magnetic field does not return to the photosphere. A simple extension of this model by Hood and Anzer (1990) has removed these weaknesses, and produced a field structure similar to the standard cartoon models of normal polarity prominences.

A cartoon model for an inverse polarity prominence (see Fig. 10.8*b*) was first proposed by Kuperus and Raadu (1974). However, the first demonstration of an inverse polarity prominence in MHD equilibrium was given by Ridgway, Amari and Priest (1990).

The next problem for theorists is to explain how such a cool plasma can form in the solar corona when the surrounding plasma is so hot. One possible mechanism is a thermal instability, originally suggested by Parker (1953). The static energy balance equation is obtained by neglecting the total time derivative term on the left-hand side of Eq. (10.1.4). Using Eqs (10.1.6) and (10.1.7), it can be written in the form

$$\boldsymbol{B} \cdot \nabla\left(\frac{\kappa_{\|}}{B^2}\boldsymbol{B} \cdot \nabla T\right) = \mathscr{L} \equiv \rho^2 Q_1(T) - h(\rho, T, B), \tag{10.6.9}$$

where the left-hand side represents thermal conduction along the equilibrium magnetic field (thermal conduction across the fields is neglected), the first term on the right-hand side represents the radiative losses due to optically thin radiation, and the second term the unknown coronal heating term. The actual dependence of the heating term on density, temperature and magnetic field strength is unknown, but is not crucial for this argument. The loss function $\mathscr{L}$ is shown in Fig. 10.9 assuming that the gas pressure is constant. Now consider a coronal plasma at 10^5 K. If the plasma cools slightly, the temperature drops, but the

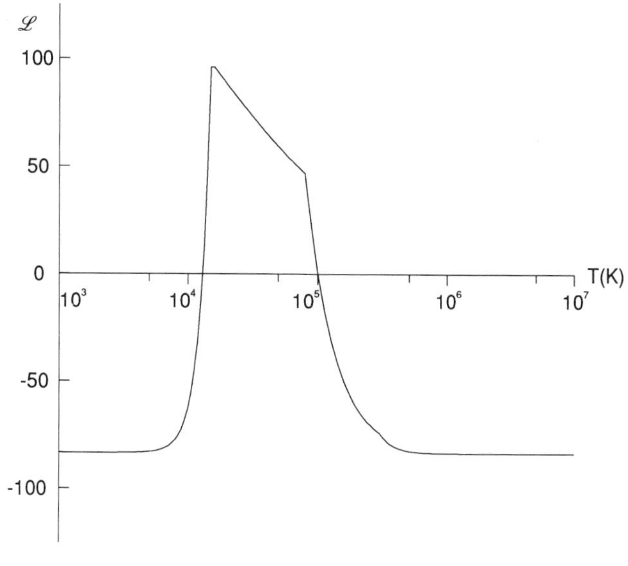

Fig. 10.9. A schematic sketch of the loss function, $\mathscr{L}$, as a function of the logarithm of the temperature. Assuming that the gas pressure is constant, the gas law, Eq. (10.1.5), has been used to eliminate ρ from Eq. (10.6.9).

radiative losses actually increase. Thus the plasma cools further, and a runaway instability can take place. It is only when the temperature drops to about 10^4 K that the instability stops and the plasma can reach a new equilibrium. In the absence of magnetic effects, the instability can exist in two different forms called the *isobaric* and *isochoric* modes.

Consider the simple model for isobaric modes. Assuming an isothermal uniform equilibrium, with $p_1 = 0$ and $\boldsymbol{B}_1 = 0$, the linearized gas law becomes

$$\frac{\rho_1}{\rho} + \frac{T_1}{T} = 0. \tag{10.6.10}$$

For disturbances of the form $f(\boldsymbol{r}, t) = f(t)\exp(i\boldsymbol{k} \cdot \boldsymbol{r})$, then the linearized form of the energy equation, Eq. (10.1.4), can be written as

$$\frac{1}{\gamma - 1}\frac{p}{T}\frac{\partial T_1}{\partial t} = -\kappa_{\|}k_{\|}^2 T_1 - \left(\frac{\partial \mathscr{L}}{\partial T}\right)_p T_1. \qquad (10.6.11)$$

The partial derivative $(\partial \mathscr{L}/\partial T)_p$ is evaluated assuming that p is held constant, and $k_{\|} \equiv (\boldsymbol{k} \cdot \boldsymbol{B})/B$ is the wave number parallel to the equilibrium magnetic field. Instability occurs when the right-hand side of the equation is positive:

$$-\kappa_{\|}k_{\|}^2 - \left(\frac{\partial \mathscr{L}}{\partial T}\right)_p > 0. \qquad (10.6.12)$$

For further details of thermal instabilities, see Hood (1992) and the references therein.

10.7 Solar flares

In simple terms, a solar flare is a rapid brightening in Hα due to the release of a large amount of magnetic energy. Large flares are amongst the most violent events in the solar system, with up to 10^{25} J of energy being released in an hour or so. A simple evaluation of all the available energy sources shows that the only available reservoir of energy is the coronal magnetic field. Theoretically, flares are divided up into two main classes, namely *two-ribbon flares* and *simple-loop flares*.

Two-ribbon flares are very large events and are nearly always associated with the eruption of an active-region filament. Two bright bands (or ribbons) of Hα emission are observed, as shown in Fig. 10.10, and as time progresses the ribbons move further apart. In addition, there is strong evidence of a restructuring and simplification of the magnetic field. *Post-flare loops* are frequently seen joining the two ribbons. On the limb, post-flare loops form, and plasma is seen draining down the loop legs. Subsequent loops are observed to form at higher heights, and their footpoints are located on the photosphere further from the filament site. So the apparent movement of the flare ribbons and the height of the post-flare loop formation are closely linked together. In fact post-flare 'loop' is a misnomer, since the main release of energy occurs during this stage.

The simple-loop flares are much smaller events, releasing 10^{23} J of energy; they consist of single or groups of loops that brighten without any major change in their overall shape.

There are many flare models but, since flares are exceedingly complex, there is no detailed model that predicts all the features present in a typical flare. Normally the overall process is divided into simpler models. First, the coronal magnetic field must be able to store the required energy. This occurs as the coronal field responds to the motion of the photospheric footpoints that inject stresses into the corona. During this energy build-up, the coronal magnetic field must be stable against ideal MHD disturbances that evolve on the local Alfvén time and could destroy the equilibrium in a matter of minutes. Next, an instability must be triggered to begin the initial release of the magnetic energy. Present theories suggest that this occurs when an ideal MHD instability is initiated and drives magnetic reconnection on an Alfvénic time scale. Other aspects of the overall flare process, including the energy release and magnetic field restructuring by magnetic reconnection, are discussed in the next section. However, an explanation of how electrons are accelerated to high velocities requires a microscopic plasma physics approach that cannot be covered by MHD theory. Thus, a solar

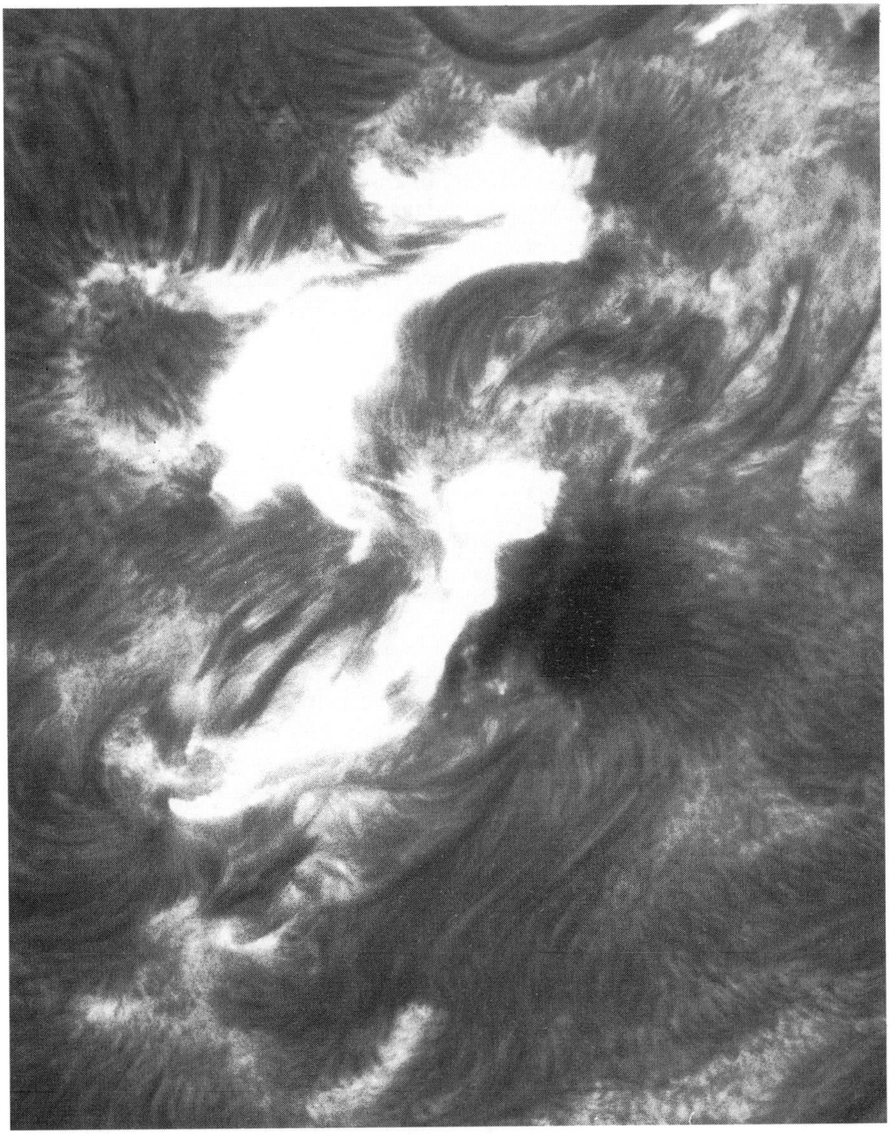

Fig. 10.10. A two-ribbon flare on the solar disc. (Courtesy BBSO.)

flare is a complex phenomenon that cannot be covered in this chapter. Further details of observations of flares are given in the books by Svestka (1976, 1981), and of MHD theories by Priest (1982) and Hood (1991).

Now we present a model for a simple loop flare. Assume the loop is a straight cylinder of length L. Therefore, the magnetic field is given by

$$\boldsymbol{B} = (0, B_\theta(r), B_z(r)), \tag{10.7.1}$$

and

$$p = p(r) \text{ and } \Theta = \frac{LB_\theta}{rB_z} \propto \frac{1}{q}. \tag{10.7.2}$$

Here Θ is the angle through which a field line is twisted in passing from one end of the loop to the other, which is related to the reciprocal of the safety factor q. The equilibrium satisfies

$$\frac{d}{dr}(\mu p) + \frac{d}{dr}\left(\frac{1}{2}B^2\right) = -\frac{B_\theta^2}{r}. \tag{10.7.3}$$

To investigate the ideal MHD stability of coronal loops, it is convenient to use the energy method of Bernstein *et al.* (1958) with a trial function for the coronal displacement ξ. If there exists a trial function, satisfying the appropriate boundary conditions, which makes the perturbed potential energy negative, then the equilibrium is unstable. On the other hand, if all allowable displacements make the perturbed potential energy positive, the plasma is stable. For an ideal plasma, the perturbed potential energy is given by

$$\delta W = \frac{1}{2\mu}\int\{|\nabla \times (\xi \times \boldsymbol{B})|^2 + \nabla \times \boldsymbol{B} \cdot [\xi \times (\nabla \times (\xi \times \boldsymbol{B}))] \\ + (\xi \cdot \nabla\mu p)\nabla \cdot \xi + \gamma\mu p(\nabla \cdot \xi)^2 + (\xi \cdot \boldsymbol{g})\nabla \cdot (\mu\rho\xi)\}dV. \tag{10.7.4}$$

The integration is over the coronal volume. The advantage of the energy method is that one can use trial functions that need not satisfy the equation of motion to test for stability. For further details see, for example, Bernstein *et al.* (1958) and Hood (1985).

When analysing the stability of coronal magnetic fields, it is assumed that the disturbances originate in the corona. Since the density rises by a factor of over 10^8 from the corona to the photosphere, the photosphere is modelled by assuming that all disturbances vanish at the footpoints, (Raadu, 1972). Thus, the coronal displacement ξ satisfies the boundary conditions

$$\xi = 0 \text{ at } z = 0, L. \tag{10.7.5}$$

These boundary conditions are frequently called *line-tying* boundary conditions, since the magnetic footpoints are thought to be tied to the dense photospheric plasma. In the original application of the energy method to cylindrically symmetric fields, but with no axial line-tying, Newcomb (1960) showed how the trial function

$$\xi(\boldsymbol{r}) = \xi(r)\exp(im\theta + ikz)$$

can be used to determine the stability of the plasma. The determination of the sign of the energy integral is reduced to establishing whether the radial component of ξ has a zero or not. Raadu (1972) modified this choice of ξ by multiplying by an amplitude factor that makes the displacement satisfy Eq. (10.7.5). Thus, he selected

$$\xi(\boldsymbol{r}) = \xi(r)\exp(im\theta + ikz)\sin(\pi z/L). \tag{10.7.6}$$

This approach was used by Hood and Priest (1979) to show that the uniform twist field (Gold and Hoyle, 1960)

$$\boldsymbol{B} = \left(0, \frac{r}{1+r^2}, \frac{\lambda}{1+r^2}\right), \quad \mu p = \frac{1}{2}\frac{(1-j^2)}{(1+r^2)^2}, \quad \Theta = \frac{L}{\lambda}, \tag{10.7.7}$$

where L is the dimensionless length of the loop, was definitely unstable when the twist exceeded 3.3π and $\lambda = 1$. The original version was for the force-free field with $\lambda = 1$. Hood and Priest (1981) solved the full partial differential equations to show that marginal stability occurs at the critical value of twist, for the force-free field, given by $\Theta_{\text{crit}} = 2.49\pi$.

Other magnetic fields have been investigated, and in most cases it is found that the field becomes unstable to an ideal kink mode when the *average* twist exceeds the Hood and Priest value of 2.5π, where the average twist is defined as

$$\langle \Theta \rangle = \frac{1}{a}\int_0^a \Theta \mathrm{d}r,$$

and a is the radius of the loop.

Having found the critical conditions for the onset of an ideal MHD instability, it is important to estimate how much magnetic energy can be released. Consider a loop of radius 10^4 km and a typical active region field strength of 100 G (10^{-2} T). From the critical twist of 2.5π, the length of the loop is approximately 8×10^4 km. The available magnetic energy is essentially due to the twisting motions that generate the poloidal field component, B_θ. Thus, the free magnetic energy is

$$\int \frac{B_\theta^2}{2\mu} \mathrm{d}V \approx 10^{24}\ \mathrm{J},$$

where the integration is taken over the volume of the coronal loop. It is apparent that there is enough energy available to drive a small loop flare but insufficient energy for a large two-ribbon flare. Note, however, that it is still necessary to release this energy since an ideal instability on its own cannot do this. It follows that nonideal effects must become important, and it is thought that resistivity is needed to allow the magnetic field to relax to a lower energy – and topologically different – configuration.

Two-ribbon flares are thought to occur in a different magnetic topology, namely a coronal arcade. Here the magnetic field lines form a tunnel-like configuration, as indicated in Fig. 10.11. If the field is approximately force-free, the line-tying effect of the dense photosphere plays an important role in stabilizing the equilibrium. The dangerous kink mode is now prohibited by the line-tying and, in all cases considered so far, force-free fields are found to be completely stable if the magnetic axis lies on or below the photosphere. Instability has only been shown if the magnetic axis exceeds a critical height above the photosphere, so that a substantial fraction of the arcade is no longer connected to it. Thus, arcades are remarkably stable, and it is possible to store a large amount of magnetic energy in such a configuration before an instability sets

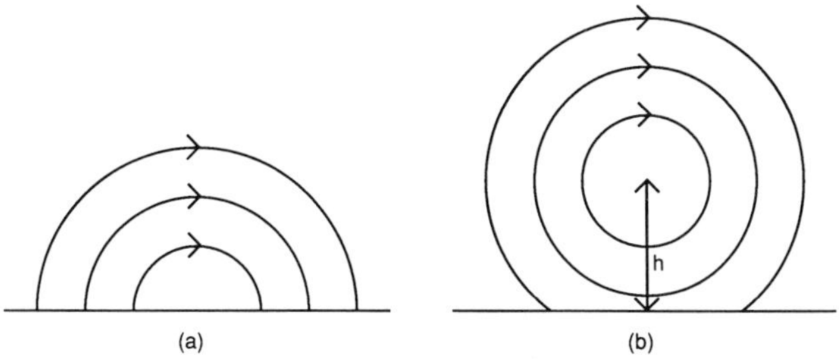

Fig. 10.11. A coronal arcade, viewed end-on with the magnetic axis (a) on or below the photosphere and (b) at a height h above the photosphere.

in. Further details of ideal and nonideal linear instabilities can be found in Hood (1992) and the references therein.

10.8 Magnetic reconnection

It has already been mentioned that an ideal instability cannot change the magnetic topology, and so the onset of an ideal instability is not the final answer to the flare problem. Current thinking is that the ideal motions push magnetic field lines closer together and cause magnetic reconnection to occur, releasing magnetic energy at a rate proportional to the Alfvén speed.

Magnetic reconnection can be defined by the simple model presented in Fig. 10.12. Fig. 10.12(*a*) shows the initial magnetic equilibrium in the form of two antiparallel fields. If the field is perturbed in the manner shown in Fig. 10.12(*b*) then in ideal MHD the magnetic pressure simply builds up near the neutral line and opposes the disturbance. However, the presence of resistivity, in Fig. 10.12(*c*), means that the magnetic field diffuses faster in the region where the current is stronger (and the field lines are closer together). This diffusion reduces the magnetic pressure and removes the original opposing force, so that the disturbance is no longer inhibited. The field lines are essentially 'cut', and they reconnect to form the topology shown in Fig. 10.12(*d*).

The release of magnetic energy occurs over several Alfvén times, so that an approximately steady state situation can develop. A simple model for reconnection is the Sweet–Parker model, originally proposed by Sweet (1958) and Parker (1963). Consider a diffusion region of length L, where L is a typical length scale of the external magnetic field, and width l. As indicated in Fig. 10.13, the inflow region has density, temperature and pressure defined by ρ_i, T_i and p_i, respectively. The outflow variables are ρ_o, T_o and p_o. Simple order of magnitude calculations describe the main characteristics of reconnection in this situation. Since the

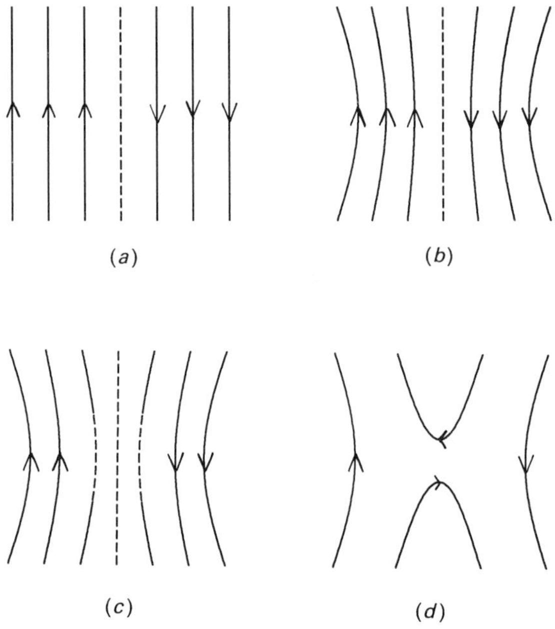

Fig. 10.12. (*a*) The equilibrium configuration. (*b*) In ideal MHD, the displacement shown is inhibited by a build-up of magnetic pressure. (*c*) In resistive MHD, the magnetic field can diffuse faster where the current is stronger. (*d*) Reconnection can therefore alter the topology of the initial configuration.

11
Gravitational plasmas

J. J. BINNEY

Star clusters, galaxies and, indeed, the universe as a whole, may to a good approximation be considered to be made up of point particles – stars, galaxies or 'inos' – whose only significant mutual interactions are provided by Newtonian gravity. The latter is an inverse-square force like the electrostatic force, so it is natural that there should be many similarities between the dynamics of these systems (conventionally termed 'stellar dynamics') and plasma physics. Moreover, we shall see that the role in an electrostatic plasma of the magnetic field is frequently mimicked by the Coriolis force that arises when we study a stellar system in a natural rotating frame. So even in this brief survey of stellar dynamics, we shall encounter several processes familiar from plasma physics, playing important roles in structuring stellar systems.

Of course, gravity, unlike the electrostatic interaction, is always attractive, and this crucial difference between the two systems invariably leaves its imprint. Consequently, there are often highly significant differences between the way a given process works in the plasma and stellar-dynamical contexts. Highlighting these differences illuminates the underlying physics in each system and such cross-fertilization should enrich each field. Probably we have not exploited these parallels as fully as we might.

Obviously, in a chapter of this length it is not possible to more than dip into stellar dynamics. Much more will be found on every topic in *Galactic Dynamics* by Binney and Tremaine (1987). A valuable mine of information on the stability of stellar systems is provided by Fridman and Polyachenko (1984). Some interesting parallels between plasma physics and stellar dynamics will be found in Lynden-Bell and Wood (1968).

For simplicity I shall refer to the particles of a gravitational plasma as 'stars'; in most applications the particles are indeed stars. But the theory applies to any objects that interact predominantly through gravity and have negligible geometrical cross-sections. So the particles might equally be massive neutrinos or massive black holes.

11.1 The universe

The cosmic microwave background provides strong evidence that the universe started out extremely homogeneous. Consequently, one of the goals of modern astrophysics is to understand how the structures we see about us grew out of the original homogeneous gravitational plasma. So let us investigate the linear stability of a homogeneous self-gravitating plasma.

As soon as one starts to think about this problem, one encounters the difficulty that with standard gravity there is no equilibrium configuration of an infinite homogeneous medium to perturb. That is, with standard attractive gravity the universe must be either expanding, or contracting, or be instantaneously in transition from expansion to contraction. Its density cannot simply be constant.

Of course there is no fundamental difficulty in studying perturbations of an expanding or a contracting homogeneous universe. But expansion or contraction does significantly complicate linear analysis, and valuable insight can be obtained by perturbing a somewhat artificial equilibrium. We imagine that the self-attraction of the homogeneous background is precisely cancelled by a repulsive force due to a 'cosmological constant'. In this case, sometimes called the Jeans swindle, the only gravitational force acting on particles is that, $-\boldsymbol{\nabla}\delta\Phi$, associated with the difference $\delta\rho$ between the density ρ and its background value. Specifically the perturbed gravitational potential satisfies

$$\nabla^2\delta\Phi = 4\pi G\delta\rho. \tag{11.1.1}$$

Let the plasma's distribution function be

$$f(\boldsymbol{x},\boldsymbol{v}) = f_0(\boldsymbol{v}) + \delta f(\boldsymbol{x},\boldsymbol{v}). \tag{11.1.2}$$

Then the translational invariance of the problem in space and time allows us to assume that $\delta\Phi$ and δf are of the form

$$\begin{aligned}\delta\Phi &= \mathrm{Re}\{\Phi_1\exp[\mathrm{i}(\boldsymbol{k}\cdot\boldsymbol{x} - \nu t)]\},\\ \delta f &= \mathrm{Re}\{f_1(\boldsymbol{v})\exp[\mathrm{i}(\boldsymbol{k}\cdot\boldsymbol{x} - \nu t)]\}.\end{aligned} \tag{11.1.3}$$

Inserting this form into the collisionless Boltzmann equation, Eq. (2.7.1) of Chapter 2, and linearizing, we easily find

$$f_1(\boldsymbol{v}) = \Phi_1\frac{\boldsymbol{k}\cdot\partial_{\boldsymbol{v}}f_0}{\boldsymbol{k}\cdot\boldsymbol{v} - \nu}. \tag{11.1.4}$$

We have to integrate f_1 over all $\boldsymbol{v}$ so as to obtain the density perturbation to which $\delta\Phi$ gives rise. For simplicity let us adopt

$$f_0(\boldsymbol{v}) = \frac{\rho_0}{(2\pi\sigma^2)^{3/2}}\exp(-v^2/2\sigma^2). \tag{11.1.5}$$

Then on orienting our axes so that $\boldsymbol{k} = k\hat{\boldsymbol{e}}_x$ we can easily perform the integrals over v_y and v_z. Poisson's equation, Eq. (11.1.1), then becomes

$$\begin{aligned}-k^2\Phi_1 &= 4\pi G\rho_1 = 4\pi G\int \mathrm{d}^3\boldsymbol{v}\, f_1\\ &= -2(2\pi)^{1/2}k\Phi_1\frac{G\rho_0}{\sigma^3}\int_{-\infty}^{\infty}\mathrm{d}v_x\frac{v_x\exp(-v_x^2/2\sigma^2)}{kv_x - \nu}.\end{aligned} \tag{11.1.6}$$

Cancelling Φ_1 from the extreme left and extreme right of Eq. (11.1.6), and performing the last remaining integral, we obtain the dispersion relation we are after.

Unfortunately, as in the case of an electrostatic plasma, the only real value of ν for which the final integral is well defined is $\nu = 0$; with any other real value of ν the integrand has a nonintegrable singularity at $kv_x = \nu$. However, the integral *is* well defined for pure imaginary v, so let us experimentally write $v = \mathrm{i}\gamma$, where γ is real. Then the dispersion relation can be written

$$k^2 = k_\mathrm{J}^2 \frac{1}{\pi^{1/2}} \int_{-\infty}^{\infty} \mathrm{d}u \frac{u \exp(-u^2)}{u - \mathrm{i}s}, \tag{11.1.7a}$$

where

$$s \equiv \frac{\gamma}{2^{1/2}\sigma k}, \tag{11.1.7b}$$

and the Jeans wave number k_J is given by

$$k_\mathrm{J}^2 \equiv \frac{4\pi G\rho_0}{\sigma^2}. \tag{11.1.8}$$

Breaking the integral in Eq. (11.1.7a) into two halves, we can show that it is real and evaluate it:

$$\begin{aligned} k^2 &= \frac{k_\mathrm{J}^2}{\pi^{1/2}} \left(\int_0^{\infty} \mathrm{d}u \frac{u \exp(-u^2)}{u - \mathrm{i}s} + \int_0^{\infty} \mathrm{d}u \frac{u \exp(-u^2)}{u + \mathrm{i}s} \right) \\ &= \frac{2k_\mathrm{J}^2}{\pi^{1/2}} \int_0^{\infty} \mathrm{d}u \frac{u^2 \exp(-u^2)}{u^2 + s^2} \\ &= k_\mathrm{J}^2 \{1 - \pi^{1/2} s \exp(s^2)[1 - \mathrm{erf}(s)]\}. \end{aligned} \tag{11.1.9}$$

It is easy to see from Eq. (11.1.9) that $k = k_\mathrm{J}$ when the growth rate γ is zero. Fig. 11.1 shows that as k decreases towards zero, γ rises to $(4\pi G\rho_0)^{1/2}$. Thus the plasma is unstable to all disturbances of wavelength greater than the Jeans length,

$$\lambda_\mathrm{J} \equiv 2\pi/k_\mathrm{J} = \sigma(G\rho_0/\pi)^{-1/2}. \tag{11.1.10}$$

Strangely, Eqs (11.1.7) cannot be solved for $k > k_\mathrm{J}$: we have seen that any pure imaginary v (real s) corresponds to $k < k_\mathrm{J}$, whereas the integral is undefined for

Fig. 11.1. Long wavelength waves in a gravitational plasma are unstable. Here the growth rate γ is plotted against wave number k.

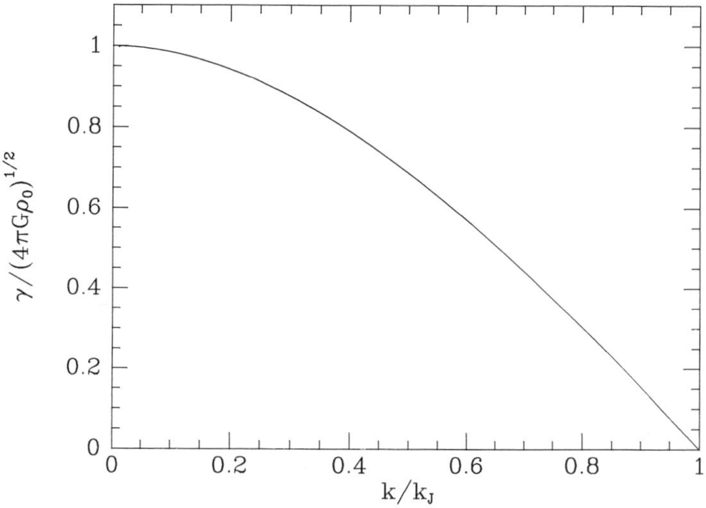

pure real v. Finally it is not hard to prove that fully complex values of v are incapable of solving Eqs (11.1.7) because they endow the integral with a nonzero imaginary part. As in the case of an electrostatic plasma, this difficulty can be circumvented by studying Landau's initial value problem rather than seeking true normal modes; that is, one solves for the evolution of a configuration that at $t = 0$ happens to be the wave $f_1 = f_s(\boldsymbol{v})\exp(i\boldsymbol{k} \cdot \boldsymbol{x})$ with $k > k_J$. One finds (for example, Ikeuchi, Nakamura and Takahara, 1974) that the disturbances initiated by such initial conditions are strongly damped and therefore of little interest.

To sum up, as in an electrostatic plasma, short wavelength disturbances in a gravitational plasma are strongly Landau damped. Long wavelength disturbances in a gravitational plasma grow exponentially as a consequence of the Jeans instability, which arises because, in the absence of adequate resistance from thermal motions, gravity causes over-dense regions to collapse to ever higher densities. By contrast, long wavelength disturbances in electrostatic plasma propagate as mildly damped waves.

11.2 Basic concepts for inhomogeneous stellar systems

The foregoing analysis essentially exhausts the interest of homogeneous gravitational plasmas; the bottom line on this analysis is that gravity can be relied upon to destroy the homogeneity of any sufficiently large system. So, if we wish to continue our study of gravitational plasmas, we have to get to grips with inhomogeneous systems such as galaxies and star clusters.

Inhomogeneous systems are enormously harder to study than homogeneous ones. Consequently many branches of physics are almost exclusively concerned with homogeneous systems – consider, for example, basic electromagnetism, thermodynamics, most of condensed-matter physics, and so on. The study of inhomogeneous plasmas requires a number of intellectual tools that are not widely used in other branches of physics. The first of these is the concept of a nontrivial mean-field orbit.

11.2.1 *Mean-field orbits*

Gravity is in a sense an even longer range force than the electrostatic force. This is because it cannot be shielded. So there is no such thing as a gravitational Debye sphere, and two test stars immersed in a swarm of randomly moving background stars will continue to attract one another no matter how far apart they are. So it is natural to expect the force on a star to be dominated by the mean gravitational field of the entire galaxy, rather than the contributions to that field from individual neighbours. Hence a natural first approximation is to suppose that stars move on orbits in the mean gravitational field calculated from a continuum approximation to the galaxy's particulate mass distribution. In the next order of approximation, we treat the difference between this idealized force-field and the actual one as a perturbation which causes the star to drift slowly from one orbit in the mean field to another.

The two-body relaxation time $t_{2\mathrm{body}}$ is the characteristic time scale of this drift between orbits in the mean field. A simple argument yields an estimate of $t_{2\mathrm{body}}$ that is valid, even though the argument itself is shot through with internal inconsistencies. Suppose the star whose orbit is being studied passes another of mass m at impact parameter b at asymptotic speed v. Then the first star's velocity changes by an amount of order

$$\delta v \simeq \frac{2Gm}{bv}. \tag{11.2.1}$$

The characteristic surface density of stars in a galaxy of half-mass radius R is of order $N/\pi R^2$, where N is the number of stars in the galaxy. So in crossing the galaxy the first star has of order

$$\delta n = \frac{N}{\pi R^2} 2\pi b db \tag{11.2.2}$$

encounters with impact parameters near b. Each encounter will produce a velocity perturbation of magnitude given by Eq. (11.2.1) but randomly oriented in direction, so $\boldsymbol{v}$ executes some sort of random walk and it is appropriate to sum each δv in quadrature. Summing over all impact parameters $b > b_{\mathrm{min}} \simeq Gm/v^{1/2}$ for which our deflection formulae Eq. (11.2.1) is valid, we find the change in v^2 during one crossing is of order

$$\begin{aligned}\Delta v^2 &= \int_{b_{\mathrm{min}}}^{R} \left(\frac{2Gm}{bv}\right)^2 \frac{2N}{R^2} bdb \\ &\simeq 8N\left(\frac{Gm}{Rv}\right)^2 \ln \Lambda,\end{aligned}$$

where

$$\ln \Lambda \equiv \ln\left(\frac{R}{b_{\mathrm{min}}}\right). \tag{11.2.3}$$

Eq. (11.2.3) gives the change in v^2 per crossing time. By the virial theorem (Binney and Tremaine, 1987, Section 4.3) $v^2 \simeq GNm/R$. Hence after $\sim (1/8)N/\ln \Lambda$ crossing times, $\boldsymbol{v}$ has wandered to a completely different region of phase space and our star cannot be said to be on anything like its original mean-field orbit. If we define t_{2body} to be the time for a star to wander in this way to a complete uncorrelated orbit, we have

$$\frac{t_{\mathrm{cross}}}{t_{\mathrm{2body}}} \simeq \frac{\Delta v^2}{v^2} = \frac{8\ln\Lambda}{N}. \tag{11.2.4}$$

Table 11.1 gives characteristic values of N, t_{cross} and t_{2body} for star clusters and galaxies, which were typically formed several gigayears ago. From the Table one sees that star clusters are many two-body times old, whereas galaxies are much younger than a two-body time. Hence, whereas collisionless dynamics should provide an adequate approximation for galaxies, when studying clusters it will be necessary to consider the cumulative effects of two-body encounters.

11.2.2 *Quasiperiodicity and action-angle variables*

What sort of orbits do typical galactic potentials support? Fig. 11.2 shows some orbits in a typical axisymmetric potential $\Phi(R, z)$. Motion in such a potential

Table 11.1. *Characteristic times of stellar systems.*

System	$\rho, M_\odot\,\mathrm{pc}^{-3}$	$\sigma,\mathrm{km\,s}^{-1}$	r,pc	t_{cross},My	$t_{2\mathrm{body}}$,Gy	age,Gy
Centre globular cluster	10 000	7	1.5	0.4	0.06	10
Centre open cluster	100	1	1	2	0.03	2
Solar neighbourhood	0.05	15	400	300	70 000	5
$r_{1/2}$ in giant elliptical galaxy	0.03	300	7000	50	8×10^8	10
Centre of giant elliptical galaxy	30	300	500	0.3	8×10^5	10

conserves L_z, the component of the star's angular momentum parallel to the potential's symmetry axis. Consequently, the star's azimuthal angular coordinate ϕ can be eliminated from the equations of motion, which then become equations for the motion in the meridional plane (R, z), where R is cylindrical radius:

$$\ddot{R} = -\frac{\partial \Phi_{\mathrm{eff}}}{\partial R},$$

$$\ddot{z} = -\frac{\partial \Phi_{\mathrm{eff}}}{\partial z}$$

where

$$\Phi_{\mathrm{eff}}(R, z) \equiv \Phi(R, z) + \frac{L_z^2}{2R^2}. \tag{11.2.5}$$

(See for example Section 3-2 of Binney and Tremaine, 1987.) Notice the beautiful regularity of the velocity vectors in Fig. 11.2; this demonstrates that at any given point (R, z) the velocity (v_R, v_z) is determined to within a simple multiplicity, such

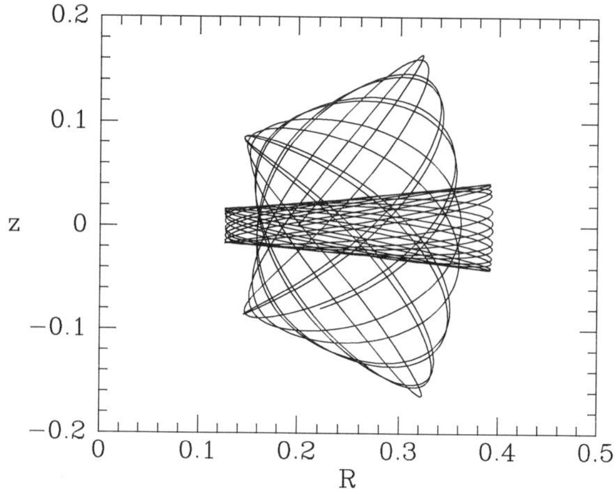

Fig. 11.2. Two orbits in the axisymmetric potential $\Phi = \frac{1}{2}\ln(R^2 + z^2/q^2)$ with $q = 0.9$. Both orbits have the same energy and angular momentum about the z-axis. At each point within a given orbit only four velocities seem to occur, implying that the orbits lie on two-surfaces in their four-dimensional phase space $(R, z, \dot{R}, \dot{z})$. This suggests that each orbit conserves an integral I_3 in addition to H and L_z, and that the orbits differ because I_3 takes different values on each orbit.

as might arise from one or more sign ambiguities. In other words, these orbits are constrained in the four-dimensional phase space (R, z, v_R, v_z) to two-dimensional surfaces. One likes to think of these two-dimensional surfaces as defined by the intersection of the level surfaces of two functions of the phase-space coordinates. One of these functions is obviously the Hamiltonian $H = \frac{1}{2}(v_R^2 + v_z^2) + \Phi_{\text{eff}}$. In general one does not know what the other function is. Conventionally it is called the *third integral*, I_3. Assuming I_3 exists, a total of three functions $H, L_z = Rv_\phi$ and I_3 are constant along the orbit. Functions with this property of being constant on an orbit are called *integrals of motion*.

Orbits that possess as many integrals of motion as they have spatial degrees of freedom enjoy the very special property of being *quasiperiodic*. This means the following. Suppose you numerically integrate equations of motion such as Eq. (11.2.5) for a d-dimensional quasiperiodic orbit. Then the value of any phase-space coordinate along the orbit is a known function of time, say $R(t)$. Fourier transform $R(t)$ and examine the resulting spectrum $\tilde{R}(\omega)$. If the orbit is quasiperiodic, you will find that (a) the spectrum is discrete, and that (b) the frequency of any line ω_α in the spectrum can be expressed as a linear combination $\omega_\alpha = \Sigma_i^d n_i \omega_i$ of d fundamental frequencies with integer coefficients n_i.

Most galaxies are probably not axisymmetric; many disc galaxies (including our own Milky Way) are clearly seen to have a stellar bar near their centres. These stellar bars are often highly elongated and are thought to be quite thin, lying nearly in the plane of the surrounding stellar disc. More indirect observational evidence and n-body simulations of galaxies forming in the expanding universe suggest that galaxies without discs (*elliptical galaxies*) and the envelopes of dark matter that are thought to surround most galaxies also have barred shapes, although less elongated ones than the conspicuous central bars of disc galaxies. Fig. 11.3 shows some orbits in (*a*) a nonrotating bar, and (*b*) a rotating bar; in Fig. 11.3(*a*) the bar's figure is stationary in inertial coordinates, whilst in (*b*) it rotates slowly with respect to inertial space. The velocity vectors of these orbits again indicate the existence of an integral of motion in addition to the Hamiltonian, and the orbits are indeed found to be quasiperiodic.

Quasiperiodic orbits are important because they are fully understood. One

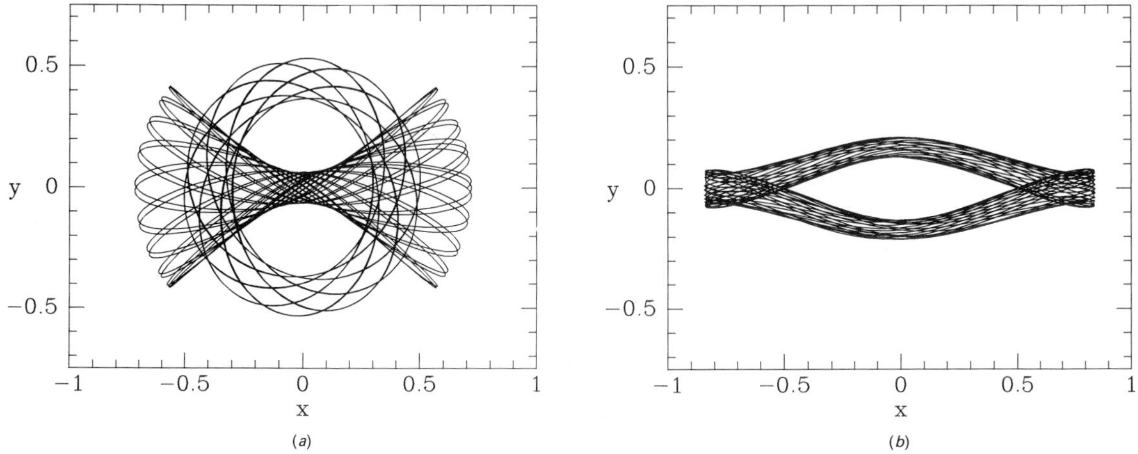

Fig. 11.3. (*a*) Two orbits in the barred potential $\Phi = \frac{1}{2}\ln(x^2 + y^2/q^2 + r_c^2)$ with $q = 0.9$ and $r_c = 0.1$. (*b*) An orbit in the same potential with $q = 0.9$ and $r_c = 0.06$ when the figure rotates with angular speed $\omega_b = 0.5$.

may show (for example, Arnold, 1978) that their integrals of motion confine them to d-dimensional tori in phase space, and that, unless the fundamental frequencies are rationally related, i.e. satisfy a relation of the type $\Sigma_i^d n_i\omega_i = 0$, a star on one of them will, over time, cover its torus uniformly. So usually all one needs to know about a star is what torus it is on – one is not very interested in *where* it is on the torus because sooner or later it will go everywhere with equal probability. In fact the situation is remarkably like that in atomic physics, where one wishes to know what orbit (n, l, m) an electron is in, but one neither knows, nor, according to quantum mechanics, *can* know, *where* the electron is in that orbit. In stellar dynamics phase information is knowable, but is frequently of no interest.

If all galactic orbits were quasiperiodic, complete information for building a steady state galaxy would consist in a list of its tori and knowledge of how many stars should be on each torus. Hence it is important to have a good way of enumerating tori. The obvious way of doing this is to label each torus by the values i_i taken by the integrals of motion $I_1 \equiv H$ etc. on that torus. However, this scheme of enumeration is not the best possible. The best labels for tori are the *actions* J_i defined as follows. On a d-torus there are d natural ways of drawing a closed path. For example, string can be wrapped around a 2-torus, such as a doughnut, either by going the short way around, or by going the long way around and thus encircling the central hole. Let $\gamma_i, i = 1,\ldots,d$, be d different paths around a d-torus drawn such that an elastic string wrapped around one path cannot be slid over the torus until it follows one of the other paths. Then the actions are given by

$$J_i = \frac{1}{2\pi}\oint_{\gamma_i} \boldsymbol{p}\cdot \mathrm{d}\boldsymbol{q}. \tag{11.2.6}$$

Here $(\boldsymbol{p}, \boldsymbol{q})$ are generalized canonical coordinates – in a nonrotating frame one can take $(\boldsymbol{p}, \boldsymbol{q}) = (\boldsymbol{v}, \boldsymbol{x})$. Paths which can be deformed into one another by sliding an elastic string over the torus from one path to the other are said to belong to the same *homotopy class*. Remarkably, the integral in Eq. (11.2.6) is the same for all paths that belong to the same homotopy class. Hence the details of the path γ_i do not matter; all that matters is its homotopy class.

The J_i are functions on phase space in the sense that through any point there is an orbit and thus a torus, and the value taken by J_i at that point is the appropriate integral Eq. (11.2.6) around that torus. Moreover the J_i are constants of motion because they take the same value everywhere on a torus and thus everywhere on the same orbit. Thus the J_i enjoy the key properties of the integrals I_i we started from.

The importance of the J_i arises because they enjoy other properties that a general integral I_i does not:

(1) They are adiabatic invariants. That is, if the Hamiltonian is changed on a time scale long compared with $\max(1/\omega_i)$, the orbital tori change too. But they do so in such a way that the J_i stay the same.

(2) They may be used as the momenta in a set of canonical coordinates. The conjugate coordinates θ_i tell you where you are on a given torus. They are called *angle variables* because increasing any one of them by 2π returns you to the same point in phase space – the trip has carried you right around the torus along one of

Somehow there is more energy associated with motion at a given speed in the tangential than in the radial direction. (Notice that this is paradoxical, given that, as we have seen, the epicycles are elongated in the tangential rather than the radial direction!)

11.5.2 *Distribution functions for discs*

By Jeans's theorem, the distribution function must be a function of the angular momentum $L_z \equiv R^2\dot{\phi}$ and the radial action given by Eq. (11.5.7). So we can now easily write down trial distribution functions for steady state discs. The most obvious such function,

$$f(x, \mathbf{v}) \propto f_0(L_z)\exp(-E_R/\sigma^2) \text{ with } \sigma \simeq 20\,\mathrm{km\,s^{-1}}, \tag{11.5.8}$$

is known as the *Schwarzschild distribution* in honour of Karl Schwarzschild of black hole fame, who deduced that the distribution of stellar velocities near the sun is significantly anisotropic. Since $20\,\mathrm{km\,s^{-1}} \approx \sigma \ll R\Omega \approx 220\,\mathrm{km\,s^{-1}}$, stars do not stray very far from the guiding centres, R_g, of their epicycles. Consequently, the function f_0 effectively determines the radial density profile of the disc. For realistic surface-density profiles, such as $\Sigma(R) \sim \exp(-R/R_d)$ with $R_d \simeq 3\,\mathrm{kpc}$, f_0 varies more slowly as a function of L_z than $\exp(-E_R/\sigma^2) = \exp(-\omega_r J_r/\sigma^2)$ does with J_r. Given this insensitivity of f_0 to L_z, Schwarzschild's observation is naturally accounted for when we use Eq. (11.5.4) to eliminate E_R from Eq. (11.5.8): the velocity distribution is seen to be approximately Maxwellian, with a radial dispersion which is larger than the tangential dispersion by a factor γ.

The factor γ in Eq. (11.5.4) helps explain another interesting observational fact. A variety of techniques make it possible to deduce the ages of stars, and it is found that the random velocities of stars increase with age. The basic mechanism by which this 'heating' of the disc comes about is easily explained. The disc's gravitational potential is not exactly smooth and axisymmetric; in addition to the irregularities due to individual stars, there are larger irregularities due to spiral arms and giant clouds of molecular gas. Let us concentrate on an irregularity generated by a giant cloud.

Fig. 11.10 is a representation of an annulus in the disc centred on the position of the cloud at $(R, \phi) = (1, 0)$, and viewed from the frame which co-rotates with the cloud. Stars on circular orbits of radius $R < 1$ overtake the cloud from the bottom left, whilst stars on circular orbits of radius $R > 1$ slip behind the cloud, from top right to upper left. As they pass the cloud, stars are tugged towards it and finish by executing nice epicycles, rather as a guitar string is set oscillating by being plucked. A few stars which approach the cloud from nearly its own radius are effectively repelled by it; this occurs because their exchange of angular

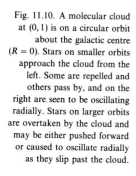

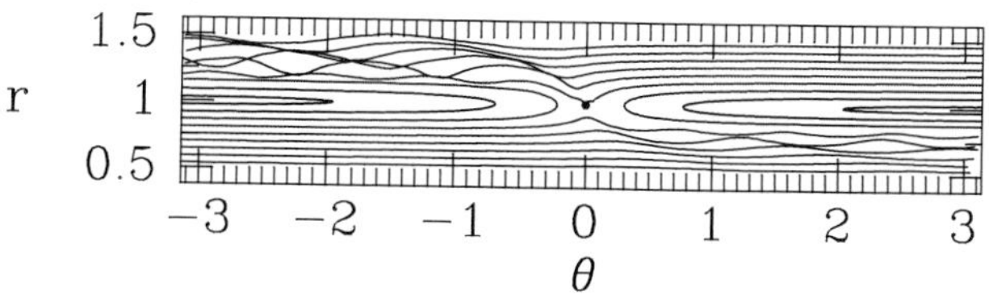

Fig. 11.10. A molecular cloud at $(0, 1)$ is on a circular orbit about the galactic centre $(R = 0)$. Stars on smaller orbits approach the cloud from the left. Some are repelled and others pass by, and on the right are seen to be oscillating radially. Stars on larger orbits are overtaken by the cloud and may be either pushed forward or caused to oscillate radially as they slip past the cloud.

momentum with the cloud causes them to switch to a guiding centre whose drift then carries them away from the cloud.

Fig. 11.10 shows that stars on initially perfect circular orbits are bound to gain nonzero E_R from an encounter with a cloud. But the situation is more complex when a star that is already executing epicycles encounters a cloud. This time its value of E_R can decrease as well as increase. To see this we work in the rotating frame in which the cloud and the galactic centre are at rest, and study an encounter between a cloud and a star that is sufficiently close to be virtually completed in a distance small compared with the star's epicycle radius. Then we can approximate the star's asymptotic in- and out-trajectories by straight lines.

Since the cloud is very much more massive than the star, the speed $|\boldsymbol{v}|$ of the star is virtually unchanged by the encounter; only the direction of $\boldsymbol{v}$ changes. From Eq. (11.5.7) and the fact that $\gamma > 1$, it now follows that E_R increases or decreases according as v_ϕ increases at the expense of v_R, or vice versa. At apocentre or pericentre, where $v_R = 0$, an encounter is bound to decrease v_ϕ and therefore E_R. Conversely, E_R will increase if the encounter occurs at $R \simeq R_g$. Thus, once stars have moved from circular orbits, it is not obvious that scattering by heavy objects will further heat the disc. Actually one can show (for example Binney and Lacey, 1988) that encounters which increase E_R swamp those that decrease it, and the disc does continue to heat. But the heating rate slows considerably as the typical value of E_R rises, and clouds alone are probably unable to drive stars to the kinds of random velocities we observe near the sun. Repeated resonant interaction with wispy spiral structure is more effective in heating stars with moderate velocity dispersions, but still seems unable to account for the fastest moving disc stars. It now seems likely that early in the life of the Milky Way, the disc was disturbed by one or more massive objects, such as a satellite galaxy, that spiralled towards the Galactic centre and eventually broke up and merged into the inner Galaxy. Such inward-spiralling objects are very effective heaters of the Galactic disc (Quinn and Goodman, 1986; Tóth and Ostriker, 1992).

11.5.3 *Stability to axisymmetric disturbances*

It is instructive to investigate the stability of the simplest disc – that in which all stars move on circular orbits. We have seen that a circular orbit of radius $R_g + x$ is represented in the sheared sheet by the trajectory $\dot{x} = 0, \dot{y} = -\frac{1}{2}Ax/\Omega_g$. How does this trajectory change when the system is perturbed by adding to the right side of the first of Eqs (11.5.2) an additional radial force $-(\mathrm{d}\delta\Phi_e/\mathrm{d}x)\exp(-\mathrm{i}\nu t)$? The changes in x and y satisfy

$$\begin{aligned}\delta\ddot{x} - 2\Omega_g\delta\dot{y} &= A\delta x - \frac{\mathrm{d}\delta\Phi_e}{\mathrm{d}x}\\ \delta\ddot{y} + 2\Omega_g\delta\dot{x} &= 0,\end{aligned} \qquad (11.5.9)$$

from which we easily find with Eq. (11.5.3)

$$\omega_r^2 - \nu^2\delta x = -\frac{\mathrm{d}\delta\Phi_e}{\mathrm{d}x}. \qquad (11.5.10)$$

$$\Sigma_1(R', \phi') \simeq \frac{\varepsilon}{R'} \exp[i(kR' + m\phi')] \left. \frac{dJ_a}{dR_g} \right|_{R_g = R'}$$
$$\times \sum_{l=-\infty}^{\infty} \exp(il\alpha) \int dJ_r J_l(\mathscr{K}X) \frac{\boldsymbol{n} \cdot \partial_{\boldsymbol{J}} f_0}{\boldsymbol{n} \cdot \boldsymbol{\omega} - \nu}$$
$$\times \int d\theta_r \exp\left[i \left(l\theta_a - m \frac{\gamma X}{R_g} \sin\theta_r - kX\cos\theta_r \right) \right]. \tag{11.5.28}$$

From Eq. (11.5.3) it is easy to show that $dJ_a/dR_g = R_g\omega_r/\gamma$ since $J_a = L_z = R_g^2\Omega_g$. For f_0 it is natural to adopt Schwarzschild's distribution function in the form $f_0(\boldsymbol{J}) = (\gamma\Sigma_0/2\pi\sigma^2)\exp(-\omega_r J_r/\sigma^2)$ (cf. Eq. (11.5.8)). Finally, the second integral in Eq. (11.5.28) involves exactly the same product of exponentials of $\cos\theta_r$ and $\sin\theta_r$ as occurs in the first line of Eq. (11.5.23a). Reducing this product to an infinite sum of exponentials as in Eq. (11.5.23a) and doing the integral over θ_r yields

$$\Sigma_1(R', \phi') \simeq \frac{\varepsilon\omega_r^2}{\sigma^4} \Sigma_0 \exp[i(kR' + m\phi')]$$
$$\times \sum_{l=-\infty}^{\infty} -l \frac{\int dJ_r |J_l(\mathscr{K}X)|^2 \exp(-\omega_r J_r/\sigma^2)}{l\omega_r + m\Omega - \nu}. \tag{11.5.29}$$

Evaluating the integral with formula 6.615 of Gradshteyn and Ryzhik (1965) we have

$$\Sigma_1(R', \phi') \simeq \frac{\varepsilon\omega_r}{\sigma^2} \Sigma_0 \exp[i(kR' + m\phi')] \sum_{l=-\infty}^{\infty} -l \frac{I_l(\chi)\exp(-\chi)}{l\omega_r + m\Omega - \nu}$$
$$= -\frac{\mathscr{K}^2\Phi_1\Sigma_0}{\omega_r^2} F(s, \chi), \tag{11.5.30a}$$

where

$$s \equiv \frac{\nu - m\Omega}{\omega_r}, \quad \chi \equiv \frac{\mathscr{K}^2\sigma^2}{\omega_r^2}, \quad F \equiv 2\exp(-\chi)\frac{1}{\chi}\sum_{i=1}^{\infty} \frac{I_l(\chi)}{1 - s^2/l^2}. \tag{11.5.30b}$$

In the tight-winding limit, the surface density that generates Φ_1 through Poisson's equation is (cf. Eq. (11.5.15))

$$\Sigma_1 = -\frac{k}{2\pi G}\Phi_1. \tag{11.5.31}$$

Equating this to the response density Eq. (11.5.30a), we obtain finally the Lin–Shu–Kalnajs (LSK) dispersion relation for stellar discs:

$$\frac{|k|}{k_{crit}} F(s, \chi) = 1, \tag{11.5.32a}$$

where

$$k_{crit} \equiv 2\pi/\lambda_{crit}$$
$$= \frac{\omega_r^2}{2\pi G\Sigma} \tag{11.5.32b}$$

by Eq. (11.5.17). A slightly more accurate version of the stability criterion Eq. (11.5.20) to axisymmetric disturbances can now be obtained by setting $\nu = m = 0$ in Eq. (11.5.32a) and summing the Bessel functions. The resulting equation for k has no solution providing Eq. (11.5.20) is satisfied.

Our next goal is to use the dispersion relation Eqs (11.5.32) to construct approximate normal modes for stellar discs. The concept of a resonant cavity provides the necessary link between Eqs (11.5.32), which describes *running* waves, and normal modes; we model the disc as a resonant cavity around which waves run back and forth, their frequency being chosen such that they always arrive back at any given spot with the same phase. A digression on the semiclassical estimation of the energy levels of the hydrogen atom will clarify this idea.

In suitable units, hydrogen's energy levels are the values of ω for which one can solve the Schrödinger equation

$$\frac{1}{r^2}\frac{\mathrm{d}}{\mathrm{d}r}\left(r^2\frac{\mathrm{d}\psi}{\mathrm{d}r}\right) - \left(\frac{l(l+1)}{r^2} - \frac{q}{r}\right)\psi + \omega\psi = 0. \tag{11.5.33}$$

We can obtain levels sufficiently above the ground state by substituting ψ of the form $\exp(ikr)$ into Eq. (11.5.33) to obtain the dispersion relation

$$k^2 = \omega - \frac{l(l+1)}{r^2} + \frac{q}{r}. \tag{11.5.34}$$

This tells us that k becomes imaginary and that the waves become evanescent at both small and large r. Let the radii at which k vanishes be $r_\pm$. Then our normal mode condition is that the total change $\Delta\phi$ in the phase of a wave that runs from r_- to r_+ and back again is $2n\pi$. So we require

$$\begin{aligned} 2n\pi = \Delta\phi &= \pi + 2\int_{r_-}^{r_+} k(r)\,\mathrm{d}r \\ &= 2\pi\left(\frac{1}{2} + \frac{q}{(2|\omega|)^{\frac{1}{2}}} - [l(l+1)]^{1/2}\right). \end{aligned} \tag{11.5.35}$$

(The extra π arises because the wave is not reflected by a sharp barrier at $r = r_\pm$; see Landau and Lifshitz, 1977, Sections 47–9, for details.) Thus our estimate of the energy levels is

$$\omega = \frac{q^2/2}{\{n - \frac{1}{2} + [l(l+1)]^{1/2}\}^2}. \tag{11.5.36}$$

For large l this is an accurate approximation to the exact values of ω.

Returning now to the LSK dispersion relation Eq. (11.5.32a), we seek the radii at which a disturbance of given frequency ν has vanishing wave number k. These are clearly the radii at which F has a pole, and from Eq. (11.5.30b) we see that F is singular whenever s is an integer other than zero. The *Lindblad resonances* are defined to be the radii of integer s. Far and away the most important resonances are the *inner Linblad resonance* (ILR), where $s = -1$, and the *outer Lindblad resonance* (OLR), where $s = 1$ (see Fig. 11.12). Does a disc consist of a resonant cavity in that waves are trapped between the ILR and the OLR?

Fig. 11.13 shows why this simple picture is not valid. This figure is a plot of $k(R)$

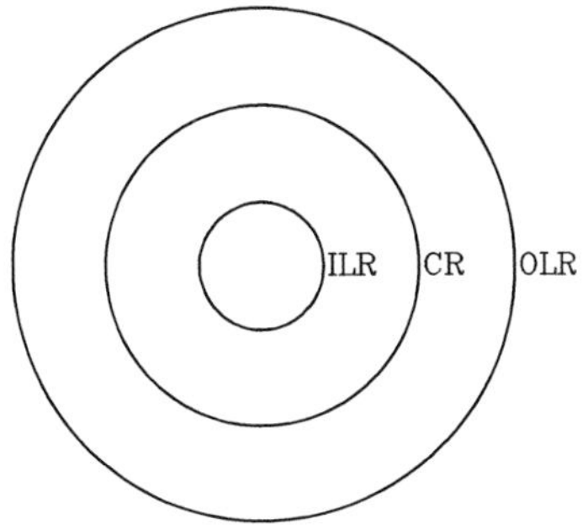

Fig. 11.12. The Lin–Shu–Kalnajs (LSK) dispersion relation Eq. (11.5.32a) indicates that the wave number of spiral density waves either vanishes or becomes arbitrarily large at the inner Lindblad resonance (ILR) and the outer Linblad resonance (OLR). A forbidden region surrounds the co-rotation resonance (CR).

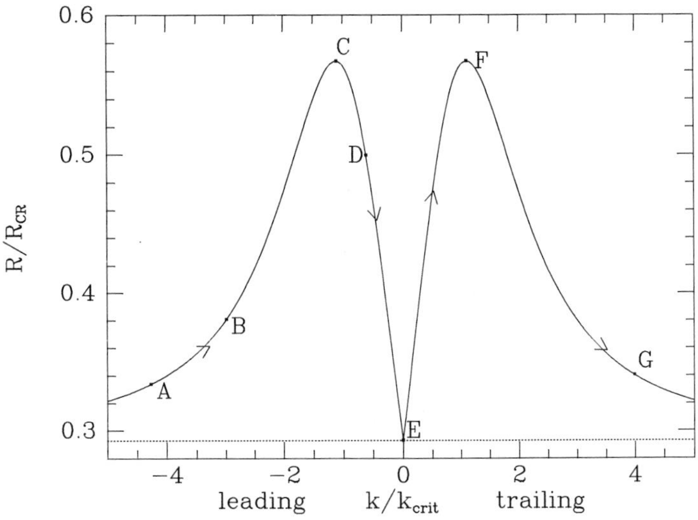

Fig. 11.13. $k(R)$ for a disc in which the circular speed is independent of R. The arrows show how a packet of tightly wound leading waves winds up into a packet of tightly wound trailing waves as it propagates from point A near the ILR ($R = 0.293R_{CR}$) to near co-rotation (at C) to the ILR at E to near co-rotation and back to the ILR. Discs in which Q is not large and the ILR lies close to the centre are effectively resonant cavities because (i) a portion of the energy of the trailing waves on the extreme right is able to excite fresh leading waves at extreme left, and (ii) waves are amplified by the 'swing amplifier' as they bounce off the co-rotation region, for example at C.

from Eq. (11.5.32a) for the simple but realistic case of a disc in which the circular speed is independent of radius; the interpretation of Eq. (11.5.32a) is complicated by the fact that Ω and ω_r are functions of radius and it is helpful to discuss the general problem of spiral structure in a particular illustrative case. Fig. 11.13 shows only the radius range just outside the ILR, which is in practice of greatest interest. The ILR is located at $R = 0.293R_{CR}$, and at this radius $k = 0$ as we have seen. At slightly larger values of R *four* values of k are allowed by Eq. (11.5.32a): two positive and two negative; two of these have small absolute values, and two have very large absolute values. Negative values of k correspond to leading waves, whereas positive values correspond to trailing waves. As one moves further away from the ILR, the two negative values of k approach one another, and likewise for the two positive values, until at a radius $\simeq 0.57R_{CR}$ they become equal. Beyond this radius Eq. (11.5.32a) has no solution. Thus there is a forbidden region around co-rotation in which spiral waves cannot propagate.

To understand the physical significance of Fig. 11.13, imagine launching a packet of tightly wound, leading waves from near the ILR. Its waves have large negative k and thus lie near point A in Fig. 11.13. It turns out that the group velocity for such waves is positive, so the packet moves outwards towards point B

and increases in wavelength (moves to smaller $|k|$). Eventually at C it reflects off the forbidden region around co-rotation, and comes back towards the ILR on the long-wavelength branch of the dispersion relation for leading waves. Eq. (11.5.32a) was obtained under the assumption that the waves were tightly wound, so it must break down somewhere near point E. But suppose the packet makes it to E and then moves onto the long-wave branch of the dispersion relation for trailing waves. Then the packet moves outwards again towards larger $|k|$, again reflects off the forbidden region around co-rotation at F and finally approaches the ILR for a second time via G, this time as a packet of tightly wound trailing waves. Thus Fig. 11.13 describes the winding up of leading waves into trailing waves, a process which is inevitable in a differentially rotating system.

How can the cycle of leading waves → trailing waves be closed, as it must be if the disc is to be a resonant cavity? Clearly tightly wound trailing waves must be converted to tightly wound leading waves at the ILR. Most of the packet's energy is Landau-damped away as its wavelength gets shorter and shorter in the second approach to the ILR. But if the ILR is sufficiently close to the centre, a little of its energy goes into exciting fresh leading waves. And when the stability parameter Q is not large, it turns out (see Toomre, 1981, for details) that the cycle can be effectively closed by quite a small transfer of energy from trailing into leading waves. The reason for this is that waves are amplified as they bounce off the forbidden region around R_{CR} by a process known as 'swing amplification'. This can be understood in the context of the shearing sheet introduced in Section 11.5.1, but want of space forbids us to pursue the matter further.

References

ARNOLD, V. I. (1978). *Mathematical Methods of Classical Mechanics.* Springer, New York.

BINNEY, J. J. and LACEY, C. G. (1988). *Mon. Not. Roy. Astron. Soc.*, **230**, 597.

BINNEY, J. J. and TREMAINE, S. D. (1987). *Galactic Dynamics.* Princeton University Press.

COHN, H. (1980). *Astrophys. J.*, **242**, 765.

FRIDMAN, K. C. and POLYACHENKO, V. L. (1984). *Physics of Gravitating Systems*, 2 vols. Springer, New York.

GRADSHTEYN, I. S. and RYZHIK, I. M. (1965). *Tables of Integrals, Series and Products.* Academic Press, New York.

IKEUCHI, S., NAKAMURA, T. and TAKAHARA, F. (1974). *Prog. Theor. Phys.*, **52**, 1807.

JULIAN, W. H. and TOOMRE, A. (1966). *Astrophys. J.*, **146**, 810.

LANDAU, L. D. and LIFSHITZ, E. M. (1977). *Quantum Mechanics.* Pergamon, Oxford.

LYNDEN-BELL, D. and WOOD, R. (1968). *Mon. Not. Roy. Astron. Soc.*, **138**, 495.

QUINN, P. J. and GOODMAN, J. (1986). *Astrophys. J.*, **309**, 472.

SAHA, P. (1991). *Mon. Not. Roy. Astron. Soc.*, **248**, 292.

TOOMRE, A. (1981). In *The Structure and Evolution of Normal Galaxies* (S. M. Fall and D. Lynden-Bell, eds), p. 111. Cambridge University Press.

TÓTH, G. and OSTRIKER, J. P. (1992). *Astrophys. J.*, **389**, 5.

13
Industrial plasmas

P. C. JOHNSON

13.1 Introduction

It is well over 100 years since plasmas were first used in technology. Many of the early applications, for example gas-filled valves, switchgear and lighting, used the properties of the ionized gas or plasma itself to achieve the required technical function. Some have developed into mature industries, whereas others are well past the end of their natural product cycle. However, since the early 1970s, growing attention has been focused on plasma as a medium for materials processing (Proud *et al.*, 1991; Vaughan, 1991), with a tremendous range of new applications being devised. In the 1990s, the use of plasmas for environmental applications such as the processing of industrial wastes is becoming important (Howlett, Timothy and Vaughan, 1992). These newer applications, covering most of the current and emerging uses of plasma in industry, can be brought together under the broad heading of *plasma processing*.

Plasma processing is now viewed as a critical technology in a large number of industries, and whilst semiconductor device fabrication for computers is perhaps the best known, it is equally important in other sectors such as bio-medicine, automobiles, defence, aerospace, optics, solar energy, telecommunications, textiles, paper, polymers and waste management (Proud *et al.*, 1991). A common theme in many current and emerging applications is plasma treatment of surfaces. For these uses, plasma is an enabling technology, allowing new products to be conceived which would be impossible to make with more traditional manufacturing processes. Even here, environmental issues are important, and environmental legislation is gradually forcing the adoption of clean technologies. Plasma is accepted as a suitable environmentally friendly process technology, producing extremely low levels of industrial waste, especially when compared with the more traditional liquid chemical treatments.

Applications of plasma processing can be grouped under the headings of *information technology*, *materials* and *the environment*, and the first part of this chapter is devoted to a discussion of the role played by plasma processing in each of these areas. In the second part, some of the main plasma sources used in the various industries are described along with aspects of the plasma science governing their behaviour. A critical parameter which often determines the choice of plasma source for a particular materials processing application is the ion energy distribution function (IEDF), describing the energy range and distribution of ions hitting the surface of the material being processed (referred to as a substrate), and particular attention will be paid to the mechanisms governing the IEDF obtained in these sources. Opportunity is taken in Section 13.5 on DC, LF and pulsed glow discharges to recall some of the basic science governing the formation of the electrostatic sheath, which is an important factor in determining the IEDF.

Of the large number of books published on plasma, those by von Engel (1965), Holohan and Bell (1974), Franklin (1976), Chapman (1980), Manos and Flamm (1989), and Rossnagel, Cuomo and Westwood (1990) are useful for different aspects of plasma applications. There are also many journals dealing with plasma, and it would be impossible to list them all here. Nevertheless the following are worth mentioning for their emphasis on the basic plasma science relevant to applications: *Journal of Applied Physics*, *Physics of Fluids B*, *Journal of Physics D: Applied Physics* and a relatively new journal, *Plasma Sources Science and Technology*.

13.2 Information technology

Electronics is a good example of an industry in which clean plasma technology has enabled rapid progress to be made. The electronics industry has grown dramatically in the last decade with annual world-wide sales reaching $705 bn in 1990 (Proud *et al.*, 1991) and doubling about every five years. Plasma reactors, with a 1990 market of $1 bn, are a small but nevertheless critical component on which the industry depends. Since the early 1980s, the drive to increase the component density, especially for larger and faster computer memory, has resulted in the displacement of many of the chemical process steps used in early device fabrication by more effective and cleaner plasma treatments. Fig. 13.1 shows that over the last 20 years the feature size has been reduced by more than a factor of 20, with corresponding increases in memory capacity of well over three orders of magnitude.

Table 13.1 lists the typical process steps used to manufacture integrated circuits (ICs), and identifies those steps for which plasma treatments can be used. Many of the steps involve plasma in one way or another, including ion implantation for which plasma sources are used to generate the ions. However,

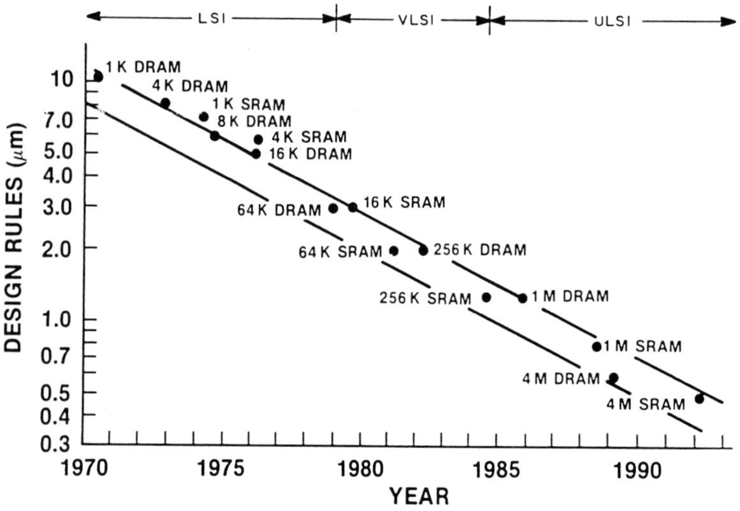

Fig. 13.1. Trends in the feature size of computer memory. DRAM, STRAM = dynamic and static random access memory; LSI, VLSI, ULSI = large, very large and ultra large scale integration. (Manos and Flamm, 1989.)

Table 13.1. *Steps in a simplified MOS integrated circuit process showing where plasma treatments are available.*

Not all of these steps are used currently in device fabrication. The ion implantation and sputtering steps shown also use plasma sources.

Process step	Implant	Plasma	Sputter
Grow thin oxide		✓	
Deposit nitride		✓	
Condition nitride			
Mask #*1 – define field*			
Pattern nitride		✓	
Implant field-boron	✓		
Remove resist		✓	
Oxidize			
Remove nitride		✓	
Remove oxide		✓	
Mask #*2 – depletion devices*			
Implant arsenic	✓		
Remove resist		✓	
Grow gate oxide		✓	
Mask #*3 – buried contacts*			
Etch oxide		✓	
Deposit polysilicon		✓	
Mask #*4 – polysilicon/gate*			
Etch polysilicon		✓	
Etch oxide		✓	
Oxidize polysilicon			
Implant arsenic	✓		
Deposit silicon oxide		✓	
Mask #*5 – contacts*			
Etch contact holes		✓	
Reflow/diffuse POCL			
Etch to clear contacts		✓	
Deposit aluminium			✓
Mask #*6 – metal*			
Etch metal		✓	
Anneal (H_2)			
Deposit glassivation		✓	
Mask #*7 – pads*			
Etch pads		✓	

the key development which has enabled such rapid progress to be made in reducing the feature size is plasma etching. Modern plasma treatments enable deep channels with vertical side-walls to be produced (Fig. 13.2), whereas ordinary chemicals attack the side-wall as rapidly as the base of the channel. Plasma etching is explored further in later sections.

Continuous improvements are being made also in plasma techniques for deposition of thin films of the materials required to build up the multilayer IC

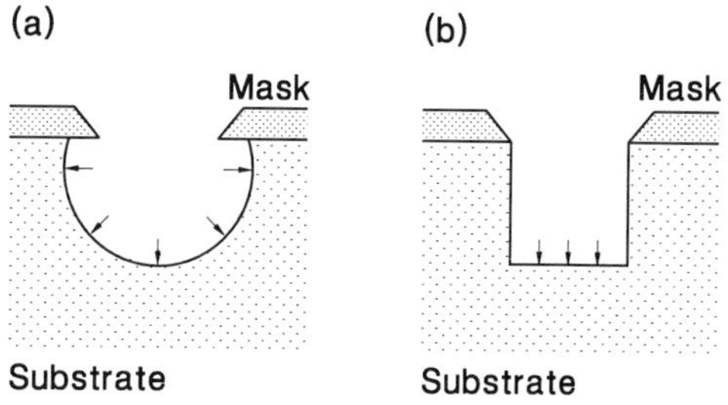

Fig. 13.2. Etching of a substrate through a mask. (*a*) Chemical etching is near isotropic giving curved walls and undercutting the mask. (*b*) Etching with plasma ions is normally directional, producing vertical side walls.

structure and for stripping of unwanted materials, such as photo-resist used as a mask for etching. Developments have reached a point where in a typical modern IC manufacturing facility, clusters of plasma tools are installed around a central substrate cassette handler, so that fully automated sequences of treatments can be performed entirely under vacuum.

Silicon remains the dominant substrate material for microelectronics, but a lot of effort is being placed worldwide on III–V and II–VI semiconductor materials such as GaAs and InP for increased speed, and for optoelectronics. Linewidths as low as 0.3 microns are now achievable, with progress towards smaller dimensions being limited by lithography rather than plasma processes. In the emerging field of nanotechnology, the limitations of optical lithography are circumvented by using finely focused electron or ion beams to write the pattern onto a photo-resist and features well below 50 nm in size have been etched (Tang, Cheung and Wilkinson, 1990). A very large effort worldwide is being devoted to further reducing the linewidth on ICs towards these levels.

13.3 Materials

Plasmas have been used for many years for welding and cutting of metals, and there is a growing market for plasma-based smelting and refining techniques for metals and glass. However, in this section emphasis will be placed on the relatively new and developing area of surface engineering of materials. This is an area in which plasma processing treatments are becoming increasingly important. For many industrial sectors, materials are required with surface properties such as corrosion or wear resistance, low friction, optical, electrical or magnetic properties, bio-medical compatibility or decorative finishes. Engineers, rather than relying on expensive bulk materials, are looking increasingly to modify the surfaces of manufactured parts to achieve the required functional performance. The materials concerned include metals and alloys, semiconductors, ceramics, glasses and polymers; applications exist or are envisaged for bulk and surface combinations of most of these types of materials.

The most common form of surface treatment is a coating, and Fig. 13.3 illustrates some of the properties required to achieve good performance from the combined system of coating and substrate (Rickerby and Matthews, 1991). Generally, the substrate is chosen for its mechanical, thermal or other properties (including cost), and a coating is applied to give the component a specific property or combination of properties. Characteristics of the coating, such as residual stress, cohesion, cracking, defects and adhesion affect the performance and lifetime of the layer. Similar importance is attached to the properties of the

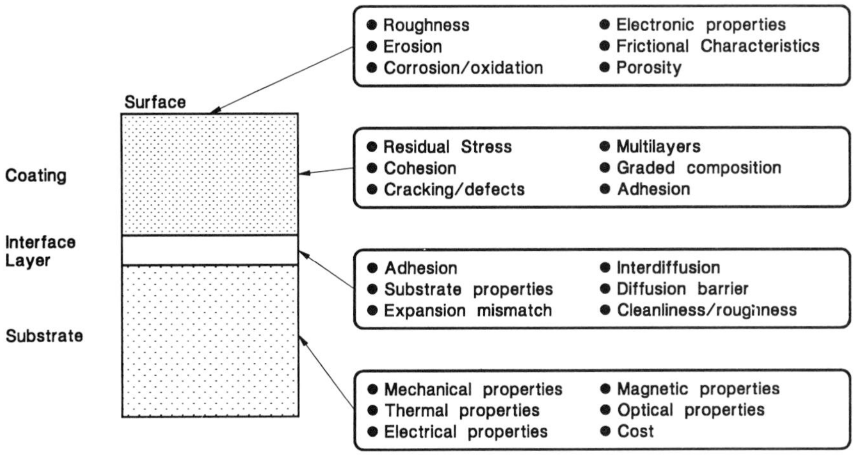

Fig. 13.3. The basis of surface engineering with a coating. The figure illustrates some of the key factors to be considered to achieve the required surface performance. (After Rickerby and Matthews, 1991.)

interface layer between the substrate and the coating, such as matching of expansion coefficients, adhesion, cleanliness, roughness or the provision of a diffusion barrier between the two.

An illustration of the coating thickness or depth of treatment for several plasma-related processes in Fig. 13.4 (Bell, 1992, University of Birmingham, private communication) shows that it is possible to achieve a range from less than a micron for ion assisted coatings to well over a millimetre for plasma and arc spraying. Physical vapour deposition processes commonly use a plasma source, such as an electron beam or magnetron, to evaporate or sputter material onto the substrate. Chemical vapour deposition and plasma assisted chemical vapour deposition are techniques for growing films using surface chemical reactions between atoms or molecules incident from a plasma or gas phase immediately above the substrate. In ion assisted coatings, energetic ions are used to impart momentum and energy to the surface to achieve increased density or to favour certain chemical pathways.

Of course there are many well-established surface treatments in use in industry, the most common being paints, electroplating and galvanizing, and the environmental damage they cause through the emission of volatile organic

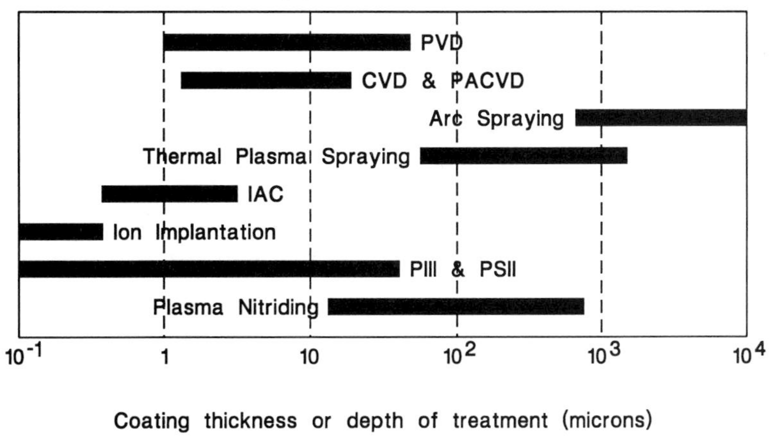

Fig. 13.4. Coating thickness or depth of treatment for a number of plasma-related processes. PVD = physical vapour deposition; CVD = chemical vapour deposition; PACVD = plasma assisted chemical vapour deposition; IAC = ion assisted coating; PIII & PSII = plasma immersion (plasma source) ion implantation. (Bell, 1992, University of Birmingham, private communication.)

solvents and the disposal of used chemicals is well known. Cost is a major factor preventing the displacement of such techniques by plasma treatments. However, plasmas are in increasing use in a number of important applications, such as hardening of steels by nitriding, coating cutting tools with titanium nitride, optical coatings and coating or hardening of hip, knee or shoulder implants. In the UK, plasma surface engineering is becoming a substantial industry, with the value of plasma treated components growing at 12–15% per year towards £150 M by the year 2005 (Matthews *et al.*, 1992).

There are many exciting new developments taking place in plasma surface engineering which benefit a wide range of industries. Synthetic diamond films are one example, with applications in cutting tools, optics and electronics, especially sensors and heat sinks. Multicomponent films, such as TiAlN, ZrAlN, TiNbN, TiCN and TiAlON are replacing TiN for cutting tools and wear parts, and are extending the available range of decorative finishes. Multilayer films, with layer thickness down to 2 nm are now becoming possible with improved process control, for example C/Si and C/Ge multilayers for optics. Graded layers with properties varying continuously from the interface to the surface, for optimum adhesion and tailored surface properties, are opening up a whole new field of *designer surfaces*. The potential for plasma technologies in surface engineering industries is evidently very large.

13.4 The environment

In addition to its use as a clean technology for materials processing, plasma is being regarded increasingly as a suitable medium for processing waste materials to reduce their impact on the environment (Howlett, Timothy and Vaughan, 1992). Techniques are being developed to process gaseous emissions such as the oxides of nitrogen and sulphur in power station emissions and volatile organic solvents, particularly from paint spraying. For solid materials, plasma treatments are being applied to bulk wastes such as tyres, household and medical refuse, and to the treatment of hazardous materials such as chemical weapons, polluted soil and dusts from industrial processes. Another important area is sterilization of medical instruments, where the plasma is used to destroy bacteria.

Atmospheric pressure plasmas are most appropriate for achieving an adequate throughput of material at low cost in these applications. Of the diverse techniques being developed, plasma torches and carbon arcs are preferred for bulk wastes and hazardous materials. In these thermal plasmas, the plasma electrons, ions and neutrals have nearly identical temperatures, and the heat transfer and process chemistry are well suited to the applications. Currently, much effort is being placed on the development of nonthermal high pressure plasma techniques such as corona discharges, high energy electron beams and pulsed microwave plasmas. A plasma in which the electrons have a high temperature gives access to higher energy dissociation processes than those achievable in a thermal plasma. This is particularly important for applications such as power station exhausts and solvents.

These applications also differ widely in scale. For power station emissions or polluted soil treatment, plasma sources in the megawatt power range are required, whereas hundreds of watts or a few kilowatts will suffice for medical sterilization or for treating the air in a small automobile paint shop. With such diversity, plasma has the potential to become a significant agent for improving the quality of the environment.

13.5 DC, LF and pulsed glow discharges

A natural starting point for a discussion of low pressure plasma sources is the classical cold-cathode DC glow discharge, illustrated in Fig. 13.5 for a cylindrical tube filled with gas at about 1 mb pressure. Also shown is the variation along the tube of the longitudinal electric field E_z and the electron and ion number densities n_e and n_i. There are two quasineutral ($n_e \sim n_i$) plasma regions, the positive column and the negative glow, in which E_z is approximately constant, and these are interfaced to each other and to the wall and electrodes by non-neutral regions with rapidly varying electric fields.

The best known feature, from devices such as the fluorescent lamp, 'neon' signs

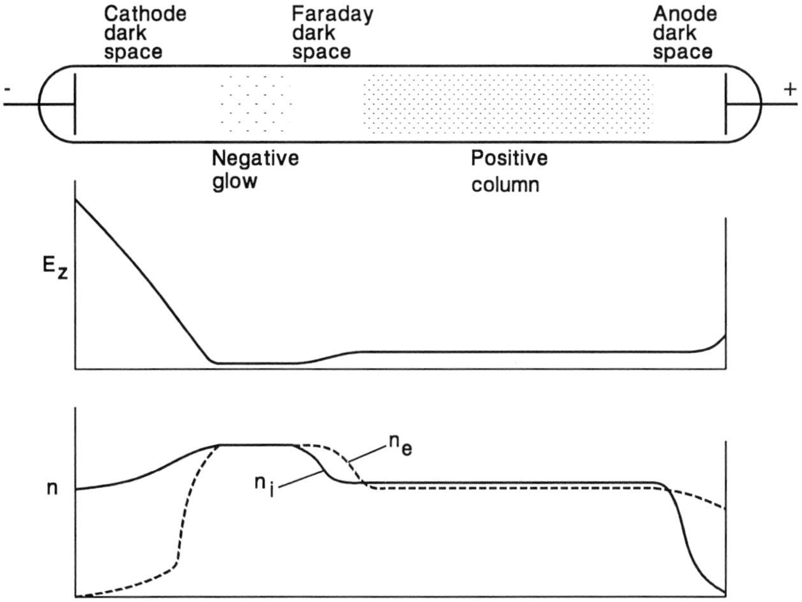

Fig. 13.5. Illustration of a cold-cathode glow discharge at a pressure $\sim$ 1 mb indicating the main regions. The variation of the longitudinal electric field, E_z, and the number densities of the electrons and ions, n_e and n_i are also shown. (After Franklin, 1976.)

and the He–Ne laser, is the positive column. Uniquely in the glow discharge, many of the properties of this region are determined by radial rather than longitudinal processes. To ensure quasineutrality, electrons and ions must flow towards the wall of the discharge tube with equal current densities. This ambipolar plasma flow is achieved through a radial electric field E_r, the main effect of which is to reduce the flux of the more mobile electrons by reducing n_e. Thus the radial profile of the plasma density is peaked in the centre of the tube.

To preserve ambipolarity out to the wall, an electrostatic (Langmuir) sheath forms between the plasma column and the wall. This region, a few Debye lengths in thickness, serves mainly to reduce further the electron flux towards the wall, so that the wall can float at a constant potential. We shall term this region the *floating sheath*. The plasma sheath plays a critical role in plasma processing of materials, and it is worth looking at the floating sheath in more detail as it forms a useful introduction to the subject.

13.5.1 *Floating sheath*

Fig. 13.6 shows the main features of the sheath and plasma regions in front of a surface floating at a constant potential. The transition from quasineutral plasma to electrostatic sheath takes place at the sheath edge and in unmagnetized plasmas – and along the magnetic field in magnetized plasmas – ions flow into a sheath from the plasma at the ion sound or Bohm speed (Franklin, 1976):

$$V_s = \left(\frac{k_B T_e}{M}\right)^{1/2}. \tag{13.5.1}$$

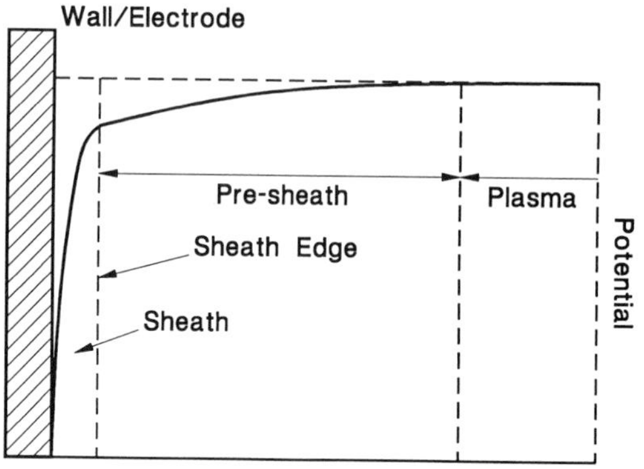

Fig. 13.6. Schematic diagram of the pre-sheath and sheath which form between a plasma and a wall or electrode drawing no net current – the floating sheath. The potential drop between the undisturbed plasma and the sheath edge is $\sim 0.5 k_B T_e$.

Here, the ions are assumed to be cold, which is generally the case in low pressure low temperature plasmas. An approximate expression for the sound speed is $V_s \approx 1 \times 10^4 (T_e/A)^{1/2}\,\mathrm{m\,s^{-1}}$, where $A = M/m_p$ and T_e is in electron-volts. Taking a typical value for T_e of 4 eV, $V_s \sim 2 \times 10^4\,\mathrm{m\,s^{-1}}$ for hydrogen ($A = 1$) and $3.2 \times 10^3\,\mathrm{m\,s^{-1}}$ for argon ($A = 40$). In front of the sheath, a quasineutral pre-sheath with a potential drop across it of approximately $0.5 k_B T_e$ gives energy to the ions so that they can satisfy the Bohm criterion at the sheath edge.

If the EEDF (electron energy distribution function) is close to Maxwellian, the electrons obey the Boltzmann law of potential:

$$n_{eb} = n_{ea} \exp\left(\frac{-e\Phi_{ab}}{k_B T_e}\right), \tag{13.5.2}$$

where n_{ea} and n_{eb} are the electron densities at two points in the plasma separated by a potential difference of Φ_{ab}. Hence for a potential drop in the pre-sheath of $0.5 k_B T_e$, the electron density at the sheath edge, $n_{es} \approx 0.6 n_{e0}$, where n_{e0} is the electron density in the undisturbed plasma beyond the pre-sheath.

For a thin sheath with no ionization, the ion flux density at the wall, Γ_{iw}, is equal to the flux crossing the sheath edge,

$$\Gamma_{iw} = n_{es} V_s. \tag{13.5.3}$$

We can make an order of magnitude estimate of this flux by taking a typical value for $n_{es} \sim 1 \times 10^{16}\,m^{-3}$, which gives $\Gamma_{iw} \sim 2 \times 10^{20}\,m^{-2}\,s^{-1}$ ($\sim 32\,A\,m^{-2}$) for hydrogen and $\sim 3.2 \times 10^{19}\,m^{-2}\,s^{-1}$ ($\sim 5.1\,A\,m^{-2}$) in argon with $T_e = 4\,eV$. Values of current density given are for singly charged ions.

From kinetic theory, the random electron flux to the wall is given by

$$\Gamma_{ew} = \frac{n_{ew} v_{Te}}{2\pi^{1/2}}. \tag{13.5.4}$$

Equating the electron and ion fluxes, and using Eq. (13.5.2) to relate n_{ew}, the electron density at the wall, to n_{es} and the potential drop across the sheath, Φ_{sw}, we obtain the following expression for the potential:

$$\Phi_{sw} = \frac{k_B T_e}{e} \ln\left(\frac{2\pi m}{M}\right)^{1/2}. \tag{13.5.5}$$

Normalizing the potentials to $k_B T_e/e$, the values obtained are -2.84 for hydrogen and -4.68 for argon. Thus for argon with a 4 eV electron temperature, the IEDF at the wall has a minimum energy of approximately 19 eV, gained in the sheath, and a width of 2 eV, due to the pre-sheath. These energies are below the sputtering threshold for most materials, and thus the floating sheath is of little use in materials processing applications which rely on physical sputtering.

Returning now to the positive column, we note that for very low gas pressures when the ion motion to the wall is collisionless, the plasma column itself performs the function of a radial pre-sheath. The potential difference between the sheath edge and the axis is again approximately $0.5 k_B T_e$. As the gas pressure is raised, E_r increases to maintain the ion flux in the presence of ion-neutral collisions and consequently the potential difference between the axis and the sheath edge becomes larger.

13.5.2 *Cathode fall sheath*

Looking now at the region around the cathode of the glow discharge, we find that the cathode fall is also an ion-rich electrostatic sheath with characteristics which are very similar to those of the floating wall sheath discussed above. The main difference is that the potential drop Φ_c across it is typically several hundred volts, and its purpose is not to maintain ambipolarity but to establish conditions for current flow between the electrodes in the discharge.

A brief description of the processes taking place in the cathode fall and negative glow regions is as follows. Starting from the cathode, ion-induced secondary electron emission or photoemission causes electrons to leave the surface, where they meet the high electric field and are accelerated by it. At very low pressures ($\approx 10^{-3}$ mbar), electrons can cross the cathode fall without making collisions with the background gas. As the pressure is increased, some collisions involving ionization can occur, increasing the electron density, as shown in Fig. 13.5. One of the first simulations of this sort of electron behaviour (Tran Ngoc An, Marode and Johnson, 1977) took a linear variation of E_z with distance and modelled the electron collisions with the background gas using the so-called Monte-Carlo technique, with the result for the EEDF shown in Fig. 13.7. A fast group of electrons crosses the cathode fall, becoming increasingly attenuated as it approaches the negative glow at $x = d_c$. The EEDF gradually fills out at lower

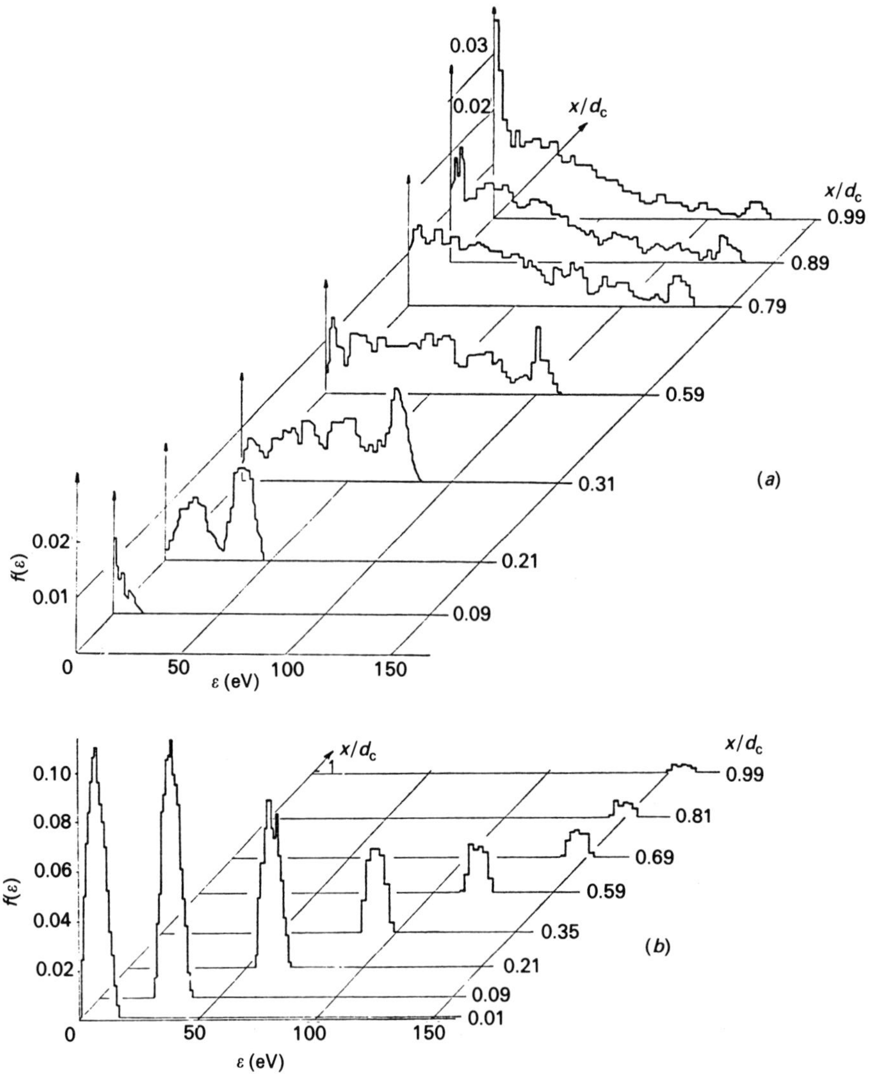

Fig. 13.7. Simulated electron energy distribution at various positions in the cathode fall of a glow discharge in helium; $p = 1.3$ mbar, $\Phi_c = 150$ V and $d_c = 13$ mm. The distribution function is divided into: (*a*) the electrons which either have suffered at least one collision or are the result of an ionizing collision; (*b*) the collision-free group. (Tran Ngoc An, Marode and Johnson, 1977.)

energies with electrons emanating from ionizing collisions which have also been accelerated by the electric field.

When the electrons enter the negative glow, where E_z is small, successive collisions slow them down until, when they emerge into the Faraday dark space, they no longer have sufficient energy to excite or ionize the background gas.

At the cathode, the current in the discharge tube is carried largely by ions whose origin is principally the electron-impact ionization in the cathode fall and negative glow. The large drop in potential between the negative glow and the cathode establishes itself, broadly speaking, so that the electron current emanating from the negative glow towards the anode is equal to the ion current hitting the cathode. In each case there are small contributions to the currents from particles of the opposite charge, but these contributions do not affect the overall picture. Thus the cathode fall and negative glow form a pair of mutually

Unlike diamond-like carbon, which is a mixture of diamond and graphite bonds in an often glassy amorphous layer, diamond films contain mainly the sp^3 diamond bond in a material which takes the form of a polycrystalline layer. Crystals in the film can vary from nanometre to micron in size, depending upon the deposition conditions. Exceptional hardness, low electrical conductivity and high thermal conductivity make this a highly desirable material for applications as diverse as cutting tools, wear parts, optical coatings, heat sinks and, with doping, as a wide band-gap semiconductor material for high temperature and radiation resistant sensors and electronic devices.

The main plasma technique for growing diamond films uses a 2.45 GHz microwave discharge (Bachmann, Leers and Lydtin, 1991). An example is depicted in Fig. 13.17. A process chamber and an antenna chamber, separated by a microwave window, together form a microwave cavity into which power is coupled from a pin antenna. The most common process gas is a $H_2/CH_4/O_2$ mixture in which H_2, at over 90%, is the main constituent. A gas pressure in the range 20 to 100 mb is used, which is between four and five orders of magnitude greater than that in an ECR source. At these higher pressures, the electron-neutral collision frequency is much greater than the electron gyro-frequency, and therefore an external magnetic field would provide no benefit.

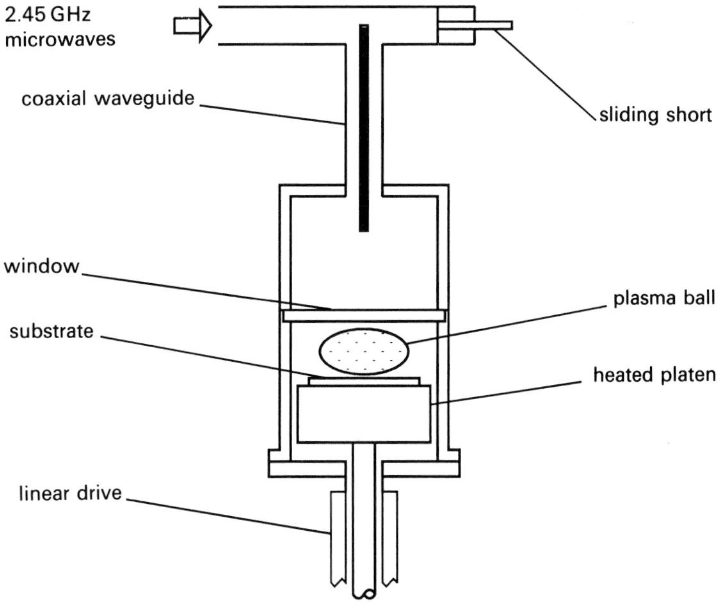

Fig. 13.17. Illustration of a microwave PACVD plasma source as used for diamond film deposition. The predominantly H_2/CH_4 process gas is fed into the process chamber close to the window.

Diamond is deposited on a substrate, usually a silicon wafer, maintained at a temperature between 600 and 950°C on a heated or cooled table. The plasma assisted chemical vapour deposition process is not fully understood, but positive ions are unlikely to play an important role at the pressures used. Carbon is provided by radicals such as CH_3, which fall onto the surface and occupy vacant bonds. Hydrogen is removed by surface recombination with free H or O atoms. The process has a strong tendency to form graphite as well as diamond, and this is removed also by H and O which etch graphite at a much faster rate than diamond, helped by the high substrate temperature. Oxygen and other etching gases are added to the process in an effort to reduce the deposition temperature, which is excessive even for tool steels.

The microwave process is favoured in many diamond applications because of the high quality and purity of the films produced and the versatility of the technique. However, even at a power of 6 kW, it is an expensive process with process times of the order of hours for micron-thick films, compared to minutes for many microelectronic plasma processes. A characteristic of the process is that the total quantity of diamond desposited (area × thickness) depends mainly on the power, the area reducing and the thickness increasing as the gas pressure is raised. Typically, at 6 kW and 80 mbar, the diameter of film produced is ~ 50 mm. Reducing film costs by increasing deposition rates and area coverage is one of the major themes of current research in this technology.

13.9 Magnetron discharges

No discussion of industrial uses of plasma would be complete without a section on magnetron discharges. They provide the principal source of sputter coatings for current applications such as metallization in microelectronics, architectural glass and food packaging. Recent developments are opening up many new applications, and it is of interest therefore to look at how these devices work and to relate aspects of their behaviour to earlier discussions of other plasma sources. More detailed descriptions can be found in two excellent reviews (Manos and Flamm, 1989; Rossnagel, Cuomo and Westwood, 1990).

Fig. 13.18 illustrates the fundamental aspects of a simple circular closed-field magnetron. A cylindrical permanent magnet (B ~ 0.1 T) imposes a magnetic field in front of a target which forms the cathode of a diode gas discharge system. Circular anodes are placed close to the poles, and applying a potential of up to 500 V between the cathode and the anode at a pressure in the region of

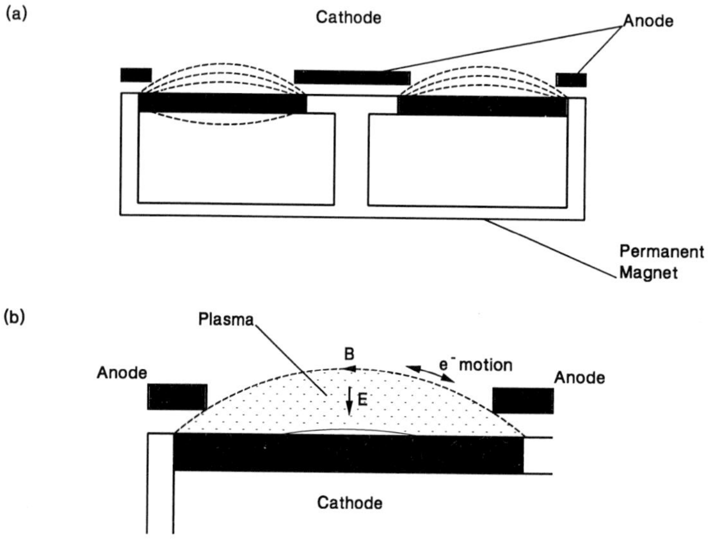

Fig. 13.18. Schematic diagram of a circular magnetron: (*a*) cross-section showing the main components and the magnetic field lines; (*b*) enlargement of the plasma region.

10^{-3} mbar, usually of argon, causes a current of several amperes to flow. The basic elements of the plasma behaviour in this system are as follows. Because of the magnetic field, electrons and ions are constrained to gyrate around field lines. For current to flow between the electrodes, electrons must cross the magnetic field, which they do at a drift speed which is consistent with Bohm diffusion. Ions, however, can reach the cathode directly, without crossing field lines. The net result is a set of linked processes which bear a remarkable similarity to those in the cathode region of the DC glow discharge.

At either end of a field line intersecting the cathode, a cathode fall is set up which drops most of the discharge voltage. Thus ions bombard the cathode at energies of several hundred electron-volts, sputtering cathode material and releasing secondary electrons. The energetic sputtered atoms escape across the plasma largely without being ionized and are then available for use in forming a coating. The secondary electrons gain energy from the cathode fall, and enter the plasma with similar energies to those of the ions bombarding the cathode. Constrained by the magnetic field to bounce between cathode sheaths, aided also by the magnetic mirror effect of the converging field lines, the fast secondaries lose energy through successive excitation and ionization collisions with the background gas in the plasma. Typical parameters for the cold electrons produced by ionization in the plasma are $T_e \sim 2$ to 20 eV and $n_e \sim 5 \times 10^{16}$ to $5 \times 10^{18}\,\mathrm{m}^{-3}$. They can also ionize the neutral gas, and in addition they carry the bulk of the cross-field current to the anode.

Field lines in contact with the anode take up a potential which is close to that of the anode, and referring to Figure 13.18 we can see that this creates an electric field perpendicular to $\boldsymbol{B}$ across the entire plasma region. Thus the plasma is subject to an additional $\boldsymbol{E} \times \boldsymbol{B}$ force which causes a drift current to flow in a path perpendicular to $\boldsymbol{B}$ and parallel to the cathode surface, this being a circular path between the magnet poles in the circular magnetron. The magnitude of this drift current is typically three to ten times the discharge current between cathode and anode.

Having looked at the plasma physics of a magnetron, we can now turn briefly to the deposition process at the substrate. The most straightforward process is sputter coating, in which usually a metal film is built up atom by atom on the surface of the substrate. Reactive sputtering is a variant in which deposition takes place in the presence of an active gas to produce, for example, metal nitride or oxide films. More advanced processes, with co-deposition from or successive exposure to different sputter targets, often in the presence of an active gas, are used to produce multicomponent or graded films, examples being multilayer optical coatings and high T_c superconductor material. The wide range of applications is reflected in the equipment supply industry, where worldwide more than 50 companies supply magnetrons varying in size from a few centimetres across to several metres in length in a 'racetrack' configuration. Use of these larger devices is growing for high performance coatings on glass and polymer films.

A relatively recent development in magnetron technology arises from the use of so-called unbalanced magnetrons (UBMs) (Munz, 1991; Monaghan *et al.*, 1993). This name arises from the fact that at the magnetron source an additional magnetic field coil or permanent magnet 'unbalances' the fluxes so that some field lines escape from the region in front of the cathode, completing their circuit by a long path into the rear of the unit. Three examples of UBM configurations are shown in Fig. 13.19.

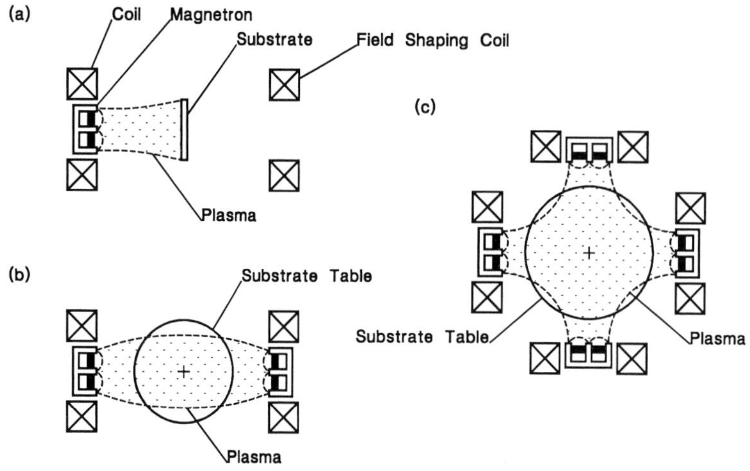

Fig. 13.19. Schematic diagrams of unbalanced magnetron configurations: (*a*) single magnetron with the substrate in the diverging field region; (*b*) two magnetrons in a mirror arrangement; (*c*) four magnetrons in a quadrupole cusp arrangement. ----- Representative magnetic field lines defining the boundary of the useful plasma.

In the simplest case, diverging field lines from a single UBM intersect a substrate placed some distance away. Similarities here with the ECR configuration are remarkable. Plasma drifting along the field lines will cause a floating sheath to form in front of the substrate, and we would expect the IEDF to be similar to that in an ECR source, with perhaps higher energies given the higher T_e values which can be obtained in the magnetron. As with the ECR source, DC or RF biasing can be added to increase the ion energy. The substrate is bombarded by a mixture of fast sputtered atoms, plasma ions, both of background gas and sputtered material, and background gas neutral atoms and molecules. Plasma ions provide the important new feature in that, through moving atoms around on the surface, they can densify films up to near-solid densities, or by adding energy they can promote surface chemical reactions which otherwise would not occur. These techniques have added considerably to the range of applications which can be addressed with magnetrons.

Further enhancements, Fig. 13.19, arise from linking the escaping field lines from two UBMs, producing a magnetic mirror configuration, or from four or more UBMs to produce multicusp configurations. With two UBMs a magnetic field is present throughout the entire plasma, and ion bombardment of a substrate or workpiece placed in the device will be preferentially from two sides, along the field. In a four-UBM system, the plasma fills a large volume in which $B \sim 0$, and plasma ion bombardment can occur all around the workpiece. Sputtered atoms from the magnetron cathodes remain directional, however, and in all cases the workpieces must be rotated to achieve uniform coatings. Developments in multi-UBM systems are focused mainly on the general engineering sector, with reactive sputtering of nitrides, such as TiN and related materials, and of oxides for applications in cutting tools, wear parts, corrosive environments and decorative coatings. This is a promising area and, unusually, it is one in which Europe retains a position as world leader.

13.10 High pressure plasmas

Finally, we shall take a brief look at applications of plasmas in the pressure range around one atmosphere. Such plasmas present a number of attractions to users through their apparent simplicity, in that ultra-high vacuum systems are not required, and their potential for high speed processing. As a consequence they are finding increasing use in surface technology and for bulk materials, including the processing of industrial wastes. The core technology here is the plasma torch, which comes in a variety of forms from systems of a few kilowatts in power for coatings to $\sim$ 1 MW for bulk materials processing.

A schematic of a torch for the production of a plasma-sprayed coating is shown in Fig. 13.20. An arc is created between a pin cathode and the torch body which forms the anode. Carrier gas forced through the torch carries plasma with it into the surrounding environment, which is normally held at a pressure somewhat below atmospheric in an inert gas. This allows the plasma to penetrate further towards the substrate and reduces contamination. The coating material in the form of a powder, often a ceramic, is fed into the plasma in the torch and is transported by the plasma onto the substrate. Provision of heat, to melt the coating material, and gas flow, to transport the material to the substrate, are thus the main functions of the torch. This process (Scott, 1992) can produce high quality coatings with micron to millimeter thicknesses (Fig. 13.4), one of the main applications being a renewable thermal barrier coating on jet engine turbine blades. Free-standing artefacts can also be produced, by coating a mandrell and then dissolving the mandrell away, an example being high temperature drawing dies for optical fibres. More recently, diamond films have been produced at a high deposition rate using a similar technology (Beulens, Buuron and Schram, 1991). Thus it is quite possible that, in time, cheaper thermal-plasma processes will begin to find wider applications in coating technology, perhaps displacing some of the lower pressure processes.

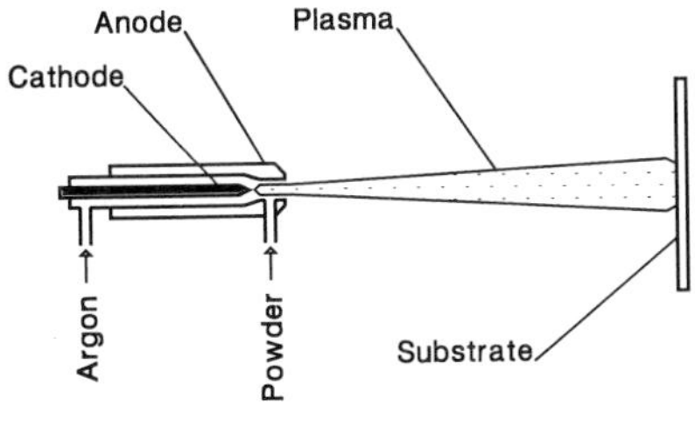

Fig. 13.20. Illustration of a plasma torch for producing plasma sprayed coatings. Powder, typically a ceramic, is added to the plasma in the torch and is conveyed as molten material or vapour to the substrate.

At the other end of the scale, a high power application to melting bulk materials (Lightfoot, Stockham and Gill, 1990) is illustrated in Fig. 13.21. The torch here is in the form of a transferred arc: the arc is struck initially between the pin cathode and the torch body, and then transferred, with the assistance of the high pressure gas flow, to the bulk material which then forms the new anode. Again, the main purpose of the torch is to transfer intense heat into the material being processed. The particular application shown in the figure is for the recovery of valuable and toxic materials, such as Ag, As, Cd, Cr, Pb and Se, from steel furnace dusts. These constituents are recovered either as a vapour or as a molten layer in the crucible. Toxic components in the remaining slag are reduced to well below the legal limits for safe disposal in landfill sites. Similar configurations are in use, or are being developed, for steel and glass smelting, the pyrolysis of used automobile tyres, the treatment of contaminated soils and many other applica-

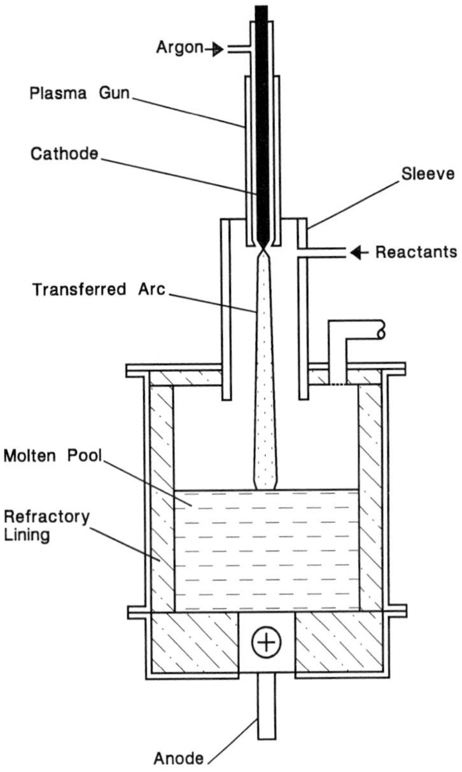

Fig. 13.21. Schematic diagram of a transferred arc process for the recovery of valuable and toxic trace metals from steel furnace dusts. Reactants are added tangentially into the sleeve around the arc. The trace metals are recovered either by draining off from the molten pool or from the hot gas in the exhaust from the vessel (Lightfoot, Stockham and Gill, 1990.)

tions. Waste and toxic material in the environment is currently a major topical issue, and thermal plasma is providing one of the core technologies helping to dispose and treat these materials.

Acknowledgements

The author would like to express his grateful thanks to colleagues in the UK Group for Plasma and Ion Surface Engineering (PISE), Prof. D. G. Armour, Prof. T. Bell, Dr I. M. Buckley-Golder, Dr N. St J. Braithwaite, Dr G. Hancock, Dr G. G. Lister, Prof. A. Matthews and Dr A. P. Webb, for many useful discussions on industrial plasma science and applications. Permission from AEA Technology to use Fig. 13.17 is gratefully acknowledged.

References

BACHMANN, P. K., LEERS, D. and LYDTIN, H. (1991). *Diamond and Related Materials*, **1**, 1.

BELL, T., DEARNLEY, P. A. and LANAGAN, J. (1988). In *The Physics of Ionized Gases* (L. Tanovic, N. Konjevic and N. Tanovic, eds). Nova Science, Commack, New York, p. 305.

BEULENS, J. J., BUURON, A. J. M. and SCHRAM, D. C. (1991). *Surface & Coatings Technol.*, **47**, 401.

CHAPMAN, B. (1980). *Glow Discharge Processes*. John Wiley & Sons, New York.

CONRAD, J. R., RADTKE, J. L., DODD, R. A., WORZALA, F. J. and TRAN, N. C. (1987). *J. Appl. Phys.*, **62**, 4591.

ENGEL, A. VON (1965). *Ionized Gases.* Oxford University Press.

FIELD, D., KLEMPERER, D. F., MAY, P. W. and SONG, Y. P. (1991). *J. Appl. Phys.*, **70**, 82.

FRANKLIN, R. N. (1976). *Plasma Phenomena in Gas Discharges.* Clarendon, Oxford.

HOLOHAN, J. R. and BELL, A. T. (1974). *Techniques and Applications of Plasma Chemistry.* John Wiley & Sons, New York.

HOWLETT, S. P., TIMOTHY, S. P. and VAUGHAN, D. A. J. (1992). *Industrial Plasmas: Focussing UK Skills on Global Opportunities.* CEST, London.

KUYPERS, A. T. (1989). *Doctorate thesis*, University of Utrecht.

LIGHTFOOT, R., STOCKHAM, J. B. and GILL, M. E. (1990). *Plasma for Industry and Environment.* British National Council for Electroheat, London, Paper 2.1.

MANOS, D. M. and FLAMM, D. L. (1989). *Plasma Etching: An Introduction.* Academic Press, San Diego.

MATTHEWS, A., ARTLEY, R. J., HOLIDAY, P. and STEVENSON, P. (1992). *The UK Engineering Coatings Industry in 2005: a Report produced for the Department of Trade and Industry.* Research Centre for Surface Engineering, Hull.

MONAGHAN, D. P., TEER, D. G., LAING, K. C., EFEOGLU, I. E. and ARNELL, R. D. (1993). *Proc. 3rd Int. Conf. Plasma Surface Engineering.* Elsevier, Amsterdam. In press.

MUNZ, W-D. (1991). *Surface & Coatings Technol.*, **48**, 81.

PROUD, J. *et al.* (1991). *Plasma Processing of Materials: Scientific Opportunities and Technological Challenges.* National Academy Press, Washington, D.C.

REINKE, P., JACOB, J. and MOLLER, W. (1991). In *Diamond and Diamond-like Films and Coatings* (R. E. Clausing, L. L. Horton, J. C. Angus and P. Koidl, eds). NATO ASI Series B, **226**, p. 661.

RICKERBY, D. S. and MATTHEWS, A. (1991). *Surface Engineering: Processes, Characterisation and Applications.* Blackie and Son, Glasgow.

RIEMANN, K-U. (1992). *Phys. Fluids B*, **4**, 2693.

RIEMANN, K-U., EHLEMANN, U. and WIESEMANN, K. (1992). *J. Phys. D: Appl. Phys.*, **25**, 620.

ROSSNAGEL, S. M., CUOMO, J. J. and WESTWOOD, W. D. (1990). *Handbook of Plasma Processing Technology.* Noyes Publications, New Jersey.

SCOTT, K. T. (1992). Proc. 15th Conference on Scanning Microscopy in Materials Testing. DVM, Kassel.

SHAMIM, M., SCHEUER, J. T. and CONRAD, J. R. (1991). *J. Appl. Phys.*, **69**, 2904.

SONG, Y. P., FIELD, D. and KLEMPERER, D. F. (1990). *J. Phys. D: Appl. Phys.*, **23**, 673.

SUZUKI, K., OKUDAIRA, S., SAKUDO, N. and KANOMATA, I. (1977). *Japan J. Appl. Phys.*, **16**, 1979.

TANG, Y. S., CHEUNG, R. and WILKINSON, C. D. W. (1990). *Electron. Lett.*, **26**, 1823.

TENDYS, J., DONNELY, I. J., KENNY, M. J. and POLLOCK, J. T. A. (1988). *Appl. Phys. Lett.*, **53**, 2143.

TRAN NGOC AN, MARODE, E. G. and JOHNSON, P. C. (1977). *J. Phys. D: Appl. Phys.*, **10**, 2317.

VAUGHAN, D. A. J. (1991). *Future of Surface Engineering in the UK.* CEST, London.

14

Transport in magnetically confined plasmas

T. E. STRINGER

14.1 Introduction

The achievement of sufficiently long particle and energy confinement is perhaps the most difficult task on the route to a fusion reactor. Energy must be confined long enough for the plasma to reach the temperature, of order 10^8°C, for thermal reactions, whilst the fuel ions must be confined long enough for a significant fraction to fuse. On the other hand, particles must not be confined so long that the spent fuel, namely alpha particles, becomes a major fraction of the plasma.

The particle and energy transport in a toroidal plasma is of two types. The first results from collisions and is referred to as neoclassical transport. Its exact evaluation is mathematically quite difficult, partly because of the complicated orbits of charged particles in a torus, but more so because of the differential nature of the Coulomb collision operator. Much elegant analysis has been done on this problem, and neoclassical loss can now be calculated accurately.

The second type of transport, known as anomalous transport, results from the fluctuating electric and magnetic fields which are generally observed in toroidal plasmas. Energy is stored in a confined plasma, both in the confining magnetic field and in the compressional energy of the plasma. Several types of wave can propagate in a plasma, and most become unstable in certain conditions, growing by extracting the stored energy. The linear theory of these instabilities has been developed over the years, and is now in good shape. Although this analysis tells one when an instability occurs, it is no longer valid when the instability approaches its saturation level. Evaluation of the associated transport requires knowledge of the saturation spectrum and the relative phase of the density and field fluctuation. To evaluate this theoretically we need a nonlinear analysis of the instability, which is much more difficult.

As an introduction to collisional transport, a straight cylindrical plasma is considered first in Section 14.2. This illustrates how currents must flow spontaneously in the plasma in order to maintain the equilibrium. In a collisional plasma this gives rise to resistive dissipation. The outward plasma diffusion releases compressional energy which maintains these currents. The additional equilibrium currents that must flow to hold a toroidal plasma in equilibrium, and the corresponding increase in resistive dissipation in a strongly collisional plasma, are discussed in Section 14.3. Most tokamak plasmas of experimental interest have long mean free paths, where some electrons and ions are reflected from regions of stronger magnetic field. This gives rise to the so-called banana orbits, described in Section 14.4. Rather than discussing the details of the resulting neoclassical analysis, a simple heuristic derivation for the loss rate is given.

Unfortunately the experimental loss rate is generally larger than neoclassical, which means that anomalous transport is dominant. Section 14.6 first derives the linear dispersion equation for drift waves, which seem responsible for at least some of the observed fluctuations. Rather than attempting to summarize the detailed theoretical predictions for the transport driven by various instabilities, Sections 14.7–14.16 outline different general methods used to study the nonlinear phase of an instability. Weak turbulence is first discussed, starting with quasilinear analysis and then the effect of mode coupling. Concepts underlying strong turbulence theory, such as the random phase approximation and renormalization, are described in Sections 14.11–14.16. Sections 14.17–14.19 give a very brief summary of the experimental observations on fluctuations and transport, and how these compare with prediction.

14.2 Transport in a cylindrical plasma

We will first discuss the simplest confined plasma – a straight cylindrical plasma in which the magnetic field is everywhere parallel to the axis. Some plasma phenomena can only be properly described by treating the plasma as an assembly of interacting particles, using kinetic equations. For this problem, however, the particle nature of the plasma plays no role, and it can be more simply described by fluid-like equations. The two-fluid model treats a plasma as two interpenetrating conducting fluids, each obeying a momentum equation of the form

$$nm_j \frac{\mathrm{d}\boldsymbol{v}_j}{\mathrm{d}t} = ne_j(\boldsymbol{E} + \boldsymbol{v}_j \times \boldsymbol{B} - \eta \boldsymbol{J}) - \boldsymbol{\nabla} p_j, \tag{14.2.1}$$

where subscript $j = \mathrm{i}$ or e for ions or electrons. Adding the ion and electron equations gives the one-fluid equation

$$nm_\mathrm{i} \frac{\mathrm{d}\boldsymbol{v}}{\mathrm{d}t} = \boldsymbol{J} \times \boldsymbol{B} - \boldsymbol{\nabla}(p_\mathrm{i} + p_\mathrm{e}). \tag{14.2.2}$$

In the confined cylindrical plasma, pressure decreases with r, the distance from the axis. To balance this pressure gradient in a steady equilibrium ($\mathrm{d}/\mathrm{d}t = 0$), we must have a poloidal current

$$J_\theta = \frac{1}{B} \frac{\mathrm{d}}{\mathrm{d}r}(p_\mathrm{i} + p_\mathrm{e}). \tag{14.2.3}$$

The source of this current may be seen from the ion and electron equations. Multiplying Eq. (14.2.1) vectorially by $\boldsymbol{B}$ gives the velocities perpendicular to the magnetic field:

$$\boldsymbol{v}_{\mathrm{i}\perp} = \frac{\boldsymbol{E} \times \boldsymbol{B}}{B^2} - \frac{\eta \boldsymbol{J} \times \boldsymbol{B}}{B^2} - \frac{\boldsymbol{\nabla} p_\mathrm{i} \times \boldsymbol{B}}{neB^2} - \frac{m_\mathrm{i}}{eB^2} \frac{\mathrm{d}\boldsymbol{v}_\mathrm{i}}{\mathrm{d}t} \times \boldsymbol{B}, \tag{14.2.4}$$

$$\boldsymbol{v}_{\mathrm{e}\perp} = \frac{\boldsymbol{E} \times \boldsymbol{B}}{B^2} - \frac{\eta \boldsymbol{J} \times \boldsymbol{B}}{B^2} + \frac{\boldsymbol{\nabla} p_\mathrm{e} \times \boldsymbol{B}}{neB^2}. \tag{14.2.5}$$

Thus the radial pressure gradients automatically give rise to ion and electron poloidal velocities (the third terms in each of the above two equations), which produce just the right poloidal current to balance the pressure gradient. This poloidal current in turn gives rise to a radial particle flux of both species (the second term):

$$nv_r = -\frac{n\eta_\perp}{B^2}\frac{d}{dr}(p_i + p_e) = -\frac{n\eta_\perp}{B^2}\left[(T_i + T_e)\frac{dn}{dr} + n\frac{d}{dr}(T_i + T_e)\right]. \qquad (14.2.6)$$

The first term has the form of a diffusive flux, $nv = -Ddn/dr$, with diffusion coefficient $D = n\eta_\perp(T_i + T_e)/B^2$.

It is a general feature of plasma confinement that any dissipation gives rise to diffusion. In the example of the plasma cylinder, poloidal current is needed to balance the pressure gradient. The energy dissipated, $\eta_\perp J_\perp^2$, to maintain this current is provided by the release of compressional energy in the expanding plasma.

The above diffusivity may be expressed in an alternative form, which suggests an alternative physical explanation. Writing the resistivity as $\eta_\perp = m_e\nu_{ei}/ne^2$, where ν_{ei} is the electron–ion collision frequency, D takes the form

$$D = \nu_{ei}r_{Le}^2(1 + T_i/T_e)/4, \qquad (14.2.7)$$

where $r_{Le} = (2T_e/m_e)^{1/2}/\omega_{ce}$ is the electron Larmor radius and ω_{ce} is the electron cyclotron frequency. This expression may be derived heuristically as follows. Each time an electron velocity suffers a 90° deflection, due to collisions with ions, its orbit restarts its gyratory motion around a different field line. The displacement of the orbit centre is comparable with the Larmor radius, and is random in direction. The diffusion produced by such a process is given by the random walk expression $D = \Delta^2/3\tau$, where Δ is the step length and τ is the average time interval between successive steps. Taking $\Delta \sim r_{Le}$ and $\tau = 1/\nu_{ei}$ gives a diffusivity similar to that in Eq. (14.2.7). This is an example of how the two alternative descriptions of a plasma, either as conducting fluids or as a distribution of gyrating particles, lead to the same final result, even when the physical processes described seem different.

14.3 Collisional transport in a resistive toroidal plasma

We now consider what happens when the cylindrical plasma is bent into a torus, as in a tokamak. The coordinates to be used are illustrated in Fig. 14.1. The magnetic field is predominantly in the toroidal direction, denoted by ϕ. The field strength falls off inversely as R, the distance from the axis of symmetry AB. This

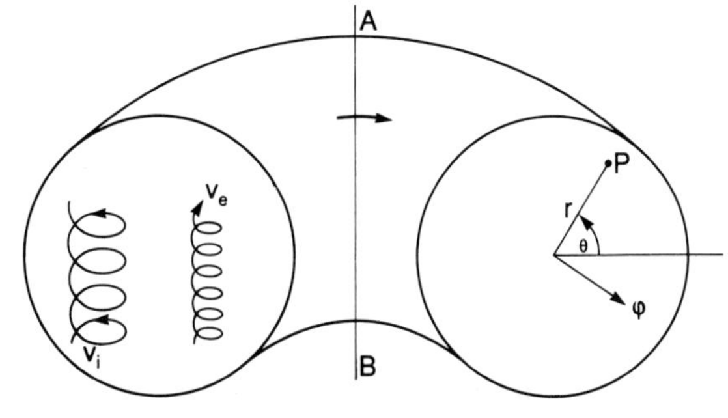

Fig. 14.1. A toroidal plasma.

gives rise to a vertical drift of the charged particles, because the curvature of the gyrating orbit on the weak field side is less than that on the strong field side, as sketched on the left-hand cross-section. Because ions and electrons gyrate in opposite directions, the vertical drifts are also opposite. If the magnetic field were purely toroidal, this would lead to an accumulation of ions at the bottom, and electrons at the top, of the torus. The resulting vertical electric field would cause the whole plasma to drift onto the outboard side of the torus wall. To prevent this happening, a poloidal magnetic field must be produced by inducing a plasma current in the ϕ-direction. Magnetic field lines then spiral around the minor axis of the torus ($r = 0$). This allows the surplus ions and electrons to neutralize each other by flowing along the magnetic field. This neutralizing current is known as the Pfirsch–Schlüter current. Its magnitude is easily shown to be

$$J_{\parallel} = -\frac{2q}{B_{\phi}}\frac{\mathrm{d}(p_{\mathrm{i}} + p_{\mathrm{e}})}{\mathrm{d}r}\cos\theta, \tag{14.3.1}$$

where $q = rB_{\phi}/RB_{\theta}$ is the safety factor. Subscripts $\parallel$ and $\perp$ denote components parallel and perpendicular to the magnetic field, respectively.

The Pfirsch–Schlüter current increases the resistive dissipation, compared to a cylindrical plasma, to

$$\frac{\mathrm{d}W}{\mathrm{d}t} = \eta_{\perp}J_{\perp}^2 + \eta_{\parallel}J_{\parallel}^2 = \eta_{\perp}J_{\perp}^2\left[1 + 2q^2\frac{\eta_{\parallel}}{\eta_{\perp}}\right]. \tag{14.3.2}$$

We may expect the plasma loss rate to be increased by a similar factor. That this is indeed the case can be shown by evaluating the poloidal variation in density and velocity over the plasma cross-section, and hence the mean particle flux. The space charge produced by the vertical ∇B drifts cannot completely escape along the field lines, since some residual electric field is needed to drive the Pfirsch–Schlüter current against resistivity. This vertical electric field produces an $\boldsymbol{E} \times \boldsymbol{B}/B^2$ plasma drift away from the axis of symmetry. The combination of the ∇B vertical drift, the $\boldsymbol{E} \times \boldsymbol{B}$ horizontal drift, and the poloidal component of the parallel flow along the helical field lines, may be visualized as a double vortex flow, as illustrated in Fig. 14.2.

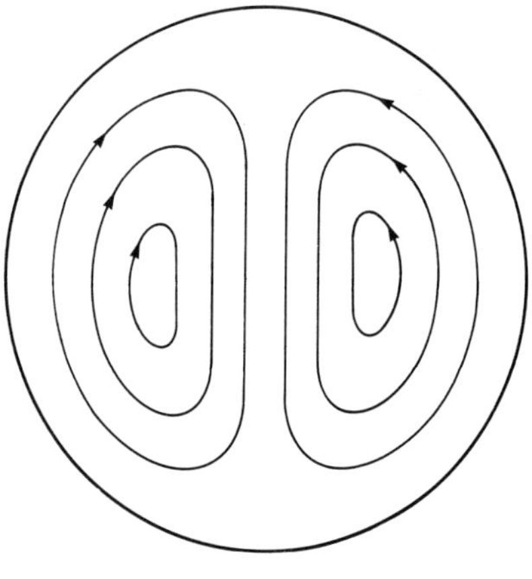

Fig. 14.2. Plasma flow pattern.

Integrating the particle flux over a magnetic surface r = const. gives the mean particle loss rate when $\mathrm{d}T/\mathrm{d}r = 0$ to be

$$\overline{nv_r} = -\frac{\eta_\perp(p_i + p_e)}{B^2}\left(1 + 2q^2\frac{\eta_\parallel}{\eta_\perp}\right)\frac{\mathrm{d}n}{\mathrm{d}r}. \quad (14.3.3)$$

The additional flux, due to toroidicity, is known as the Pfirsch–Schlüter flux. Since a typical value for q is 3, and $\eta_\parallel/\eta_\perp = 1/2$, the collisional diffusion in a resistive toroidal plasma is typically an order of magnitude greater than that in a comparable cylinder.

14.4 Diffusion in a torus with long collisional mean-free path

The foregoing discussion assumed that the plasma behaviour can be described by resistive fluid equations. However, these equations are no longer valid when the collisional mean free path, λ_{mfp}, is longer than the distance a particle must travel along a field line to get from top to bottom of the torus. This latter distance, the so-called connection length, equals qR. Over the central region of most tokamaks, $\lambda_{\mathrm{mfp}} > qR$, and the kinetic description must be used for the plasma.

The most important new effect to occur at long mean free paths is *magnetic trapping*. This results from the conservation of the two adiabatic invariants of the plasma motion, i.e. total particle energy, $W = m(v_\parallel^2 + v_\perp^2)/2$, and magnetic moment $\mu = mv_\perp^2/2B$. The variation in toroidal magnetic field strength may be written as

$$B = B_0\frac{R_0}{R} = \frac{B_0R_0}{R_0 + r\cos\theta} \approx B_0(1 - \varepsilon\cos\theta). \quad (14.4.1)$$

Here R_0 is the radius of the minor axis of the torus, B_0 is the field strength on this axis, and the inverse aspect ratio, $\varepsilon = r/R_0$ is assumed small. As a particle follows a field line over a magnetic surface r = const., the variation in its parallel velocity is

$$v_\parallel^2 = 2W/m - v_\perp^2 = 2W/m - v_{\perp 0}^2(1 - \varepsilon\cos\theta), \quad (14.4.2)$$

where subscript 0 denotes the value of a variable at $\theta = 0$. When $2W/m - v_{\perp 0}^2 < \varepsilon v_{\perp 0}^2$, i.e. when $v_{\parallel 0}^2 < 2\varepsilon v_{\perp 0}^2$, $v_\parallel^2$ becomes negative when θ exceeds $\theta_r = \cos^{-1}(1 - v_{\parallel 0}^2/\varepsilon v_{\perp 0}^2)$, which is physically meaningless. What happens is that the parallel velocity becomes zero when it reaches $\theta = \theta_r$, and it then reverses sign. Referring to Fig. 14.3, the particle bounces to and fro between the reflection points $\pm\theta_r$. Only particles whose velocity vector is nearly perpendicular to the magnetic field, i.e. $|v_{\parallel 0}|/v_0 < (2\varepsilon)^{1/2}$, are trapped. Scattering collisions with other particles produce a gradual change in the velocity pitch angle. The average time before a trapped particle is scattered out of the narrow trapped velocity band is ε/ν_e, where ν_e is the scattering frequency for 90° deflection.

A charged particle follows a field line only in first approximation. Magnetic gradients or electric fields cause it to drift perpendicular to the field line. The important drift in a torus results from the variation in toroidal magnetic field strength, discussed in Section 14.3. Consider the case where this drift is vertically downwards. When the particle is in the upper half of the cross-section, $0 < \theta < \pi$, the ∇B drift carries it closer to the plasma centre, whereas in the lower half it

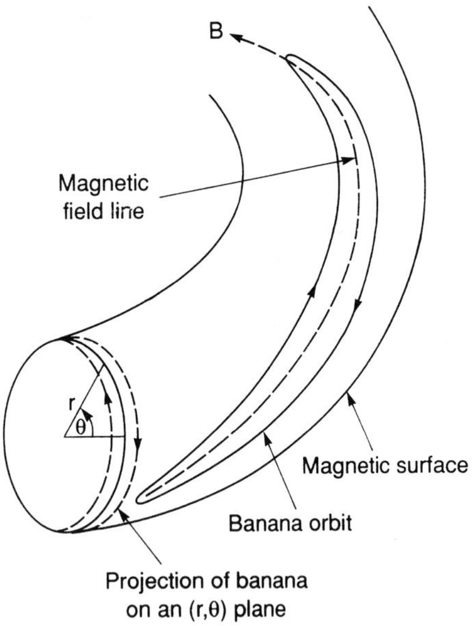

Fig. 14.3. A trapped particle orbit.

carries it away from the centre. The effect is to give a finite width to a trapped particle orbit, as illustrated in Fig. 14.3. Since the particle spends an equal time in the upper and lower halves, the net radial displacement over a complete bounce orbit is zero, i.e. the orbit closes on itself. Such trapped particle orbits are referred to as banana orbits, and the parameter range where trapping occurs is called the banana regime. The average width of a trapped particle orbit may be obtained by multiplying the bounce time, $\tau_b = qR/\varepsilon^{1/2}v$, by the vertical drift velocity, $V_D = v^2/2\omega_c R$. This width, $\Delta_b = qv/2\varepsilon^{1/2}\omega_c$, can be rewritten as $\Delta_b = \varepsilon^{1/2} r_L B_\phi/B_\theta$, where $r_L = v/\omega_c$ is the Larmor radius for the species considered. Since $B_\phi/B_\theta = O(10)$, and $\varepsilon^{1/2}$ is typically 0.5, Δ_b is significantly larger than the Larmor radius.

We will now discuss the transport of trapped particles. The problem can be solved analytically, starting from the kinetic equation. The collision operator plays an important role, and must be included in its differential form. This is because particles do not contribute to the outward flux whilst they are trapped, but only when they are scattered into, or out of, the trapped velocity band. Their mean radius then suffers a discrete change, of order Δ_b, whose sign depends on where in the banana orbit it is scattered into, or out of, the trapped velocity band. The analysis is rather complex, but the approximate loss rate can be deduced heuristically, by applying the random walk argument, already used for the resistive cylindrical plasma. Taking $\tau = \varepsilon/v$ and $\Delta = \Delta_b$ gives

$$D = \frac{vq^2 r_L^2}{4\varepsilon^{3/2}}. \tag{14.4.3}$$

In deriving Eq. (14.4.3) the diffusion coefficient for the trapped particles has been multiplied by $\varepsilon^{1/2}$, because only that fraction of the particles is trapped and hence takes part in the random walk. The diffusion coefficient in Eq. (14.4.3) thus applies to the whole plasma.

The above derivation applies to both ions and electrons but, of course, the appropriate values must be taken for ν, ω_c, and r_L. Eq. (14.4.3) predicts that $D_i \sim D_e(m_i/m_e)^{1/2}$. However, the ion and electron loss rates must be almost equal – otherwise an enormous space charge field rapidly builds up. (This is referred to as the ambipolar condition.) This contradiction is resolved in the analysis, where it is shown that the particle fluxes depend on the radial electric field (or, to be more exact, on $E_r - B_\theta u_{j\parallel}$, where $u_{j\parallel}$ is the mean velocity of the jth species parallel to the magnetic field). E_r quickly reaches an ambipolar value, at which the ion particle flux is reduced to the electron rate. Thermal diffusivities are also given approximately by Eq. (14.4.3), but now the ion thermal diffusivity really is larger than that of the electrons by the factor $(m_i/m_e)^{1/2}$. The energy dissipation which, we have said, always accompanies an increase in diffusion, is less obvious in the banana regime. It appears as a parallel viscosity, introduced by the presence of trapped particles. Comparing the particle diffusivities in the plasma cylinder and the resistive and long mean free path toroidal plasmas we obtain the ratios:

$$D_{\text{cyl}} : D_{\text{res}} : D_{\text{ban}} = 1 : q^2 : q^2 \varepsilon^{-3/2}.$$

Between the Pfirsch–Schlüter and banana regimes there is an intermediate range, known as the plateau regime, defined by the condition $1 < \lambda_{\text{mfp}}/qR < \varepsilon^{-3/2}$. In this regime, the resistive fluid equations are invalid, and on average each particle is scattered out of the trapped velocity band before it has time to complete one bounce orbit. The particle flux is predominantly due to particles with $v_\parallel \approx 0$ (or, when a radial electric field is present, $v_\parallel \approx E_r/B_\theta$). Such particles have very slow poloidal rotation. Hence the radial ∇B drift produces a relatively large displacement before changing sign as the particle moves from the upper to the lower half sections. These particles play a role analogous to resonant particles in kinetic instability analysis, the energy dissipation being Landau damping. Combining the diffusivities for the three regimes gives the well-known variation with collisionality illustrated in Fig. 14.4.

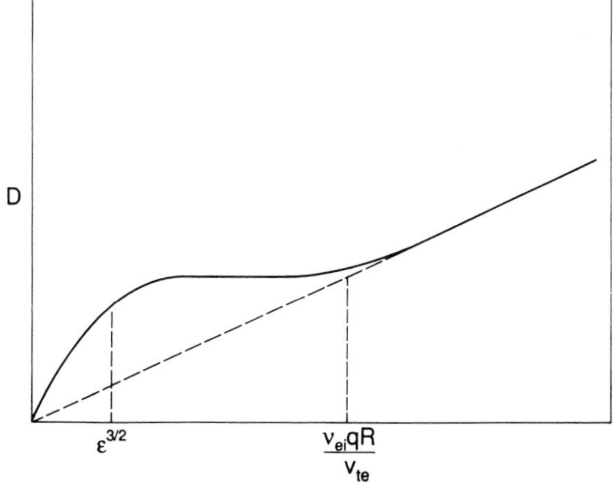

Fig. 14.4. Variation of neoclassical diffusivity with collisionality.

14.5 Anomalous transport and weak turbulence theory

The most conspicuous instabilities are the low-m MHD modes, where the mode variation is written as $\exp[i(m\theta - n\phi - \omega t)]$. However, there appears to be no correlation between the intensity of these modes and the measured electron loss. Although they are clearly responsible for disruptions, and may greatly enhance the local transport near their resonant surfaces, they do not appear to contribute to the steady background loss. Thus we are interested only in shorter wavelength fluctuations, known as microturbulence.

There are two obvious ways that particles and energy can escape – by $\boldsymbol{E} \times \boldsymbol{B}$ drift across the confining field, or by parallel motion along magnetic field lines whose trajectories diffuse radially. As will be discussed later, it is not possible with present diagnostics to establish conclusively which is responsible for the observed transport, and so both should be considered. Most attention has been given to the $\boldsymbol{E} \times \boldsymbol{B}$ flux resulting from different forms of electrostatic drift modes. However, stochastic magnetic fields due to electromagnetic instabilities, e.g. rippling or resistive ballooning modes, provide an alternative path for energy escape.

14.6 Drift waves

Since drift waves are the most popular explanation for anomalous transport, we will give here an elementary account of these waves. To derive the dispersion equation from the two-fluid equations, a more exact form than that given in Eq. (14.2.1) must be used. Because of the gyratory motion of the ions, their momentum balance cannot be completely described by scalar pressure, but requires a pressure tensor. The off-diagonal components of this pressure tensor are generally small, but, because of cancellation of the larger terms, they become important in the drift wave analysis. This analysis is much simpler when the guiding-centre equations are used, in which pressure plays no role.

Since drift waves are electrostatic, we may write $\boldsymbol{E} = -\nabla\phi$. The dispersion equation can be derived from the following equations:

$$\boldsymbol{v}_{i\perp} = -\frac{\nabla\phi \times \boldsymbol{B}}{B^2} - \frac{m_i}{eB^2}\frac{d}{dt}\nabla_\perp\phi, \tag{14.6.1}$$

$$n_0 m_i \frac{dv_{i\parallel}}{dt} = -\nabla_\parallel(n_0 e\phi + \gamma T_i n_i), \tag{14.6.2}$$

$$\frac{\partial n_i}{\partial t} + \nabla \cdot (n\boldsymbol{v}_i) = 0, \tag{14.6.3}$$

$$\frac{\tilde{n}_e}{n_0} = \frac{e\phi}{T_e}. \tag{14.6.4}$$

Eq. (14.6.1) is the ion guiding-centre velocity perpendicular to the magnetic field, including the polarization drift. Eq. (14.6.2) is the component of the ion equation of motion parallel to the magnetic field, where the ion pressure variation is assumed to be $\tilde{p}_i = \gamma T_{i0}\tilde{n}_i$. Eq. (14.6.3) is the ion continuity equation, and Eq. (14.6.4) is the Boltzmann distribution for the electrons; n_0 denotes the unperturbed density and $\tilde{n}$ is the perturbation.

In slab geometry, where density varies in the x-direction and the z-axis coincides with the magnetic field direction, these linearized equations are readily solved for a simple wave

$$\phi = \phi(x)\exp[\mathrm{i}(k_y y + k_{\parallel} z - \omega t)]. \tag{14.6.5}$$

From quasineutrality, $\tilde{n}_{\mathrm{i}} = \tilde{n}_{\mathrm{e}}$. Hence $\tilde{n}_{\mathrm{i}}$ can everywhere be replaced by ϕ, using Eq. (14.6.4). Eq. (14.6.2) gives $v_{\mathrm{i}\parallel}$. Substituting in the continuity equation for $v_{\mathrm{i}\parallel}$ and $\boldsymbol{v}_{\mathrm{i}\perp}$ immediately gives the dispersion relation

$$\omega^2(1 + \tau k_y^2 r_{\mathrm{Li}}^2/2) - \omega\omega_* - V_{\mathrm{s}}^2 k_{\parallel}^2 = 0. \tag{14.6.6}$$

Here $r_{\mathrm{Li}} =$ is the ion Larmor radius, $\tau = T_{\mathrm{e}}/T_{\mathrm{i}}$, $\omega_* = -(k_y T_{\mathrm{e}}/eBn_0)(\mathrm{d}n_0/\mathrm{d}x)$ is the electron diamagnetic drift frequency, and $V_{\mathrm{s}} = [(T_{\mathrm{e}} + \gamma T_{\mathrm{i}})/m_{\mathrm{i}}]^{1/2}$ is the sound speed.

A drift wave is just the familiar ion acoustic wave propagating in an inhomogeneous plasma. The restoring force in an ion acoustic wave is the pressure gradient and the electric force, see Eq. (14.6.2). In a homogeneous plasma, the pressure gradient results solely from the plasma compression due to its parallel motion. This gives rise to the $V_{\mathrm{s}}^2 k_{\parallel}^2$ term in Eq. (14.6.6). In an inhomogeneous plasma, however, density is also perturbed by the $\boldsymbol{E} \times \boldsymbol{B}$ motion, which convects plasma along the gradient in the equilibrium density, $n_0(x)$. This gives rise to the $\omega\omega_*$ term. It becomes important only for waves propagating nearly perpendicular to the magnetic field, $k_{\parallel}/k_y \leq r_{\mathrm{Li}}/L_n$, where $L = n_0/(\mathrm{d}n_0/\mathrm{d}x)$ is the density scale length. In this limit one of the two solutions of Eq. (14.6.6), the drift wave branch, is approximately

$$\omega_{\mathrm{D}} = \frac{\omega_*}{1 + k_y^2 r_{\mathrm{Li}}^2/2}. \tag{14.6.7}$$

The $k_y^2 r_{Li}^2$ term in Eq. (14.6.6) comes from the ion polarization drift, and is important only at short wavelengths.

When the drift wave becomes unstable, it derives its energy from the plasma compressional energy. This is transferred to the wave via the electron current flow along the magnetic field. In a homogeneous plasma this current extracts energy from the wave to support the ohmic or Landau dissipation, thus damping the wave. In an inhomogeneous plasma, however, $\boldsymbol{E} \times \boldsymbol{B}$ convection converts some of the radial pressure gradient into a gradient along the magnetic field. These gradients can drive an electron current whose phase is such that it does work against the wave electric field. This increases the wave energy. Any dissipation introduces an out-of-phase addition to the electron Boltzmann distribution,

$$\frac{\tilde{n}_{\mathrm{e}}}{n_0} = \frac{e\phi}{T_{\mathrm{e}}}(1 - \mathrm{i}\delta); \tag{14.6.8}$$

δ is known as the nonadiabatic electron response, and is the term responsible for instability or damping. It can result from collisional resistance, resonant electrons, or trapped electrons. Since the nonlinear behaviour of the wave is independent of the origin of this term, it will not be discussed in detail.

14.7 Weak turbulence theory

In weak turbulence analysis the fluctuation level is assumed low enough that nonlinear effects can be treated as perturbations. The solution is expanded with the fluctuation level as the small parameter:

$$f(\boldsymbol{r},\boldsymbol{v},t) = f^{(0)}(\boldsymbol{r},\boldsymbol{v}) + f^{(1)}(\boldsymbol{r},\boldsymbol{v},t) + f^{(2)}(\boldsymbol{r},\boldsymbol{v},t) + \cdots. \tag{14.7.1}$$

Here $f^{(1)}$ is the usual solution of the linearized equations. The assumption that this is everywhere a reasonable first approximation to the nonlinear solution is implicit in the perturbation expansion; $f^{(2)}$ includes all terms which are second order in the fluctuation amplitude.

14.8 Quasilinear analysis

To first order in the fluctuation amplitude the particle and energy fluxes vary sinusoidally with y and z, similarly to Eq. (14.6.5). Hence when averaged over an x = constant plane, the mean fluxes vanish. When evaluating the second order fluxes, the products of first order quantities must be retained, and some of these have a nonvanishing average. Thus the second order fluxes may be evaluated using the first order solution for the fluctuations. Since the fluctuations are described by the linearized equations, they can be expressed as a sum of independent normal modes. For example,

$$\tilde{f}(\boldsymbol{r},\boldsymbol{v},t) = \textstyle\sum_k f(\boldsymbol{k},\boldsymbol{v})\exp[\mathrm{i}(\boldsymbol{k}\cdot\boldsymbol{r} - \omega(\boldsymbol{k})t)], \tag{14.8.1}$$

where $\omega(\boldsymbol{k})$ satisfies the linear dispersion relation. Consider, for simplicity, electrostatic modes in a cylindrical plasma where the magnetic field lines are helices, lying on r = const. surfaces. The net radial particle flux across a magnetic surface is

$$\Gamma = \frac{1}{2\pi}\int_0^{2\pi} n\left[\frac{\boldsymbol{E}\times\boldsymbol{B}_0}{B_0^2}\right]_{\mathrm{r}} \mathrm{d}\theta = -\frac{1}{2\pi r B_0}\textstyle\sum_k \displaystyle\int_0^{2\pi} \tilde{n}(\boldsymbol{k})\frac{\partial\phi(\boldsymbol{k})}{\partial\theta}\mathrm{d}\theta. \tag{14.8.2}$$

Here subscript r denotes the radial component and, as usual, θ is the poloidal angle. It is straightforward to evaluate the quasilinear fluxes for any specific instability, once the linear analysis is complete. A general form will now be derived for the particle transport due to electron drift waves.

As discussed in Section 14.6, the electron density perturbation can be written as the Boltzmann distribution with a nonadiabatic contribution due to resonant or trapped electrons, or collisions:

$$\tilde{n}_{\mathrm{e}}(\boldsymbol{k}) = \frac{n_0 e}{T_{\mathrm{e}}}\phi(\boldsymbol{k})[1 - \mathrm{i}\delta(\boldsymbol{k})]. \tag{14.8.3}$$

Substituting for ϕ in Eq. (14.8.2), and assuming $\delta \ll 1$, gives

$$\Gamma = \frac{n_0 T_{\mathrm{e}}}{eB}\textstyle\sum_k k_\theta \delta(\boldsymbol{k})\left[\dfrac{\tilde{n}(\boldsymbol{k})}{n_0}\right]^2. \tag{14.8.4}$$

The linear growth rate of the drift wave, $\gamma(k)$, is determined by the electron nonadiabatic response, $\gamma(k) = \delta(k)\omega_*$. Hence Eq. (14.8.4) can be written as a diffusive flux, $-D\mathrm{d}n_0/\mathrm{d}r$, with

$$D = L_n^2 \textstyle\sum_k \gamma(\boldsymbol{k})\left[\dfrac{\tilde{n}(\boldsymbol{k})}{n_0}\right]^2. \tag{14.8.5}$$

Quasilinear analysis cannot determine the stationary fluctuation level. However, when this can be measured, Eq. (14.8.5) is frequently used to estimate the resulting cross-field transport, for comparison with the observed confinement. It also provides an upper limit on the transport. Thus it seems unlikely that the density perturbation could grow beyond the amplitude at which it reverses the local density gradient (since it is the density gradient which drives the instability). This gives the upper bound $k\tilde{n} \leq dn_0/dr$, i./e. $\tilde{n}/n_0 \leq 1/k_\perp L_n$. The same limit results from the condition that the fluctuating $\tilde{\boldsymbol{E}} \times \boldsymbol{B}$ drift shall not exceed the wave velocity ω_*/k_y. Substituting in Eq. (14.8.5) gives the well-known upper limit, first given by Kadomtsev,

$$D \lesssim \frac{\gamma(\boldsymbol{k})}{k_\perp^2}. \tag{14.8.6}$$

This result is open to several criticisms. A wave cannot grow indefinitely, and the above result might suggest that, when it reaches saturation, the transport ceases. Most analyses of the saturation mechanism find that the important nonlinear effects come from the ion equations, whereas the electrons still behave linearly. Thus the phase difference between $\tilde{n}$ and $\tilde{\phi}$ is still given by δ. This can be expressed in terms of the linear growth rate, even though the nonlinear growth rate vanishes.

14.9 Mode coupling

A more serious criticism is the neglect of second order terms in the derivation of the fluctuations. For the simple example of electrostatic waves in a homogeneous plasma, the kinetic equation is

$$\frac{\partial f}{\partial t} + \boldsymbol{v} \cdot \frac{\partial f}{\partial \boldsymbol{r}} + \frac{e\boldsymbol{E}}{m} \cdot \frac{\partial f}{\partial \boldsymbol{v}} = 0. \tag{14.9.1}$$

Substituting the expansions of Eqs (14.7.1) and (14.8.1), in first order the $\boldsymbol{k}$-component is the linearized equation for $f^{(1)}(\boldsymbol{k})$. In second order, collecting all terms which vary as $\exp i(k_y y + k_z z - \omega t)$ gives an equation for $f^{(2)}(\boldsymbol{k})$:

$$\frac{\partial f^{(2)}(\boldsymbol{k})}{\partial t} + \boldsymbol{v} \cdot \frac{\partial f^{(2)}(\boldsymbol{k})}{\partial \boldsymbol{r}} = \frac{ie}{m} \sum_{\boldsymbol{k}_1 + \boldsymbol{k}_2 = \boldsymbol{k}} \phi^{(1)}(\boldsymbol{k}_1)\boldsymbol{k}_1 \cdot \frac{\partial f^{(1)}(\boldsymbol{k}_2)}{\partial \boldsymbol{v}}, \tag{14.9.2}$$

where the right side is summed over all pairs of waves which satisfy the selection rules $\omega_1 + \omega_2 = \omega$, $\boldsymbol{k}_1 + \boldsymbol{k}_2 = \boldsymbol{k}$. Only for certain types of dispersion curves, such as curve 2 in Fig. 14.5, can we satisfy the selection rules. Such a dispersion curve is referred to as the decay type. Curve 1 is an example of nondecay dispersion.

Decay-type dispersion allows resonant coupling to transfer energy between different normal modes. An obvious saturation mechanism operates when resonant coupling transfers energy from a wave number which is unstable to one which is damped. This provides the energy sink needed to balance the input over the unstable range of wave numbers. Most drift-type instabilities have their maximum growth when $k_\perp r_L = O(1)$. It can be seen from Eq. (14.6.7) that the propagation is highly dispersive here, and hence is the nondecay type. Beat waves are still produced by mode interaction, but they are forced, as opposed to natural waves. These forced waves interact in turn, and contribute to the normal modes

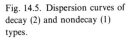

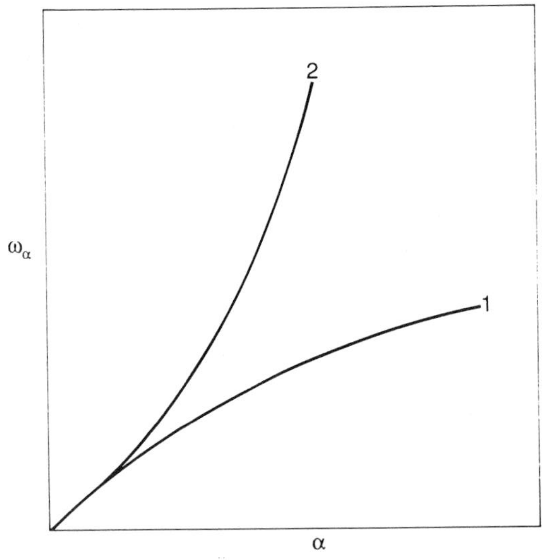

Fig. 14.5. Dispersion curves of decay (2) and nondecay (1) types.

in third order, since it is easy to satisfy the selection rules with four waves (for example, $\omega_1 + \omega_2 = \omega_3 + \omega_4$).

The transfer of energy by mode coupling into a forced wave is less effective than when the beat wave is a natural plasma wave. These forced waves may, however, have the following important effect. Because drift wave phase velocities lie between the ion and electron thermal velocities, there are few resonant ions. Among the forced waves, however, are some whose phase velocity $(\omega_1 - \omega_2)/(\boldsymbol{k}_1 - \boldsymbol{k}_2)_{\|} \sim v_{\mathrm{Ti}}$, the ion thermal velocity. Although their amplitude may be low, this is more than made up for by the higher density of resonant ions. As well as giving energy transfer between the waves and resonant ions (nonlinear ion Landau damping), this interaction also results in energy transfer between the participating natural waves. This can be regarded as wave scattering on ions, and is often referred to as ion Compton scattering. Energy is transferred from short to longer wavelengths. Since the more usual effect of resonant wave coupling is to feed energy to shorter wavelengths, this is referred to as inverse energy cascade.

14.10 Limitations on weak turbulence theory

In principle, the foregoing expansion procedure can be extended to higher order in the fluctuation amplitude. However, because it is a perturbation method, it necessarily results in modes whose character is not too different from the linear prediction. In particular, it leads to a precise relation between ω and k. Although such well defined modes are sometimes observed when the fluctuations are weak, in most conditions of interest the spectrum is very diffuse, i.e. for each k-component the time variation corresponds to a wide frequency content, $\Delta\omega \sim \omega$.

Such a diffuse spectrum is characteristic of strong turbulence. Here wave coupling continuously transfers energy between different wave-number components, so that each component grows and decays on a time scale not much longer than its period. This is illustrated by Fig. 14.6, which shows the computed variation of two k-components in a turbulent drift wave spectrum, obtained by

Waltz (1983) by numerically solving the nonlinear wave-coupled equation. Fig. 14.6(*a*) shows the amplitude and phase of a smaller component and Fig. 14.6(*b*) shows the amplitude of the largest component. If the time variation in Fig. 14.6(*a*) is Fourier analysed, the spectrum width $\Delta\omega_\kappa \sim \omega_k$. It is perhaps more meaningful to refer to $\Delta\omega_\kappa$ as the decorrelation rate.

Criteria for the validity of weak turbulence theory cannot readily be derived. The strength of interaction between different wave-number components depends both on their amplitudes and their dispersion. As the dispersion becomes stronger, different components remain in phase for shorter times, and so have less time to interact. As mentioned in Section 14.8, nonlinearity usually affects the ions first, whereas the electrons can still be described by linearized equations. Thus, even when mode coupling is strong enough to give diffuse spectra, the electron transport for a known fluctuation level may still be given by the quasilinear result in Eq. (14.8.5).

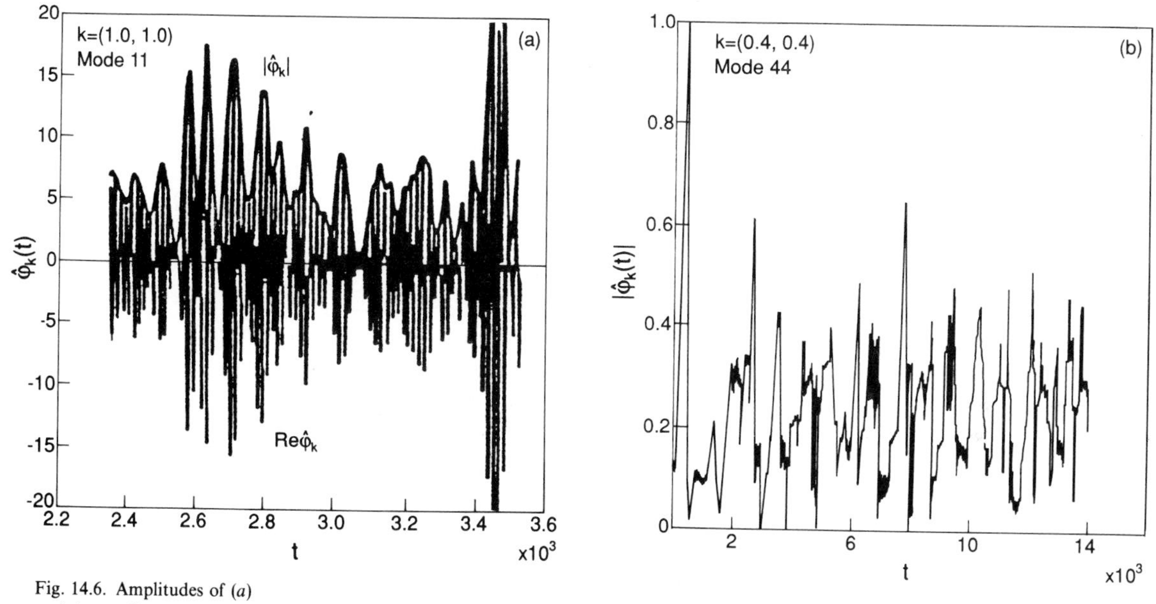

Fig. 14.6. Amplitudes of (*a*) one of the smallest and (*b*) the largest modes versus time. (From Waltz, 1983.)

14.11 The random phase approximation (RPA)

With increasing fluctuation intensity, the effect of coupling between different modes becomes too strong to be treated simply as a perturbation to the linear equation. New techniques must be developed to extend the straightforward expansion in powers of the intensity. We will consider the situation where the coupling between any pair of waves is not so strong that their phases become locked together, as happens for example in shock waves. In this condition we can use the random phase approximation, which states that, in lowest order, there is no phase correlation between components of a turbulent spectrum. Thus each mode propagates at near its natural velocity. Coupling with other components must, of course, have some effect on the wave, but this effect is assumed to be of second order in the fluctuation amplitude.

The fluctuations can be expressed as an integral over their Fourier spectrum:

$$f(\boldsymbol{r},\boldsymbol{v},t) = \int_{-\infty}^{\infty} \mathrm{d}\boldsymbol{k} \exp(\mathrm{i}\boldsymbol{k}\cdot\boldsymbol{r}) f_{\boldsymbol{k}}(\boldsymbol{v},t).$$

Substituting this into the nonlinear kinetic equation, of which Eq. (14.9.1) is the simplest example, gives an equation of the form

$$\left[\frac{\partial}{\partial t} + L_{\boldsymbol{k}}\right] f_{\boldsymbol{k}} = \int_{-\infty}^{\infty} V_{\boldsymbol{k},\boldsymbol{k}_1} f_{\boldsymbol{k}_1} f_{\boldsymbol{k}-\boldsymbol{k}_1} \mathrm{d}\boldsymbol{k}_1, \tag{14.11.1}$$

where the linear operator $L_{\boldsymbol{k}}$ is the same as occurs in the usual linearized analysis, and $V_{\boldsymbol{k},\boldsymbol{k}_1}$ will be referred to as the coupling operator. Multiplying this equation by $f_{-\boldsymbol{k}}$, the complex conjugate of $f_{\boldsymbol{k}}$, gives an equation for the wave amplitude

$$\frac{1}{2}\left[\frac{\partial}{\partial t} + L_{\boldsymbol{k}}\right] |f_{\boldsymbol{k}}|^2 = f_{-\boldsymbol{k}} \int V_{\boldsymbol{k},\boldsymbol{k}_1} f_{\boldsymbol{k}_1} f_{\boldsymbol{k}-\boldsymbol{k}_1} \mathrm{d}\boldsymbol{k}_1. \tag{14.11.2}$$

The assumption of weak coupling allows us to expand the solution in powers of $V_{\boldsymbol{k},\boldsymbol{k}_1}$. Write $f_{\boldsymbol{k}} = f_{\boldsymbol{k}}^{(0)} + f_{\boldsymbol{k}}^{(1)}$, where $f_{\boldsymbol{k}}^{(1)}$ represents the effect of the coupling with other waves:

$$\left[\frac{\partial}{\partial t} + L_{\boldsymbol{k}}\right] f_{\boldsymbol{k}}^{(1)} = \int V_{\boldsymbol{k},\boldsymbol{k}_1} f_{\boldsymbol{k}_1}^{(0)} f_{\boldsymbol{k}-\boldsymbol{k}_1}^{(0)} \mathrm{d}\boldsymbol{k}. \tag{14.11.3}$$

The corresponding equation for $|f_{\boldsymbol{k}}^{(1)}|^2$ must include all second order terms on the right of Eq. (14.11.2):

$$\frac{1}{2}\left[\frac{\partial}{\partial t} + L_{\boldsymbol{k}}\right] |f_{\boldsymbol{k}}^{(1)}|^2 = \int V_{\boldsymbol{k},\boldsymbol{k}_1} [f_{\boldsymbol{k}_1}^{(1)} f_{\boldsymbol{k}-\boldsymbol{k}_1}^{(0)} f_{-\boldsymbol{k}}^{(0)} + f_{\boldsymbol{k}_1}^{(0)} f_{\boldsymbol{k}-\boldsymbol{k}_1}^{(1)} f_{-\boldsymbol{k}}^{(0)} + f_{\boldsymbol{k}_1}^{(0)} f_{\boldsymbol{k}-\boldsymbol{k}_1}^{(0)} f_{-\boldsymbol{k}}^{(1)}] \mathrm{d}\boldsymbol{k}_1. \tag{14.11.4}$$

Now substitute for each first order component using the solution of Eq. (14.11.3). Invoking the RPA, the only products which survive the integration contain either $|f_{\boldsymbol{k}_1}|^2 |f_{\boldsymbol{k}}|^2$ or $|f_{\boldsymbol{k}_1}|^2 |f_{\boldsymbol{k}-\boldsymbol{k}_1}|^2$. Terms of the first type are called coherent, and the second type incoherent.

This may all seem rather tedious, and is no different from the standard perturbation expansion. The new step, referred to as renormalization, is now to transfer terms proportional to $|f_{\boldsymbol{k}}|^2$ to the left side, and treat them as zero order. The purpose behind this may be illustrated by a linear example, since here we know the exact answer.

14.12 Renormalization

Consider a simple harmonic oscillator with weak damping:

$$\ddot{x} + 2\nu\dot{x} + \omega_0^2 x = 0. \tag{14.12.1}$$

We choose as initial conditions $\dot{x}(0) = 0, x(0) = 1$. If we were to treat the small damping term as a perturbation, then in lowest order $x^{(0)} = \cos\omega_0 t$, and in next order

$$\ddot{x}^{(1)} + \omega_0^2 x^{(1)} = 2\nu\omega_0 \sin\omega_0 t. \tag{14.12.2}$$

This gives a component, still having frequency ω_0, which increases linearly with time. The exact solution is, of course, known to be

$$x = \exp(-\nu t)\left[\cos\Omega t + \frac{\nu}{\Omega}\sin\Omega t\right], \tag{14.12.3}$$

where $\Omega = (\omega_0^2 - \nu^2)^{1/2}$. The perturbation solution, which could also be obtained by expanding Eq. (14.12.3) in powers of ν, is correct at small times, but gives quite the wrong behaviour at larger times ($\nu t > 1$).

Weak turbulence analysis breaks down because the linearized solution ceases to be a good first approximation. If an expansion is to be used, the effect of turbulence must somehow be included in the lowest order approximation. The above example gives a clue as to how this may be done – any small terms which *can* be included in the zero order operator, *should* be included. Thus the coherent terms in Eq. (14.11.4) are denoted, for example, by $\nu_{\boldsymbol{k}} |f_{\boldsymbol{k}}|^2$, and combined with the linear operator on the left.

An example of renormalization which has an obvious physical interpretation is the scattering of particle orbits by waves, studied by Dupree in connection with the collisionless drift wave instability. This renormalization changes the resonant denominator in the kinetic equation to $\omega - k_{\parallel} v_{\parallel} + \mathrm{i} D k_{\perp}^2$, where D expresses the diffusion of the particle orbit due to the fluctuating $\tilde{\boldsymbol{E}} \times \boldsymbol{B}$ drift. The value of D is determined later in the analysis by the self-consistency argument that the resulting fluctuation level produces the orbit diffusion. This led to fluctuation levels and diffusion rates which are not inconsistent with observation.

Including the renormalized term with the linear terms gives rise to a nonlinear dielectric function. The incoherent terms in Eq. (14.11.4) act as a source, giving a relation of the form

$$\varepsilon(k, \omega, |f_{\boldsymbol{k}_1}^2|) |f_{\boldsymbol{k}}|^2 = S_{k\omega,\,\mathrm{inch}}. \tag{14.12.4}$$

If the incoherent terms are ignored, the analysis gives a nonlinear dispersion relation, including terms in $|f_{\boldsymbol{k}_1}|^2$ introduced by the coherent coupling term. The linear part of the dielectric function generally contains imaginary terms, giving rise to the linear instability. The nonlinear terms may also include imaginary terms, expressing the damping caused by coupling of wave energy into other components. The intensity at which these imaginary terms cancel yields the saturation level for the instability. Such analysis is called *one point renormalized theory*.

For a given spectrum intensity the above nonlinear dispersion equation still yields a discrete value of ω for each wave number k, whereas experimentally waves with any specified wave number are found to cover a frequency range. To describe this behaviour the incoherent terms in Eq. (14.12.4) must be retained. The spectrum then becomes diffused. Such analysis is referred to as *two point renormalized theory*.

14.13 The direct interaction approximation (DIA)

Renormalization is sometimes done in a phenomological way, without going through the foregoing expansion procedure. The nonlinear terms can be replaced by a term of the form $\nu_k f_k$, where ν_k is an unspecified function of the turbulence and is determined later by self-consistency as, for example, in Dupree's particle scattering operator just described. However, it is more reliable to derive the renormalized equation systematically, using the direct interaction approximation. This was developed for fluid turbulence by Kraichnan. It was first used in plasma physics by Kadomtsev, who called it the *weak coupling approximation*.

The DIA assumes the interaction between any pair of waves to be weak.

However, because of the large number of waves, the overall effect is not small, and should be included in the lowest order solution. The basic analysis has already been done in Section 14.11. For clarity we Fourier-transform Eq. (14.11.1) in time, write the linear operator as $\omega_k f_k$, and the coherent mode coupling terms as $-\mathrm{i}\nu_k f_k$. The equation then takes the form

$$(\omega - \omega_k + \mathrm{i}\nu_{\boldsymbol{k}})f_{\boldsymbol{k}}(\omega) = \mathrm{i}\int V_{\boldsymbol{k},\boldsymbol{k}_1} f_{\boldsymbol{k}_1}(\omega_1) f_{\boldsymbol{k}-\boldsymbol{k}_1}(\omega - \omega_1)\,\mathrm{d}\boldsymbol{k}_1\,\mathrm{d}\omega_1. \tag{14.13.1}$$

We now explicitly evaluate all terms on the right of Eq. (14.11.3) as integrals over $\boldsymbol{k}_1$ and ω_1. Those proportional to $|f_{\boldsymbol{k}}(\omega)|^2$ provide the definition of ν_k, while the rest give $S_{\boldsymbol{k}\omega}$. When the steady state spectrum $|f_{\boldsymbol{k}}|^2$ is found, the resulting anomalous transport is usually derived from Eq. (14.8.4) for the $n\boldsymbol{E} \times \boldsymbol{B}$ flux. An example of this is given in the next section.

14.14 The Hasegawa–Mima model

This model was used in one of the first applications of the DIA method to drift wave turbulence. Many subsequent analyses use extensions of this model. Electron dissipation effects (collisional or resonant particle) are neglected, and the electrons are described by the simple adiabatic response $\tilde{n}_e = n_0 e\phi/T_e$. Expressing $\phi(\boldsymbol{r}, t)$ as a spatial Fourier series, $\phi(\boldsymbol{r}, t) = \sum_{\boldsymbol{k}} \phi_{\boldsymbol{k}}(t)\exp(\mathrm{i}\boldsymbol{k}\cdot\boldsymbol{r})$, the ion density $n_{\mathrm{i}\boldsymbol{k}}$ is readily obtained from the continuity equation, Eq. (14.6.3). Equating this to $n_{\mathrm{e}\boldsymbol{k}}$ gives the Hasegawa–Mima equation

$$\frac{\partial \phi_{\boldsymbol{k}}}{\partial t} + \mathrm{i}\omega_D \phi_{\boldsymbol{k}} = \sum_{\boldsymbol{k}=\boldsymbol{k}_1+\boldsymbol{k}_2} V_{\boldsymbol{k}_1,\boldsymbol{k}_2}\phi_{\boldsymbol{k}_1}\phi_{\boldsymbol{k}_2}, \tag{14.14.1}$$

where

$$V_{\boldsymbol{k}_1,\boldsymbol{k}_2} = \frac{r_{LS}^2}{(1 + k^2 r_{LS}^2)B}(\boldsymbol{k}_1 \times \boldsymbol{k}_2)\cdot \hat{e}_z [k_2^2 - k_1^2];$$

$r_{LS} = (T_e/m_i)^{1/2}/\omega_{ci}$ is the ion Larmor radius at $T = T_e$, and ω_D is defined by Eq. (14.6.7).

The mode coupling on the right of Eq. (14.14.1) results solely from the nonlinearity in the ion polarization drift. The relative magnitude of the mode coupling and linear terms is

$$\frac{k^4 r_{LS}^2}{B\omega_{*k}}\phi_{\boldsymbol{k}} \sim (k r_{LS})^2 k L_n \frac{e\phi_{\boldsymbol{k}}}{T_e}.$$

Since we expect $e\phi/T_e \sim 1/kL_n$, the mode coupling term becomes comparable when $kr_{LS} = O(1)$.

Applying a DIA-type analysis to Eq. (14.14.1), Hasegawa and Mima obtained a differential equation for the steady state spectral distribution $|\phi_{\boldsymbol{k}}|^2$ at higher $\boldsymbol{k}$. This gives a distribution peak at $kr_{LS} \sim 1$, with a width similar to that observed experimentally.

14.15 Numerical solution

An alternative approach is to solve the coupled equations, Eq. (14.11.1), for the time evolution of the different Fourier components. An example of this is the study, by R. E. Waltz and colleagues, of the evolution of drift wave turbulence in a collisionless plasma. They included linear and nonlinear ion Landau damping, magnetic shear, and the nonadiabatic electron response in the form $n_k = (e\phi_k/T_e)(1 - i\delta_k)$. An analytic expression is assigned to δ_k, which models the trapped electron response.

Rather than directly solving the coupled wave equations for the time evolution of the individual components $\phi_k(t)$, Waltz used the DIA form of the equations. This reduced the computing time by a factor of between 10 and 100. For example, the number of modes required to approach the infinite mode number limit was 30×30. In typical experimental conditions, he found a stationary turbulent state whose fluctuation level $\tilde{n}/n$ and transport coefficient D were both within a factor of two of typical experimental values. In both magnitude and scaling, the numerical results agree with the simple mixing length model of Section 14.8; for example, $\tilde{n}/n_0 \sim 1/k_x L_n$, $D \sim \gamma/k_x^2$. The decorrelation rate $\Delta\omega_k - (0.1 - 0.5)\omega$ is less than generally observed experimentally. The wave-number spectrum is isotropic in k_x and k_y, in agreement with experiment, and peaks at $k_\perp r_L \sim 0.3$.

14.16 Effect of magnetic fluctuations

The radial particle flux is

$$\Gamma = \int dS \frac{\tilde{n}\tilde{E}_p}{B_\phi} + \int dS \frac{\tilde{B}_r}{B_\phi} \int dv\, v_\parallel \tilde{f}(\boldsymbol{r}, \boldsymbol{v}, t), \tag{14.16.1}$$

where subscripts r, p, and ϕ denote radial, poloidal and toroidal components, respectively. So far we have considered only the first component, since the drift wave is predominantly electrostatic. However, there are several instabilities, such as the resistive ballooning and microtearing modes, which produce fine scale magnetic fluctuations. Even the drift wave produces a small magnetic perturbation when finite β effects are included. In a sheared magnetic field, each mode is localized around a mode rational surface. Because of the close proximity of these surfaces at high m (poloidal mode number), the magnetic perturbations from modes centred on different surfaces may overlap. A magnetic field line then performs a random walk as it wanders from the vicinity of one rational surface to another, leading to an ergodic field. The very much higher particle mobility and thermal conductivity along the magnetic field may result in a radial loss from parallel flow which exceeds the cross-field loss even when the magnetic perturbation is relatively small.

Even when the magnetic field fluctuations are specified, it is not simple to derive the resulting heat loss. Rechester and Rosenbluth showed the key parameter to be L_F, the distance travelled along a pair of neighbouring magnetic field lines before they diverge significantly. It is defined by the relation $dr(s) = dr(0)\exp(s/L_F)$, where dr is the perpendicular separation, and s is the distance along a field line. The particle radial transport will obviously depend on the mean free path. In the limit where $\lambda_{mfp} > L_F$, they deduce the electron thermal diffusivity to be

$$\chi_e = v_{T_e} D_M, \tag{14.16.2}$$

where $D_M = 1/B^2 \int_0^\infty \tilde{B}_r(0)\tilde{B}_r(s)\,ds$, and v_{T_e} is the electron thermal velocity. D_M is called the field line diffusion coefficient. The collisional regime, $\lambda_{mfp} < L_F$, is divided into three subregions, each with a different expression for χ_e. In the least collisional,

$$\chi_e = v_{T_e} D_M \frac{\lambda_{mfp}}{L_F}. \tag{14.16.3}$$

The saturation level and transport have been evaluated for several MHD modes, including the microtearing and the rippling mode (this is unstable near the plasma edge). Here we will discuss only the nonlinear analysis for the resistive ballooning mode by Carreras *et al.*

Resistive ballooning modes are a short wavelength form of the resistive interchange mode, driven by pressure gradient and field line curvature. They are called ballooning because the mode amplitude is largest on the outboard side, where the toroidal curvature is destabilizing. Renormalizing the equations leads to a saturation amplitude and an electron diffusivity due to the stochastic magnetic field given by

$$\chi_e = \frac{3}{2} v_{T_e} a \frac{q}{S}\left[\frac{\beta q^2}{\varepsilon L_p g}\right], \tag{14.16.4}$$

where q is the safety factor, S is the usual ratio of resistive to poloidal Alfvén times, $L_p = p/|dp/dr|$, g is the shear $(d \ln q/d \ln r)$, and $\varepsilon = r/R$ is the inverse aspect ratio.

14.17 Techniques for measuring LF fluctuations

Since probes are restricted to the cool edge region, where $T \leq \sim 1$ eV, the main tool for measuring density fluctuations elsewhere in fusion plasmas is microwave and laser scattering. For an incident beam of wave number $\boldsymbol{k}_i$ and frequency ω_i, the scattered wave along a direction defined by $\boldsymbol{k}_s$, at frequency ω_s, is proportional to $n_e S(\boldsymbol{k}_s - \boldsymbol{k}_i, \omega_s - \omega_i)$, where $S(\boldsymbol{k}, \omega)$ is the spectral power density of the fluctuations. The frequency spectrum of the fluctuations at a particular wave number is obtained by Fourier analysis, while the k spectrum is determined by pivoting the incident and scattered beams from shot to shot. For example, varying the poloidal wave number from 3 cm^{-1} to 40 cm^{-1}, spans the range $kr_{Li} \sim 1$.

The magnetic fluctuations inside the limiter shadow can be measured by pick-up coils. The magnetic signal from an MHD mode decays with distance from its rational surface, roughly as r^{-m-1}, where m is the poloidal mode number. Thus low-m MHD modes can be satisfactorily detected, and their mode amplitude deduced. High-m MHD modes, however, are detectable only if they are localized near the edge.

14.18 General description of the measured fluctuations

Because the short-wave magnetic fluctuations in the plasma interior cannot be measured with existing diagnostics, one cannot say *a priori* whether electrostatic cross-field diffusion, or parallel flux along stochastic magnetic fields, is the dominant loss mechanism. The short-wave density fluctuations generally follow a similar pattern in all tokamaks. The spectral energy, $S(\boldsymbol{k}) = \int S(\boldsymbol{k}, \omega)\,d\omega$, covers a broad range of wave numbers. It commonly has a maximum at around $kr_L \sim 0.5$, and falls off sharply for $kr_L > 1$. The radial and poloidal wave numbers are comparable, whereas $k_\parallel$ is much smaller. For each k the frequency spread is wide, $\Delta\omega \sim \omega_k$, and the spectrum peaks at a frequency comparable to the diamagnetic drift frequency. However, radial electric fields are generally present, giving poloidal rotations of this same order. It is difficult to separate Doppler shift from rest-frame frequency.

The relative density fluctuation, $\tilde{n}/n(r)$, tends to increase with radius, ranging from 0.1 to 1% in the centre, to 10 to 100% at the edge. The shapes of the wave-number and frequency spectra, however, appear to vary little with radius. Magnetic fluctuations are low at the edge ($\tilde{B}/B \sim 10^{-5}$ to 10^{-4}) and appear to increase into the plasma.

14.19 Measurement versus theory

The characteristics described above are similar to those for drift wave turbulence. The most unstable drift wave typically has $k_\theta r_L \sim 0.1$ to 0.3. Its frequency is consistent with the observed peak frequency, bearing in mind the Doppler shift could change the apparent frequency by a factor of two or three. The diffuse frequency spectrum observed, and the isotropy in $\boldsymbol{k}_\perp (k_\theta \sim k_r)$, are predicted by several strong turbulence analyses of the drift wave.

The measured fluctuation level is comparable with the upper limit predicted by Kadomtsev's mixing length argument (Section 14.8). Fig. 14.7, taken from Liewer (1985), plots the $\tilde{n}/n_0$ measured on several tokamaks against r_L/L_n. The dotted band shows $\tilde{n}/n = (3 - 5)r_{Li}/L_n$. Evaluating the mixing length result, $\tilde{n}/n = 1/k_\perp L_n$, for the most unstable wave number, $k_\perp r_L \sim 0.2$, gives the prediction $\tilde{n}/n \sim 5r_{Li}/L_n$.

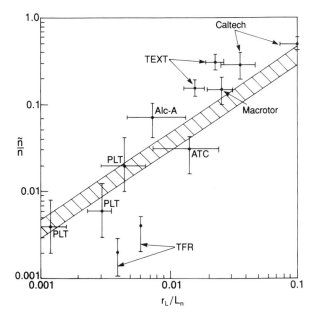

Fig. 14.7. Measured fluctuation level versus r_L/L_n for several tokamaks.

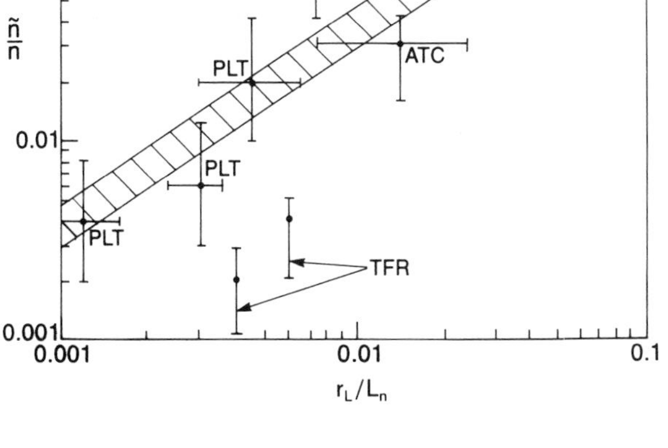

The density fluctuations thus show the characteristics expected for turbulent drift waves. However, the evidence that they are responsible for anomalous transport is ambiguous. For example, the quasilinear $\boldsymbol{E} \times \boldsymbol{B}$ transport expected from the measured fluctuations in TFR, assuming them to be dissipative trapped electron modes, is on average an order of magnitude less than the observed loss rate. In addition, the radial variation in the measured and calculated thermal diffusivities is quite different. Whereas the experimental diffusivity has a minimum at mid-radius, the quasilinear diffusivity has a maximum there. At mid-radius, the calculated diffusivity is about one-third of that measured, which could be within the uncertainties in the comparison. Near the centre and edge, however, the ratio is much smaller, and some other loss mechanism must be dominant.

In some other tokamaks, however, the calculated quasilinear flux appears to be consistent with observation. In the CALTECH tokamak, where lower temperatures permitted the use of probes, $\tilde{n}$, $\tilde{\phi}$ and $\tilde{B}_r$ could all be measured. The quasilinear $\boldsymbol{E} \times \boldsymbol{B}$ flux, due to the electrostatic fluctuations, is consistent with the measured confinement time, whereas the transport due to magnetic fluctuations, estimated using the Rechester and Rosenbluth result in Eq. (14.16.2), is an order of magnitude less than that observed.

The strongest evidence for magnetic fluctuations as the source of electron energy loss is from ISX-B. In high-β discharges ($\beta_p > 1$), the higher frequency magnetic fluctuations (100–200 kHz), measured in the limiter shadow, were found to be well correlated with the electron diffusivity. The experimental χ_e, deduced from transport code analysis, agreed both in magnitude, and in its parameter scaling, with the theoretical estimate for the resistive ballooning mode given in Eq. (14.16.4). When $\beta_p < 1$, the experimental χ_e did not follow the scaling of Eq. (14.16.4). This is not surprising since the resistive ballooning mode becomes less important at lower β. Presumably, some other instability then becomes dominant.

14.20 Summary

Charged particle transport across a plasma slab, due solely to collisions, is the simplest example of plasma diffusion. It illustrates the connection between dissipation and transport, and the concept of a random walk process, where each collisional scattering produces a discrete radial displacement in either direction.

Additional parallel currents must flow in a toroidal plasma to balance the perpendicular drifts in the toroidal magnetic field. In a resistive plasma the balance is not complete, and the resulting space charge produces electric drifts which increase the plasma loss. At higher temperatures, the mean free path is so long that particles can travel many times around the torus before colliding. Particles with velocity nearly perpendicular to the magnetic field are then reflected by the higher magnetic field on the inboard side of the cross-section, performing banana orbits. Because of the relatively large radial width of these orbits, their random walk loss rate is high, dominating the total diffusion.

Neoclassical transport is the minimum possible loss rate, which occurs even if there are no fluctuations or field imperfections. In practice, the measured loss rate is higher, and the excess over neoclassical is called anomalous transport because it is not yet fully understood. It is almost certainly produced by the electric and magnetic fluctuations which are always observed experimentally. Although the small amplitude analysis for most instabilities which can occur in tokamaks is well established, in real situations these instabilities generally reach a saturated turbulent state. Their study requires nonlinear analysis, which is more difficult. Several nonlinear techniques have been described in general form.

The comparison of theory and experiment is still inconclusive. The density fluctuations observed within the plasma are consistent with drift wave instabilities. Their mean frequency is comparable with the diamagnetic frequency, the lack of more precise agreement being attributable to Doppler shift. Their k-spectrum peaks at around $k_{\perp} r_{\mathrm{Li}} \sim 0.5$ and the overall level is commonly of order $\tilde{n}/n \sim r_{\mathrm{Li}}/L_n$, both in agreement with prediction. The fluctuation amplitude frequently, but not always, correlates with changes in confinement. Although recent progress in analysing the nonlinear phase of instability growth has improved the agreement between the predicted and measured transport, further improvement in understanding is needed before the transport can be predicted with confidence.

References

LIEWER, P. C. (1985). *Nucl. Fusion*, **25**, 543.
WALTZ, R. E. (1983). *Phys. Fluids*, **26**, 169.

15
Radio-frequency plasma heating

R. A. CAIRNS

15.1 Introduction

One of the problems associated with magnetically confined fusion is that of heating the plasma to the 10–12 keV temperature range that is necessary. The resistivity of a plasma varies with temperature, going as $T^{-3/2}$, and the result of this decrease in resistivity as the temperature increases is that Ohmic heating becomes less effective at high temperatures and that some form of auxiliary heating is generally necessary. Most large experiments use neutral beam heating, radio-frequency heating or, very often, a combination of the two.

Radio-frequency heating, the subject of this chapter, involves launching a high power electromagnetic wave from an antenna or waveguide at the edge of the plasma. Some absorption will take place as a result of collisions between particles, converting the ordered motion of the oscillation in the field of the wave into random thermal motion. However, the collision frequency in a high temperature fusion plasma is not high enough for this to be effective, and it is necessary for the wave to be tuned to some resonant frequency at which collisionless absorption can take place. A viable scheme requires us to identify a wave which can be launched from the edge and propagate to the central region of the plasma, there to be absorbed. An additional feature is that the resonant nature of the absorption makes it possible to control the absorption profile to some extent. By using radio-frequency heating to modify the temperature profile, it may be possible to suppress some of the instabilities driven by temperature or density gradients at rational surfaces.

Four main frequency ranges, in which heating can take place have been investigated. In describing the main features and parameters associated with them, we shall have in mind applications to a typical tokamak experiment. However, it should be borne in mind that similar principles apply to other magnetic confinement devices. The lowest frequency scheme involves Alfvén waves, and typically requires waves at a few megahertz. Going to frequencies of a few tens of megahertz, we get into the ion cyclotron range of frequencies, where the wave frequency is equal to some low multiple of the cyclotron frequency of one or more of the ion species present. This is the scheme which is most widely used at the moment, and, in JET for example, has been successful in coupling substantial power into the plasma and raising the temperature to the 10 keV range. Increasing the frequency to a few gigahertz brings us to the lower-hybrid range, and finally, at frequencies of around 30 GHz and above, we reach the electron cyclotron range.

The importance of this topic to fusion research means that it has attracted a good deal of attention and produced a large body of literature. The object of this chapter is to give an introduction to the basic processes involved in these various schemes and some indication of how they work in practice. For readers who wish for more details, some suggestions for further reading are given, and these, in turn, will provide further references to the research literature.

15.2 Wave propagation

The propagation of waves through an inhomogeneous plasma is obviously a crucial consideration for any radio-frequency heating scheme. The wave modes which exist in a plasma are discussed in Chapter 3, and the aim of this section is to give an outline of some features which are important in the present context.

The simplest treatment of an inhomogeneous plasma arises if the inhomogeneity is along only one direction, say the x-axis, and we can use the WKB method. The electric field and all other wave quantities are taken to vary as

$$\boldsymbol{E}_0(x)\exp\left(\mathrm{i}\int_0^x k_x(x')\mathrm{d}x' + \mathrm{i}k_y y + \mathrm{i}k_z z - \mathrm{i}\omega t\right),$$

with k_x determined by the local dispersion relation. The amplitude $\boldsymbol{E}_0(x)$ varies on a scale much longer than the wavelength. This type of approximation can be extended to more than one dimension, when it is known as the eikonal or ray tracing method. The amplitude in this case is assumed to vary as

$$\boldsymbol{E}_0(x)\exp(\mathrm{i}\Phi - \mathrm{i}\omega t),$$

where Φ is a function of position with $\nabla\Phi = \boldsymbol{k}$. The direction of energy flow is given by the group velocity

$$\boldsymbol{v}_\mathrm{g} = \frac{\partial\omega}{\partial\boldsymbol{k}} = -\frac{\partial D/\partial\boldsymbol{k}}{\partial D/\partial\omega}, \tag{15.2.1}$$

where $D(\omega, \boldsymbol{k}) = 0$ is the dispersion relation. This single equation does not determine the vector $\boldsymbol{k}$, and the second equation

$$\frac{\mathrm{d}\boldsymbol{k}}{\mathrm{d}t} = \frac{\partial D/\partial\boldsymbol{r}}{\partial D/\partial\omega} \tag{15.2.2}$$

is needed to give its variation along the ray path. From the antenna at the plasma edge, where waves are generated with a known wave-number spectrum, the pair of equations Eqs (15.2.1) and (15.2.2) is solved to find the path of the ray in space and the accompanying change in wave number. The damping of the wave is calculated from the non-Hermitian part of the dielectric tensor. Numerical codes incorporating this type of calculation are generally necessary if the toroidal geometry of a tokamak is to be treated in a realistic way, and are especially useful in the higher frequency ranges, where the condition that the wavelength be small compared with the scale of the machine is more readily satisfied.

Even if the assumptions of ray tracing theory are generally true within the plasma, there are local regions in which they break down and where a more detailed solution of the full wave equations is necessary. The best known such regions are where the wave-number component along the direction of inhomogeneity goes to zero or infinity. The first case is known as a cut-off, and is associated with reflection of the wave, whereas the second, a resonance, is associated with absorption of the wave.

A third case in which the WKB approximation fails arises from the fact that a plasma can support a variety of different wave modes. If they have different wave numbers then they propagate independently, but there may be localized regions in an inhomogeneous plasma where two waves have almost the same wave number. The behaviour is typically as shown in Fig. 15.1.

Fig. 15.1. The variation of wave number with position in a region where mode conversion occurs.

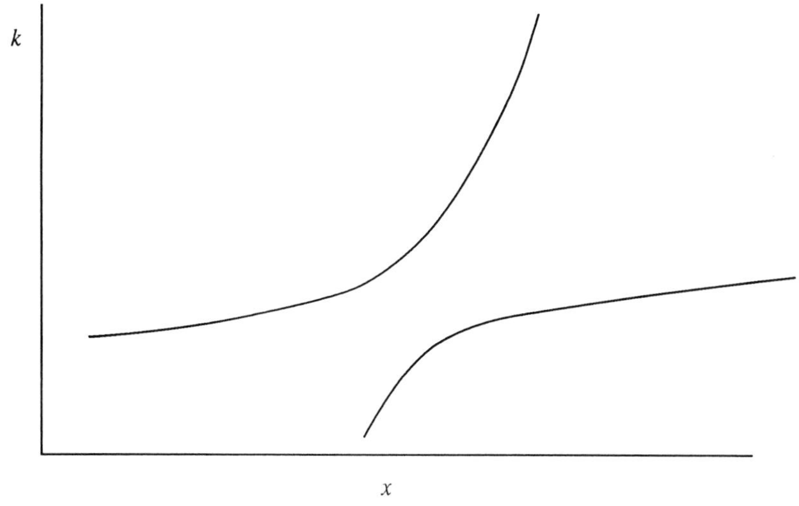

If energy is incident on this region in one of the modes shown, then in general it will be divided between the two modes on emerging from this region. This phenomenon is known as mode conversion. Very often one of the modes has a cut-off in the neighbourhood of the mode conversion, the reflection at which allows energy to couple to the waves going in the opposite direction. Further, the methods of heating we consider generally involve a localized dissipation mechanism, where cyclotron or other resonance occurs. Usually this is precisely the region in which mode conversion also occurs, so that we have the complication of having to take account of the two effects at the same time. To do so, it is necessary to find some approximation to the full set of wave equations. If the ratio of the Larmor radius of the species present to the wavelength is small, then expansion in this ratio is possible and yields a system of differential equations. If it is not small, then a set of integro-differential equations results, solution of which is a rather formidable task.

The two basic collisionless absorption mechanisms are Landau and cyclotron damping. The former occurs when the resonance condition

$$\omega - \boldsymbol{k} \cdot \boldsymbol{v} = 0 \tag{15.2.3}$$

is satisfied, whereas the latter involves the interaction of the wave with the cyclotron motion of one of the particle species in the plasma at frequency ω_c. Looking at this as a linear oscillator problem, we might expect to find resonance only when the driving frequency, that is the wave frequency ω, matches the natural frequency of the oscillator, that is ω_c. However, if we take account of the thermal motion of the particle, we must consider the time variation of the field as we follow the unperturbed orbit of the particle.

In the laboratory frame, the wave may be taken to vary as

$$\exp(\mathrm{i}k_\perp x + \mathrm{i}k_\| z - \mathrm{i}\omega t), \tag{15.2.4}$$

with the perpendicular component of the wave vector taken to be in the x-direction. The particle orbit is

$$x = \frac{v_\perp}{\omega_c}\sin(\omega_c t), \tag{15.2.5}$$

$$z = v_\| t, \tag{15.2.6}$$

assuming, as will suffice for our purposes, that at $t = 0$ the particle is at the origin. Substituting Eqs (15.2.5) and (15.2.6) into Eq. (15.2.4) we get the time variation of the field along the particle orbit to be

$$\exp\left(\mathrm{i}\frac{k_\perp v_\perp}{\omega_c}\sin(\omega_c t) + \mathrm{i}k_\| v_\| t - \mathrm{i}\omega t\right),$$

which equals

$$\sum_{l=-\infty}^{\infty} J_l\left(\frac{k_\perp v_\perp}{\omega_c}\right)\exp(\mathrm{i}l\omega_c t + \mathrm{i}k_\| v_\| t - \mathrm{i}\omega t). \tag{15.2.7}$$

Here J_l is the Bessel function of order l, and we have used the identity

$$\sum_{l=-\infty}^{\infty} J_l(a)\exp(\mathrm{i}lx) = \exp(\mathrm{i}a\sin x).$$

We see from this that the field seen by the particle is Doppler shifted because of the parallel velocity of the particle, and that it also contains frequency components shifted from the wave frequency by integer multiples of the cyclotron frequency. The coefficients of these Fourier components are the Bessel functions, which are a familiar feature of the dispersion relation of a magnetized plasma. This decomposition means that there is a resonant interaction between the wave and the plasma whenever the Doppler-shifted frequency is an arbitrary multiple of the cyclotron frequency, that is the condition

$$\omega = l\omega_c + k_\| \omega_\| \tag{15.2.8}$$

is satisfied. It is in regions where this or the Landau resonance condition is satisfied that a strong wave–particle interaction occurs and there can be strong damping of the wave.

15.3 Particle diffusion

When a particle is in resonance with a wave, as described in the previous section, there is a steady transfer of energy from one to the other, rather than the oscillatory response of a nonresonant particle. However, whether a particular particle gains or loses energy depends on its phase with respect to the wave, and it is certainly not the case that when a wave is being damped all the resonant particles are gaining energy. Rather, each particle follows a random walk in velocity space, sometimes gaining energy and sometimes losing energy as its phase relation with the wave changes. The net effect is that particles diffuse in velocity, and the velocity distribution tends to spread out. A spreading of the velocity distribution is equivalent to an increase in temperature, so heating results, assuming the initial distribution to be peaked about the origin. This effect is in competition with energy loss mechanisms, and the final temperature and rate of heating result from the balance between these effects.

Sometimes it is important to know the direction of diffusion in velocity space, which can be deduced from simple momentum conservation arguments as follows. We note first that particles are free to move in the direction parallel to the steady magnetic field, so that any parallel momentum deposited by the wave will appear as a change in the parallel momentum of the particle. We note in passing that a force perpendicular to the field results in motion perpendicular to the field and the force. Thus we may expect the perpendicular component of any wave momentum that is absorbed to produce such a transverse drift. For the wave,

$$\frac{\text{momentum}}{\text{energy}} = \frac{\mathbf{k}}{\omega}$$

and so the changes in parallel momentum and energy of a particle interacting with the wave must satisfy

$$\frac{\Delta p_z}{\Delta E} = \frac{k_z}{\omega}.$$

For a Landau resonance this implies that

$$\Delta E = v_z \Delta p_z,$$

which can only be the case if the velocity perturbation is in the parallel direction. Thus a wave which is damped by Landau damping will produce a tail on the velocity distribution function pulled out along the field direction.

In the case of a cyclotron resonance, the resonance condition implies that

$$\frac{\Delta p_z}{\Delta E} = \frac{k_z}{l\omega_c + k_\parallel v_\parallel},$$

and if we combine this with the fact that

$$\Delta E = m v_\parallel \Delta v_\parallel + m v_\perp \Delta v_\perp$$

we find that the direction of the velocity change is as indicated in Fig. 15.2, that is along a circle centred at the parallel phase velocity of the wave. Generally, the phase velocity is much greater than the particle thermal velocity, so this direction is almost perpendicular to the magnetic field.

15.4 Alfvén wave heating

Alfvén waves occur in the frequency range below the ion cyclotron frequency, and are generally encountered in the context of magnetohydrodynamics. In MHD, three wave modes are found – the Alfvén wave and the fast and slow magnetosonic waves. The first two of these also occur in the low frequency limit of

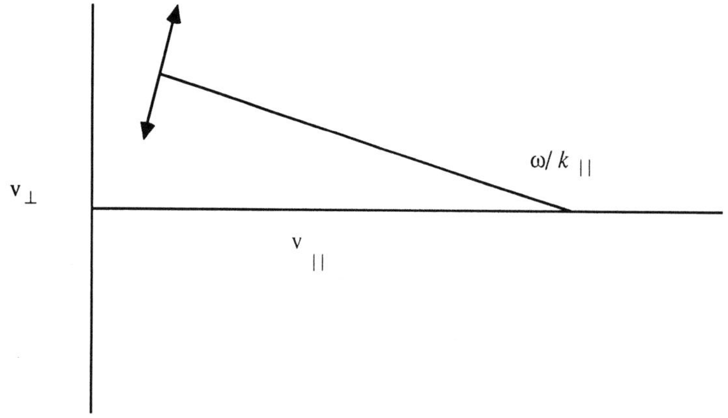

Fig. 15.2. The velocity change in cyclotron resonance is along a circle centred on the parallel phase velocity of the wave.

cold plasma wave theory, where they may be referred to as the slow and fast Alfvén waves, referring to their relative phase velocities, or the shear and compressional Alfvén waves, referring to the direction of the velocity perturbation in them. In the present context it is these which are of interest, since the slow magnetosonic wave has a phase velocity of the order of the ion thermal velocity, under typical tokamak conditions, and is so strongly Landau damped that it cannot penetrate any appreciable distance into the plasma.

The dispersion relations of the slow and fast Alfvén waves are

$$\omega = k_{\parallel} V_A \tag{15.4.1}$$

and

$$\omega = k V_A, \tag{15.4.2}$$

where V_A is the Alfvén speed. The main features of these waves which are relevant to the present discussion are their group velocities. The fast wave propagates isotropically, and its group velocity is simply parallel to its phase velocity. The slow wave, on the other hand, has a dispersion relation which depends only on the parallel component of the wave number, with the result that its group velocity is in the parallel direction.

These differences in group velocity are essential to Alfvén wave heating, the basic idea of which is that a fast wave is excited near the plasma edge and carries energy across the field lines into the plasma, until a layer is reached at which the Alfvén velocity, which varies across the inhomogeneous plasma, satisfies Eq. (15.4.1). Energy is then transferred to the slow mode. Since the latter has no component of group velocity across the field, this energy is trapped on the magnetic surface where the resonance occurs and is eventually converted into thermal energy by damping. A more detailed treatment, including kinetic effects, shows that an extra wave mode, the kinetic Alfvén wave, exists and carries energy across the field to some extent. This affects the details of the energy absorption profile, but makes little difference to predictions of the overall energy absorption.

Since the Alfvén speed varies continuously across the plasma, the values of ω and $k_{\parallel}$ which can satisfy the resonance relation (15.4.1) also vary continuously, and there is a continuous absorption spectrum. If the toroidal geometry of the system and extra physical effects, such as the two-fluid behaviour of electrons and ions, are taken into account, the continuous spectrum may be found to split up into narrowly spaced lines, though this makes little difference in practice. More importantly, other discrete modes appear at frequencies below the continuous range. These, known as *global modes*, are determined by the boundary conditions on the torus, rather than by a local resonance condition.

The existence of the continuous spectrum and of the global modes has been confirmed in experiments, and there is good agreement with the theoretical predictions of the absorption spectrum. However, when it comes to coupling substantial power levels into the plasma, results have been somewhat disappointing, and there are some doubts as to the extent to which Alfvén waves can, in reality, transfer energy efficiently from an external antenna into the bulk of the plasma.

15.5 Ion cyclotron heating

Ion cyclotron heating is the scheme which has, so far, been most successful in producing high temperatures in experimental devices. It has been widely used in large tokamaks. In JET, for example, ion cyclotron power of the order of 20 MW has been injected and temperatures in excess of 10 keV reached.

In a plasma with a single ion species there are no cold plasma resonances in the ion cyclotron range of frequencies. For parallel propagation, certainly, the left-hand circularly polarized wave has a resonance at the ion cyclotron frequency, but in tokamak geometry, where the dominant gradients in equilibrium quantities are across the field, there is no resonance at a frequency below the lower hybrid, which is well above the ion cyclotron frequency in the centre of the plasma. As the density goes to zero, the lower-hybrid frequency approaches the ion cyclotron frequency, so there may be edge effects in which it plays a role. Absorption near the centre of the plasma is the result of finite temperature effects. At the fundamental resonance in a single species plasma, the electric field in the wave is almost entirely circularly polarized so as to rotate in the opposite direction to the particles. The result is that there is only a weak interaction between the wave and the particles and only weak absorption.

At the second harmonic, that is when the wave frequency is twice the cyclotron frequency, the situation is different. Cyclotron absorption now takes place quite strongly, and is proportional to $(k_\perp v_T/\omega)^2$, at least when the Larmor radius is much less than the wavelength. Absorption at the fundamental is, on the other hand, proportional to $(k_\perp v_T/\omega)^4$, which is much smaller. These results are obtained from the standard dispersion relation for a hot plasma, expanded in powers of $(k_\perp v_T/\omega)$.

In analysing the dispersion relation, one property which allows simplification is the fact that, at the frequencies in question, electrons can flow along the magnetic field and neutralize any electric field in this direction. More exactly, the parallel field turns out to be smaller than the perpendicular field by the ratio of the electron mass to the ion mass. To a good approximation the parallel component can be neglected and the dispersion relation reduced to a 2×2 determinant. When we go to an inhomogeneous system, with coupled differential equations for the field components, the simplification is even more worthwhile.

With the hot plasma terms included, the variation of wave number across the plasma is as sketched in Fig. 15.3, showing the fast wave coupled to a Bernstein

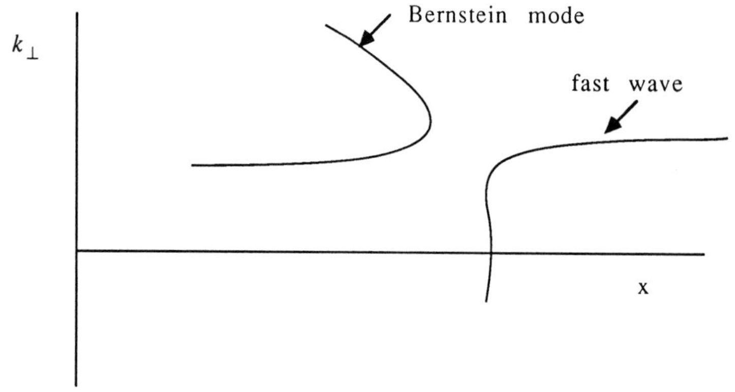

Fig. 15.3. The variation of wave number with position in the vicinity of the second harmonic resonance, showing the coupling of the fast wave to the Bernstein mode.

mode. The fast wave is just the continuation of the fast magnetosonic wave to higher frequencies, whereas the Bernstein mode is a wave which only exists in a hot plasma and which propagates, with much smaller group velocity than the fast wave, in the vicinity of a cyclotron harmonic.

A detailed analysis of propagation through the resonance layer, where there is strong cyclotron damping as well as mode conversion, is too complicated to give here. However, we can obtain some important qualitative information about the wave from the behaviour of the curves in Fig. 15.3. The first thing to note is that the Bernstein wave is backward propagating, that is its group velocity is in the opposite direction to its phase velocity. If energy is incident from the high field side (the left), then it couples to the branch of the Bernstein mode going off to higher wave number, and the result is partial transmission, partial mode conservation to this branch, and partial absorption. Coming from the low field side, the energy in the fast wave couples to the branch of the Bernstein mode which goes to a cut-off. Here it can be reflected and interacts once more with the fast wave going in the opposite direction to the incident wave. The result is partial reflection. In this case the incoming energy is divided amongst transmission, reflection, mode conversion, and local absorption. Energy reflected or transmitted in the fast wave can leave the plasma, but the Bernstein wave does not connect to a vacuum mode and its energy must eventually go into the plasma. What it does is carry energy away from the cyclotron resonance and contribute to a broadening of the absorption profile.

Often there are two species of ion present. This would certainly be the case in a deuterium–tritium reactor, and is also the case in many experiments. Then there is a cold plasma resonance at the two-ion hybrid resonance, which lies between the cyclotron frequencies of the two species. In a hot plasma, the wave number does not tend to infinity at the resonance, but instead connects to a Bernstein mode, so that there is once more mode conversion. An important special case is minority heating, when one species is present in a small concentration. In this case the two-ion hybrid resonance is close to the fundamental cyclotron resonance of the minority species. The argument about the polarization of the wave being in the wrong direction to give strong absorption at the fundamental does not apply in this case, since the polarization is largely determined by the majority species. There is strong damping on the minority ions and, because there is a lot of energy going into a small fraction of the ions, a hot tail on their velocity

then a tail can be drawn out along one direction only. Such an asymmetry in the parallel distribution of electron velocities produces a flow of electrons along the field, that is an electric current. It has, in fact, proved possible to create and sustain a tokamak discharge using lower-hybrid heating alone.

For the accessibility condition to be satisfied and, if desired, to produce the asymmetry required for current drive, it is necessary to control the spectrum of parallel wave numbers which is launched into the plasma. This is done by launching the wave from an array of waveguides, the relative phases of which can be adjusted. six or eight waveguides were common in early experiments, but some recent experiments, for example the large JT-60 machine in Japan, have used many more. The larger the number of waveguides, the narrower the spectrum can be made.

15.5 Electron cyclotron heating

With this method of heating we arrive at frequencies of the order of tens of gigahertz upwards. The frequency appropriate to absorption at any harmonic may be estimated by noting that the electron cyclotron frequency is 28 GHz in a 1 T field. (Note that this is not the angular frequency, but $\omega_{ce}/2\pi$. This is the number which is normally meant if the frequency of a wave heating system is quoted.) At lower frequencies, it may well be possible for the wave which is generated to be evanescent at the plasma edge, and to tunnel through to a region in which it propagates. However, the high frequency, and resultant short wavelength, makes this impossible in the electron cyclotron range, except in very small devices. Thus, it must be possible for the wave to travel from the edge to the absorption region without meeting any cut-off at which it would be reflected. As we shall see, this imposes density limits on the range of operation. However, it does have the advantage that there are no problems with the coupling of energy into the plasma, and the wave can simply be launched from a waveguide aperture at the edge.

The mechanism of absorption involves hot plasma effects in the vicinity of the layer where the wave frequency equals the cyclotron frequency, or one of its low harmonics. In a cold plasma, the right-hand circularly polarized component of the field vanishes, and there is no absorption, in the same way as was explained for the ion cyclotron case. Absorption depends, once again, on thermal effects. An important difference between electron and ion cyclotron heating is that the parallel field can be just as large as the perpendicular field at electron cyclotron frequencies. The other main difference between the two cases is that relativistic effects are important in the electron cyclotron case. The absorption depends on a wave–particle resonance, in which the difference between the wave frequency and the cyclotron harmonic is the Doppler shift resulting from the parallel motion. However, the cyclotron frequency depends on the mass of the particle, and so there is a relativistic shift in this frequency which depends on the energy of the particle. Because we are dealing here with small shifts, the effect is important at typical tokamak temperatures of a few kilo-electron-volts, where a simple comparison with the electron rest energy (511 keV) would not lead one to expect relativistic effects to be particularly important.

If we return to the question of wave propagation from the plasma edge to the centre, we can use the cold plasma dispersion relation, which is a good approximation except near the cyclotron resonance. Two modes exist, the ordinary (O) mode and the extraordinary (X) mode. For simplicity we shall consider perpendicular propagation, although similar results, with slightly different density limits, hold for oblique propagation. For the O mode the dispersion relation is

$$n^2 = 1 - \omega_p^2/\omega^2, \tag{15.6.1}$$

where n is the refractive index kc/ω. The wave then propagates so long as the density is below the value at which $\omega_p^2 = \omega^2$. Higher densities can be reached if the wave frequency is increased, for example by using absorption at the second harmonic rather than the fundamental. In typical tokamak operating conditions, the plasma and electron cyclotron frequencies are of the same order, so the restrictions imposed by this accessibility condition are of some consequence.

For the X mode, the dispersion relation is

$$n^2 = \frac{(\omega_p^2 - \omega^2)^2 - \omega^2\omega_{ce}^2}{\omega^2(\omega^2 - \omega_p^2 - \omega_{ce}^2)}. \tag{15.6.2}$$

This implies that there is a cut-off when

$$(\omega_p^2 - \omega^2)^2 = \omega^2\omega_{ce}^2, \tag{15.6.3}$$

and the upper-hybrid resonance when

$$\omega^2 - \omega_p^2 - \omega_{ce}^2 = 0. \tag{15.6.4}$$

To see how these affect the propagation of the wave across the field we construct a Clemmow–Mullally–Allis (CMA) diagram, as in Fig. 15.8. In the diagram the curves defined by the cut-off condition Eq. (15.6.3) and the resonance Eq. (15.6.4) are plotted on a diagram whose axes are $X = \omega_p^2/\omega^2$ and $Y = \omega_{ce}^2/\omega^2$. In the context of a wave of given frequency in an inhomogeneous plasma, X is a measure of the density and Y of the magnetic field. As the wave travels through the plasma it traces out a curve on the CMA diagram determined by the variation of the plasma parameters along its path. If the wave starts at the edge of the plasma, where the density is zero, then the corresponding curve on the diagram starts on the Y-axis. In a tokamak, the magnetic field increases from the outside to the inside of the torus, so a wave launched from the outside will follow a curve along which Y increases, whereas one launched from the inside will follow a curve along which Y decreases.

The problem of accessibility then reduces to the question of whether there is a curve satisfying the above constraints and reaching the point at which the wave is absorbed. Absorption can take place at the fundamental, which lies on the line $Y = 1$, or at the second harmonic, on $Y = 1/4$. In the regions A and B of the diagram, the wave does not propagate. It then becomes clear that to reach the absorption region at the fundamental, the wave must be launched from the inside, whereas to reach the second harmonic it may be launched from either side. Absorption can also occur at the upper-hybrid resonance, which can be approached from the high field side. However, before reaching the hybrid resonance, the wave must cross a hot plasma resonance, and in large hot plasmas little energy penetrates through to the hybrid resonance.

For absorption at the fundamental, with inside launch, the wave must reach

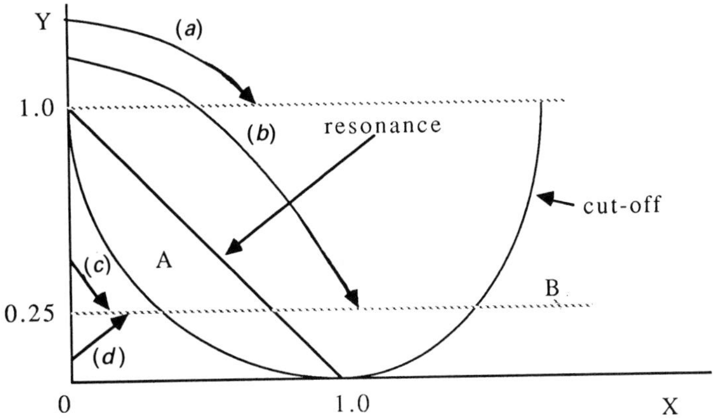

Fig. 15.8. The CMA diagram for waves in the electron cyclotron frequency range. The regions A, between the cut-off and the resonance, and B, to the right of the cut-off, are those in which the wave does not propagate. The possibilities for accessing a resonance are (*a*) fundamental from high field side, (*b*) second harmonic from high field side, (*c*) second harmonic from high field side, (*d*) second harmonic from low field side. Note that for (*b*) to occur the wave would have to be at least partially transmitted through the fundamental resonance, and the magnetic field would have to vary by a factor of more than two across the plasma. The more likely method of second harmonic heating from the high field side is (*c*).

the line $Y = 1$ before it meets the high density branch of the cut-off. This gives the condition that X has to be less than two when $Y = 1$, that is

$$\omega_p^2/\omega_{ce}^2 < 2. \tag{15.6.5}$$

For a given magnetic field, this yields a density limit which is twice that which applies to the O mode.

For second harmonic absorption, following the curves (*c*) or (*d*) in Fig. 15.8, the wave must reach the value $Y = 1/4$ before it reaches the low density branch of the cut-off, that is before X reaches 1/2. This gives the same density limit as Eq. (15.6.5). For inside launch at the second harmonic, following curve (*b*), we require that $X < 3/2$ when $Y = 1/4$, which imposes the density limit

$$\omega_p^2/\omega_{ce}^2 < 6. \tag{15.6.6}$$

For a given magnetic field, there are advantages in going to the second harmonic since it raises the density limit. On the other hand, it may lead to technical problems in obtaining a high power source at the required frequency. Inside launch, for the second harmonic using the X mode, also has advantages as regards the density limit, but is less convenient than outside launch from an engineering point of view.

If we turn to the absorption process, the particles which are in resonance are those for which

$$\omega = l\omega_{ce}/\gamma + k_\parallel v_\parallel, \tag{15.6.7}$$

where γ is the usual relativistic factor and the cyclotron frequency is calculated using the rest mass. In general, the plasma may be assumed to be weakly relativistic, with particle velocities much less than c, in which case the resonance condition may be approximated by

$$\omega = n\omega_{ce}\left(1 - \frac{1}{2}\frac{v^2}{c^2}\right) + k_\parallel v_\parallel. \tag{15.6.8}$$

This is a quadratic equation in $v_\parallel$ which can be solved to give the resonant values of $v_\parallel$ for given values of $v_\perp$ and of the other parameters of the plasma. The result is

$$\frac{v_\parallel}{c} = n_\parallel \pm \left[n_\parallel^2 - \frac{v_\perp^2}{c^2} + \frac{2(l\omega_{\text{ce}} - \omega)}{\omega_{\text{ce}}} \right]^{1/2}, \tag{15.6.9}$$

where $n_\parallel = k_\parallel c/\omega$. In contrast, the nonrelativistic equivalent is

$$\frac{v_\parallel}{c} = -\frac{l\omega_{\text{ce}} - \omega}{n_\parallel \omega_{\text{ce}}}. \tag{15.6.10}$$

We can obtain Eq. (15.6.10) from Eq. (15.6.9) if we take the minus sign and assume that $n_\parallel^2$ is much greater than both of the other terms in the bracket. If the temperature is small then there is only a very small range of values of $n_\parallel$ for which this condition is not met, and for which the velocity components are not far out on the tail of the distribution. However, it only takes a temperature of a few kilo-electron-volts for the range to become significant.

An important consequence of the relativistic effect is that no resonant value of the velocity exists if

$$-\frac{2(l\omega_{\text{ce}} - \omega)}{\omega_{\text{ce}}} > n_\parallel^2. \tag{15.6.11}$$

This occurs if we go far enough to the low field side, where the wave frequency exceeds the cyclotron harmonic frequency. Beyond the point at which the two sides in Eq. (15.6.11) become equal, the absorption is strictly zero. In the nonrelativistic approximation, and on the high field side in the relativistic case, the absorption falls off gradually away from the resonance as the resonant velocities move out into the tail of the distribution. Thus the relativistic mass shift has an important effect on the absorption profile. In the limit of perpendicular propagation, there is no absorption at all on the low field side. Detailed calculations of the absorption profile require the use of dielectric tensor elements calculated from the relativistic version of the Vlasov equation, details of which we do not have space to discuss here.

At the fundamental, the O mode is strongly absorbed at perpendicular incidence and the absorption falls off as the angle of incidence increases. The fact that cyclotron heating, which we have already shown to involve an increase in the energy in the perpendicular degree of freedom, is associated with a wave whose electric field is in the parallel direction, is quite simply explained by the fact that the heating is done by the $\boldsymbol{v} \times \boldsymbol{B}$ force. In the case of the X mode at the fundamental, absorption is weak at perpendicular incidence and increases as the incidence becomes more oblique. At the second harmonic, the X mode is strongly absorbed over a wide range of angles, whereas the O mode is much more weakly absorbed.

Experiments with electron cyclotron heating have shown that it is an effective mechanism, and that, in a medium size tokamak, a wave can be almost completely absorbed in a single pass through the resonance, in line with theoretical expectations. One result of the short wavelength of these waves is that the absorption is well localized in the region of the resonance. This may be useful in allowing some measure of control of the radial temperature profile, and hence also of the current profile, since the resistivity depends on temperature. This may allow control of plasma stability, by modifying the profiles in the neighbourhood of rational-q surfaces.

15.7 Current drive

In addition to the use of radio-frequency waves to augment the Ohmic heating of a confined plasma, there is also a great deal of interest in the possibility of using waves to drive the current in a tokamak. This is because the standard method of driving current by having the plasma as the secondary of a transformer circuit is, by its very nature, confined to pulsed operation. A viable method of producing steady state current drive would get round the engineering and operational difficulties posed by pulsed operation of a reactor.

The most obvious way to produce a current is to exploit the Landau resonance, which accelerates particles along the field lines. We have described how lower-hybrid heating, in the regime where absorption takes place via Landau damping, produces a tail on the parallel distribution function. If the spectrum of waves is asymmetric, so that the tail is only pulled out in one direction, then a current is produced. Such a spectrum is readily produced by suitable phasing of the waveguides used for lower-hybrid launch, and the effectiveness of this scheme has been amply demonstrated in a large number of experiments.

Cyclotron absorption, producing an increase in the energy associated with the perpendicular degree of freedom, is not such an obvious candidate, but it is possible to use it to drive a current. The mechanism for this depends on the velocity dependence of the collision frequency. If, by adjusting the parallel wave-number spectrum of electron cyclotron waves, we heat electrons travelling to the right, say, then these electrons, whilst they have not gained any parallel momentum from the wave, will collide less often with ions than electrons moving to the left. The result of the asymmetry in collision frequency is that electrons acquire a net drift to the right and ions to the left. The total momentum remains zero, but because the ion mass is so much larger than the electron mass the ion velocity is correspondingly small and the ion contribution to the current is negligible.

It might be thought at first sight that this would be a very inefficient way of driving current compared with direct acceleration along the field via Landau damping, but it turns out to have a theoretical efficiency of the same order. The parameter by which the efficiency is measured is the ratio of current to power. In a reactor this will have to have a value high enough that the amount of output power which has to be recirculated back into the plasma to keep the current going is at an acceptable level.

A rough estimate of the efficiency can be made very easily as follows. Taking first the case of electrons being pushed along the field as a result of Landau damping, a change in parallel velocity $\delta v_{\parallel}$ produces a current

$$J = q\delta v_{\parallel}, \tag{15.7.1}$$

and the change in energy is

$$E = mv_{\parallel}\delta v_{\parallel}. \tag{15.7.2}$$

If the collision frequency for the electrons is ν, then the current will persist for a time of the order of $1/\nu$. To maintain a steady current, it will be necessary to push the electron again after this time has elapsed, and the average power requirement per electron is $E\nu$. In this way we can arrive at an estimate of the ratio of current density to absorbed power density

$$\frac{J}{P_d} = \frac{q}{mv_{\parallel}\nu}. \tag{15.7.3}$$

Now, ν depends on the electron energy, going as $1/v^3$ for velocities in excess of the thermal velocity. From Eq. (15.7.3) we see that the limit of large velocity, when ν becomes small, is a favourable one for current drive, and this is the regime which is exploited in lower-hybrid current drive.

In the case of cyclotron heating, we can make a similar crude estimate. The rate of loss of momentum by fast electrons is

$$\frac{dp}{dt} = mv_{\parallel}\nu. \tag{15.7.4}$$

The current drive arises because an increase in the perpendicular velocity of particles travelling in one direction reduces ν, and so decreases the rate of loss of momentum. In the steady state, when the electron and ion drifts have adjusted to this asymmetry, we may expect the change in ν to be compensated for by a change in $v_{\parallel}$. Thus,

$$mv_{\parallel}\nu \approx m(v_{\parallel} + \delta v_{\parallel})(\nu + \delta\nu)$$

so that

$$\delta v_{\parallel} \approx -\frac{v_{\parallel}\delta\nu}{\nu}.$$

Again, if the current is to be maintained continuously, the change in $v_{\perp}$ which produces the change in collision frequency must be renewed at intervals of $1/\nu$, and so we obtain an estimate of the ratio of current density to power density:

$$\frac{J}{P_d} = -\frac{qv_{\parallel}\delta\nu}{mv_{\perp}\nu^2\delta v_{\perp}}.$$

Now, since ν goes as $1/v^3$,

$$\frac{\delta\nu}{\nu} = -\frac{3v_{\perp}\delta v_{\perp}}{v^2},$$

and so

$$\frac{J}{P_d} \approx \frac{3qv_{\parallel}}{mv^2\nu}. \tag{15.7.5}$$

Comparing Eq. (15.7.5) with Eq. (15.7.3) we see that they are of the same order of magnitude if we insert the same typical values of velocity.

These values tell us the efficiency in terms of current and power densities, but a more useful figure is the ratio of total current around the tokamak to the total absorbed power. To obtain this, consider a ring of major radius R and cross-section A. Then the current in this ring is AJ and the total power absorbed is $2\pi RAP_d$. Thus we have

$$\frac{I}{P} \approx \frac{1}{2\pi R}\frac{J}{P_d}. \tag{15.7.6}$$

The collision frequency scales as $nT^{-3/2}$, where n is the plasma density, and the weak temperature dependence of the Coulomb logarithm can be neglected for our purposes. We can then see from Eq. (15.7.6) and Eq. (15.7.3) or Eq. (15.7.5) that I/P scales as T/Rn, and so efficient current drive is favoured by high temperature and low density. The efficiency appears to decrease with the size of the machine, but in a large machine it is easier to ensure that all the energy is absorbed at a single pass by particles on the tail of the distribution. For a plasma at 10 keV, 1 m major radius and density 10^{20} m^{-3}, the numerical value of I/P is of the order of $0.03(v_{ph}/v_T)^2$ A/W, for lower-hybrid current drive with waves of phase velocity v_{ph}.

The estimates of efficiency given here are obviously just simple first approximations, but they are sufficient to show that a substantial amount of power is required to drive the current in a tokamak and that a successful current drive scheme will require careful design to optimize the efficiency. More exact calculations rely on the same principle of looking at the balance between the driving effect of the wave and collisions. The effect of the wave is generally taken to be represented by a quasilinear diffusion term, and collisions by a Fokker–Planck term. Numerical or analytical estimates of the distribution function and the resulting current can then be made, using a variety of techniques. For a complete description of the behaviour, we must also include studies of the coupling of the antenna to the plasma, ray tracing calculations of the wave propagation in the toroidal geometry of the tokamak, and the effect of the current drive on the plasma equilibrium. If we use current drive to increase the current, the self-inductance of the plasma and its inductive coupling to the Ohmic heating coils will have to be considered. As always, the effect of inductance is to oppose any change in the current. When radio-frequency waves are being used to ramp up the current an electric field is set up which may act to limit the rate of increase that is possible.

Experimentally, current drive by lower-hybrid waves has been most successful so far, and has been shown to be capable of driving up and maintaining the current in a tokamak. Electron cyclotron heating has also been demonstrated, but has been found to be less efficient than lower hybrid, and rather less efficient than the simple theory given here would suggest. This may be the result of losses of fast electrons, in which case the theoretical value may be approached in larger machines.

While lower-hybrid current drive has been very successful in present-day machines, there are doubts about its applicability in the reactor regime. These arise because of the limits on the spectral range. An upper limit to the parallel refractive index is set by the accessibility condition, and a lower limit is set by the fact that we do not wish strong Landau damping to take place in the cooler plasma near the edge. As the plasma becomes larger and hotter, these two limits approach each other, and it may not be possible to launch waves which get beyond the edge of the plasma. For this reason, interest in the fast wave has developed. It is polarized so that the parallel component of electric field is smaller than for the slow wave, so it is less strongly Landau damped. Another advantage is that it does not have to be launched with a frequency above the lower-hybrid frequency, as is evident from Fig. 15.4. Going to lower frequencies relaxes the constraints of accessibility and allows the use of larger and more efficient waveguides. Technically, however, it is somewhat more difficult to launch.

The fact that radio-frequency current drive is most efficient at low densities, while Ohmic current drive works well at the high densities needed for fusion, has led to suggestions that the two be combined in schemes involving 'transformer recharge'. The idea here is that the current is driven inductively by a changing current in the Ohmic coils at high plasma density. When the maximum current swing in the Ohmic coil has been reached, the density is dropped and the radio-frequency current drive switched on. With enough power, it is possible to maintain the current in the plasma and also, through the inductive coupling between the plasma and the Ohmic heating coil, drive the current in the latter back to its original value, at which point the density can be increased and the cycle started again. By using each current drive method under favourable conditions, it is possible to keep the current at a steady value with a lower average radio-frequency power than would be needed for continuous current drive. In a reactor this would be at the expense of pulsed output.

The current drive theory discussed above involves resonant interaction between the wave and fast particles. Recently there has been a good deal of interest in low frequency current drive, the essential idea of which can be described by a simple fluid theory. At frequencies well below the particle cyclotron frequencies, the electric and magnetic fields are related by the Ohm's law,

$$\boldsymbol{E} + \boldsymbol{v} \times \boldsymbol{B} = \eta \boldsymbol{J}. \tag{15.7.7}$$

If a wave is set up with a nonzero average value of $\boldsymbol{v} \times \boldsymbol{B}$, then a steady current can be driven. For the phase relation between $\boldsymbol{v}$ and $\boldsymbol{B}$ to be such that the average is nonzero, the wave must be damped. A viable scheme of this sort needs a wave which is sufficiently strongly damped and whose polarization is such that the average value of $\boldsymbol{v} \times \boldsymbol{B}$ is along the direction of the steady magnetic field.

Since anomalous transport properties are important in various aspects of plasma physics, readers may question the validity of our estimates based on classical collision frequencies. There is now ample experimental evidence that these collision frequencies do give a good description of the current drive process, since the efficiencies obtained agree with those calculated on this basis. Much work remains to be done in understanding the details and in optimizing the efficiency, but the essential mechanisms appear to be well understood.

The use of radio-frequency waves for both heating and current drive has been a very successful part of fusion research. It is now possible to inject high power levels into a plasma without producing impurity problems or catastrophic decreases in confinement time. Ion cyclotron heating is used on a large number of machines, and lower-hybrid current drive is now routinely used to extend the time scale of experiments beyond what would be possible with inductive current drive alone.

Bibliography

BORNATICI, M., CANO, R., DE BARBIERI, O. and ENGLEMANN, F. (1983). *Nucl. Fusion*, **23**, 1153.

CAIRNS, R. A. (1991). *Radiofrequency Heating of Plasmas*. Adam Hilger, Bristol.

FISCH, N. J. (1987). *Rev. Mod. Phys.*, **59**, 175.

GOLANT, V. E. and FEDEROV, V. I. (1989). *Plasma Heating in Toroidal Fusion Devices*. Consultants' Bureau, New York.

LITVAK, A. G. (1992). *High Frequency Plasma Heating*. American Institute of Physics, New York.

VACLAVIK, J. and APPERT, K. (1991). *Nucl. Fusion*, **31**, 1945.

16

Boundary plasmas

G. McCRACKEN

16.1 Introduction

In many studies of plasmas there is no necessity to consider the boundary of the plasma in any detail, for example in astrophysical plasmas and inertial confinement fusion. In the case of laboratory plasmas, the boundary is normally the walls of the vessel in which the plasma is formed. However, there are other common cases, such as arcs at atmospheric pressure, where there is a boundary between the plasma and a cold gas. In the case of magnetic confinement for nuclear fusion, the plasma boundary has assumed an increasingly important role as the power in the plasma has increased towards the conditions required for net power generation. In this chapter we shall consider the principal features of the plasma in the region close to the vacuum vessel.

In magnetic confinement, the plasma is held within closed magnetic flux surfaces, normally generated by a combination of fields due to external conductors and to currents flowing in the plasma. Such fields can only be generated within a restricted volume, and there is therefore a boundary determined by the *last closed flux surface* (LCFS). The shape of the LCFS is determined by the magnetic fields, but the position may be determined by the presence of a solid surface. Such a surface is called a *limiter*. Alternatively the closed surface may be determined entirely by the magnetic fields, so that when the plasma ions and electrons diffuse across the LCFS they flow predominantly along the magnetic field until they reach a solid surface. This is the basic geometry of a divertor. The two situations are compared in Fig. 16.1. The essential difference between the two is that with a limiter the LCFS is in contact with a solid surface, whereas with a divertor the solid surface is removed some distance from the LCFS. The region outside the LCFS, in both divertor and limiter configurations, is commonly referred to as the scrape-off layer (SOL), see Fig. 16.1.

In this chapter we will first consider the factors which control the boundary conditions and, in particular, the plasma transport along and across the magnetic flux surfaces. In Section 16.5 the particular case of parallel transport in a divertor configuration is discussed. Impurity production and transport are considered in Section 16.6 and the physical process governing recycling of plasmas at solid surfaces is discussed in Section 16.7. Finally, in Section 16.8, the methods used to minimize impurity production are outlined.

16.2 The plasma–solid boundary

A solid surface in contact with a plasma formed from ions and electrons in thermal equilibrium will be initially subjected to a much greater flux of electrons

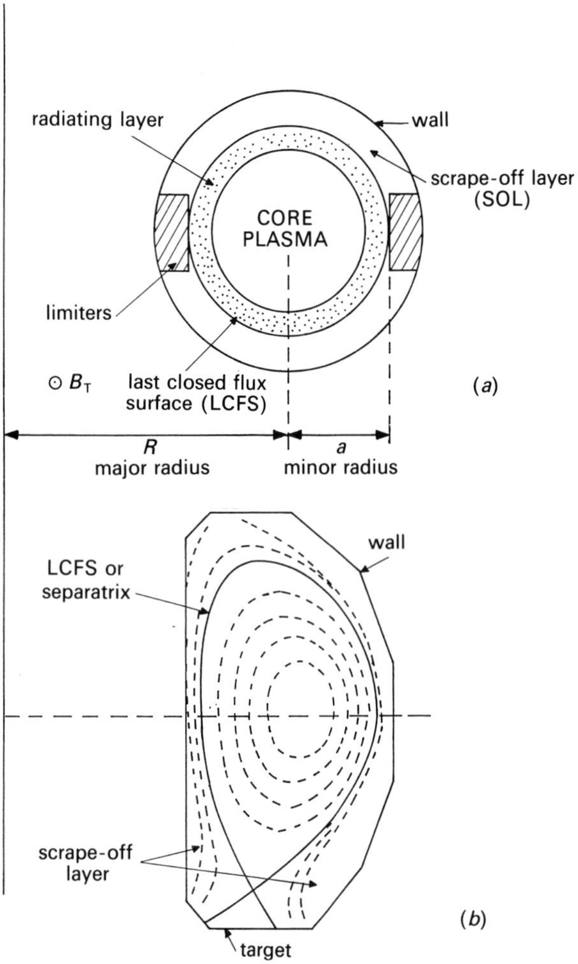

Fig. 16.1. Schematic diagram of flux surfaces in a tokamak (*a*) with a limiter and (*b*) with a divertor.

than of ions, due to the higher velocity of the electrons. This flux results in a potential being built up between the plasma and the surface until sufficient electrons are retarded for charge balance to be established. The condition for equal flux densities of ions and electrons, J^+ and J^-, can be stated as

$$J^+ = en_e V_s = J^- = 1/4 en_e v_e (1 - \delta) \exp\left(-\frac{\Phi_s}{T_e}\right), \tag{16.2.1}$$

where n_e is the plasma density, T_e is the electron temperature in electron-volts, Φ_s is the equilibrium potential of the surface with respect to the plasma, δ is the secondary electron yield due to ions and electrons, v_e is the mean random electron velocity, and V_s is the ion sound speed. v_e and V_s are defined as follows:

$$v_e = \left[\frac{8k_B T_e}{\pi m_e}\right]^{1/2}, \tag{16.2.2}$$

$$V_s = \left[\frac{k_B(T_e + T_i)}{m_i}\right]^{1/2} \le v_e, \tag{16.2.3}$$

where m_e and m_i are the electron and ion masses, and T_i is the ion temperature. The reason for the ion flux being proportional to V_s will be discussed in Section 16.3.

From these equations we can express the potential, Φ_s, known as the sheath potential, as

$$\Phi_s = -0.5k_BT_e\ln\left[\frac{2\pi m_e}{m_i}\frac{(1+T_i/T_e)}{(1-\delta)^2}\right]. \tag{16.2.4}$$

For hydrogen ions, if we assume $\delta = 0$ and $T_i = T_e$, we get the simple result

$$\Phi_s = 2.5k_BT_e. \tag{16.2.5}$$

In practice, a surface acts as a sink both for ions and electrons. An ion arriving at the surface has a high probability of picking up an electron and becoming neutral. Thus there needs to be some method of continually generating the plasma. This is commonly produced by ionization of the neutral gas in the vessel. Atoms or molecules return from the walls into the plasma and are ionized by collisions with the plasma electrons. Such a process is termed *recycling*. Some external source of energy is required to keep the electrons hot, since the continuous flux to the surface is a loss of energy, which has to be replaced.

The energy of incident ions can cause erosion of the surface by sputtering, evaporation, desorption and other mechanisms. The resulting impurities are transported into the plasma and cause further energy loss by radiation. We consider these processes in more detail in Section 16.6. The fact that there is a sheath potential means that if the plasma is in contact with a solid surface at a fixed potential, the plasma potential is determined by the sheath potential. Electric fields can, however, be applied in the boundary by having two or more surfaces in contact with the plasma at different potentials. Electric fields, either externally induced or naturally occurring in the boundary, can have a significant effect on the transport.

The energy of the plasma ions reaching the surface is controlled by the sheath. The ions are accelerated by the sheath so that when they reach the surface their energy distribution is approximately an accelerated Maxwellian, that is a Maxwellian distribution in which the velocity of each ion has been increased by a constant value. The electrons are decelerated by the sheath. The distribution remains a Maxwellian but the flux is reduced. Examples of experimentally measured distributions are shown in Fig. 16.2. It can be seen that T_i is considerably larger than T_e. Although this is an extreme example, experimental data frequently show this trend. In our models we will however generally assume $T_i = T_e$ for convenience.

We now consider the flow of power to a surface as a result of the sheath. The energy transported by a Maxwellian distribution of ions or electrons is $2kT$ per particle. In addition, there is the energy resulting from the acceleration of ions by the sheath. The flux of electrons is limited by ambipolarity to the same value as the flux of ions. Thus we have the power P given by

$$P = J_sT_e\left[\frac{2T_i}{T_e} + \frac{2}{1-\delta} - 0.5\ln\left\{\frac{2\pi m_e(1+T_i/T_e)}{m_i(1-\delta)^2}\right\}\right]\mathrm{W\,m^{-2}}, \tag{16.2.6}$$

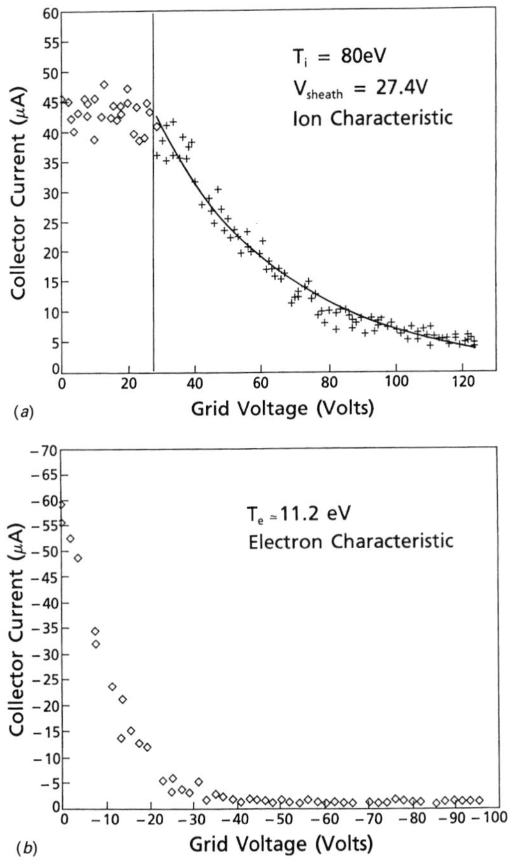

Fig. 16.2. (*a*) Integral ion energy distribution measured in the plasma boundary of the DITE tokamak. (*b*) Integral electron energy distribution measured in DITE.

where J_s is the ion current density in A m^{-2} and T_e is again in electron-volts. For $T_i = T_e, \delta = 0$, in a deuterium plasma we obtain

$$P \simeq 7.8 \times 10^{-16} \gamma n_e T_e^{3/2} \text{ W m}^{-2}, \tag{16.2.7}$$

where $\gamma = P/J_s T_e$ defined by Eq. (16.2.6) is the sheath power transmission coefficient.

In practice δ is close to unity and γ is significantly enhanced above the value for $\delta = 0$. As δ increases, the sheath potential drops, Eq. (16.2.4); however, the effective value of δ cannot exceed ~ 0.8 due to space-charge effects. For any practical surface under given cooling conditions there is a maximum tolerable power, P_m, determined by its material properties. Hence the maximum density which can be in contact with a surface without destroying it is

$$n_e(\max) = 1.3 \times 10^{15} \frac{P_m}{\gamma T_e^{3/2}} \text{ m}^{-3}. \tag{16.2.8}$$

16.3 Radial distribution of n_e and T_e in the scrape-off layer

Outside the LCFS there is parallel flow along the field lines towards the limiter or target plate. There is also perpendicular flow across the magnetic field at a much slower rate, Fig. 16.3. In toroidal devices the diffusion of both power and particles is at an 'anomalous' rate, higher than predicted by neoclassical theory. In steady state, provided there are no other sources (such as ionization) or sinks in the SOL, the particle loss along the field from any given flux tube must be balanced by the net flow across the field into the flux tube. We can thus write

$$\frac{\mathrm{d}}{\mathrm{d}r}\left[D_{\perp}\frac{\mathrm{d}n_{\mathrm{e}}}{\mathrm{d}r}\right]=\frac{n_{\mathrm{e}}V_{\mathrm{s}}}{L_{\mathrm{c}}}, \tag{16.3.1}$$

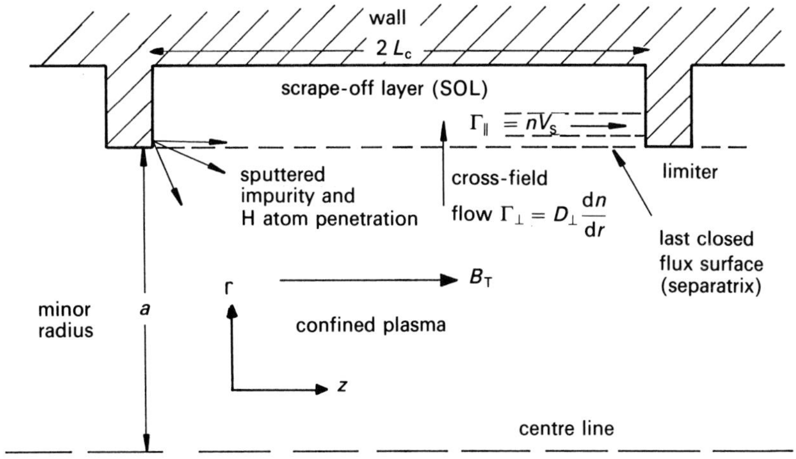

Fig. 16.3. A two-dimensional schematic diagram of the plasma flow from the core plasma into the SOL by cross-field diffusion and parallel flow in the SOL to the limiters.

where $2L_{\mathrm{c}}$ is the connection length along the flux tube to the solid surface, see Fig. 16.3, and $D_{\perp}$ is the cross-field diffusion coefficient. With the assumption that V_{s} is independent of radius in the SOL, Eq. (16.3.1) is readily integrated to obtain

$$n_{\mathrm{e}}(r)=n_{\mathrm{e}}(a)\exp\{-(r-a)/\lambda_{n}\}, \tag{16.3.2}$$

where the density e-folding length λ_n is given by

$$\lambda_{n}=\left[\frac{D_{\perp}L_{\mathrm{c}}}{V_{\mathrm{s}}}\right]^{1/2}. \tag{16.3.3}$$

The electron heat balance can be considered in an analogous way, and a similar result is obtained:

$$T_{\mathrm{e}}(r)=T_{\mathrm{e}}(a)\exp\{-(r-a)/\lambda_{T}\}, \tag{16.3.4}$$

where the electron temperature e-folding length, λ_T, is given by the relation

$$1+\frac{\lambda_{n}}{\lambda_{T}}=\frac{\delta}{5/2+\psi_{\perp}\lambda_{n}/D_{\perp}\lambda_{T}}, \tag{16.3.5}$$

and $\psi_{\perp}$ is the cross-field thermal conduction coefficient. Using Eq. (16.2.7) we can also write an equation for power flux similar to those for n_{e} and T_{e}.

We now have the e-folding lengths in the SOL in terms of the cross-field transport parameters. In practice, since we generally do not have any *a priori* knowledge of the cross-field transport, we can use measured values of λ_n and λ_T to deduce $D_\perp$ and $\psi_\perp$. Measurements of the e-folding length for n_e and T_e can be made with electrical probes, provided the plasma impurity content is small. Values are typically 10 mm. From this measurement, together with the assumption that $T_e = T_i$, V_s can be calculated and a value for $D_\perp$ deduced. This is typically $1\,\mathrm{m^2\,s^{-1}}$. There is experimental evidence that $D_\perp$ varies inversely proportional to density. However, as density and temperature changes in the boundary are correlated, this behaviour can also be interpreted as a temperature dependence of $D_\perp$.

We can also make rough estimates of the values of the density and temperature at the LCFS by making global particle and energy balances. The integrated particle flux to the limiter must be equal to the total flux diffusing out across the LCFS from the confined plasma. If we define a particle replacement (or global recycling) time τ_p then the flux out of the plasma Γ_n can be written

$$\Gamma_n = \frac{\bar{n}_e V}{\tau_p}, \tag{16.3.6}$$

where V is the plasma volume.

The flux to the limiter, Γ_L, for a single full aperture circular limiter is

$$\Gamma_L = \pi a \int_a^\infty n(a)\exp(-(r-a)/\lambda_n)V_s\mathrm{d}r, \tag{16.3.7}$$

where it is assumed that λ_n is small compared with the distance between the limiter and the wall. Integrating Eq. (16.3.7) and equating Eqs (16.3.6) and (16.3.7) we obtain a value for the density at the limiter $n(a)$,

$$n(a) = \frac{\bar{n}V}{\tau_p}\frac{1}{2\pi a\lambda_n V_s}. \tag{16.3.8}$$

Using a simple edge transport model for τ_p, taking into account the ionization mean free path of atoms entering the plasma, we can derive

$$n_e(a) = \frac{\overline{\sigma v}_i}{6\bar{v}_n}\lambda_\Gamma \bar{n}_e^2 \approx 5\times 10^{21}\bar{n}_e^2\,\mathrm{m^{-3}}, \tag{16.3.9}$$

where $\overline{\sigma v}_i$ is the ionization rate coefficient, $\bar{v}_n$ is the average atomic velocity, and λ_Γ is the flux e-folding length. This simple expression turns out to give a remarkably good prediction of the edge density over a wide range of tokamaks and a range of densities; see Fig. 16.4.

16.4 Parallel transport in the SOL

The simplest model capable of describing the main features of transport along the magnetic field in the SOL is the isothermal fluid model. We consider the general features of parallel flow for a boundary with a limiter, as in Fig. 16.3. Steady state, inviscid, isothermal one-dimensional flow is determined by conservation of particles and momentum:

$$\frac{\mathrm{d}}{\mathrm{d}z}(nv) = S \tag{16.4.1}$$

and

$$nm_i v\frac{\mathrm{d}v}{\mathrm{d}z} = -\frac{\mathrm{d}p}{\mathrm{d}z} + enE - mvS, \tag{16.4.2}$$

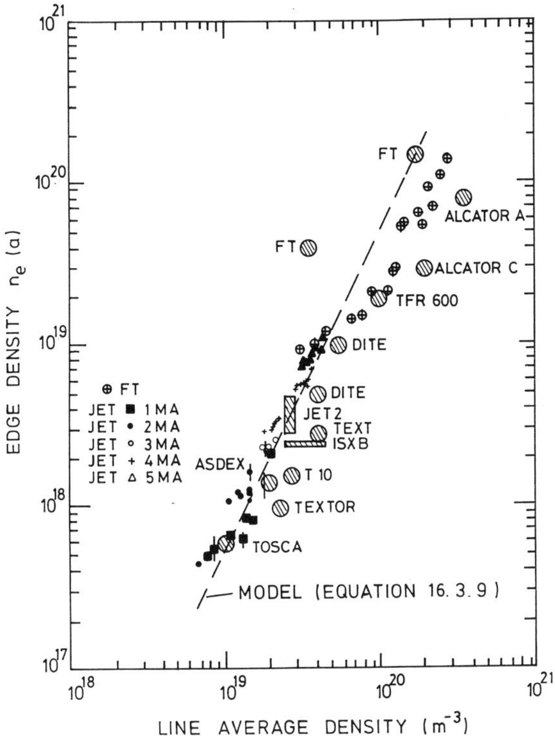

Fig. 16.4. Relationship between edge density $n_e(a)$ and line average density $\bar{n}_e$ for a wide range of tokamaks. The line is obtained from the Stangeby model, Eq. (16.3.9).

where $p = nkT_i, E = -\,d\Phi/dz$, S is the source rate of particles due to cross-field transport or ionization, and the ion charge is taken as unity (no impurities). The first term on the right-hand side of Eq. (16.4.2) is the pressure gradient, the second is the electric field, and the third is the drag due to acceleration of the cold ions from the source. Using the Boltzmann relation, $n = n_0\exp(eV/kT_e)$, Eqs (16.4.1) and (16.4.2) can be combined to give

$$\frac{dM}{dz} = \frac{S}{nc_s}\frac{(1 + M^2)}{(1 - M^2)}, \qquad (16.4.3)$$

where $M = v/V_s$ is the Mach number of the flow.

From Eq. (16.4.3) we can see that, as M tends to unity, dM/dz tends to infinity and the plasma solution breaks down. $M = 1$ indicates the state of the plasma sheath. The fact that the ion velocity must reach the sound speed at the start of the plasma sheath is known as the Bohm criterion. If we make the further assumption that S is proportional to n then we obtain as $M \rightarrow 1$

$$\frac{n(M)}{n(0)} = \frac{1}{1 + M^2} \rightarrow 0.5, \qquad (16.4.4)$$

and

$$\Phi(M) = T_e \ln(1 - M^2) \rightarrow 0.69T_e, \qquad (16.4.5)$$

where $n(0)$ is the density at $z = 0$, that is at the stagnation point, halfway between the limiters.

Many other models, both fluid and kinetic, have been proposed for parallel flow. However, for the most important parameters, the predictions of the different models are very similar. A schematic presentation of the results of the model for the density, potential and flow velocity is given in Fig. 16.5. We have already shown that the potential drop in the sheath is given by Eq. (16.2.4).

Up to now we have considered only the plasma ions. Impurities which are produced at the limiter and flow into the SOL are also subject to the electric field and their own pressure gradient. In addition, they are subject to the frictional force of the ion flow towards the limiter. It can be shown that they are typically accelerated up to a velocity in the range of 0.1–0.5V_s.

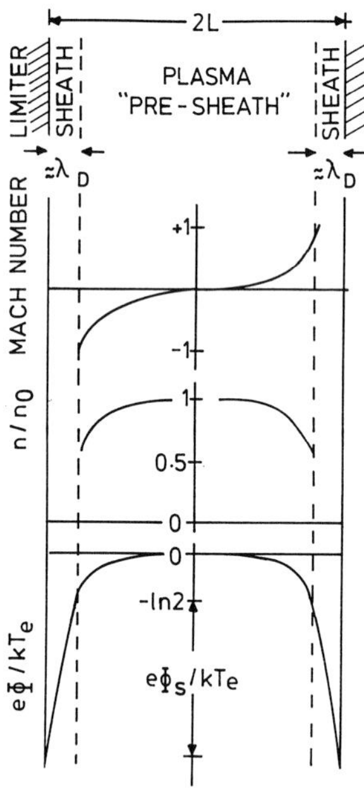

Fig. 16.5. Schematic of the distribution of parallel velocity, normalized density and plasma potential between two semi-infinite plates, based on fluid models. The thickness of the sheath is exaggerated for clarity.

16.5 Divertors

As discussed in Section 16.1, the position and shape of the LCFS may be formed solely by the magnetic field structure, allowing the first solid surface exposed to plasma to be remote from the confined plasma. In the case of a divertor, the LCFS is frequently referred to as the *separatrix*, Fig. 16.1. The production of impurities, recycling and energy losses may be kept remote from the confined plasma. Such an arrangement is in principle a very attractive method of reducing the effect of plasma–surface interactions. The key feature which can be achieved with a divertor is a temperature and density gradient along the field lines, which allows a high temperature, low density plasma at the separatrix to be converted to a high density, low temperature plasma at the target plate.

The basic plasma parameters for a divertor plasma can be represented by a simple model which expresses pressure balance, electron heat conduction and power transport across the sheath. If we assume that $T_i = T_e$ everywhere, we may write

$$n_d T_d = 0.5 n_u T_u, \qquad (16.5.1)$$

to flow. The presence of a plasma can produce a 'unipolar' arc where the applied potential is produced by the plasma sheath. When the arc is initiated, electrons from a local 'cathode spot' are accelerated by the sheath potential. The flow of current reduces the sheath potential so that high energy plasma electrons, in the tail of the Maxwellian distribution, can surmount the barrier of the sheath potential and reach the surface, thus completing the circuit. The minimum current required to sustain an arc is $\sim 10\,\text{A}$. Arcs can be readily detected by examination of surfaces as they leave characteristic tracks, and such tracks are routinely observed on components in fusion devices. However, at least in tokamaks, such arcs and the consequent impurity production seem only to occur during the initial plasma production or during instabilities. There is no evidence that arcing is a serious impurity generating process under stable operating conditions in present fusion devices. This may be due to the fact that in tokamaks the plasma diffuses principally across the magnetic field. Under these conditions the limiter and wall tend to charge positively with respect to the plasma. Thus electron emission from the solid is suppressed and the unipolar arc cannot be sustained.

(*e*) *Chemical reactions*

These reactions have been studied in detail and, as for physical sputtering, yields are known as a function of ion energy. Unlike physical sputtering, the yields are very dependent on surface temperature. Typical yields are shown for graphite surfaces in Fig. 16.16(*a*) for hydrogen bombardment and in Fig. 16.16(*b*) for oxygen bombardment. The higher yield obtained at 820 K compared with 300 K in Fig. 16.16(*a*) is due to the formation of methane. Such yields are sensitive to surface conditions in a way that is not yet fully understood. Yields measured under tokamak conditions are generally much lower than under clean laboratory conditions. A possible reason is that the presence of metals on the graphite surface may inhibit the chemical reaction rate.

16.6.3 *Impurity transport*

In order to discuss cross-field transport of impurities, we must first consider the source function. This depends on ionization of the incoming neutrals and therefore on the local plasma density and temperature. Collisions with electrons and ions result in excitation, ionization, dissociation and heating. If we consider a flux of atoms $\Gamma(r)$ starting from the wall and moving with a velocity v_0 across the magnetic field into a plasma with a radial profile of density $n_e(r)$ and temperature $T_e(r)$, then considering only transport in the radial direction, the flux will decay with distance as

$$\Gamma(r_w - r) = \Gamma_0 \exp\left(-\int_{r_w}^{r} \frac{\overline{\sigma v}}{v_0} n_e(r_w - r)\mathrm{d}r\right). \qquad (16.6.9)$$

Here $\overline{\sigma v}$ is the rate coefficient for ionization of impurities, r_w is the wall radius, and Γ_0 is the initial impurity flux entering the plasma. If we assume the density profile to be exponential with an e-folding value of λ_n, and a value of $n_e(\omega)$ at the boundary, and that $\overline{\sigma v}$ is independent of radius, we can write the neutral atom density as

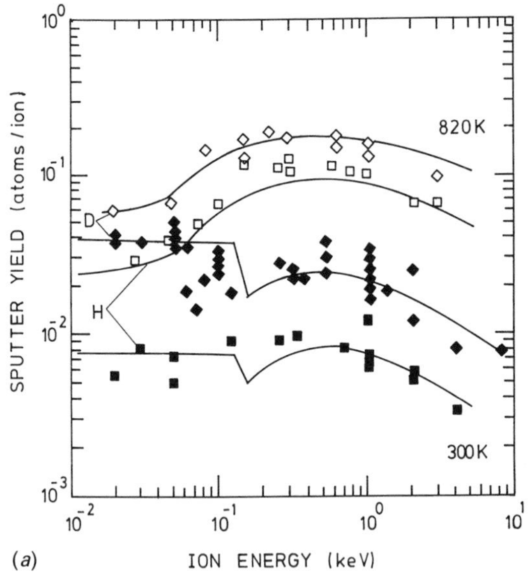

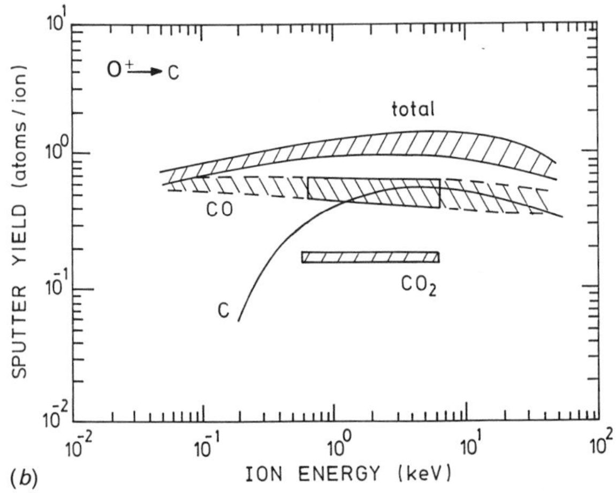

Fig. 16.16. Energy dependence of total sputter yield of graphite by (*a*) hydrogen ions and (*b*) oxygen ions. The upper curves in (*a*) are taken at a surface temperature which corresponds to the maximum in the chemical sputter yield.

$$\Gamma(r_w - r) = \Gamma_0 \exp\left(-\frac{\lambda_n}{d}[\exp((r_w - r)/\lambda_n) - 1] \right), \tag{16.6.10}$$

where

$$d = v_0/\overline{\sigma v} n_e(w). \tag{16.6.11}$$

The ionization rate at any point r is given simply by

$$S(r) = \mathrm{d}\Gamma/\mathrm{d}r. \tag{16.6.12}$$

For arbitrary experimental profiles of n_e and T_e, similar calculations of $\Gamma(r)$ and $S(r)$ can be carried out numerically.

Given the source function, we must consider the transport of ions. Inside the LCFS the ions are by definition on closed field lines. If we assume toroidal uniformity then we only need consider cross-field diffusion. A simple one-dimensional model has been proposed, which gives the central impurity concentration $n(0)$ in terms of $D_\perp$, the impurity influx Γ_m, and the ionization mean free path, λ_i:

$$n(0) = \frac{\Gamma_m(\lambda_i + \lambda_n)}{A_p D_\perp}, \tag{16.6.13}$$

where A_p is the plasma surface area. The source is in this case considered to be a delta function rather than the more complex function given in Eq. (16.6.12). The impurity concentration distributions from this model are shown schematically in Fig. 16.17. The case of a distributed source has also been analysed. This is basically the same model that is used to derive Eq. (16.3.9) for the plasma density.

The velocity distribution of sputtered atoms is an important factor in determining the depth of penetration into the plasma. The effect of different penetration depths on the central concentration for the same influx rate is shown in Fig. 16.18. It is seen that if the impurities are ionized in the SOL, then due to parallel transport many are rapidly returned to the limiter by parallel flow. If the impurity penetrates across the LCFS before ionization, then it makes a much larger contribution to the central concentration. It is seen that the central

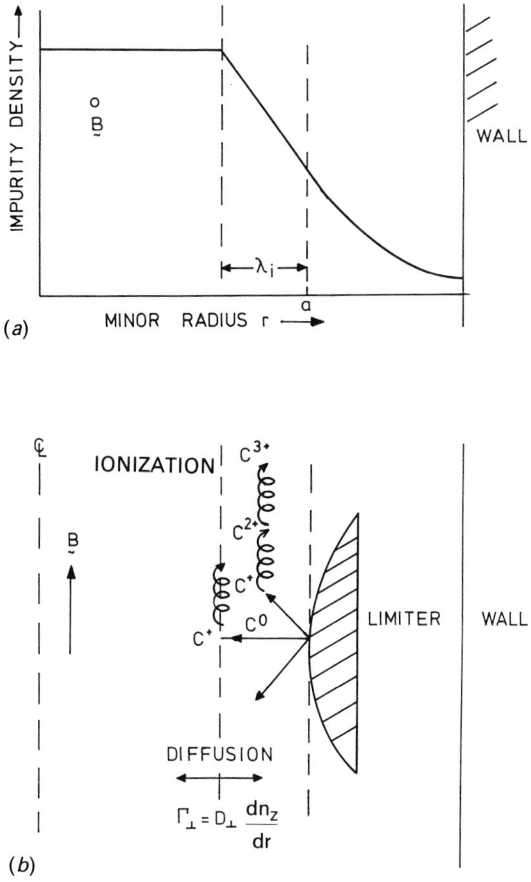

Fig. 16.17. Schematic of impurity transport in the boundary layer. (*a*) The one-dimensional (radial) steady state distribution. The penetration distance of the neutral λ_i corresponds to one ionization mean free path. (*b*) The two-dimensional cylindrical approximation illustrating successive ionization stages.

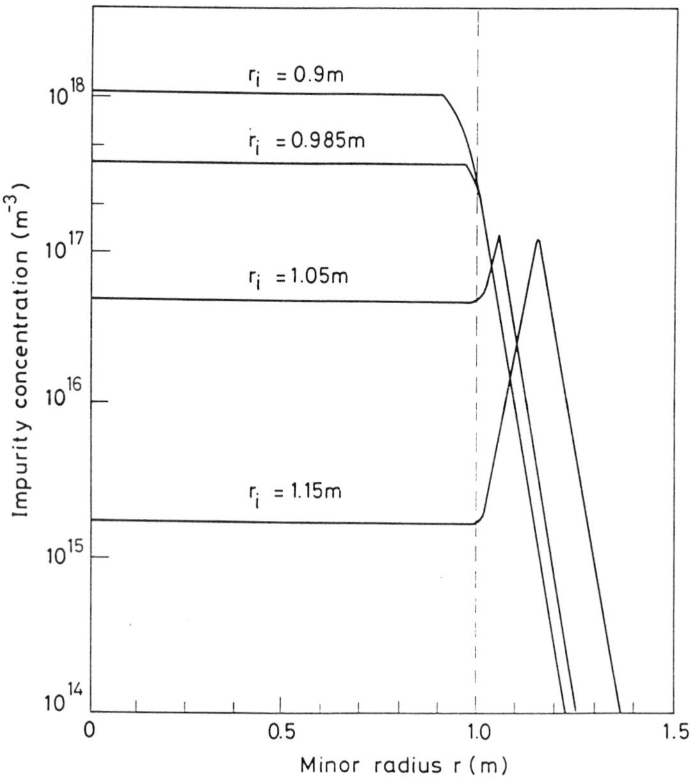

Fig. 16.18. Model calculation of steady state distributions of impurities illustrating the effect of different ionization mean free path. Cross-field diffusion is assumed constant $= 0.5\,m^2\,s^{-1}$. Influx rate $= 5 \times 10^{20}$ atoms s^{-1} over $160\,m^2$.

impurity concentration can vary by orders of magnitude for the same impurity production rate, depending on the penetration of the neutrals. This simple analysis has been carried out with a constant radial diffusion coefficient as the only transport mechanism. In some circumstances the concentration profile is quite peaked. This is attributed to an anomalous inward velocity term. The ratio of central to edge value of the concentration depends on the inward velocity, but is typically a factor of two.

Impurities entering the plasma undergo excitation, ionization, heating and diffusion. Once ionized, their transport is fastest along the field lines. The behaviour is illustrated by Fig. 16.19, which shows the toroidal distribution of carbon ions resulting from the deliberate introduction of methane gas at a point source. The distribution of neutral carbon atoms CI, from the gas source, is shown in the Figure. The singly and doubly ionized species (CII and CIII) are found to move consistent with classical (Spitzer) collisions. While the ions are moving along the field, they also move slowly across it, as shown schematically in Fig. 16.17(*b*).

16.6.4 *Ionization and heating*

The incoming impurity atoms are ionized by electron impact and then heated by ion–ion collisions. We consider the time scales of the different processes. The characteristic time for ionization is

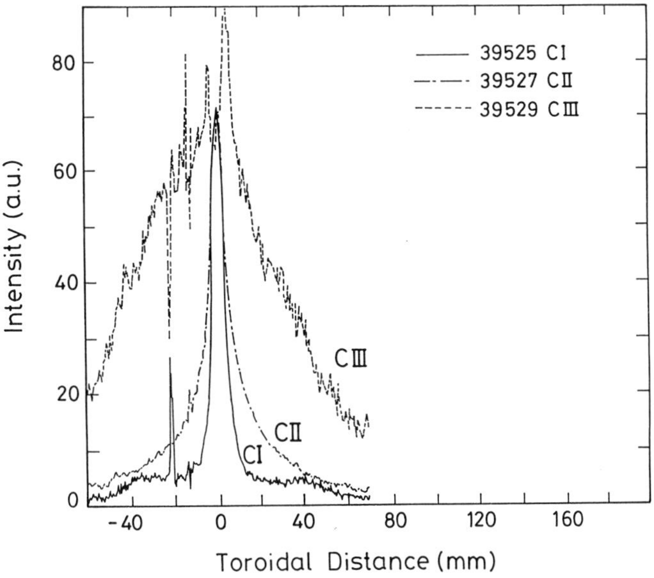

Fig. 16.19. Toroidal distribution of emission of carbon in different charge stages during puffing from a limiter at the LCFS in TEXTOR.

$$\tau_{iZ} = \frac{1}{n_e \overline{\sigma v}}. \tag{16.6.14}$$

The classical (Spitzer) thermalization time is given by

$$\tau_{th} = \frac{m_I T_B^{3/2}}{1.4 \times 10^{-13} m_B^{1/2}, n_B Z_B^2 Z_I^2 \ln\Lambda}, \tag{16.6.15}$$

where m_I and m_B, Z_I and Z_B are the masses and charge states of the impurity and background plasma ions, n_B and T_B are the background plasma density and temperature, and $\ln\Lambda$ is the Coulomb logarithm. The thermalization rate is

$$\frac{dT}{dt} = \frac{T_B - T_{imp}}{\tau_{th}}, \tag{16.6.16}$$

where T_{imp} is the impurity temperature. Using Eq. (16.6.16) we can calculate the impurity ion temperature attained in a given charge state, before ionization to the next higher state

$$T_{imp} = T_B - (T_B - T_0)\exp(-\tau_{iZ}/\tau_{th}), \tag{16.6.17}$$

where T_0 is the initial impurity temperature. The important factor determining T_{imp} is thus the ratio τ_{iZ}/τ_{th}, which from Eqs (16.6.14) and (16.6.15) is independent of density and dependent only on local temperature.

Results for oxygen and carbon as a function of background plasma temperature (assuming $T_e = T_i$) are shown in Fig. 16.20. At low plasma temperature, the collision rate is high, the ionization rate is low, and the impurities are rapidly thermalized. As the plasma temperature increases, the ionization rate increases and the thermalization rate decreases, so that ionization is likely to occur before thermalization is complete. Measurements of low charge state impurity ion temperatures using Doppler broadening have confirmed this picture.

The charge state distribution of impurities is important not only for determining the radiation but also for calculating the impurity production rate.

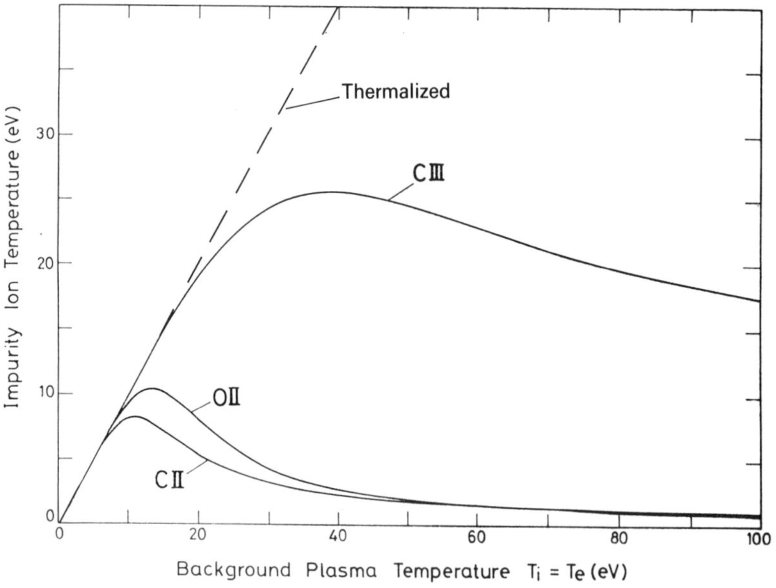

Fig. 16.20. Calculation of impurity ion temperature of CII, OII and CII ions as a function of background plasma temperature ($T_i = T_e$, initial ion temperature $T_0 = 0$) calculated using Eq. (16.6.16).

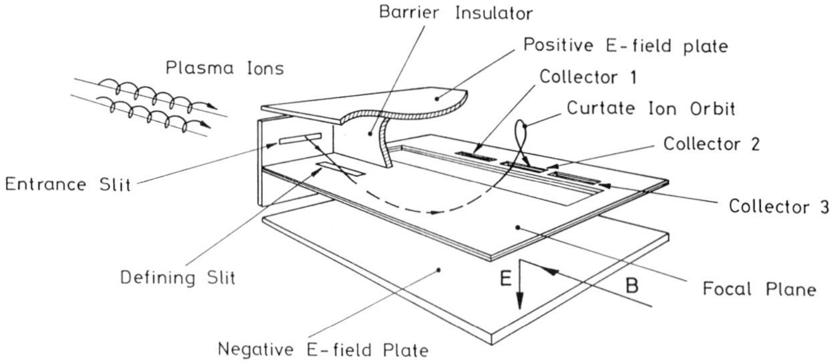

Fig. 16.21. Schematic showing the internal structure of the plasma ion mass spectrometer used in the DITE tokamak.

Multiply-charged impurities will be accelerated by the sheath potential to an energy proportional to their charge. This can increase their sputter yield. A technique known as *plasma ion mass spectrometry* (PIMS) has recently been developed for measuring the charge state distribution in the SOL. The principle is shown in Fig. 16.21. Ions are selected by a small slit and then subjected to an electric field normal to the total tokamak magnetic field. They perform cycloidal orbits in the combined electric and magnetic fields and are detected electrically at a collection. The configuration produces both spatial focusing and focusing of ions of different velocities. Results are shown for the distribution of charge states of oxygen and carbon in the SOL of the DITE tokamak in Fig. 16.22. It is found that the complete spectrum of charge states from singly charged to fully stripped ions is present. The most probable charge for low mass impurities is three to four.

16.6.5 *Impurity feedback mechanism*

Under some conditions, when there are high self-sputter yields or high evaporation rates, it might be thought that the impurity concentration would

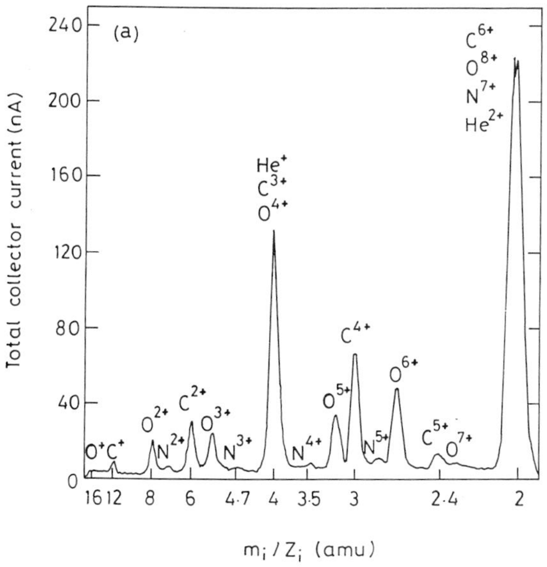

Fig. 16.22. Mass spectrum of ions in the SOL of the DITE tokamak using the plasma ion mass spectrometer.

increase indefinitely, causing the plasma to be extinguished. However, there is a simple energy balance argument, which shows that for slowly changing conditions this does not happen. The global power balance between input heating power P_T, radiation P_R and conduction and convection losses P_c is given by

$$P_T = P_R + P_c. \tag{16.6.18}$$

Thus if the impurity influx increases, the power radiated increases, and the power conducted to the edge must decrease. The argument is presented schematically in Fig. 16.23. Since we have shown in Section 16.3 that the edge density is not, to first order, affected by the power balance, the reduction in edge T_e will reduce both the power flux and the ion energy of particles arriving at the limiter or divertor. It thus reduces the impurity influx again. For a given surface material, the overall equilibrium level of impurities is determined by the total input power and the average density. The cleanliness of surfaces and the absolute sputter yield of the limiter material, both of which determine the impurity influx rate for a given ion energy, are of course of primary importance.

16.7 Recycling processes

Recycling is the general term used for those processes that result in the plasma ions interacting with a surface and returning to the plasma again. When an energetic ion or atom interacts with a surface, it undergoes a series of elastic and inelastic collisions with the atoms of the solid. It may in principle either be backscattered after one or many collisions, or it may slow down in the solid and become implanted in it. Both these processes have been studied in some detail in recent years, although there are still insufficient data at very low energies, < 0.1 keV, due to difficulty in detection of the backscattered particles, which are mainly neutral.

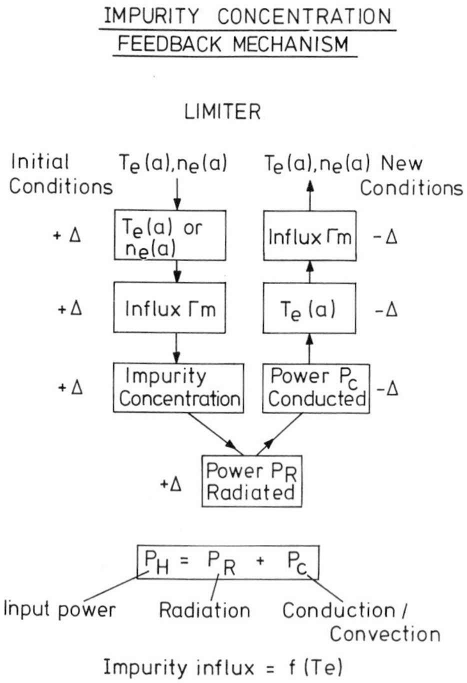

Fig. 16.23. Schematic of the impurity feedback mechanism which controls the absolute level of impurities in the plasma for a given power input.

16.7.1 *Backscattering*

The range and energy loss of fast ions in solids can be described in general terms for all ion-target combinations using a reduced energy ε given by

$$\varepsilon = \frac{32.5 m_2 E}{(m_1 + m_2) Z_1 Z_2 (Z_1^{2/3} + Z_2^{2/3})^{1/2}} \text{(keV)}, \tag{16.7.1}$$

where Z_1, m_1, Z_2 and m_2 are the atomic numbers and mass numbers for the ion and target atoms, respectively. Analytical calculations of the reflection coefficient are difficult. A more promising technique is to use Monte-Carlo calculations, but the interaction potential in the elastic collisions is still not known reliably at low energies. Collating the experimental data and plotting them in terms of reduced energy, we get an estimate of the particle reflection coefficient R_P versus energy as shown in Fig. 16.24. The energy reflection coefficient R_E defined as the ratio of the mean reflected energy to the incident energy is also shown in Fig. 16.24. The backscattering increases rapidly for increasing target atomic number and decreases with increasing incident ion energy. Above 10 keV the backscattering coefficient is less than 10% for most targets.

The backscattered particles are primarily neutral atoms, due to electron capture from the solid. Experimental results indicate that the probability of being backscattered as an ion is almost independent of target material and depends only on the energy of the *backscattered* particle. This probability is $\sim 10\%$ at 5 keV. The energy distribution of the backscattered particles has been measured for incident ion energies down to ≈ 0.1 keV and calculated, again using Monte-Carlo techniques, down to even lower energies. Some results are shown in Fig. 16.25. The distribution is continuous, typically having a maximum, the

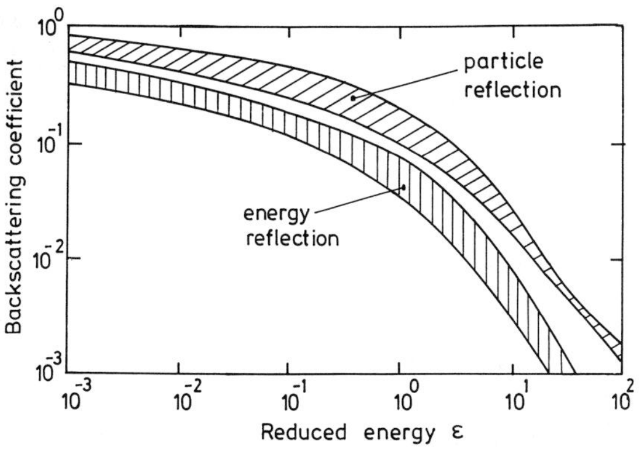

Fig. 16.24. Particle and energy reflection coefficients R_p and R_E of various ions incident on various surfaces as a function of the reduced energy (cf. Eq. (16.7.1)).

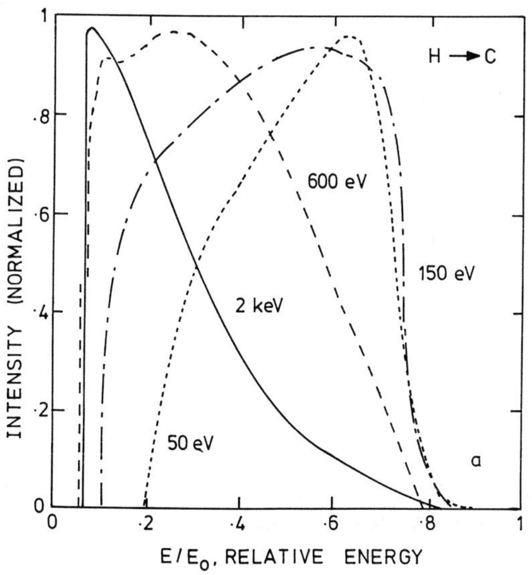

Fig. 16.25. Measured energy distribution of hydrogen backscattered from carbon, for different incident ion energies. The distributions are normalized to the same maximum intensity.

position of which depends on both the energy and the angle of incidence of the primary ion. The maximum energy of the backscattered particles is close to that of the primary energy. Thus the component of incident ions which are energetically backscattered from surface consists predominantly of neutral atoms whose mean energy is roughly 30 to 50% of the incident energy.

16.7.2 *Re-emission of gas*

Incident ions which are not backscattered will slow down to thermal velocities in the solid. Most species of implanted ion become trapped in vacancies or interstitial sites, and no further transport occurs when they have slowed down to the temperature of the lattice. However, hydrogen diffuses relatively rapidly in many metals, even near room temperature. Moreover, under many plasma conditions the flux of ions to the surface is so large that the concentration of hydrogen in the surface can easily approach the atomic concentration in the lattice itself. It is not surprising therefore that experimentally a diffusive flow is observed from the solid back into the vacuum, which eventually equals the incident flux. Since the surface acts as a sink and the range is normally small compared with the sample thickness the concentration gradient will be much greater towards the surface than towards the interior. Thus the flow to the surface will approximately equal the incident flux. The gas which desorbs thermally is molecular and at the same temperature as the metal surface.

The form of the re-emitted gas flux as a function of time has been shown to be approximately described by the diffusion model outlined above, the theoretical flux being given by:

$$\frac{J}{J_0} = \operatorname{erf}\left[\frac{R}{(4Dt)^{1/2}}\right], \qquad (16.7.2)$$

where J_0 is the incident flux, R is the mean ion range, D is the diffusion coefficient of the gas in the solid, t is the time and, erf is the error function defined by $\operatorname{erf}(x) = (2/\pi^{1/2})\int_0^x \exp(-x^2)dt$. The form of the curves is shown in Fig. 16.26. Since diffusion coefficients vary with temperature in the well known way, $D = D_0\exp(-E_d/T_s)$, where E_d is the activation energy for diffusion in electron-volts, an increase of temperature results in an increase of D and hence the more rapid attainment of equilibrium. However, ion implantation produces damage sites in the lattice at which the diffusing atoms become trapped and the effective diffusion coefficient can be as much as two orders of magnitude lower than for diffusion in annealed samples, where no damage is present. The damage sites can be annealed out by heating to temperatures close to the melting point.

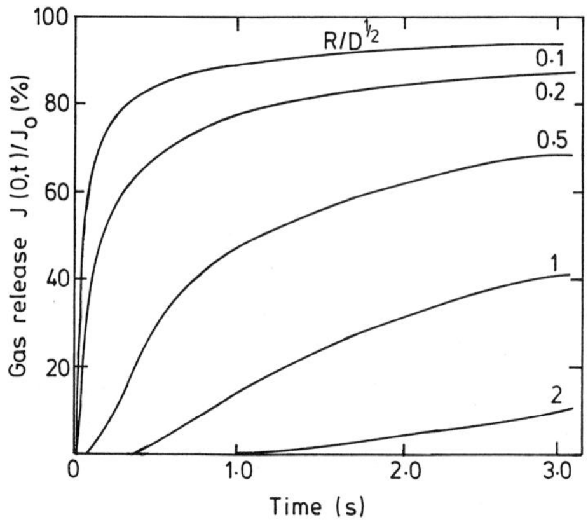

Fig. 16.26. Time dependence of gas re-emission, calculated from the simple diffusion model, Eq. (16.7.2).

The rate at which implanted hydrogen ions will be released is in practice difficult to determine, depending on the damage history of the lattice, the surface temperature, and the concentration of gas left in the surface. Under the conditions of many plasma physics experiments, with high fluxes and regular repetition rates, it is probable that the surface will reach saturation, one hydrogen molecule being released for every two ions which slow down in the lattice. In tokamaks, where there will be rapid recycling during the rising current phase when the plasma is not well contained, it is possible that saturation may be reached during a single pulse.

16.7.3 *Desorption*

Under most practical conditions, some of the working gas is adsorbed on the wall and thus it is possible that it may be desorbed by particles and radiation. The processes by which this can happen have been discussed in Section 16.6.2 on impurities. Although perhaps not strictly recycling, these desorbed particles cannot easily be distinguished from those produced by the other two recycling processes. Since this is an additional source of gas, it can lead to effective recycling rates greater than unity. Desorption coefficients as high as 10 atoms/ion have been observed. The major difference between this effect and backscattering or re-emission is that it will decay with time as the surface concentration is depleted. This depletion may not be significant during the period of a single discharge pulse. There is clear evidence that desorption can contribute to the recycling process from the fact that in some tokamaks the density increases steadily throughout the discharge and is unrelated to the initial gas pressure in the system. It has been shown that recycling can be reduced by evaporation of fresh metallic films in the torus as discussed in Section 16.8.1.

To summarize, the recycling behaviour consists of three processes. In backscattering, the recycled particles are predominantly neutral atoms with energies smaller than, but comparable with, the incident energy. In gaseous re-emission, those atoms which are trapped in the solid diffuse out to produce a flux of cold molecules which varies with time. Finally, if the working gas can adsorb on the wall before the start of a pulse, a time-dependent flux of desorbed neutral atoms or molecules with a mean energy of a few electron-volts will be released.

16.8 Minimizing impurities

16.8.1 *Vessel conditioning*

The first step requited for minimizing impurities is to remove the loosely bound impurities adsorbed on surfaces. This can be done by the desorption processes discussed in Section 16.6.2. The most effective techniques are baking, to thermally desorb the most weakly bound adsorbed species, and glow discharge cleaning (GDC), to remove most of the rest. A positively biased electrode is inserted and the vessel used as the cathode. No magnetic field is used so that the bombardment of the vessel is fairly uniform. The glow discharge provides ion and electron bombardment of surface, which releases the gas and allows it to be pumped out. Optimized glow discharge conditions are, typically, low operating pressure $\sim 10^{-3}$ mbar to maximize ion energies, and current density $\sim 10^{-5}$ A cm^{-2}. In order to maintain the discharge at low pressure, an auxiliary electrode with an RF coil operating typically at 13.45 MHz is advantageous. Although a range of gases can be used for glow discharge cleaning, the heavier rare gases tend to cause excessive sputtering, which produces metallic films on windows and insulators. It has been found preferable to use hydrogen as the discharge gas and to rely as much on chemical as on physical sputtering. It is generally found that more than 100 h of GDC is necessary before tokamak discharges can be started. It is beneficial to combine GDC and baking in order to minimize the chance of desorbed gases being pumped out of the vessel.

GDC has the advantage that it bombards most of the vessel uniformly, but the current and power densities are rather low. Use of short, higher power, tokamak-like discharges with a magnetic field allows a higher flux density of power and particles on the limiters and other surfaces close to the plasma, so that they can be cleaned up more thoroughly than the walls of the vessel.

Evaporating reactive metal layers such as titanium, chromium, aluminium, and more recently beryllium, has been effective in further reducing the problem with adsorbed gases. These active metals are particularly good at removing oxygen from the system by adsorption. This process is commonly referred to as *gettering*.

Metallic impurities can only be tolerated at very low concentrations, as discussed in Section 16.6.1. Even when the limiters are made of graphite, metallic impurities sputtered from the walls have been a problem. This has been minimized by putting a thin film of low Z materials on the wall, processes known as *carbonization* and *boronization*. Boron films have the advantage that they act as getters and also that they do not adsorb water vapour when the vessel is exposed to air. A disadvantage of these films is that they contain high concentrations of hydrogen. This can lead to difficulties in density control.

16.8.2 *Impurity screening*

The impurities released from a surface, either of the limiter or of the divertor target, may get ionized in the SOL and thus flow directly back to the surface without entering the confined plasma. If this happens, little effect on the plasma should be apparent. In the case of the divertor in particular, impurity transport is still not fully understood. There is a balance between frictional forces due to the flowing plasma, diffusion due to density gradients and temperature gradient terms. This is a very active area of research. The impurity screening is difficult to assess. The divertor configuration lends itself to improved impurity screening, as the impurity source can in principle be a considerable distance from the separatrix. However, two-dimensional flow circulation has been shown to occur in computational models which may reduce the effective screening. In the case of the limiter, considering the simple one-dimensional model discussed in Section 16.6.3, it is clear that the impurities will be screened more effectively when ionization occurs further from the LCFS.

16.8.3 *Materials selection*

For a given power flow to the target the plasma edge temperature is reduced as density is increased. Thus the sputter yield must decrease faster, than in proportion to energy in order to reduce the overall impurity influx, $\Gamma = Ync_s$.

We have discussed in Section 16.6.1 how the relative importance of impurities depends on atomic number. However, if the effective yield is high, even low Z impurities can be deleterious. A figure of merit R has been proposed as follows:

$$R = \frac{C}{Y_D(1 - Y_m)}, \tag{16.8.1}$$

where C is the maximum tolerable impurity concentration, and Y_D and Y_m are the hydrogen and self-sputter yields, respectively.

This figure of merit is plotted as a function of plasma temperature for a range of materials in Fig. 16.27. For very low plasma temperatures the ion energy is below the sputter threshold and R increases. At very high temperatures the sputter yield decreases and so R also increases. Because of their relatively high energy threshold for sputtering, the high mass refractory metals are good at low plasma temperature. However, this relatively simple criterion does not take into account the possibility of $T_i \gg T_e$, or of low mass impurities being present and sputtering the high mass impurities. These factors complicate the assessment of optimum conditions for fusion reactor design.

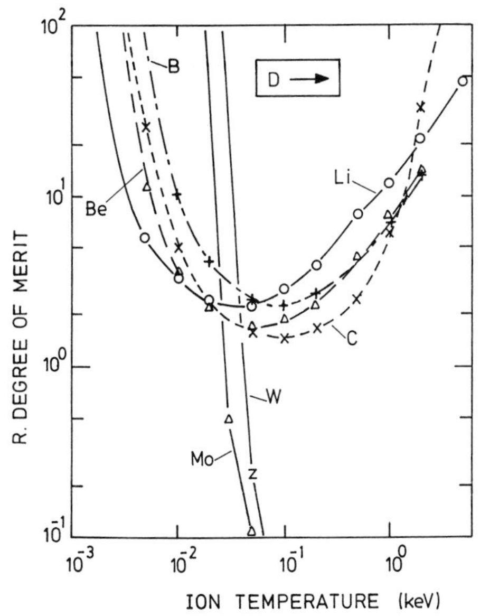

Fig. 16.27. The figure of merit for limiter materials as a function of plasma temperature, calculated from Eq. (16.8.1).

Even if sputtering can be controlled by keeping plasma temperatures low, the most serious problem in large fusion devices is the power density. For a given power output this is controlled by the physical size of the reactor, by the fraction of the fusion power which it is possible to radiate, and by the detailed magnetic design of the divertor itself. Possible ways of overcoming this difficulty are magnetic sweeping of the separatrix across the target and increasing the radiation in the divertor region itself. The ultimate limitation on the power density which can be tolerated is basically determined by materials' properties themselves. While some progress in the development of materials can be expected, this is recognized as difficult, expensive and time consuming. It is unlikely that power levels above 10 MW m^{-2} can be maintained under steady state operating conditions.

A further factor related to materials selection is the erosion of surfaces. Some transfer of the impurities back to the surface is to be expected so only net erosion is relevant. Unfortunately, this too is difficult to assess because of the lack of full understanding of impurity transport. Net erosion will eventually cause rupture of the coolant channels; equally, net deposition will reduce the heat transfer rate.

Bibliography

The most comprehensive and up-to-date reviews of the plasma boundary and of plasma surface interactions are:

BEHRISCH, R. and POST, D. E. (eds) (1985). *Physics of Plasma Wall Interactions in Controlled Fusion.* Proceedings of a NATO Advanced Study Institute. Plenum Press, New York.

JANEV, R. K. (ed.) (1991). *Nucl. Fusion Suppl.*, **1**.

JANEV, R. K. (ed.) (1992). *Nucl. Fusion Suppl.*, **2**.

LANGLEY, R. A. *et al.* (1984). *Nucl. Fusion*, Special Issue.

STANGEBY, P. C. and McCRACKEN, G. M. (1990). *Nucl. Fusion*, **30**, 1225–379.

There has been a series of biennial conferences covering plasma surface interactions and plasma boundary physics starting in 1974. The proceedings of these conferences are published in the *Journal of Nuclear Materials* under the title 'Plasma surface interactions in controlled fusion devices'. The last four conferences are:

- Princeton, 1987: Vols 145–7;
- Aachen, 1989: Vols 162–4;
- Bournemouth, 1990: Vols 176–7;
- Monterey, 1992: to be published.

17

How to build a tokamak

T. N. TODD

17.1 Introduction

The talk which gave rise to this chapter was subtitled 'Engineering issues for physicists'. The purpose of the following pages is to allow a physicist to understand the mix of physics and engineering which has to come together to achieve some desired level of machine performance, or to match some resource investment, if that is a more important constraint. Although the discussion proceeds with the initial assumption that no equipment exists *a priori*, for those with an existing device there are many possible upgrades or conditioning techniques which will be apparent.

The main content is divided into two areas: first generic issues, including scalings for size, power, cost, etc.; and secondly specific issues, such as machine topology, design choices, conditioning processes and control system options. I have tried to ensure that my suggestions cover a wide range of possible resource limitations, albeit with a bias towards the smaller end of this spectrum.

17.2 Project aims and constraints

The level of machine sophistication required varies with the real purpose of the device. At the bottom of the scale might well be the requirement to establish a national plasma physics centre, in which case it is of little consequence whether or not the device is a tokamak, and the size and parameters are essentially irrelevant. The next evolutionary step could be to enhance the training of the nation's plasma physicists. This necessitates a large enough machine to have interesting plasma parameters compared with glow discharges, multipoles and similar low temperature devices, for example a tokamak of $\gtrsim$ 5kA plasma current.

Further along the scale, an aim might be to create national expertise in tokamaks, now requiring something big enough to clearly exhibit normal tokamak behaviour, perhaps $\gtrsim$ 10 kA. Ignoring flagship-class machines, the top of the range will be a device intended to address certain key tokamak problems. This will have to be adequate to achieve credibility in the world community, say $\gtrsim$ 150 kA, probably with some specific design features.

The allowable sophistication is also a function of the desired time scale for design, construction and operational life (possibly including upgrades) and of course the envisaged budgetary constraints (noting that in 'first world' countries we expect to spend $\approx$ \$150k per professional woman-year*).

*Or man-year.

17.3 Machine size

Consideration of the 'real purpose' and broadly estimated resource constraints will have suggested the likely plasma current range for the prospective new machine. This section will relate the current to the other leading parameters and describe some costing algorithms. Iteration will allow a detailed combination of the key parameters of the machine to be determined, consistent with the budgetary constraint in mind (with care to leave a substantial margin for uncertainties!).

Tokamaks nearly all operate with an edge safety factor q (i.e. circumference-normalized field line pitch) ≥ 2. There are various ansatzes for q, such as

$$q = \frac{5a^2B}{RI}\left(\frac{1+\kappa^2}{2}\right)\left(1+\left(1+\frac{\Lambda^2}{2}\right)\varepsilon^2\right)[1.24 - 0.54\kappa + 0.3(\kappa^2+\delta^2) + 0.13\delta], \tag{17.3.1}$$

where a = minor radius (m), B = toroidal field (T), R = major radius (m), I = plasma current (MA), κ = plasma elongation, $\Lambda = \beta_p + l_i/2$ = poloidal beta + 0.5 × internal inductance, $\varepsilon = a/R$, which can be crudely approximated by

$$q \approx 6a^2B\kappa/RI. \tag{17.3.2}$$

It is difficult to make positionally stable plasmas with κ much above two, so this becomes

$$q \approx (6 \to 12)a^2B/RI \text{ or } (6 \to 12)\varepsilon^2RB/I,$$
$$\Rightarrow \varepsilon^2RB \approx qI/(6 \to 12); \tag{17.3.3}$$

therefore

$$\varepsilon^2RB \gtrsim I/(3 \to 6) \tag{17.3.4}$$

for conventional tokamaks. The cost of the device will obviously rise with R and B, but also rises with ε as the central region of the machine becomes an engineering nightmare due to the conflicting demands of the transformer core, toroidal field coils, vacuum vessel wall and inboard limiters or armour-tile.

A set of scalings developed from JET cost analysis and benchmarked against the LHD and SIE large stellarator proposals is as follows:

load assembly: \$17M × $aR(\kappa)^{1/2}$ [m] (17.3.5)

pumps and cooling system: \$16M × $aR(\kappa)^{1/2}$ [m] (17.3.6)

power supplies: \$2.6M × $a^2B^2\kappa$ [m, T] (17.3.7)

auxiliary heating: \$2M × P_{aux} [MW] (17.3.8)

buildings: \$0.2M × number of professional staff (17.3.9)

These can be considered to give a '± 3 dB' cost estimate for any proposed 'first world' device, with obvious caveats regarding existing building availability and any problematical special design features.

The number of professional people-years required to implement the device can be very roughly estimated by dividing the resulting total by \$150k, as indicated earlier.

17.4 Auxiliary heating

Auxiliary heating is applied to tokamaks to alter the intrinsic Ohmic heating behaviour, typically to achieve one or more of the following effects. The most common aim is simply plasma heating, for example to approach the β (i.e. normalized plasma pressure) limit, or to change the collisionality of the electrons or ions. Another popular motivation is to generate bulk current drive, perhaps to extend the plasma pulse duration or to address current drive as an issue. A more precise requirement is MHD instability control (by changing the current or pressure profiles). Each of these will now be addressed in more detail in order to allow the choice of a suitable scheme and to determine the likely power requirements.

Turning first to *bulk plasma heating*, the required power is of course given by the stored energy divided by the energy confinement time, both of which can be estimated. Conventional tokamaks to date generally demonstrate a limit to β (plasma pressure divided by magnetic field pressure) given both experimentally and theoretically by

$$\beta \approx 3.5I/aB[\mathrm{MA, m, T, \%}], \tag{17.4.1}$$

where

$$\beta = <p>_{\mathrm{vol}}/(B^2_{\phi_{\mathrm{vac}}}/2\mu_0)\,[\mathrm{Pa, T, scalar}]. \tag{17.4.2}$$

The coefficient 3.5 in Eq. (17.4.1) lies between the ideal MHD kink mode and ballooning mode limits and is fairly representative of the experimental limit. Tokamaks also exhibit an energy confinement time scaling similar to, for example, ITER-89P:

$$\tau_{\mathrm{E}} = 0.038(M/Z)^{0.5}I^{0.85}R^{1.2}a^{0.3}\bar{n}_{19}^{0.1}B^{0.2}(\kappa/P_{\mathrm{tot}})^{0.5} \tag{17.4.3}$$

[ion mass and charge, MA, m, m, $10^{19}\,\mathrm{m}^{-3}$, T, MW, s].

It can therefore be seen that since the total power required to attain a given β is given by

$$P_{\mathrm{tot}} = \frac{a^2R\kappa B^2\beta}{8.49\tau_{\mathrm{E}}}[\mathrm{m, T, \%, s, MW}], \tag{17.4.4}$$

the required maximum power consistent with Eqs (17.4.1) and (17.4.3) becomes:

$$P_{\mathrm{tot}} \approx \frac{118a^{1.4}\kappa B^{1.6}I^{0.3}Z}{R^{0.4}\bar{n}_{19}^{0.2}M}[\mathrm{m, T, MA, m, 10^{19}\,m^{-3}, MW}] \tag{17.4.5}$$

including an Ohmic heating contribution P_{Oh} (MW) $\lesssim$ 1 volt per turn $\times$ I (MA).

There is little point in having more than (say) twice the power necessary to reach the β limit: either the plasma stored energy saturates or a disruption is induced.

If the scheme chosen is either ion or electron cyclotron resonance heating (ICRH or ECRH), there will be a constraint on B (strictly on $|\boldsymbol{B}|$ at the desired heating location, allowing for plasma paramagnetism and diamagnetism) to match the favoured heating frequency f at a chosen harmonic number l (or vice versa):

$B \approx (M/Z)f/15l$ [ion mass no., ion charge, MHz, harmonic no., T] for ICRH, (17.4.6)

$B \approx f/28l$ [GHz, harmonic no., T] for ECRH. (17.4.7)

In the case of ECRH the plasma density has to be kept low enough for wave access to the resonance from either the high field side (HFS) or low field side (LFS), typically yielding (with margins for additional refractive effects)

$$\bar{n}_{e19} \lesssim 6(f/60)^2 \quad (17.4.8)$$

for X mode fundamental (HFS);

$$\bar{n}_{e19} \lesssim (f/60)^2 \quad (17.4.9)$$

for O mode or

$$\bar{n}_{e19} \lesssim 1.3(f/60)^2 \, [10^{19}\,\mathrm{m}^{-3}, \mathrm{GHz}] \quad (17.4.10)$$

for X mode second harmonic (LFS).

If the scheme chosen is lower-hybrid heating, the access criterion for the lower-hybrid waves to reach the core of the plasma is (approximately)

$$B/(n_{e19})^{1/2} \gtrsim 2N_{\parallel}/(N_{\parallel}^2 - 1) \, [\mathrm{T}, 10^{19}\,\mathrm{m}^{-3}] \quad (17.4.11)$$

where $N_{\parallel} = c/v_{\mathrm{phase}}$ is dictated by the antenna design, but is typically ≈ 3.

If neutral beam heating is chosen then the plasma line averaged density has to be large enough to stop the majority of the beam, i.e.

$$\text{beam path length} \times \bar{n}_{e19} \approx 22(E_{\mathrm{beam}}/M_{\mathrm{beam}})^{2/3}(2.3 - \ln(\mathscr{F}/0.1)) \quad (17.4.12)$$
$$[\mathrm{m}, 10^{19}\,\mathrm{m}^{-3}, \mathrm{MeV}, \mathrm{scalar}]$$

(where M_{beam} is the mass of the beam particles in units of the proton mass, $\mathscr{F}$ is the shine-through fraction, and I have roughly summed and fitted the cross-sections for charge exchange and ionization due to electrons and ions). In addition, the plasma current has to be large enough to confine the resulting fast ions, i.e. to keep the full banana orbit widths at some prescribed small fraction of the minor radius:

$$I \geq 2.0(\varepsilon M_{\mathrm{beam}} E_{\mathrm{beam}})^{1/2}/(Z(\Delta_{\mathrm{banana}}/a)) \quad (17.4.13)$$
$$[\text{beam ion mass, MeV, beam ion charge, MA}].$$

Hence with $\Delta_{\mathrm{banana}}/a \leq 3$, where ε is the local inverse aspect ratio $\approx 0.5a/R$ say,

$$I \gtrsim 6(\varepsilon M_{\mathrm{beam}} E_{\mathrm{beam}})^{1/2}/Z. \quad (17.4.14)$$

It should also be borne in mind that tokamaks so far all lie in the Hugill diagram, i.e. they exhibit (more or less) a minimum q of 2.0 (occasionally 1.0) and density limits given by

$$\underset{\text{disruptions}}{1.5 \times 10^{-14}} \lesssim \frac{I}{\pi a^2 \kappa \bar{n}} \lesssim \underset{\text{runaway electrons}}{1.5 \times 10^{-13}} \, [\mathrm{A}, \mathrm{m}, \mathrm{m}^{-3}]. \quad (17.4.15)$$

One therefore has to juggle the required operating parameters of the machine to ensure that all the appropriate conditions are satisfied simultaneously; it can be quite hard to find consistent parameter sets for neutral beam heating in a small machine or density limit studies with second harmonic ECRH, for example.

Another important effect that can be produced by auxiliary heating is *bulk plasma current drive*, in principle possible with all the types of heating scheme considered above but most reliably demonstrated with lower-hybrid waves and neutral beams. Consideration of electron scattering processes gives rise to theoretical current drive scaling of the form

$$I = k \frac{PT_e}{Rn \ln\Lambda}, \quad (17.4.16)$$

with $k \propto 1/N_{\parallel}^2 T_e$ for directed lower-hybrid waves with an adequate $N_{\parallel}$ to gain access, and $k \approx$ constant for neutral beams allowed to reach the plasma core (with beam energy $\approx 40T_e$). Here $\ln\Lambda$ is the Coulomb logarithm, which typically has a value close to 16. Experimentally the power required to drive a given plasma current is

$$P_{\mathrm{LHCD}} \approx (0.3 \rightarrow 2) I R n_{19} [\mathrm{MA,\ m,\ 10^{19}\,m^{-3},\ MW}], \tag{17.4.17}$$

and (for $Z_{\mathrm{eff}}/Z_{\mathrm{beam}} \approx 2$ and typical ε)

$$P_{\mathrm{NBCD}} \approx 5 I R n_{19}/T_e \cos\phi_{\mathrm{beam}} [\mathrm{MA,\ m,\ 10^{19}\,m^{-3},\ keV,\ MW}], \tag{17.4.18}$$

where ϕ_{beam} is the angle between the beam and the magnetic field.

Finally, some measure of *MHD instability control* might be required of the heating system. Since this should only be an exercise in perturbing the plasma pressure or current profile by localized heating or current drive, the necessary power can be estimated as a small fraction (10 or 20% say) of that required to perform the gross heating or current drive, deduced using the scalings and constraints given above (and noting that the values of magnetic field strength and plasma parameters used should refer to the point of interest). In the special case of stabilizing a low-m mode with current drive localized to the region of the mode resonance and synchronized to the mode phase, a still smaller fraction should in principle suffice, since $\lesssim 1\%$ of the total plasma current needs to be so driven.

Consideration of the effects required of the auxiliary heating system and the overall consistency of the machine parameters will lead, via the equations above (or their more accurate variants), to the selection of one or more schemes and the necessary power in each one. No auxiliary heating system is cheap, with the possible exceptions of Alfvén wave heating (simple remote antennae, ≈ 20 MHz operation), which has significant problems in implementation and interpretation, and of small scale ECRH installations using microwave cooker magnetrons ($\lesssim 20$ kW, 2.45 GHz). Apart from these, construction costs can be expected to fall in the general area of \$(1 → 4)/watt-at-the-plasma. In addition, the size of the team required to run the auxiliary heating system can be substantial, and such a team is usually dedicated to this function, representing a considerable resource commitment. The physical size of the associated plant should also be considered, as well as the instantaneous power demand (typically a factor $\approx 2 \rightarrow 5$ greater than the power-at-the-plasma).

The sections of this chapter so far have all been aimed at finding a machine size, plasma current, heating power and so on consistent with the broad objectives and overall budget available. The remaining sections concentrate on design options for the key components of the load assembly and power supplies, permitting the physicists and engineers to converge upon a workable design from often rather different concepts (Fig. 17.1).

17.5 Topology

An early design option, generally with far-reaching consequences, is to choose which coils are nearest to the plasma: those producing the toroidal or the poloidal field. Table 17.1 lists the relative advantages of each option when fully pursued. There are of course machines with poloidal field coils both within and external to the toroidal field coil 'solenoid', typically featuring fast position control coils close to the plasma and (optionally iron-cored) solenoid windings and main equilibrium field coils placed remotely.

Fig. 17.1. 'The ideal tokamak' as perceived by a physicist and an engineer.

Table 17.1.

Coils nearest plasma	Advantages	Disadvantages
Toroidal field	Smallest possible stored magnetic energy No interlinking of coils	Many coils needed to avoid severe ripple Restricted OH solenoid diameter if air-cored Difficult to get strong plasma shaping
Poloidal field	Easy to shape plasma Possible gain in plasma vertical stability Largest possible (air-cored) OH solenoid diameter Can use fewer TF coils Good access for (small) diagnostics	Interlinked coils, therefore joints somewhere Larger TF stored energy

The choice of basic topology leads to a variety of design solutions, some of which are shown in Figs 17.2–17.5. Although in principle adopting the solution with the toroidal field coils closest to the plasma would allow all the windings to be constructed and assembled as complete unjointed rings (which will be a virtual necessity for superconducting coils), sometimes other considerations lead to jointing the poloidal field coils as well. In the case of TEXTOR, the requirement for ready access to the plasma facing components inside the vacuum vessel led to the vertical-plane split shown in Fig. 17.2. The JET assembly (Fig. 17.3) was intended

Fig. 17.2. TEXTOR, showing the vertical split in the vessel and poloidal field coils. By kind permission of KFA Jülich.

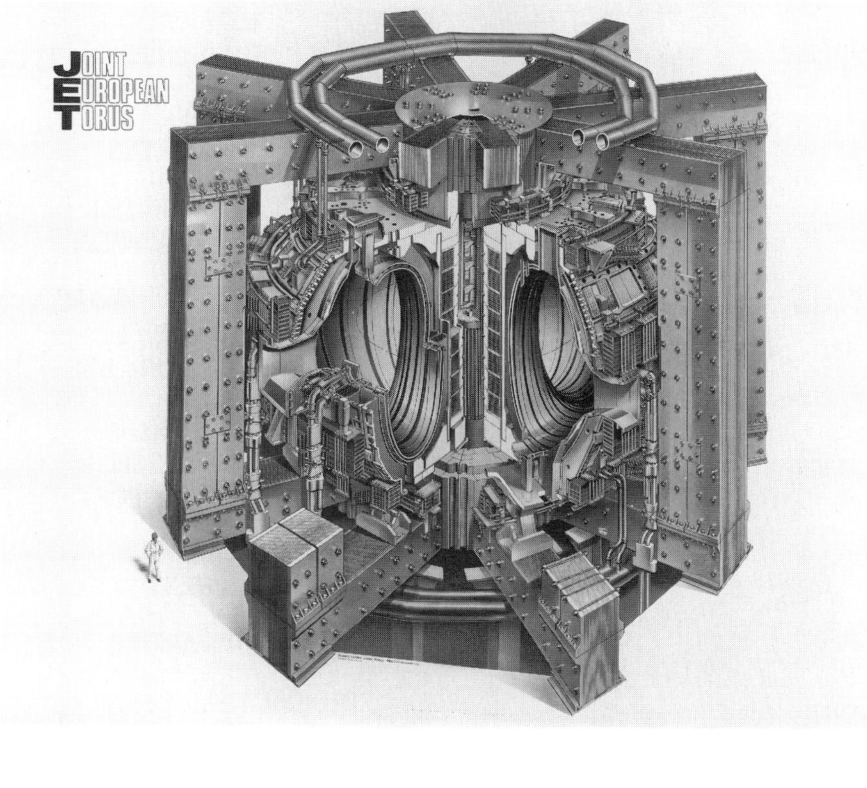

Fig. 17.3. JET, showing the unlinked poloidal and toroidal field coils. By kind permission of JET Joint Undertaking.

to be compatible with remote handling of octants of the vessel and the associated toroidal field coil subassembly, leading to several joints being provided in each outer poloidal field coil (which also facilitated the road transport of these larger coils to the JET site).

DIII-D (Fig. 17.4) and COMPASS are fully representative of the conventional air-cored class of tokamak, with the poloidal field coils all interlinking the toroidal field coil set. START, although similar in concept, is a design extremum with no solenoid and a 'single turn' toroidal field coil arrangement (Fig. 17.5) driven by the pursuit of minimum possible plasma aspect ratio.

17.6 Toroidal field coils

The toroidal field (TF) coil system has to be strong enough to tolerate the forces imposed on it, which are primarily the self- (i.e. bursting) force,

$$F_{\text{self}} \approx 2.5 B_{\phi 0}^2 R_0^2 / NR \text{ [T, m, m, Nm}^{-1}] \quad (17.6.1)$$

in each of the N limbs, and the toppling force due to the interaction with the poloidal fields. Often this is worst immediately following a disruption, due to the vertical field (B_v) in the absence of toroidal current in the plasma (and vessel):

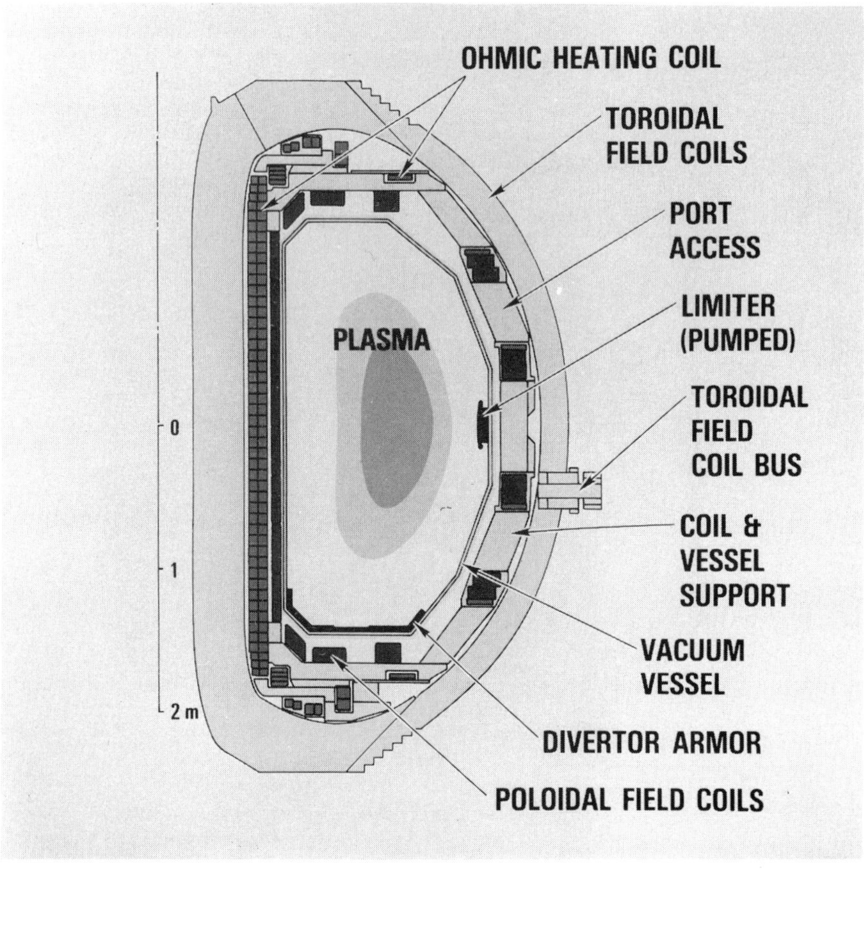

Fig. 17.4. DIII-D, showing the jointed toroidal field coil and interlinked poloidal field coils. By kind permission of General Atomics, funded by the US Department of Energy.

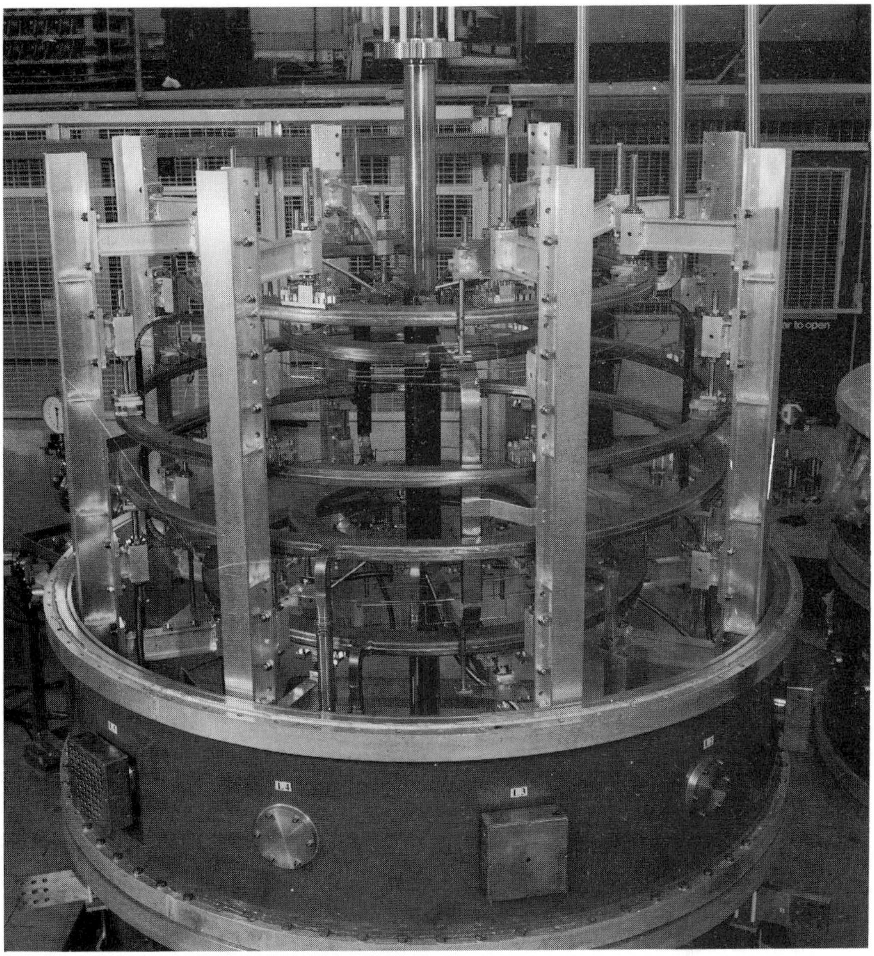

Fig. 17.5. START, showing the compact single turn toroidal field coil and jacketed poloidal field coils inside the vacuum vessel.

$$F_{\text{toppling}} = \frac{5B_0R_0}{N} \times B_v\,[\text{T, m, T, Nm}^{-1}]. \qquad (17.6.2)$$

$$\approx \frac{B_0I}{2N}\ln(6R_0/a)\,[\text{T, MA, Nm}^{-1}]. \qquad (17.6.3)$$

The integral of these forces around the coil gives rise to tensile and bending stresses (which can be reduced to pure tension for the self-force on an appropriately D-shaped coil). Finite element modelling to determine the stress distribution will generally be necessary for all but the simplest coil geometries, with the aim of obtaining an adequate cycle lifetime (say ten times the number of machine shots envisaged). This typically results in keeping below $\approx$ 30 MPa for conventional copper (generally known as oxygen-free, high conductivity or OFHC) (Fig. 17.6), ranging up to $\approx$ 200 MPa for special copper alloys.

However, heating in the coils is often more of a problem than the mechanical stresses imposed, and frequently a compromise between strength and conductivity is required, since alloying the copper to produce a high strength material

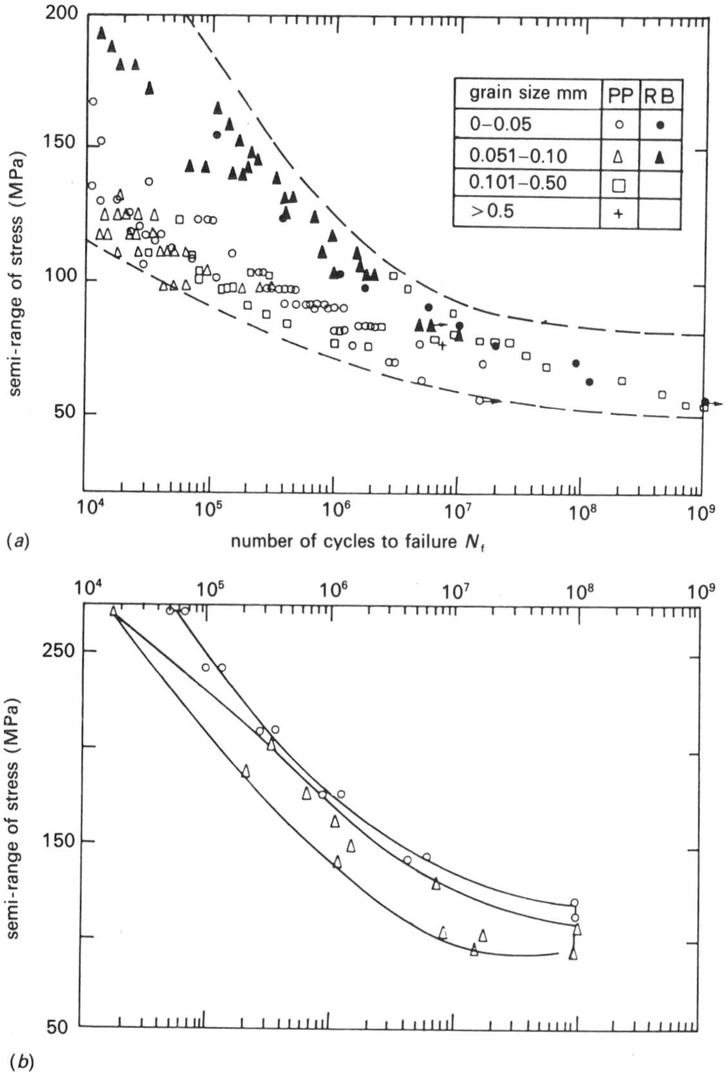

Fig. 17.6. Fatigue curves of copper. (*a*) Effect of grain size for oxygen-free and tough pitch annealed copper. (PP = push–pull test; RB = reverse bend test.) (*b*) Effect of type of copper (70% cold worked). ○ Oxygen-free copper; △ tough pitch copper. After Murphy (1981).

increases its resistivity (see Fig. 17.7). Interestingly, there is a school of thought which asserts that the state of work-hardening of the delivered material may not be very critical. This is because long periods of vibration and heating during operation have an annealing effect while cyclic stressing generates work hardening itself, so asymptotically some more or less 'half-hard' condition (or at least an equilibrium independent of the starting condition) is achieved.

Two disadvantages accrue if a high resistivity conductor is chosen: first the extra power and energy demanded to provide the specified toroidal field, and secondly the various thermomechanical and thermohydraulic problems caused by the temperature rise of the coils and their insulation.

The Ohmic heating rate in a conductor is given by

$$\dot{\theta} = \eta J^2/\rho S, \tag{17.6.4}$$

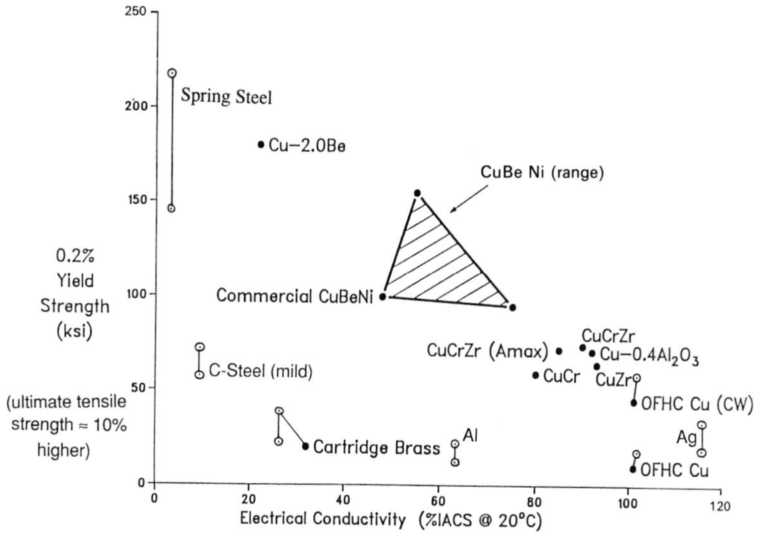

Fig. 17.7. Properties of conducting materials; strength versus conductivity. Inconel has a yield strength of ≈75 ksi and conductivity of 1.3% IACS. After Kaye and Laby (1973).

which for warm copper becomes

$$\dot{\theta} \approx 0.6J^2 \, [\mathrm{kA\,cm^{-2}, K\,s^{-1}}]. \tag{17.6.5}$$

A survey of typical designs shows that when $J \lesssim 2 \to 3\,\mathrm{kA\,cm^{-2}}$ it is possible to run copper coils in steady state using water cooling, while otherwise intershot cooling is usually required. If the vault of inner TFC limbs has a copper fraction f_c, and inner and outer radii r_1 and r_2, respectively, then

$$J = \frac{0.5B_0R_0}{f_c\pi(r_2^2 - r_1^2)} \, [\mathrm{T, m, m, kA\,cm^{-2}}]. \tag{17.6.6}$$

Insulation of the epoxy-resin–glass type is often used, imposing an upper temperature limit $\lesssim 100°\mathrm{C}$, i.e. a rise during the shot of $\lesssim 80°\mathrm{C}$. In this case $\int J^2 dt$ is limited to $133\,\mathrm{kA^2\,cm^{-4}\,s}$, i.e.

$$(r_2^2 - r_1^2) \geq \frac{0.5B_0R_0(T_{\mathrm{pulse}})^{1/2}}{f_c\pi(133)^{1/2}} \, [\mathrm{T, m, m^2}], \tag{17.6.7}$$

hence

$$(r_2^2 - r_1^2) \geq \frac{B_0R_0(T_{\mathrm{pulse}})^{1/2}}{73f_c} \, [\mathrm{T, m, s, m^2}]. \tag{17.6.8}$$

This determines the necessary extent of the toroidal field coil vault in major radius under these temperature rise constraints (with a margin of safety given by the thermal capacity of the insulation and the water in the cooling channels, both neglected here). One reason for making the START toroidal field coil a single turn structure was to avoid the reduction in f_c incurred by inter-turn insulation. Multiple parallel return paths are provided via the outer limbs of the coil set, as shown in Fig. 17.8.

An alternative, albeit technologically demanding, approach is to use superconducting coils. The advantage of a modest power supply requirement (stored

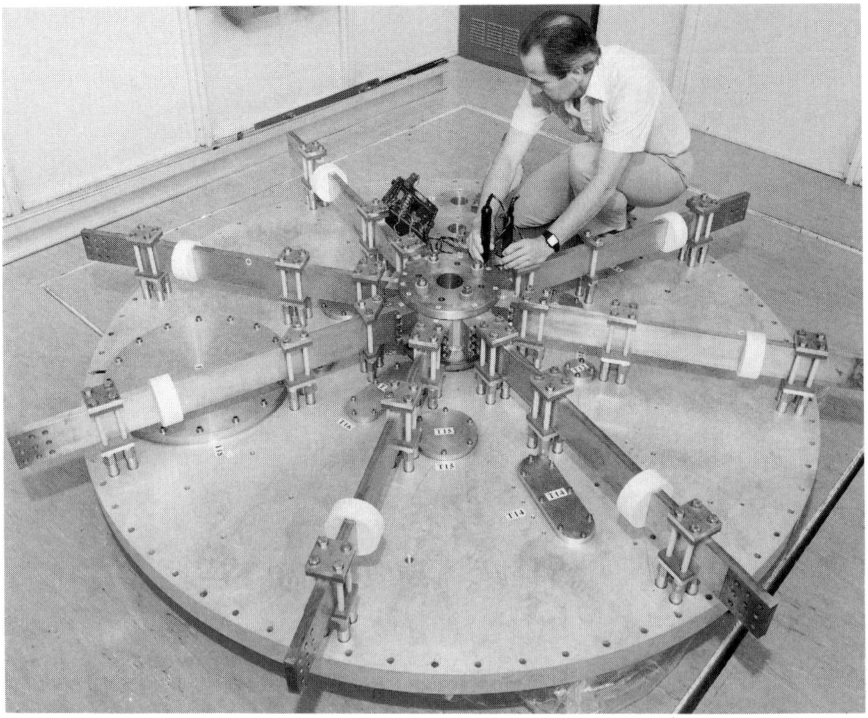

Fig. 17.8. The electrically parallel toroidal field coil return limbs of START.

energy ÷ desired rise time) is offset by the need for liquid He and liquid N_2 refrigeration plant and its power consumption (10–20 MW for a big machine), and the space demands in the overall conductor pack for the cryostats and thermal insulation.

Although the critical current density at some transverse magnetic field in the superconducting material itself can be very large, the requirement for quench stabilization adds a great deal of copper (or aluminium) and liquid helium to the conductor cross-section, so that critical current densities averaged over the whole conductor are typically only a few kiloamps per centimetre squared at 10 T. A recent NbTi example (for operation at a peak field of 6.1 T) tolerates 122.5 kA/cm^{-2} in the NbTi alone, corresponding to 21.0 kA cm^{-2} in the cable but only 6.8 kA cm^{-2} in the overall conductor.

Rewriting Eq. (17.6.6) to represent the required radial build of the superconductor given some critical current density results in

$$(r_2^2 - r_1^2) \geq \frac{0.5 B_0 R_0}{f_c \pi J_{\text{crit}}}, \qquad (17.6.9)$$

noting that J_{crit} is a function of the peak magnetic field seen by the conductor, which is $\gtrsim B_0 R_0/(R_0 - a)$, and of the operating temperatures, as shown in Fig. 17.9.

Layering the winding one way or another to produce lower current density in the high field region can help significantly, but is costly due to the increased complexity of coil manufacture or number of power supplies. The detail of preparing an actual design for a superconducting coil solution is beyond the scope of this work, necessitating analysis of all sources of eddy current heating,

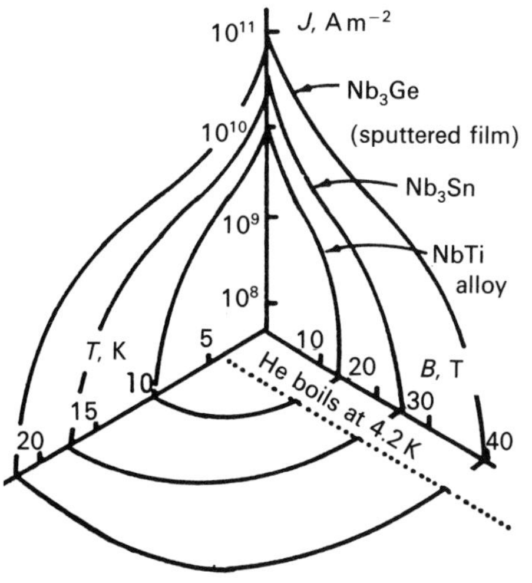

Fig. 17.9. Properties of common superconductors; critical current density versus imposed magnetic field strength and temperature. After Dolan (1982).

neutron heating etc., as well as the more obvious thermomechanical problems.

The power demanded by a coil system comprises the reactive and resistive elements and usually peaks at the end of the ramp-up phase, when both terms are large. In the case of a toroidal field set, it is usually dominated by the reactive power producing the stored magnetic energy in the required rise time:

$$P_{\text{reactive}} = \frac{2.5 B_0^2 R_0^2 h \ln(R_2/R_1)}{T_{\text{rise}}} [\text{T, m, m, s, MW}], \qquad (17.6.10)$$

where h, R_2, and R_1 are, respectively, the height and the major radii of the outer and inner toroidal field coil limbs (assuming rectangular coils). An upper limit to the resistive power can be estimated by assuming a constant conductor cross-section around a rectangular coil, here assuming the resistivity of warm copper:

$$P_{\text{resistive}} \approx \frac{0.3 B_0^2 R_0^2 (h + R_2 - R_1)}{f_c (r_2^2 - r_1^2)} [\text{T, m, m, m, MW}], \qquad (17.6.11)$$

with parameters as defined above. Clearly the resistive power is proportional to the resistivity of the conductor, which is a function of the material selected and the operating temperature, as exemplified in Fig. 17.10. This leads some designers to opt for liquid nitrogen cooled coils, particularly if they are required to make use of modest power supplies already available to them. Where the power supply is not 'given', a type has to be chosen which is capable of meeting the pulse (or steady state) specification. The advantages and disadvantages of the most common types available are shown in Table 17.2. Homopolar generators are not included, since their very low voltage, high current output is not readily accommodated in the overall design of a small tokamak.

17.7 Stray fields

An important issue which has sometimes been overlooked by tokamak designers is the minimization of stray fields created by various parts of the machine and its ancillaries. The stray fields fall into two categories:

Table 17.2. *Power supply options*

Type	Advantages	Disadvantages
Capacitor banks	Very low mains power demand; simple and very cheap	Very poor control of current waveform
Inductive storage	Cheaper than capacitor bank systems at very large energies	Switching problems
Flywheel motor generators	Low (pony motor) mains demand; optional stator field control; very large powers and energies readily available	Output frequency falls during pulse
Steady state motor generators	Some flywheel effect to accommodate transients; clean DC output with optional feedback control by stator field	Severe mains demand; feedback response time can be slow
Thyristor rectifier	Fairly fast if multiphase (e.g. 12 or 24 phases); natural inversion available	Ripple and noise in load current; worst possible option for mains demand; large signal BW poor, particularly for turn-off; four-quadrant versions 'messy'
Thyristor chopper	Very fast; can run from capacitor bank(s); readily made multiple-quadrant	Ripple and noise in load current (and/or complexity if multirail)
Transistor amplifier	Linear, clean output, very fast four-quadrant feedback control	Limited to a few hundred kilowatts per unit; expensive

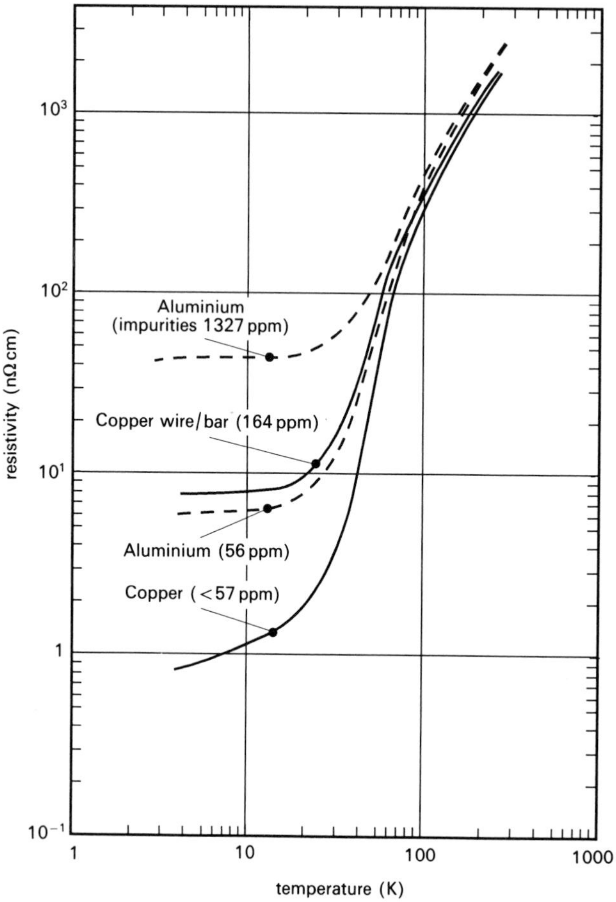

Fig. 17.10. Properties of conducting materials: resistivity of copper and aluminium versus temperature.

(i) average perpendicular fields, which inhibit plasma breakdown and may cause plasma motion during the plasma discharge;

(ii) resonant magnetic perturbations which tend to create magnetic islands in the plasma region, possibly affecting plasma confinement and stability (for example via 'mode lock' phenomena).

The perpendicular fields usually originate from systematic (or net) tilt errors in the placement of the coils, or from the interconnection scheme if this is not well thought out. The toroidal field coil interconnections of COMPASS are so arranged that viewed vertically or horizontally the dipole loop formed by the interconnecting and return bars cancels that formed by the 'joggle' into each rectangular coil and its return bar. Conventionally one tries to achieve $B_\perp/B_\phi \ll 10^{-3}$, which directly maps to a constraint on the permissible construction error tilt angles in radians. In COMPASS the coils have been aligned during assembly to ≈ 0.4 mm compared to a coil size $\approx 1.5 \times 1.0\,\mathrm{m}^2$, and the feeder arrangement provides an axisymmetric field $\lessapprox 0.1$ mT at an operating field strength of 1.75 T.

Resonant magnetic perturbations primarily originate from random differences in placement from coil to coil, and from nonaxisymmetric stray fields generated by the feeder bars (especially if the number of turns in the toroidal field coil set is small, so that the feeder bars carry a large current and are physically massive). If a resonant magnetic perturbation penetrates the plasma it will form a magnetic island at the relevant resonance (where $q \equiv m/n$). The island width formula is

$$\frac{W}{a} = \left(\frac{16}{n}\frac{\tilde{b}_r}{B_0}\frac{R}{a}\frac{r_q}{a}\right)^{1/2}, \tag{17.7.1}$$

where W is the full width, n is the toroidal mode number, $\tilde{b}_r$ is the perturbation field, B_0 is the toroidal field, and $r_q = q/\nabla q$. It follows that to achieve a maximum island width of $W/a \leq 0.1$ (ignoring any possible plasma amplification or attentuation of the perturbation) typically requires

$$\frac{\tilde{b}_r}{B} \leq 0.1^2 \times \frac{1}{16} \times \frac{1}{3} \times 1 \approx 2 \times 10^{-4}. \tag{17.7.2}$$

This is quite demanding (unless mode-locking and island formation are inhibited by rapid plasma rotation, as is usually the case in small machines) but can be achieved by careful optomechanical alignment of the coils during assembly, or magnetically sensing the low n asymmetries of the coils and arranging some corrective mechanism.

17.8 Poloidal field coils

The functions of the poloidal field (PF) coils are to produce the magnetizing flux, main equilibrium field, shaping fields and fast position feedback. A major design option, requiring careful consideration, is the inclusion of an iron core, which makes everything about the magnetizing system much simpler and everything about the stray fields and equilibrium modelling much more difficult.

Some or all of the poloidal field coil functions can be combined using appropriate feedback control techniques, generally resulting in simpler mechanical engineering and more complicated power supplies and feedback systems. Some significant operational advantages (and engineering disadvantages) accrue if a poloidal divertor is included in the design to improve impurity control (as originally intended) or energy confinement times (as serendipitously discovered). Plasma shaping fields (including those for a divertor configuration) are of the same magnitude as, indeed are part of, the main equilibrium fields and have comparable power requirements and stresses. Hence, in the analyses expanded below, the only 'worked example' of the equilibrium coils is the main vertical field set.

Truly circular and aligned poloidal field coils do not 'feel' the main toroidal field, only its small ripple. The principal forces are therefore self- (hoop) and vertical and radial forces arising from other PF coils and the plasma current. Often the most highly stressed PF coil is the Ohmic heating (OH) solenoid, because experimentalists always seek the largest possible volt-seconds swing in the pursuit of the largest possible pulse duration or plasma current. The rest of this section outlines the electromechanical design requirements for the magnetizing and vertical field windings, with attention to their functions, power demands, stresses and alignment requirements.

The magnetizing winding provides the flux swing necessary to produce and sustain the plasma current for the desired pulse duration. The volt-second consumption seems to be reasonably well approximated for small machines by

$$\Delta\Phi \approx 1.5IL + t_{\text{pulse}}V_{\text{loop}}\,[\text{MA}, \mu\text{H}, \text{s}, \text{Wb}], \tag{17.8.1}$$

where the 1.5 is an empirical factor accounting for all loss processes, or for large machines

$$\Delta\phi \approx 2 + IL + t_{\text{pulse}}V_{\text{loop}}, \tag{17.8.2}$$

where the 2 is now the empirical term.

The resistive loop voltage of tokamaks is remarkably invariant, being ≈ 2 volts per turn for pure Ohmic heating in a moderately 'dirty' device (i.e. one with a high Z_{eff} resulting from a large impurity fraction) and ≈ 1 volt per turn in a well conditioned machine with a clean plasma. Auxiliary heating (raising the electron temperature), current drive, or the presence of a significant runaway electron population can greatly reduce the loop voltage. In the equations above, L is the total inductance of the plasma loop, comprising the inductance internal and external to the plasma cross-section:

$$L = L_{\text{int}} + L_{\text{ext}}, \tag{17.8.3}$$

which can be approximated by

$$L \approx \mu_0 R(l_i/2 + \ln(1.3R/a\kappa^{1/2}))\,[\text{m}, \text{H}], \tag{17.8.4}$$

where l_i is the normalized internal inductance (typically ≈ 0.8–1.6). Alternatively, very crudely,

$$L \approx 2R\,[\text{m}, \mu\text{H}]. \tag{17.8.5}$$

The required flux swing can be produced with an air-cored or iron-cored transformer design. The iron-core option makes the transformer very simple to design (if strong saturation is avoided), since the net ampere-turns required for magnetizing the iron are very small and the iron core produces good coupling between the primary and the plasma. The total primary ampere-turns in this case are thus nearly equal to the plasma current, allowing a 'constant current' (i.e.

high impedance) primary circuit to more or less dictate the plasma current. Another advantage is that the primary windings can be placed almost anywhere – an elegant choice is a Helmholtz pair larger than the plasma radius, such that the stray vertical field produced is approximately that required by the plasma (having allowed for the inward attraction of the iron). Since soft iron saturates at around 2 T, the required cross-sectional area of the core is simply $\approx \Delta\Phi/4\,(\mathrm{m}^2)$, assuming a bidirectional flux swing.

The major disadvantages of an iron core are the toroidal asymmetries it introduces, including resonant magnetic perturbations, and the loss of equilibrium when it saturates and the stray fields sharply increase. These effects are overcome by moving to an air-cored system, where the major disadvantages are (generally) extremely poor coupling, leading to large primary ampere-turns, and strong constraints on the primary winding distribution to avoid the generation of stray fields in the plasma. An example of the magnetizing winding distribution in a typical air-cored tokamak is shown in Fig. 17.11 – the aim is to produce a discretized approximation to the toroidal equivalent of a skin current, which produces no internal net field, that is to say no net field in the plasma region.

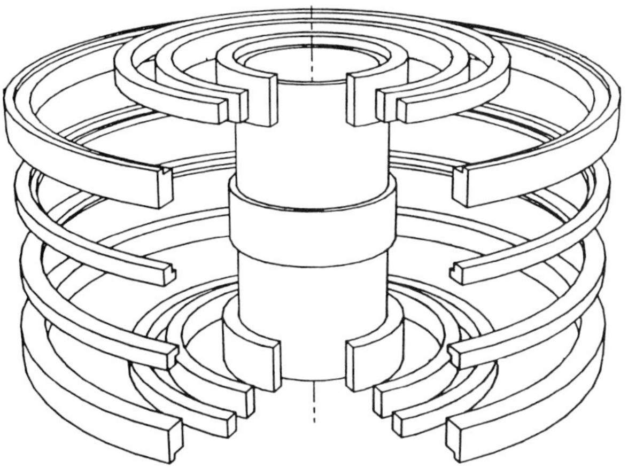

Fig. 17.11. Typical magnetizing winding arrangement in an air-cored tokamak.

The volt-seconds produced by a simple long solenoid are (for a bidirectional swing B_{max}, radius r_{sol}, thickness δr_{sol}, packing fraction f_{sol}) given by

$$\Delta\Phi = 2B_{max}\pi r_{sol}^2\ [\mathrm{T,\ m,\ Wb}], \tag{17.8.6}$$

or

$$\Delta\Phi = 8\pi^2 J r_{sol}^2 \delta r_{sol} f_{sol}\ [\mathrm{kA\,cm^{-2},\ m,\ m,\ Wb}]. \tag{17.8.7}$$

The average hoop stress in such a winding is

$$\sigma = 20\pi J^2 r_{sol} \delta r_{sol} f_{sol},\ [\mathrm{kA\,cm^{-2},\ m,\ m,\ MPa}], \tag{17.8.8}$$

which should be kept below the fatigue–failure limit for the envisaged life (σ_{max} $\approx$ 30 MPa for OFHC Cu, 200 MPa for special alloys). Thus the stress-limited flux swing is

$$\Delta\Phi(\sigma) \approx 10(r_{sol}^3 \delta r_{sol} f_{sol} \sigma_{max})^{1/2}\ [\mathrm{m,\ m,\ MPa,\ Wb}]. \tag{17.8.9}$$

The heating rate is as given above for the toroidal field coils, so that for an epoxy-resin insulated OFHC copper solenoid, $\int J^2 dt \lesssim 133\,\mathrm{kA^2\,cm^{-4}\,s}$. For a triangular current waveform $\int J^2 dt = J_{max}^2 T_{pulse}/3$, hence

$$J_{max}(\theta) \lesssim 20(t_{pulse})^{1/2}\,[\mathrm{kA\,cm^{-2}, s}], \tag{17.8.10}$$

and accordingly

$$\Delta\Phi(\theta) \approx 1600 r_{sol}^2 \delta r_{sol} f_{sol}/(t_{pulse})^{1/2}\,[\mathrm{m, m, s, Wb}].$$

Thus one has to juggle with the r_{sol} and δr_{sol} to obtain the desired volt-seconds swing without breaking or overheating the magnetizing solenoid. This is particularly difficult in tight aspect ratio devices.

The vertical field requirement of a tokamak is approximated by

$$B_v \approx \frac{I}{10R_0}\{\ln(8R_0/a) + \beta_p + l_i/2 - 3/2\}, \tag{17.8.11}$$

where I is the plasma current, so that

$$B_v \approx \frac{I}{10R_0}\ln(6R_0/a)\,[\mathrm{MA, m, T}]. \tag{17.8.12}$$

The vertical field produced by a Helmholtz coil pair is

$$B_v \approx I_{coil}/1.1R_{coil}\,[\mathrm{MA\,turns, m, T}]. \tag{17.8.13}$$

Thus,

$$I_{coil} \approx 0.11\frac{R_{coil}}{R_0} I \ln(6R_0/a)\,[\mathrm{m, MA, MA\,turns}] \tag{17.8.14}$$

for this coil arrangement, and the same caveats on J_{max}, $\int J^2 dt$, etc. apply. The minimum cross-sectional area for this coil (if constructed with epoxy-resin insulated copper) is accordingly

$$A_{coil} \gtrsim 10(t_{pulse})^{1/2} R_{coil} I \ln(6R_0/a)/R_0 f_{coil}\,[\mathrm{s, m, MA, m, cm^2}]. \tag{17.8.15}$$

In the early days of tokamaks, the equilibrium vertical field requirement was simply lumped into 'shell stabilization' of all positional and kink instabilities, partly as a result of experimental antecedents with no toroidal field which were strongly kink unstable. In these systems any small motion of the plasma (or some element of it) created image currents in a closely surrounding conductive shell, generating a restoring force. Clearly the equilibrium plasma position was not necessarily centred in the shell, and would drift outwards to compensate the resistive decay of the image currents, so that the maximum plasma aperture could not be sustained for the full shot duration. Other disadvantages were nonaxisymmetries created by toroidal gaps (necessary to allow the plasma current induction) and diagnostic ports, potentially creating resonant magnetic perturbations. This technique dropped out of favour as tokamak pulse lengths increased and kink stability in normal operation was reliably demonstrated without any close conducting shell.

In an unsaturated iron-cored system the primary voltage is essentially given by $V_{pri} = N_t V_{loop} + I_{pri} R_{pri}$ and the primary current is only marginally higher than the approximation $I_{pri} = I/N_t$, where N_t is the number of primary turns. At all times (while the core is not saturated) the power demand $I_{pri}V_{pri}$ is therefore usually dominated by IV_{loop}, with V_{loop} given by

$$V_{loop} = V_{resistive} + (I\dot{L}) \qquad (17.8.16)$$

i.e.

$$V_{loop} \approx (1 \to 2) + 2\dot{I}R_0\,[\text{MA, m, V}]. \qquad (17.8.17)$$

The power demand for the vertical field coils is of course part of the (usually negligible) $I^2_{pri}R_{pri}$ term if these coils are also the primary winding as suggested above.

In an air-cored system the power supply has to drive a large current swing in the magnetizing winding, which typically has an inductance $\approx$ 50% greater than the central solenoid alone. Thus

$$L_{tot} \approx 1.5 \times 3.9N^2_{sol}r^2_{sol}l_{sol}\,[\text{m, m, }\mu\text{H}], \qquad (17.8.18)$$

where N_{sol} and l_{sol} are the number of turns and length of the solenoid, respectively. The reactive voltage is dominated by $\dot{I}_{pri}L_{tot}$, since coupling to the plasma is usually very poor.

Resistive power (allowing an additional 25% for the magnetizing coils other than the solenoid) is

$$P_{resistive(sol)} \approx 5\pi r_{sol}f_{sol}\delta r_{sol}l_{sol}J^2\,[\text{m, m, m, kA cm}^{-2}\text{, MW}]. \qquad (17.8.19)$$

The vertical field coils are usually operated with a flat-top pulse tracking the plasma current, with resistive power (for a Helmholtz pair) given by

$$P_{resistive}(B_v) \approx J^2f_{coil}A_{coil}R_{coil}/400\,[\text{kA cm}^{-2}\text{, cm}^2\text{, m, MW}], \qquad (17.8.20)$$

with some addition for reactive power during the ramp-up;

$$P_{reactive}(B_v) \approx 3I_{coil}\dot{I}_{coil}R_{coil}\ln\left(\frac{1.1R_{coil}}{a_{coil}}\right) \qquad (17.8.21)$$

[MA turns, MA turns per second, m, MVA],

where I_{coil} is the total ampere-turns in one of the pairs.

Feedback systems are usually relatively low power but fast, typically based on thyristor choppers or linear amplifiers of 10–500 kW (Fig. 17.12). It is sometimes necessary to incorporate a decoupling transformer in the feedback amplifier circuit to cancel large voltages imposed upon it by mutual inductance with other poloidal field systems in the load assembly.

The PF coils have to be circular and well aligned to the TFC to avoid producing resonant magnetic perturbations and possible islands – see for example Fig. 17.13. The nonaxisymmetric alignment errors of the PF coils can be measured (and optionally corrected) using an accurate array of matched pick-up coils connected in opposite pairs, as implemented in DIII-D and COMPASS (Fig. 17.14). The PF coils also have to be positioned in radius and height so as to minimize stray perpendicular fields (particularly important for the magnetizing winding since it may be the only poloidal field coil set carrying current at the time of plasma initiation). The required positional tolerance is $\approx 10^{-3}$ of the major radius of the machine for each type of error. Coils with small numbers of turns are

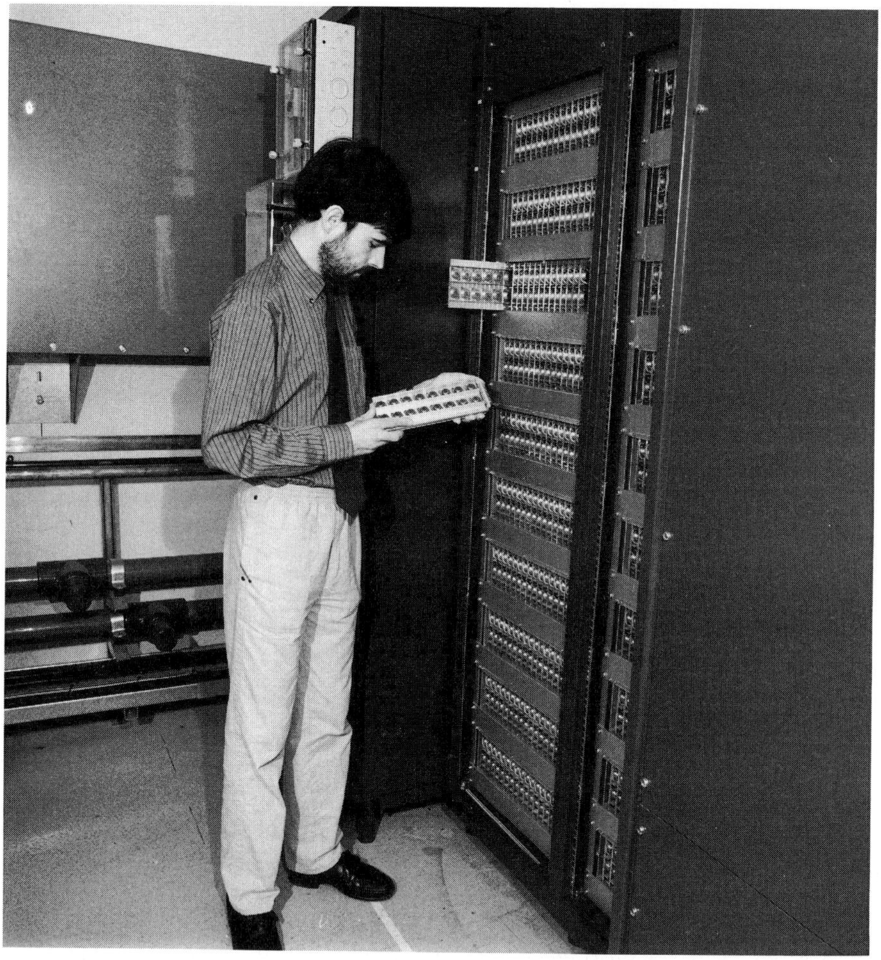

Fig. 17.12. Representative linear transistor amplifier: many thousands of bipolar transistors in parallel giving a specification of $\pm 60\,V$, $\pm 5\,kA$, DC – 20 kHz.

troublesome because of the effective dipole errors associated with the feeder bars (unless the feeders are coaxial, quadrupole or a higher multipolarity), and the turn-to-turn joggles of the winding.

17.9 Support structure

Some structure has to accommodate the toppling forces on the TF coils and the vertical forces on the PF coils. The support structure is responsible for maintaining the alignment of all the TF and PF coils, with the accuracy requirements discussed earlier. There are many geometrical options. Three strikingly different ones are shown in Figs 17.15(*a*), (*b*) and (*c*). Since the support structure has to be strong, stainless steel is commonly used in order to obtain high strength with low magnetic permeability. However, stainless steels increase in permeability where worked, cut or welded, and so (sometimes even after heat treatment) it is easy to generate nonaxisymmetric and potentially resonant perturbation fields.

Any volume of unsaturated magnetic material 'sucks in' the ambient magnetic

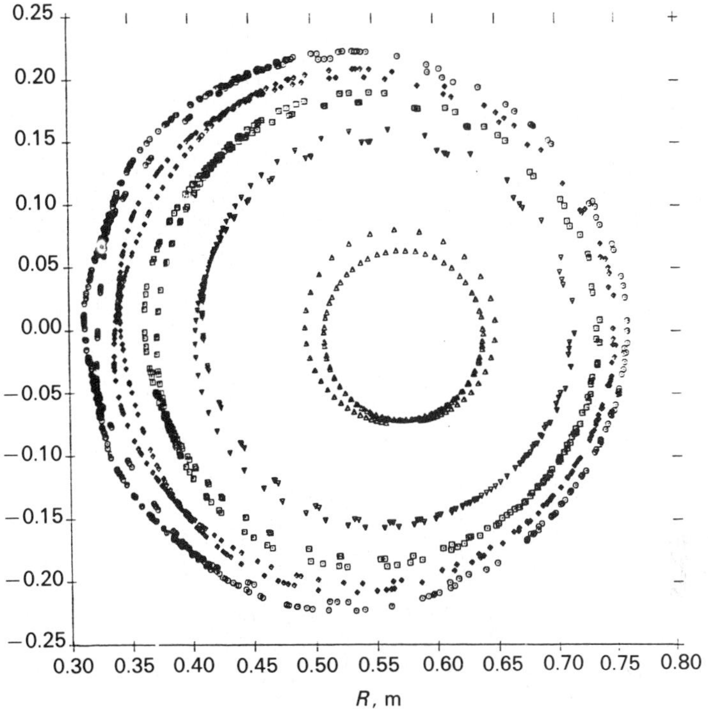

Fig. 17.13. Example of magnetic islands in a tokamak cross-section caused by an alignment error in a poloidal field coil.

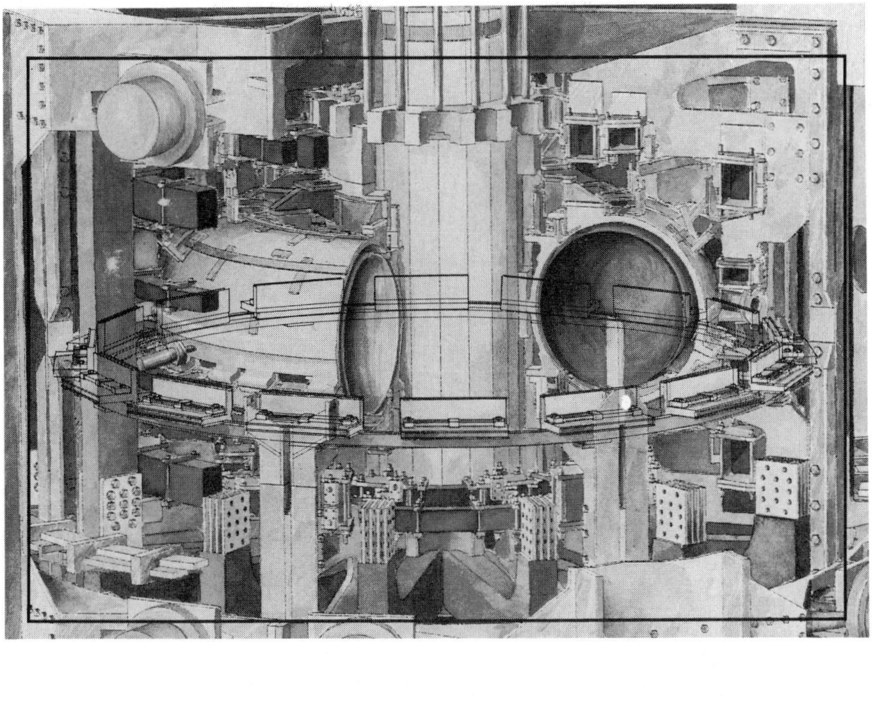

Fig. 17.14. The poloidal field coil aligment sensor array of COMPASS.

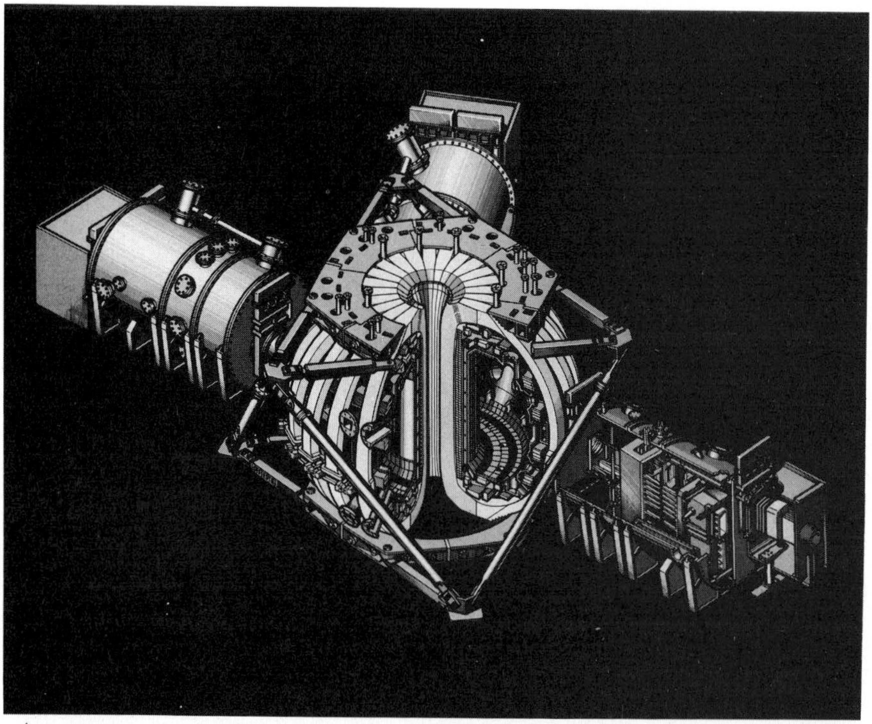

(*a*)

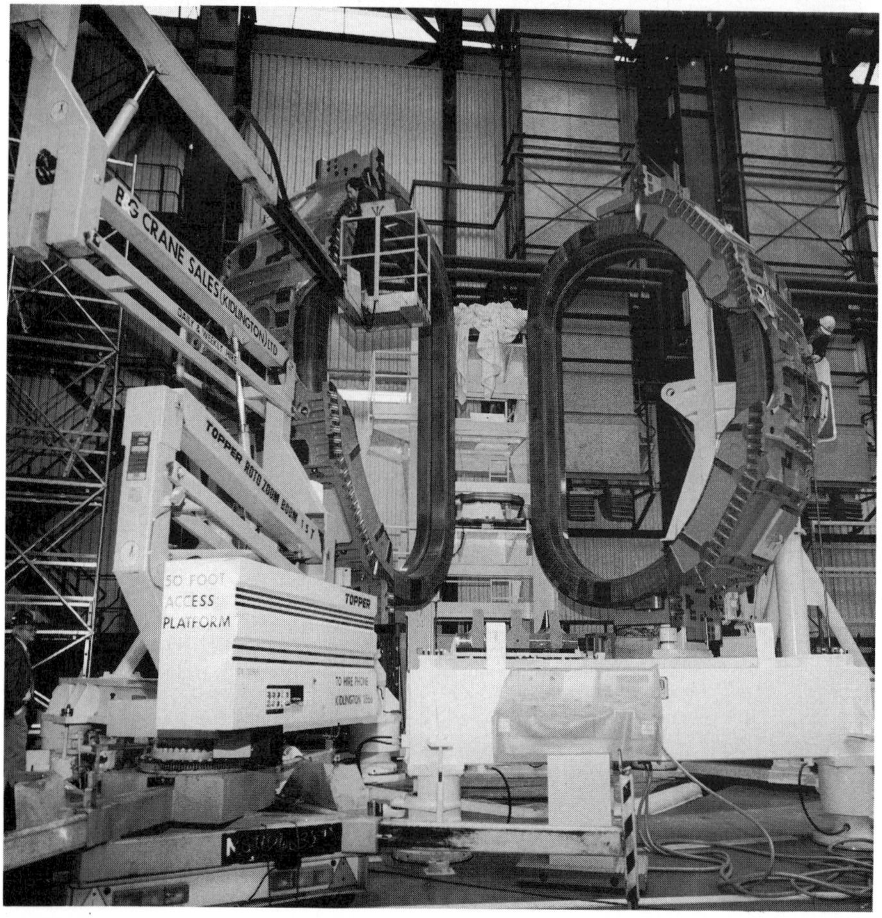

(*b*)

Fig. 17.15. Support structure options: (*a*) DIII-D (geodesic) (by kind permission of General Atomics, funded by the US Department of Energy; (*b*) JET (shell) (by kind permission of JET Joint Undertaking); (*c*) COMPASS (shear plate).

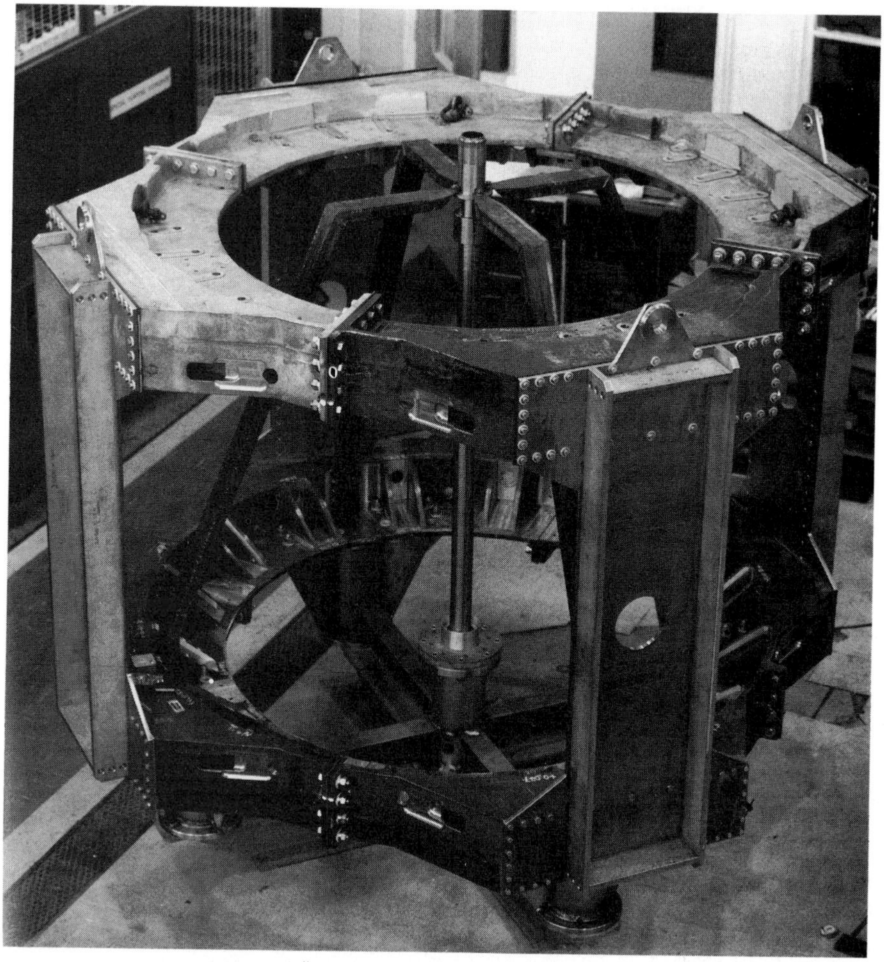

(c)

field, creating a disturbance in the field akin to a dipole source. This dipole source produces a field back in the plasma region, which will usually be nonaxisymmetric and is therefore likely to generate islands. Outside the TF coils, the ambient field is dipole-like, and the critical volume to generate 10^{-4} of the poloidal field at the plasma edge is given by

$$V_c \lesssim 300\mathscr{R}^6/R_0^2 a(\mu_R - 1)\,[\mathrm{m, m, m, cm^3}] \tag{17.9.1}$$

for $\mu_R \approx 1$, where $\mathscr{R}$ is the range of the offending volume from the machine centre. Inside the TFC the ambient field is B_0R_0/R, and the same criterion (at the machine centre) yields

$$V_c \lesssim 250IRR_0^2/aB_0(\mu_R - 1)\,[\mathrm{MA, m, m, m, T, cm^3}], \tag{17.9.2}$$

where R is the major radius of the offending piece of material.

17.10 Vacuum vessel

Tokamak vacuum vessels are subject to a number of significant stresses, arising from the following sources (in decreasing order of importance):

- disruption-induced currents;
- thermal gradients;
- air pressure;
- human malpractice;
- diagnostic loads;
- currents induced during plasma start-up;
- diagnostic flange bolting.

The largest of these, caused by disruption induced currents, arises when the plasma terminates rapidly, commutating its current into the vessel. This current then 'stands off' the externally imposed poloidal fields, giving rise to inward pressures on the vessel which can easily reach a bar or so (that is, several hundred kilopascals). Worse than this, the commutated current has to get around port-holes, causing regions of high current density and crossing of the toroidal field, producing very large local stresses.

It has been shown that rapid plasma displacement events (during vertical instability) can 'scrape off' up to $\approx 20\%$ of the toroidal plasma current where it flows helically (termed the halo current), resulting in large currents crossing the toroidal field within components mounted inside the vessel, severely stressing the mountings.

The apparently trivial solution to achieve a high strength vessel is to make it toroidally continuous and of thick metal. However, the loop resistance is then so low that large currents will be induced during plasma start-up, creating a vertical field which inhibits the gas break-down unless compensated. This can be overcome by gapping the vessel, but then one has to design a robust insulating joint of high vacuum specification (and worry about the plasma disruption effects on coils and structural components further out). An alternative solution is to put the vacuum vessel outside the poloidal field coils, but then there are difficulties with maintaining the vacuum standard, given the large number of components and the range of materials selected for items mounted inside the vessel. START is an example of this approach, whereas COMPASS-C represents a common compromise between strength and loop resistance, employing a deeply corrugated stainless steel bellows (Fig. 17.16). Inconel is frequently adopted in smooth vessel designs due to its high resistivity and mechanical strength and reliably low permeability, as in COMPASS-D; see Fig. 17.17.

17.11 Vessel conditioning

A high vacuum standard is needed in a tokamak to keep the radiated power and loop voltage (and hence runaway electron population) low. Conventional practice reveals that operation is very difficult with base pressures above about 2×10^{-7} torr of $H_2O + CO$, whereas, with better than about 5×10^{-8} torr of such residual 'gases', the plasma behaviour becomes dependent on the conditions of the walls and limiters. This is particularly true of operation with hydrogen isotopes, because plasma chemistry at the wall and limiters causes H_2O, OH, CH_4 and so on to be produced and then dissociated in the bulk plasma.

Some routes to a high vacuum standard include the following:

- the achievement of an all-metal set of vacuum seals, welded wherever reasonable;
- the avoidance of materials with high vapour pressures (for example, brass, plastics, ...);

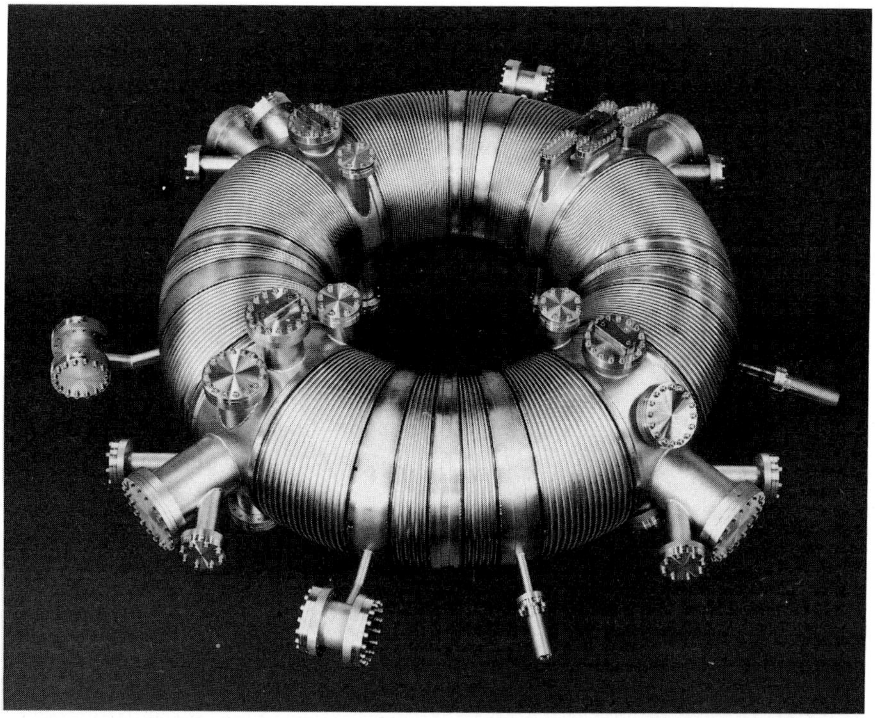

Fig. 17.16. COMPASS circular cross-section vacuum vessel: corrugated stainless steel 0.7 mm thick.

Fig. 17.17. COMPASS D-shaped vacuum vessel: continuous welded inconel 3 mm thick.

• the avoidance of trapped volumes (for example, in screw threads and between mating surfaces);

• the banning of marker pens and crack-detecting dye penetrants for use inside the vessel;

• the avoidance of the use of cleaning fluids with tightly binding high Z elements (for example, the ubiquitous chlorine);

• the prohibition of direct skin contact with anything destined to go in the vacuum system;

• the baking of components of the vessel itself during manufacture (for example to 400°C);

• the baking of everything that goes inside the vessel to whatever temperature it will tolerate (for example to 200°C);

• the specification and preservation of mirror-bright interior surfaces (perhaps including electropolishing the vessel interior surfaces);

• the avoidance, or protection of, plastics to which the plasma has a line of sight.

Assuming that most of the guidelines for vessel construction have been followed, one will have achieved an apparently hygienically clean vessel, probably of stainless steel, inconel or aluminium. However, all these materials stay clean because they form very thin stable oxide layers which inhibit further atmospheric corrosion. The oxides are readily reduced by energetic hydrogen, however, causing the problems outlined above.

Many of the established cures, following baking of the vessel *in situ* to accelerate the outgassing of water and hydrocarbon solvents from the structure, feature glow discharge cleaning at $\approx 15\,\mu\mathrm{A\,cm^{-2}}$ in various gases:

• in hydrogen (creating water which is pumped away) – this loads the vessel walls with hydrogen;

• in methane (or methane-hydrogen, etc.) to create a graphitic or carbide layer smothering any oxides, etc. – this also loads the walls with hydrogen;

• in helium, to get the hydrogen out (the helium diffuses out subsequently since it is chemically inert);

• in trimethyl boron $B(CH_3)^3$ to create a carbon–boron-carbide layer (apparently even better than simple carbonization because of the lower Z);

• in diborane (with care since this is a highly toxic and explosive gas) to lay down a high boron content layer, claimed by some to be more effective than the trimethyl boron approach.

Alternatively one can just getter titanium or chromium over a large fraction of the vessel interior to a thickness of about a monolayer every shot or so – these metals have a strong affinity for oxygen and many other impurity species and an almost magical effect on the plasma behaviour, even in a 'dirty' system (meaning one with a natural base pressure of water, carbon monoxide and various hydrocarbons of up to $\approx 5 \times 10^{-7}$ torr). Gettering is however like an addictive drug – once started it is hard to stop, and there are many significant long term disadvantages (such as coating over windows and flaking of loose titanium).

Long ago, diffusion pumps were very popular in fusion research devices since they were fairly cheap, available in many sizes, and could pull a fairly good vacuum. However, they tend to backstream oil vapour so that cool baffles and or liquid nitrogen cold traps have to be added to obtain acceptable behaviour, resulting in undesirable system complexity and occasional oily failures. Today, turbomolecular pumps (TMP) are widely favoured, producing excellent base pressures (if kept clean), easily down to the 10^{-10} torr region if the system is clean and leak free. Various bearing designs are employed, and some have been known to break up, allowing oil vapour back into the system. The high rotational speed vanes suffer strong eddy currents (and hence drag) if the TMP is exposed to magnetic fields, however, so siting needs consideration. Diffusion and turbomolecular pumps are generally backed by mechanical pumps, which can also serve to rough pump the vessel from atmosphere. A typical small machine would have a pair of 450 torr l s^{-1} TMPs backed by a 400 $m^3 h^{-1}$ Rootes pump and a 40 $m^3 h^{-1}$ rotary pump.

Cryopumps are used by some and are hard to beat where enormous pumping speeds are required (such as in ion beam neutralizer cells). They are expensive and can give operational problems (such as spontaneous dumping if overloaded). Getter pumps are available in standardized cartridge form, giving limited lifetime but good performance with very simple installation requirements (just a local heater to regenerate the active surface).

17.12 Limiters

Limiters are required in tokamaks to define the place and material of the principal plasma edge interaction, to stop runaway electrons from damaging the vacuum vessel, to protect the vessel from neutral beam shine-through, and to shadow in-vessel components from the plasma edge. Historically, poloidal ring (even 'diaphragm') limiters were popular, but, as power input and pulse duration have risen, these have become somewhat less favoured, giving way to various toroidal ring options; see Fig. 17.18. Some machines have also fared well with horizontal rails or vertical outboard rail limiters, often with fairly sophisticated front surface profiles to optimize the power handling capability.

Some of the plasma input power is lost as radiation and charge exchanged energetic neutral atoms, but a substantial fraction generally flows to the limiters. At the plasma edge the charged particles flow rapidly along the field lines and diffuse slowly across them. This gives rise to a 'fuzzy' edge region characterized by an exponential fall-off of power flux, density and temperature:

$$P_{\parallel} = P_{\parallel 0}\exp[-(r - r_{\text{lim}})/\lambda] \qquad (17.12.1)$$

with

$$\lambda \approx (\chi_{\perp}\tau_{\parallel})^{1/2}, \qquad (17.12.2)$$

where $\chi_{\perp}$ is the cross-field diffusion coefficient ($\gtrsim 1\ \mathrm{m^2\,s^{-1}}$), and $\tau_{\parallel}$ is the time the ions take to explore the field lines between limiter intersections, $\lesssim 2\pi Rqf/v_{\text{th(i)}}$, where f is a geometrical factor ≈ 1, related to the distribution of the limiters. For typical tokamaks this gives rise to $\lambda \approx 0.5$–2 cm, while $P_{\parallel 0}$ is $\gtrsim 10\ \mathrm{kW\,cm^{-2}}$.

A high surface heat flux for a short duration causes a temperature gradient in the limiter material, leading to differential expansion resisted by the elasticity of the material but limited by the yield point. The figure of merit for the thermal shock capability of materials is thus

$$\frac{\text{yield stress}}{\text{elastic modulus}} \times \frac{\text{thermal conductivity}}{\text{expansion coefficient}} \times (\text{density} \times \text{heat capacity}). \qquad (17.12.3)$$

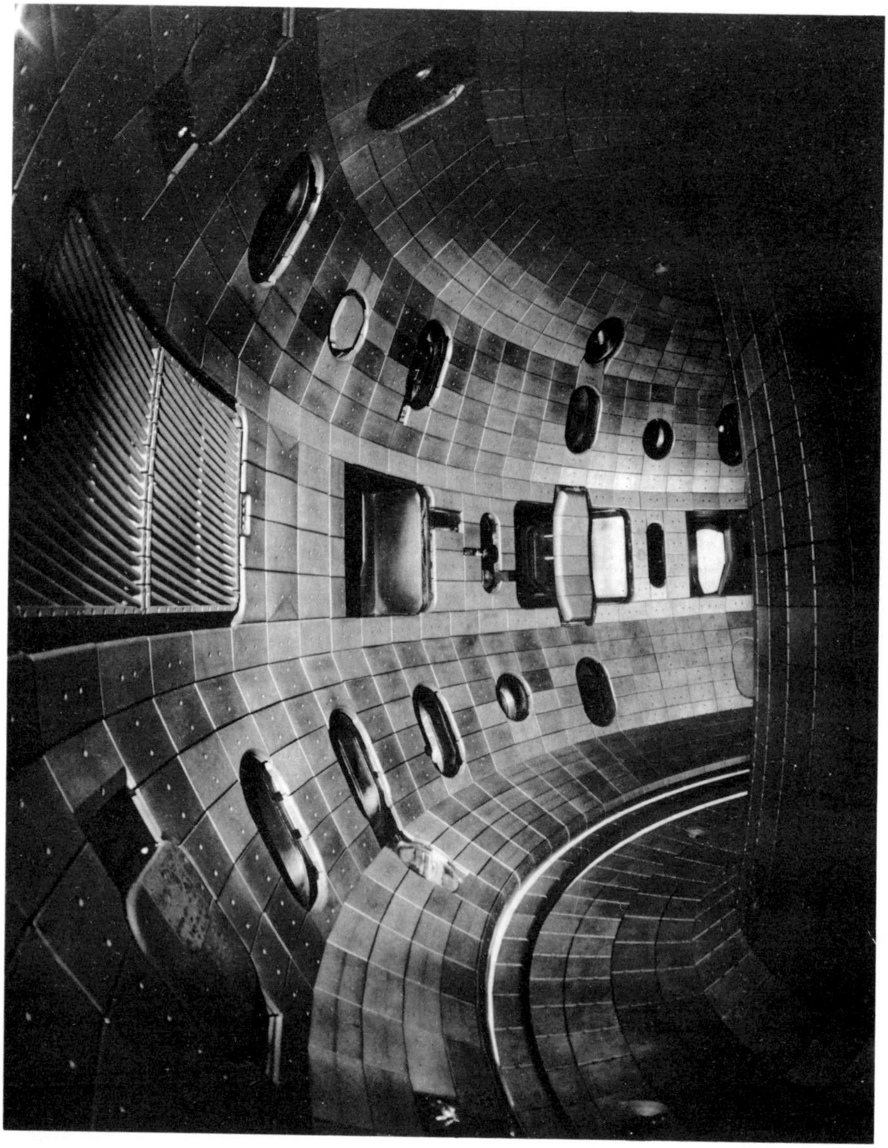

Fig. 17.18. Toroidal limiter in DIII-D. By kind permission of General Atomics, funded by the US Department of Energy.

The thermal shock figure of merit, combined with a requirement for high melting point, allows a small range of suitable limiter materials to be identified. Typical examples are refractory metals such as tungsten, tantalum, molybdenum, titanium, carbon (graphite), certain carbides, particularly those of titanium, silicon, boron (usually as coatings) and beryllium.

The early machines mostly used refractory metals, but graphite is now very popular, particularly as it sublimes rather than melts and allows runaway electrons to penetrate (and thus dissipate their energy) deeply. Ordinary isostatically compressed graphite can be improved by densification using chemical vapour deposition techniques (which also seal off the porosity), and by incorporating carbon fibres to raise the strength and conductivity. Large quantities of graphite can result in gas absorption/desorption problems, and at very high thermal loads (and hence surface temperature) a type of 'cluster sputtering' arises which rapidly contaminates the plasma. In addition, it is hard to match the expansion coefficient of graphite to metals, so active steady state cooling is complicated and expensive. Few carbides are available in block form, and thin coatings of these materials are rapidly eroded. Beryllium has considerable toxicity and therefore handling problems, particularly in finely particulate beryllia forms, such as arise during machining or when venting a torus where beryllium has been plasma-eroded.

17.13 Control and data acquisition

Successful resolution of all the issues described so far should lead to a convergence of the physicists and engineers involved, albeit not necessarily to what either group initially had in mind (Fig. 17.19). The load assembly and power supplies have accordingly been defined, but of course some form of central control system is also necessary in order to run the machine with all subsystems appropriately synchronized. The minimal control system possible for a very simple tokamak would comprise a slow electronic timer driving relays to control the slow plant, with a fast timer controlling ignitrons or thyristors switching capacitor banks, etc., and triggering oscilloscopes, electronic integrator resets and other diagnostic equipment.

Computer-driven systems are now of course becoming the vogue, allowing

Fig. 17.19. A realistic tokamak.

TOKAMAK DESIGN PROCESS:- STAGE 37

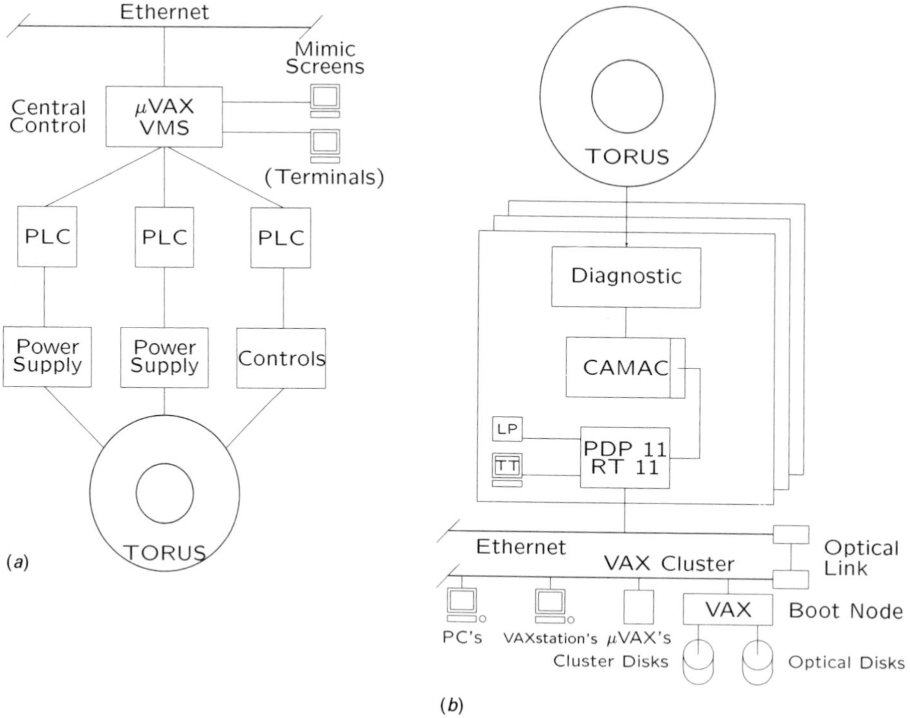

Fig. 17.20. (*a*) Example of computer control system, showing a hierarchy of computers. (*b*) Example of data acquisition and management system.

increasingly sophisticated sequence control, error checking and optional parameter-combination limits (for example to avoid probable q-limits, density limit or runaway effects (Eqs (17.3.1, 17.4.14)). A common type of layout is shown in Fig. 17.20(*a*). In addition, a computer-driven system can be given extensive real-time plant-monitoring capability, arbitrarily recorded with the data from the plasma physics diagnostics and available for display on mimic diagrams.

Whatever the system chosen, there will be many dozens of triggered events during a typical shot, and bitter experience shows that one should insist upon a unique definition of 'time zero' in *all* the software, thumbwheels and so on, to avoid confusion!

The purpose of this chapter is to describe how to *build* a tokamak, and so the selection and provision of suitable diagnostics and the *modus operandi* of the entire facility will not be addressed here. The final area to be considered is data acquisition, which is essentially part of the control system, indeed sometimes is indistinguishable from it in hardware terms. The most primitive form of data acquisition is, I think, a manually wound-on film camera with a remotely controlled shutter, looking at a simple CRT oscilloscope and requiring off-site processing – yes, people used to work that way! (And I think a few still do.) Historically, the advent of Polaroid film enormously improved the situation by allowing almost immediate inspection of the data. Analogue (i.e. screen) storage oscilloscopes came along in the 1970s, allowing instant inspection but usually compromising bandwidth or writing speed and often suffering from stray poloidal fields distorting the traces. The screen was (is) optionally photographed with a Polaroid camera if the stored traces seemed to be meaningful.

Digital sampling has, like all electronic technology, advanced out of all recognition in just the 15 years I have experienced it. Early systems sampled ≈ 8

channels at $\lesssim$ 10 kHz (sampling rate) for typically 1024 (i.e. 1 kilobyte) samples per channel. Today we sample many hundreds of channels at anything up to several megahertz for typically 16–128 kilobyte per channel, generating many megabytes of data per shot. This is typically all archived between shots on a networked computer, allowing general access from many terminals for subsequent analysis. Fig. 17.20(*b*) shows a typical system. The processing and printing of this volume of data between shots (even when intelligently 'sparsed' using special processing routines) is, I have found, invariably adjusted by the diagnosticians to dictate the acceptable repetition rate of the machine.

Data management is now an important issue, with a trend towards uniformity of data array formats and standardized raw data processing routines to avoid problems with staff continuity.

17.14 Summary

This chapter has attempted to outline the broad-brush thinking underlying the specification of a generic tokamak and its control systems. Most of the concepts considered are essentially fundamental, resulting from topological arguments and the properties of materials, but inevitably as the years go by and new innovations in composite materials, high temperature superconductors and so on appear, there will be some changes (perhaps radical) in tokamak design philosophy. Semiconductor technology advances inexorably, promising a greater range of feedback controllable, high speed power devices and ever greater computer power for control and data acquisition systems. The requirements for 'good vacuum practice' are probably least likely to change over the next few years, but I would imagine that even this area will have advanced significantly or changed emphasis by the end of the century, requiring some new author to outline 'how to build a tokamak' as the new one begins.

References

DOLAN, T. J. (1982). *Fusion Research*, vol. 3. Oxford: Pergamon Press, p. 637.

KAYE, G. W. C. and LABY, T. H. (1973). *Tables of Physical and Chemical Constants*, 14th edn. Longman.

MURPHY, M. C. (1981). *Fatigue of Engineering Materials and Structures* **4** (3), 199.

18

Survey of fusion plasma physics

R. S. PEASE

18.1 What is a plasma?

The term *plasma* was introduced by Langmuir (1928) to describe the state of matter in the positive columns of glow discharge tubes. This state is now recognized to occur throughout space and the stars, and in a wide range of laboratory and industrial devices. This chapter reviews the basic physical features which characterize plasmas, including plasmas in magnetic fields, and reviews briefly their application to magnetic fusion experiments.

A plasma is primarily an ionized gas, that is a gas of positive ions and electrons, both of them being free in the sense that their mean kinetic energy is much larger than the Coulomb potential energy, $\sim e^2 n^{1/3}/4\pi\varepsilon_0$. Here e is the electronic charge and n_e is the electron number density. Quantum mechanically both the ion and the electron states form a continuum. Other atomic and ionic species may be present: negative ions, neutral atoms and molecules, molecular ions, metastable species; and the positive ions can have a multiplicity of charges, $Z_i e$. Generally, these other species introduce complexity rather than new plasma physics; here we usually assume a single species of positive ion, whose mass m_i greatly exceeds that of the electron, m_e, and the influence of any other species is neglected.

The chief characteristic, identified by Langmuir, is that the plasma is electrostatically neutral, that is the mean number density n_i of ions and of electrons n_e ensures that only a small net electric charge occurs in macroscopic volumes of plasma. To a high degree of approximation,

$$n_e = \sum n_i Z_i, \tag{18.1.1}$$

where the summation is over all species of ion. The question of what constitutes a 'macroscopic volume' and 'high degree of approximation' is discussed below. Plasmas have a number of other important derived characteristics; here we deal first in more detail with the two principal features already introduced.

18.2 Ionization

The ionization potential of an atom or ion is the minimum potential through which an electron must be accelerated in order to dislodge an electron from that atom or ion. Examples of this potential for isolated ground state neutral atoms are:

- Cs, 3.9 V (the lowest potential)
- He, 24.6 V (the highest neutral atom potential)
- H, 13.6 V (related directly to the Rydberg energy $E_R = 13.6\,\text{eV}$).

The ionization potential χ_r for removing an electron from r-times ionized atoms is generally much higher than for neutral atoms, and $e\chi_r$ ranges up to Z^2E_R for hydrogen-like ions, subject to relativistic correction for the highest values of atomic number Z. The various particles in a plasma interact ergodically through their mutual Coulomb interactions. Thus, after a sufficient time undisturbed by external forces, the plasma can be characterized by a temperature. The ionization energy, $e\chi_r$, is related to temperature T by Boltzmann's constant $k_B = 0.861\,682 \times 10^{-4}\,\text{eV}\,\text{K}^{-1}$. A temperature of 1 eV is 11 610 K. Thus plasmas are generally characterized by temperatures that are high compared with room temperature. The degree of ionization of a gas can be calculated simply only on the assumption of thermal equilibrium, using equilibrium thermodynamics; it is expressed by the Saha equation, Eq. (18.2.1) below.

18.2.1 *Thermal equilibrium*

Let number densities be n_r of r-times ionized atoms, and n_{r+1} of $(r+1)$-times ionized atoms; then, when the system is in thermodynamic equilibrium,

$$n_{r+1} = n_r\left(\frac{U_{r+1}}{U_r}\right)\left(\frac{2}{n_e}\frac{(2\pi m_e k_B T)^{3/2}}{h^3}\right)\exp(-e\chi_r/k_B T). \tag{18.2.1}$$

Here, U_{r+1}, U_r, are the partition functions of the ions, m_e is the electron rest mass, h is Planck's constant,

$$U_r = \sum g_{rl}\exp(-\varepsilon_{rl}/k_B T), \tag{18.2.2}$$

where g_{rl} is the statistical weight and ε_{rl} is the energy above ground of the lth excited state.

Where the ground states of the ions are the predominant contribution to the ion partition functions, the ratio U_{r+1}/U_r is of order unity because both ion species are assumed to be freely moving in a field-free region. The other factors on the right-hand side of Eq. (18.2.1) represent the partition function of the free electron; $(h^2/2\pi m_e k_B T)^{3/2}$ is the cube of the de Broglie wavelength. For classical, highly nondegenerate plasmas $n_e \ll (2\pi m_e k_B T/h^2)^{3/2}$; and so the degree of ionization n_{r+1}/n_r is very high when $k_B T \sim e\chi_r$. Fig. 18.1 shows the percentage ionization $\alpha \equiv 100n_1/(n_1 + n_0)$, computed from Eq. (18.2.1) for a hydrogen plasma as a function of temperature. Here $n_{r+1} \sim n_r$ when $k_B T \sim (1/10)e\chi_r$ a result which is a useful rule-of-thumb for a wide range of situations, because of the dominance of the exponential term.

The thermal equilibrium assumed in Eq. (18.2.1) breaks down most usually from the loss of thermal equilibrium with radiation. Typically, when the dimensions of the plasma L are such that

$$n_e\sigma_T L < 1, \tag{18.2.3}$$

where σ_T is the Thomson cross-section $(e^2/4\pi\varepsilon_0 m_e c^2)^2 8\pi/3 = 6.65 \times 10^{-29}\,\text{m}^2$, this equilibrium with the radiation field may be lost, subject of course to the reflectivity and temperature of the surrounding material. Because of the small value of σ_T, it is necessary to adopt special means to get laboratory plasmas into thermal equilibrium with radiation. For the solar corona, the plasma surrounding the sun, Eq. (18.2.3) is far from satisfied, despite the large value of L (typically, $n < 10^{15}\,\text{m}^{-3}$ and $L \sim 10^8$ m). Very often, plasmas are not in equilibrium; for

velocity is small compared with V_s and the waves are no longer damped; however, when the wavelength becomes less than λ_D, Eq. (18.3.3), the waves cease to propagate and adopt a fixed frequency, the ion plasma frequency

$$\omega_{pi}^2 = n_i(Z_i e)^2/\varepsilon_0 m_i. \tag{18.3.10}$$

18.4 Experimental evidence for Debye length

The main experimental evidence for the magnitude of the screening distance λ_D in a plasma comes from experiments on:

(i) the edges of plasma: electron sheath thickness;
(ii) the damping of waves: Landau damping at short wavelengths;
(iii) the scattering of radiation by plasma;
(iv) the measurements of (screened Coulomb) collision cross-sections.

The agreement between experiment and theory of the damping of electron plasma oscillations as the wavelength approaches the Debye length is shown in Fig. 18.3. Moreover, it was additionally shown experimentally that, as predicted, this Landau damping is due to those electrons in the Maxwellian distribution whose velocity is close to the phase velocity of the wave.

The spectra and amplitudes of electromagnetic waves scattered by the plasma reveal fundamental information about the magnitude and velocity of thermal fluctuations in the electron density. Provided the scattering is weak, the wave number of the incident radiation $\boldsymbol{k}_i$ is related to the scattered wave number $\boldsymbol{k}_s$ by the generalized Bragg's law:

$$\boldsymbol{k}_i - \boldsymbol{k}_s = \boldsymbol{k}_d,$$

where $\boldsymbol{k}_d$ is the wave number of the Fourier component of the electron density. When the incident wavelength is chosen appropriately, $|\boldsymbol{k}_d|$ can be near to $2\pi/\lambda_D$, the wave number corresponding to the Debye length. Forward scattering provides the amplitude and frequency of Fourier components with wavelengths longer than λ_D; backward scattering looks at shorter wavelength Fourier components. Fig. 18.4 shows the dramatic difference between the spectra in the forward and backward directions in the scattering of ruby laser light by a plasma. In the forward direction, Fig. 18.4(*a*), there are two peaks corresponding to electron plasma oscillation, at frequency $\pm\omega_{pe}$ given by Eq. (18.3.8), and a central peak, with zero mean shift, but broadened by sound waves, Eq. (18.3.9). In the backward direction the frequency spectrum is largely the Doppler broadening due to scattering from randomly moving electrons with a Maxwellian velocity distribution. A continual transition between these two spectra can be obtained by varying the scattering angle, θ. These experiments confirm, for a nonperturbed plasma, the magnitude and velocities of the electron density fluctuation and their correlations with the ion motions, predicted by the fundamental statistical theory. This experimental observation of the plasma microstructure can be compared to X-ray diffraction showing the arrangement of atoms in condensed matter. The collective scattering of radiation (in the forward direction in Fig. 18.4(*a*)) was first observed in the backscatter of radar signals from the ionospheric plasma (Bowles, 1958; see also Sheffield, 1975), where the 7.3 m radar wavelength used substantially exceeded the Debye length of about 2 mm.

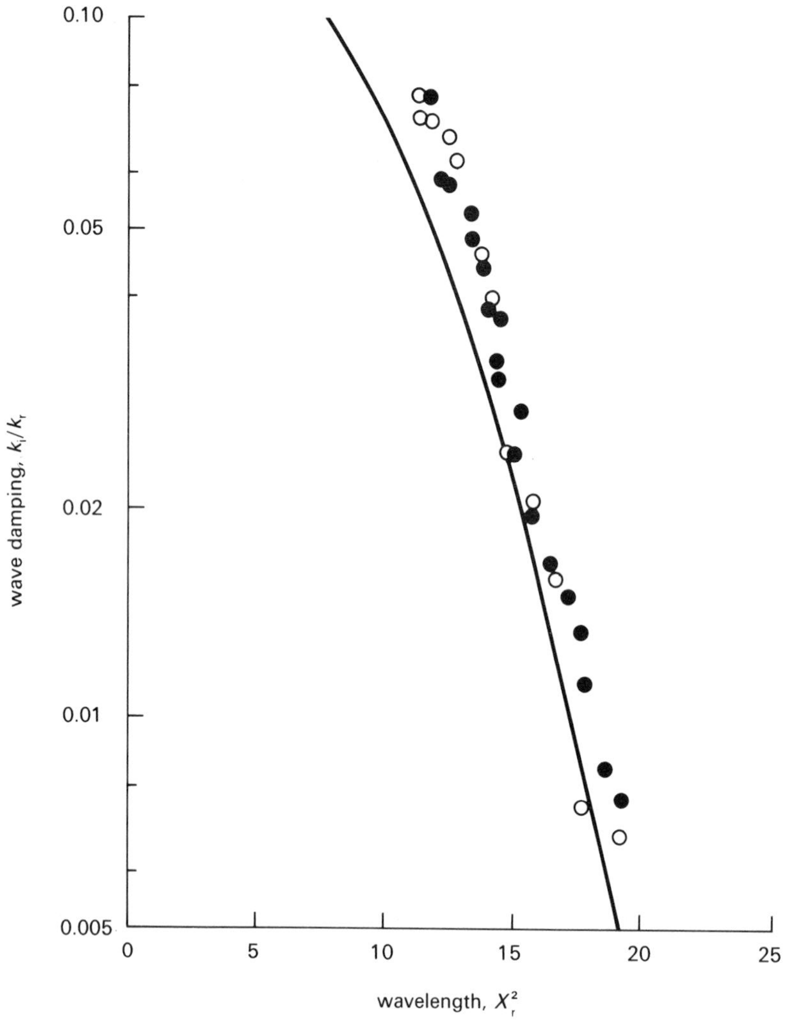

Fig. 18.3. Observed and calculated damping of electron plasma oscillations. k_i/k_r is the ratio of imaginary to real parts of the wave vector; X_r is the ratio $2\pi/\boldsymbol{k}_r\lambda_D$; ● $k_BT_e = 5.9\,\text{eV}$; ○, $k_BT_e = 8.6\,\text{eV}$; —— prediction of Landau. From Malmberg and Wharton (1966).

18.5 Collisions in a plasma

The interaction between the individual electrons and ions in a plasma is characterized by the screened electrostatic Coulomb potential given in Eq. (18.3.6). The interactions take the form of numerous small deflections rather than the single large ones familiar from the kinetic theory of gases.

In this circumstance, the mean free path L_c is the average distance travelled by a particle in randomizing its direction of motion. The ultimate quantitative definition of L_c is given by

$$-\frac{\mathrm{d}v_{\parallel}}{\boldsymbol{v}} = \frac{\mathrm{d}x}{L_c}. \tag{18.5.1}$$

Here $\mathrm{d}v_{\parallel}$ is the mean change in velocity parallel to $\mathrm{d}x$ when moving along the element $\mathrm{d}x$ of its trajectory. To fix ideas, if we consider an electron of velocity v interacting with individual stationary heavy ions, each surrounded by the

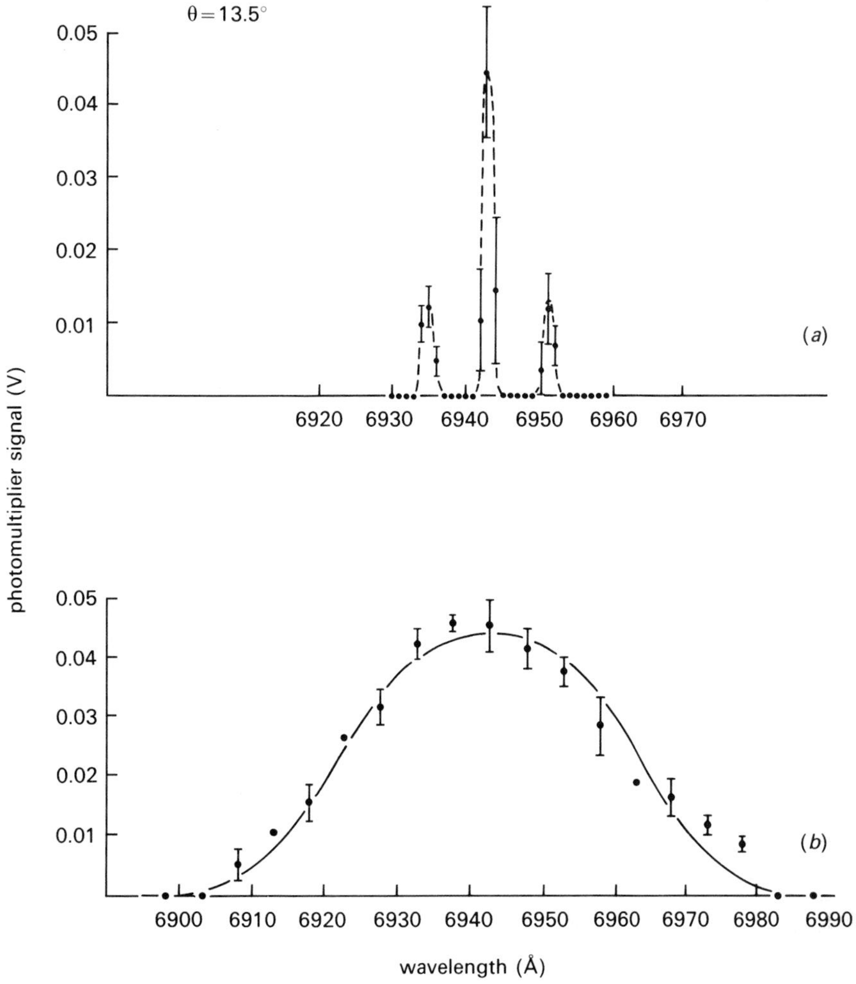

Fig. 18.4. Spectra of ruby laser light scattered from a plasma in (*a*) forward direction and (*b*) backward direction The solid line is the theory. (Ramsden and Davies, 1966.)

potential Eq. (18.3.6), then L_c can be defined in terms of a cross-section $\sigma_{ei}n_i \equiv L_c^{-1}$ where σ_{ei} is given by

$$\sigma_{ei}(v) = \pi b^2 \ln(\lambda_D/b), \tag{18.5.2}$$

and $b = Z_i e^2/2\pi\varepsilon_0 m_e v^2$ is the distance of closest approach. This cross-section is sometimes known as the *momentum transfer cross-section*. Eq. (18.5.2) illustrates two main points: first, the cross-section varies strongly with the particle velocity, and consequently any calculation of the transport involves accurate averaging of $\sigma_{ei}(v)$ over the particle velocity distribution (generally obtainable only by using Boltzmann's equation or its derivative); secondly, the general trend is that collision cross-sections decrease as the square of the electron energy, so that the plasma becomes more collisionless as the temperature increases, until relativistic velocities are reached. To get the orders of magnitude, the mean free path L_c can be averaged over the Maxwellian velocity distribution, and the electron–ion mean free path λ_{ei} is then estimated to be

$$\lambda_{ei} = 4 \times 10^9 T^2/(Z_i^2 n_i \ln\Lambda)\,[\mathrm{m, K}], \tag{18.5.3}$$

where $\Lambda = (\lambda_D/b)$, and T is in kelvin; $\ln\Lambda$ is called the Coulomb logarithm, and has the value 10–20 in most cases. Taking $\ln\Lambda = 15$, we have, for T_e in kelvin,

$$\sigma_{ie} = (\lambda_{ei} n_i)^{-1} \sim 4 \times 10^{-9} Z_i^2/T_e^2 \;[\mathrm{m^2, K}]. \tag{18.5.4}$$

The collision frequency ν_{ei} is given by

$$\nu_{ei} \sim 2.5 \times 10^{-5} n_i Z_i^2/T_e^{3/2} \;[\mathrm{m, K, s^{-1}}]. \tag{18.5.5}$$

The mean free paths are illustrated for $Z_i = 1$ in Table 18.1, showing these to be much larger than the other physical lengths so far considered.

The times needed for electron–electron and ion–ion collisions to establish a Maxwellian distribution with temperatures T_e and T_i can be calculated by these considerations, and are about $\lambda_{ee}/(k_B T_e/m_e)^{1/2}$ and $\lambda_{ii}/(k_B T_i/m_i)^{1/2}$, respectively, where λ_{ee} and λ_{ii} are the mean free paths for the appropriate collisions analogous to Eqs (18.5.1) and (18.5.2). The time needed for the ion and electron temperatures to come into equilibrium, τ_{eq}, is defined as

$$\frac{dT_i}{dt} = (T_e - T_i)/\tau_{eq}, \tag{18.5.6}$$

where

$$\tau_{eq} = \frac{3 m_i m_e (k_B T_i/m_i + k_B T_e/m_e)^{3/2}}{8(2\pi)^{1/2} n_e Z_i^2 e^4 \ln\Lambda} \tag{18.5.7}$$

and $(k_B T_i/m_i)^{1/2} < (k_B T_e/m_e)^{1/2}$.

This equalization of temperature is, generally speaking, the slowest of the collision processes. It is exponential only when the term $(k_B T_i/m + k_B T_e/m_e)^{3/2}$ does not change greatly during the process. Values of $\ln\Lambda$ and a discussion of the quantum effects when the distance of closest approach, b, is less than the electron de Broglie wavelength, see Spitzer (1962).

18.5.1 *Electrical resistivity*

The electrical resistivity η has been calculated for a wide variety of circumstances. Taking account of electron–electron collisions in the calculation of the electron velocity distribution in the presence of an electric field yields

$$\eta = 65.3 Z_i \ln\Lambda/T_e^{3/2}\,[\Omega\mathrm{m\,K}]. \tag{18.5.8}$$

The coefficient 65.3 applies to the case $Z_i = 1$; as $Z_i \to \infty$ it should be replaced by 39.0; exact values are given by Spitzer (1962).

This formula is valid only when the mean electron drift velocity $J/n_e e$ is less than about 10^{-2} of the mean electron thermal velocity $(k_B T_e/m_e)^{1/2}$. If this condition is violated, the high energy end of the electron Maxwellian spectrum (where the mean free path, given by Eq. (18.5.2), becomes very long) is accelerated without limit. The phenomenon of runaway electrons is observed experimentally, and was calculated in detail by Dreicer (1959).

Agreement between experiment and theory of plasma resistivity at high temperatures is indicated by recent experiments on the tokamaks JET and TFTR. The observed and calculated resistivity on JET are shown in Fig. 18.5. For TFTR, the average ratio of calculated to observed resistivity is given by Zarnstorff *et al.*

(1990) as 1.014 ± 0.06. In both cases, theory and experiments are dealing with a very complex situation; yet the agreement, with no fitted parameters, is much better than that achieved in the case of ordinary condensed matter. The experiments substantiate the calculated electron–ion and electron–electron transport coefficients and the underlying cross-sections and coefficients up to energies of 10 keV or so.

These are the best data we have so far in support of collisional transport theory in ordinary plasmas; less accurate but more direct data are available for the slowing down of ion beams in plasmas, for the particle velocity scattering (in so-called magnetic mirror machines), and for runaway electrons. There are good observational data also supporting Eq. (18.5.7) for τ_{eq}. Thus far, there are no grounds for supposing that there are any major errors in collisional transport theory. The evidence for the classical–quantum transition at $k_B T_e \simeq E_R Z_i^2$ does not seem to exist for a plasma: the Coulomb logarithm is too insensitive to the change, and small-angle scattering experiments analogous to those carried out for scattering by atoms do not seem to have been reported. In sufficiently high magnetic fields, the Debye length has to be replaced by the electron Larmor radius in the Coulomb logarithm; and large departures from Eq. (18.5.2) occur

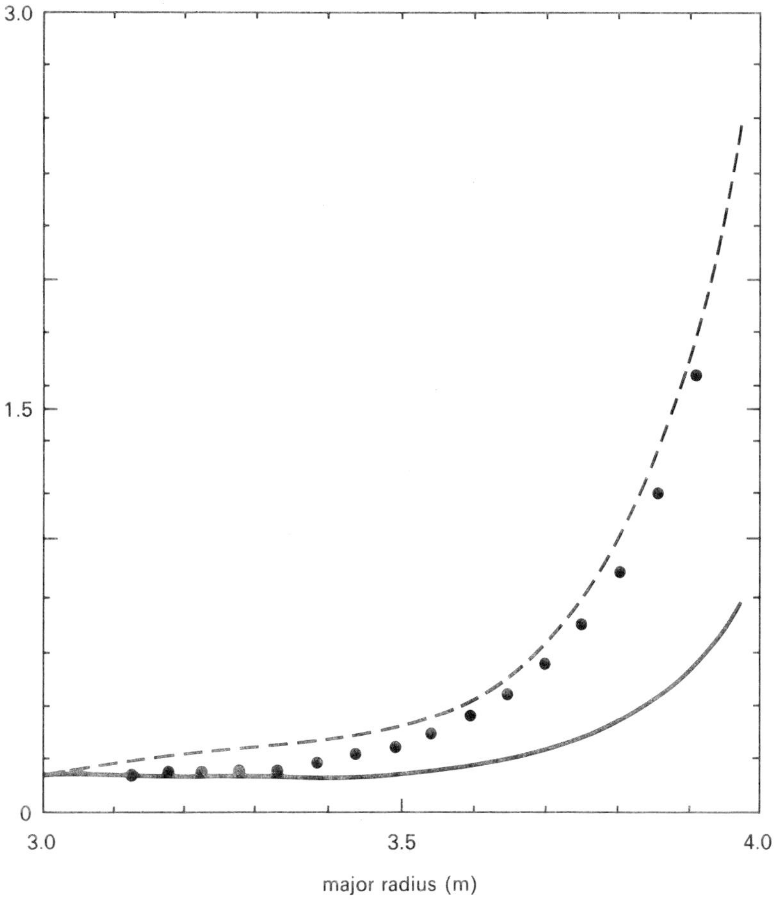

Fig. 18.5. Observed and calculated resistivity of the plasma in the JET tokamak, as a function of position. The major radius of 3.0 m is the centre of the minor cross-section, and is the hottest part of the plasma (T_e up to 4 keV); the major radius of 3.9 m corresponds to the edge of the plasma column (Bartlett *et al.*, 1988). • Experiments; ——— calculated from Spitzer (1962); ---- calculated using neoclassical theory (see Connor, p. 75 of Wesson, 1987).

when the Larmor radius becomes comparable with b, the distance of closest approach (Beck *et al.*, 1992).

However, plasma properties often depart from the predictions of collisional theory, as discussed in Section 18.10. Such departures arise, it seems, because the plasma is subject to large perturbations, such as shock waves, imperfect equilibria, instabilities, or turbulent plasma flow. The consequent convection and nonlinearities overwhelm the abovementioned collisional effects.

18.6 Plasma in magnetic fields

In a magnetic field, the charged particles constituting the plasma tend to move in helical paths, the axis of the helix being parallel to the magnetic field $\boldsymbol{B}$. The angular frequency of rotation about the lines of force is known as the cyclotron frequency, denoted by ω_{ce} and ω_{ci} for electrons and ions, respectively.

For nonrelativistic velocities in a magnetic field with strength B,

$$\frac{\omega_c}{2\pi} = \frac{ZeB}{2\pi m} = 1.54 \times 10^7 \frac{ZB}{A} \text{ [Hz, T]}, \tag{18.6.1}$$

where Ze is the electric charge on the particle and A is its atomic weight. Because plasma is electrically conducting, the magnetic field can dominate its fluid dynamics. This dominance in astrophysical plasmas was brought to prominence by Alfvén in the 1940s (Alfvén, 1950); it subsequently became important for laboratory studies of magnetically confined plasmas for thermonuclear fusion. As an indication of the potential size of the forces, when $B = 0.5\,\text{T}$, the magnetic energy density $B^2/2\mu_0$ corresponds to a pressure of about 1 atm. This pressure is exerted only if currents can flow in the plasma.

The simplest case of low frequency kinetics, where the rates of change are less than both the electron and cyclotron frequencies, and the plasma can be treated as a fluid, is known as the magnetohydrodynamic (MHD) approximation.

18.6.1 *Simple magnetohydrodynamics*

The equation of motion of the plasma is

$$\rho \frac{d\boldsymbol{v}}{dt} = \boldsymbol{J} \times \boldsymbol{B} - \nabla p. \tag{18.6.2}$$

Here ρ is the mass density, generally taken to be $n_i m_i$; $\boldsymbol{v}$ is the velocity of the centre of mass of the fluid element; $\boldsymbol{J}$ is the current density and the pressure p is

$$p = n_e k_B (T_e + T_i/Z_i). \tag{18.6.3}$$

Ohm's law in a moving fluid is, in the simplest approximation,

$$\boldsymbol{E} + \boldsymbol{v} \times \boldsymbol{B} = \eta \boldsymbol{J}, \tag{18.6.4}$$

where η is the plasma resistivity. The electric field $\boldsymbol{E}$, the magnetic field $\boldsymbol{B}$ and the plasma velocity are those observed by the same (stationary) observer. When the plasma is highly conducting, the approximation $\eta\boldsymbol{J} = 0$ can often be used, and the resulting theory is known as ideal MHD.

The above three equations, together with the equations of conservation of matter and of electric charge, combined with Maxwell's equation, are the set of MHD equations in the single-fluid model.

18.6.2 *Plasma motion across the lines of force*

Alfvén introduced the concept that the magnetic lines of force move with the plasma and are, so to speak, frozen to it by the high electrical conductivity. Using the induction Maxwell equation

$$\nabla \times \boldsymbol{E} = -\frac{\partial}{\partial t}\boldsymbol{B}, \tag{18.6.5}$$

it can be shown that the magnetic flux Φ_s through a surface defined by a closed contour which moves with the plasma, varies as follows:

$$\frac{\mathrm{d}}{\mathrm{d}t}(\Phi_s) = \iint \nabla \times (\boldsymbol{E} + \boldsymbol{v} \times \boldsymbol{B})\mathrm{d}\boldsymbol{S}. \tag{18.6.6}$$

When the resistivity is small, $\eta \boldsymbol{J} \to 0$, and thus by Eq. (18.6.4), the magnetic flux Φ_s is unchanged.

Alfvén's concept, though vivid, is not altogether reliable: for example, an isolated volume of plasma can move across a magnetic field by becoming polarized, the resulting charge separation producing the electric field $\boldsymbol{E}$ to offset the $e(\boldsymbol{v} \times \boldsymbol{B})$ force.

18.6.3 *Skin depth effect*

The validity of Alfvén's concept depends also on the dimension L of the plasma region being considered, and the time scale. Thus in a time t, an inhomogeneity in the magnetic field will diffuse a distance δ_s normal to $\boldsymbol{B}$ in a medium:

$$\delta_s = (\mu_0 \eta t)^{\frac{1}{2}}, \tag{18.6.7}$$

which defines the resistive skin depth, δ_s.

If $\delta_s \ll L$, then one can speak of a highly conducting plasma and use ideal MHD theory. Even in this case, at singularities, the product $\eta \boldsymbol{J}$ cannot always be taken to be zero and neglected: resistive MHD theory is needed to study low frequency stability involving singularities even when $\delta_s \ll L$.

18.6.4 *Waves in magnetized plasma*

The most important wave motions which develop as a consequence of the magnetic field are the Alfvén waves. They have low frequencies $\omega < \omega_{ci}$ and are generally characterized by a phase velocity V_A, the Alfvén velocity, given for the case $V_A \ll c$ by:

$$V_A^2 = B^2/\mu_0 n_i m_i. \tag{18.6.8}$$

This is the phase velocity of shear Alfvén waves, which propagate along the lines of force; the plasma velocity and the electric field are orthogonal, and are normal to B. Fig. 18.6 shows the agreement between theory and experiment for such waves in a long quasiuniform column of argon plasma embedded in a magnetic field. These shear Alfvén waves can be thought of as arising from a tension B^2/μ_0 per unit area of the line of force.

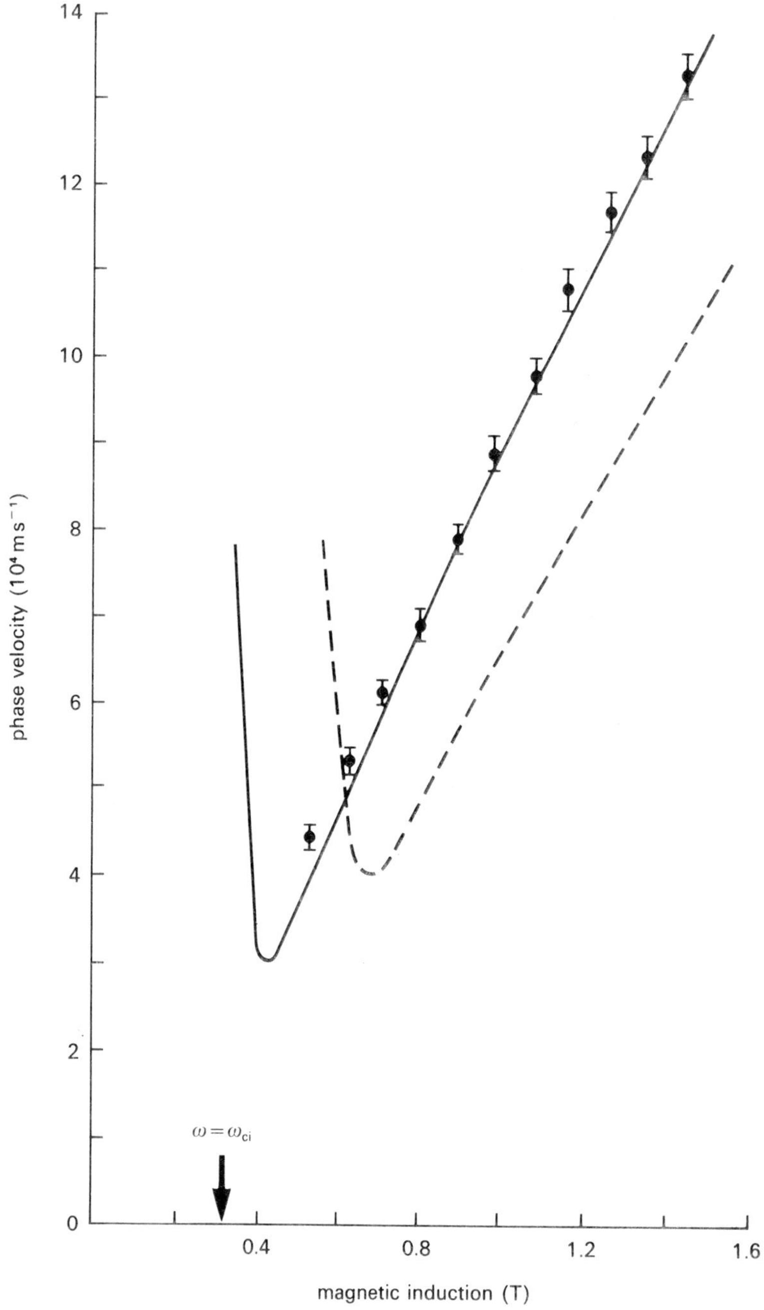

Fig. 18.6. Observed and calculated phase velocities of shear Alfvén waves (Jephcott and Stocker, 1962). ----,—— Theoretical, with different adjustments for collisions between ions and neutral atoms; ⧲, experimental.

up to infinitely conducting walls, then the current-carrying plasma would be wholly stable (in a state of minimum energy) for all nonzero values of B_ϕ and I provided that everywhere

$$\mu_0 \boldsymbol{J} = \mu \boldsymbol{B}, \tag{18.9.2}$$

where μ is a scalar constant which is everywhere the same; the solution of Eq. (18.9.2) chosen has to be the one with the smallest magnetic energy inside the torus (Taylor, 1976).

The quantity μ, which has the dimension (length^{-1}), describes the strength of the stabilizing field. In a torus of circular minor cross-section and radius a, with a/R small, for a symmetrical discharge,

$$\mu = 2B_\theta(a)/a\overline{B_\phi}, \tag{18.9.3}$$

where $B_\theta(a)$ is the azimuthal field generated by the pinch current I_ϕ at the walls, and $\overline{B_\phi}$ is the mean value within the torus of the stabilizing field B_ϕ.

In practice, the ideal conditions leading to Eq. (18.9.2) can be only approximately met: there are three defects which limit the useful range of I_ϕ and B_ϕ. These defects are:

(i) the metal walls have a finite conductivity;

(ii) the conducting plasma cannot extend out to the walls – indeed a vacuum-like region tends to surround the plasma ring;

(iii) if plasma pressure is to be contained, then there must be a component of current density J normal to the magnetic field, Eq. (18.6.11).

The two main types of toroidal pinch now under study, namely the tokamak and the reverse field pinch (RFP) are the least affected by these departures from ideal conditions. In the tokamak the condition needed is:

$$q(r) \equiv \left(\frac{B_\phi}{B_\theta}\right)\left(\frac{r}{R}\right) \geq 1 \text{ everywhere;} \tag{18.9.4}$$

$q(r)$ is the pitch length of the helical lines of force, normalized to the major circumference; condition (18.9.4) requires $\mu R < 2$. The RFP requires that

$$1.6 > \frac{\mu a}{2} > 1.2; \tag{18.9.5}$$

then the solution of Eq. (18.9.2) requires that the sign of B_ϕ is, in the outer regions, opposite to that of the mean value $\overline{B_\phi}$. These two configurations are illustrated in Fig. 18.12. Nowadays most effort in fusion research is concentrated on the tokamak configuration, because the greatest departures from Eq. (18.9.2) with the greatest degree of stability is obtained with the strongest stabilizing field B_ϕ.

The configuration described by Eq. (18.9.2) is sometimes known as the *relaxed state*, because it is the state to which the magnetic configuration of the pinch tends to relax as a result of MHD instability. Its most striking experimental verification is the spontaneous appearance of the reversed B_ϕ when condition (18.9.5) is satisfied (Fig. 18.13); indeed this was one of the observations which compelled the development of theory (Butt *et al.*, 1958). For a detailed discussion of tokamaks see Wesson (1987); for the reverse field pinch, see Prager (1990).

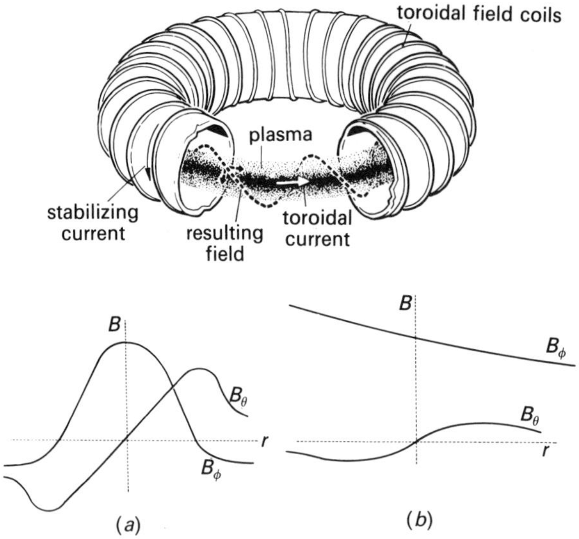

Fig. 18.12. Magnetic field components as a function of minor radius, r, in (a) the reverse field pinch and (b) the tokamak confinement systems. B_ϕ is the toroidal field and B_θ is the poloidal field.

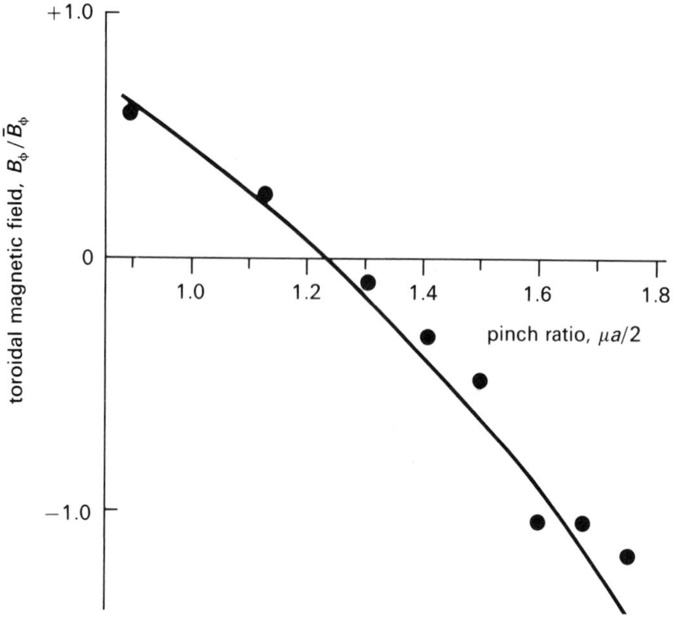

Fig. 18.13. Observed and calculated toroidal magnetic field in the Zeta toroidal pinch, normalized to mean value, as a function of $\mu a/2$.
—— relaxation theory; ● experimental. From Taylor, (1976).

18.9.3 *Confinement properties*

The confinement properties of the pinch stabilized with B_ϕ depend mostly on the toroidal current I_ϕ (the 'pinch' current) and plasma energy. The energy per unit length of the confined plasma is given by Eq. (18.6.13) modified with a coefficient β_p:

$$8\pi^2 a^2 \bar{n} k \bar{T} = \mu_0 \beta_p I^2. \tag{18.9.6}$$

Here $\bar{n}$ is the mean particle density and $\bar{T}$ is the mean temperature. The coefficient β_p is known as the poloidal beta, and, in a steady state pinch driven by a steady toroidal electric field, has the approximate value $\eta_{\parallel}/\eta_{\perp} \simeq \frac{1}{2}$; here $\eta_{\parallel}$ and $\eta_{\perp}$ are plasma resistivities parallel and perpendicular to $\boldsymbol{B}$. β_p can exceed 1/2 when the plasma energy is enhanced by additional heating provided by external means, as is common in many contemporary experiments. In a tokamak, the upper theoretical limit of β_p for linear ideal MHD stability is about R/a, thus showing a potential enhancement of the pinch effect by the stabilizing field.

If the confined plasma with Ohmic heating suffers any losses of energy additional to that due to radial collision diffusion, then $\beta_p < 1/2$ in a steady state. When radiation losses become high, Ohmic heating can be insufficient to maintain plasma pressure. The minor radius of the plasma ring starts to contract, which destabilizes the current ring, and a sudden instability and relaxation – known as a disruption – involving loss of energy and current to the walls, can terminate the discharge. Especially at mega-ampere currents, the control of energy balance and particle density to avoid disruption is an important feature of experimentation. Even if the only radiation is bremsstrahlung, this effect becomes important when

$$\beta_p I_\phi > 1.5\,\text{MA}. \tag{18.9.7}$$

Thermal insulation provided by the pinch in the case where plasma extends out to the walls is determined ideally by ion thermal conduction losses. The ion thermal diffusivity κ_i has, in a toroidal pinch, the approximate value

$$\kappa_i \simeq r_{L\theta}^2 \nu_{ii} (r/R)^{1/2}, \tag{18.9.8}$$

where r is the minor radius, R is the major radius, ν_{ii} is the ion–ion collision frequency, and $r_{L\theta}$ is the mean ion Larmor radius in the poloidal field component B_θ. A rough value of the required quantity $n_e\tau_E$ can be calculated

$$n_e\tau_E \simeq I^2 V_{Ti} (R/a)^{1/2}/Z_{\text{eff}}, \tag{18.9.9}$$

where V_{Ti} is the ion thermal velocity ($\sim 10^6\,\text{m}\,\text{s}^{-1}$), R/a is the aspect ratio of the torus, and Z_{eff} is an effect mean ion charge. Assuming an ohmically heated ideally stable pinch, the numerical values give for Lawson's criterion

$$I \simeq 1.5\,\text{MA for } n\tau_E \sim 10^{20}\,\text{m}^{-3}\,\text{s}.$$

18.9.4 *Alpha particle confinement*

For the confinement of alpha particles, it is highly desirable that the Larmor radius in the poloidal field shall be less than the minor radius of the plasma ring. This requires

$$I \simeq 3\,\text{MA}.$$

This latter condition thus requires a somewhat higher current than is suggested by the ideal thermal insulation, Eq. (18.9.9).

18.10 Experimental observations

Techniques now exist to determine experimentally almost every important parameter as a function of time and position in the plasma. The direct measurement of $q(r)$ has been carried out in a number of experiments (Fig. 18.14), which confirm the conditions $q(r) > 1$ in a tokamak, except near the centre where, it appears, a marginally lower value, about 0.8, is found in a number of experiments. Fig. 18.14 exhibits a quite wide departure from Eq. (18.9.2), especially at the edges where the effective plasma resistivity becomes high. If Eq. (18.9.2) were followed exactly, $q(r)$ would be roughly constant. Thus tokamaks have a significantly more peaked current density than given by Eq. (18.9.2).

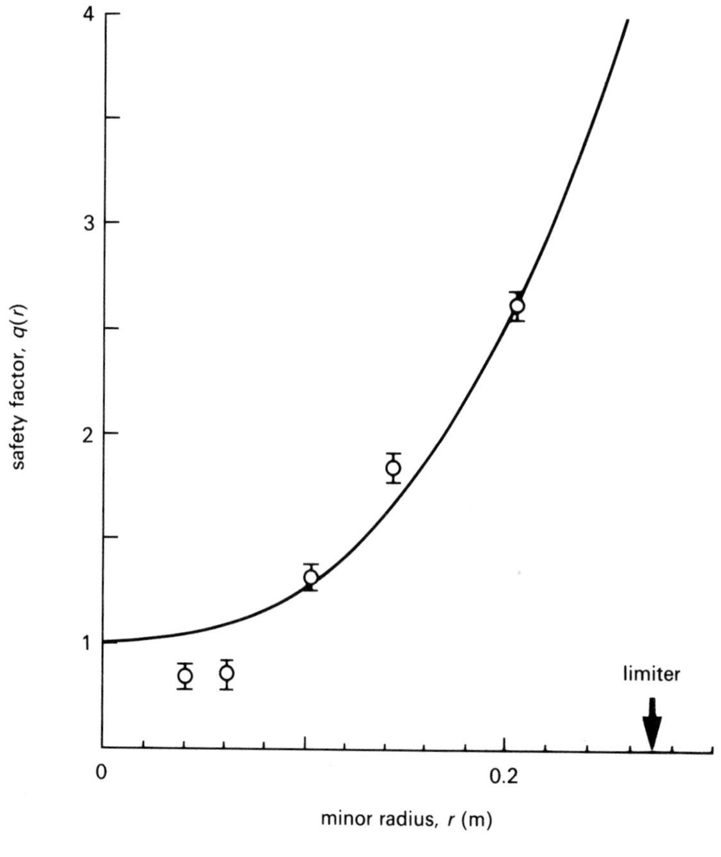

Fig. 18.14. The tokamak configuration: values of the safety factor $q(r)$, measured by optical means (Ṫ) ——; inferred from measured T_e and Eq. (18.5.8) in the DITE tokamak (Forrest *et al.*, 1977).

The most striking experimental confirmation of Eq. (18.9.6) was obtained on the Soviet T-3 tokamak. Fig. 18.15 shows the observed energy of the confined plasma increasing as the square of the current I_ϕ. Moreover, the constant of proportionality is quite near to the theoretical value $\beta_p \sim 1/2$. The individual variations of T with n were also measured. These encouraging observations are characteristic of currents I_ϕ in the range $\sim 100\,\text{kA}$. In today's higher current experiments with ordinary Ohmic heating, β_p falls to values of 0.05, indicating a significant defect in the thermal insulation properties of the pinch. With the powerful additional heating now available, values of β_p in excess of unity are

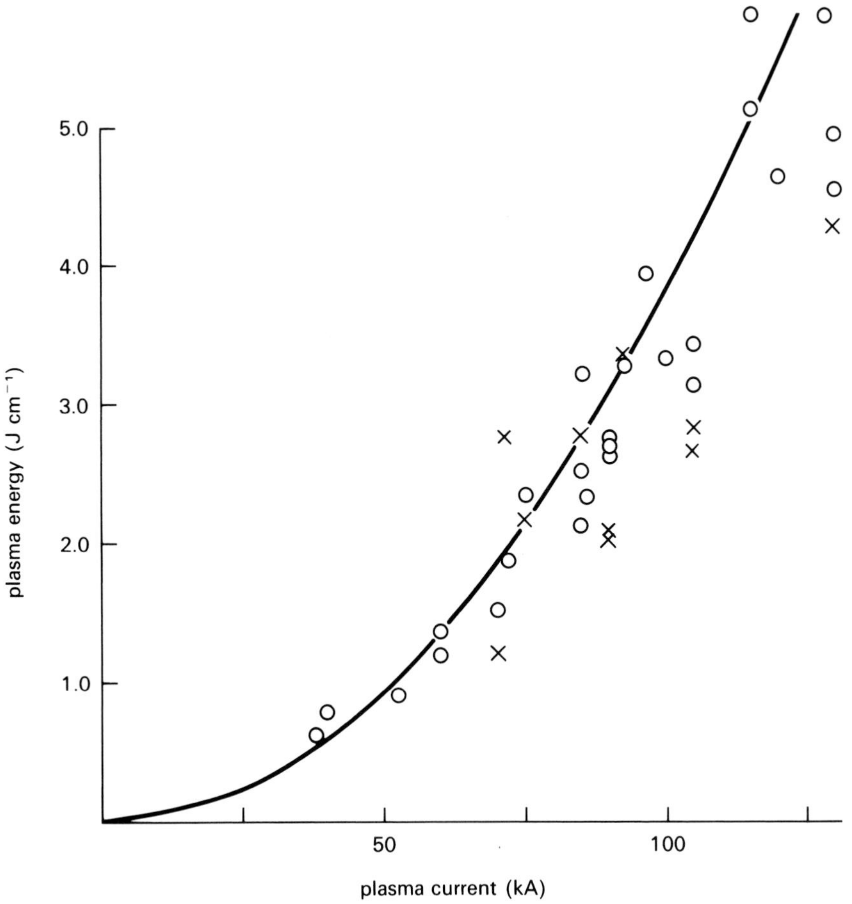

Fig. 18.15. Confined plasma energy as a function of toroidal current. No additional heating. ——Eq. (18.9.6) with $\beta_\theta = 0.5$; ○ Thomson scattering measurements; × diamagnetic measurements. T–3 tokamak (Anashin *et al.*, 1971).

readily obtained, and the pressure limit of ideal MHD linear theory is reached (Fig. 18.16). But the deficiency of the thermal insulation remains: indeed it is increased by additional heating. The values of the confinement parameter $n\tau_E$ have been measured in many experiments. Some authors use the peak number density of deuterium ions, others use the more easily measured mean-line-of-sight electron density $\bar{n}_e$, as is determined by simple interferometry. Eq. (18.9.9) corresponds approximately to the latter. Fig. 18.17 shows the trend of $\bar{n}_e\tau_E$ as a function of I_ϕ^2 derived from a wide variety of experiments, and is compared with a rough theoretical line $n\tau_E \propto I_\phi^2$ fitted at the 1.5 MA point to the original calculations for JET. This line therefore does not take account of the variations of the quantity

$$V_{Ti}(R/a)^{1/2}/Z_{eff}$$

from experiment to experiment, which are required together with profile factors if the exact collisional confinement time is to be calculated. But this quantity does not vary widely.

Thus one can see that experiment follows theory in that $n\tau_E$ increases broadly as I^2; however, in the mega-amp current range, the observed confinement parameter $n_e\tau_E$ is below ideal theory by a factor between about 10 and 100. The empirical scaling law found from a large number of results is

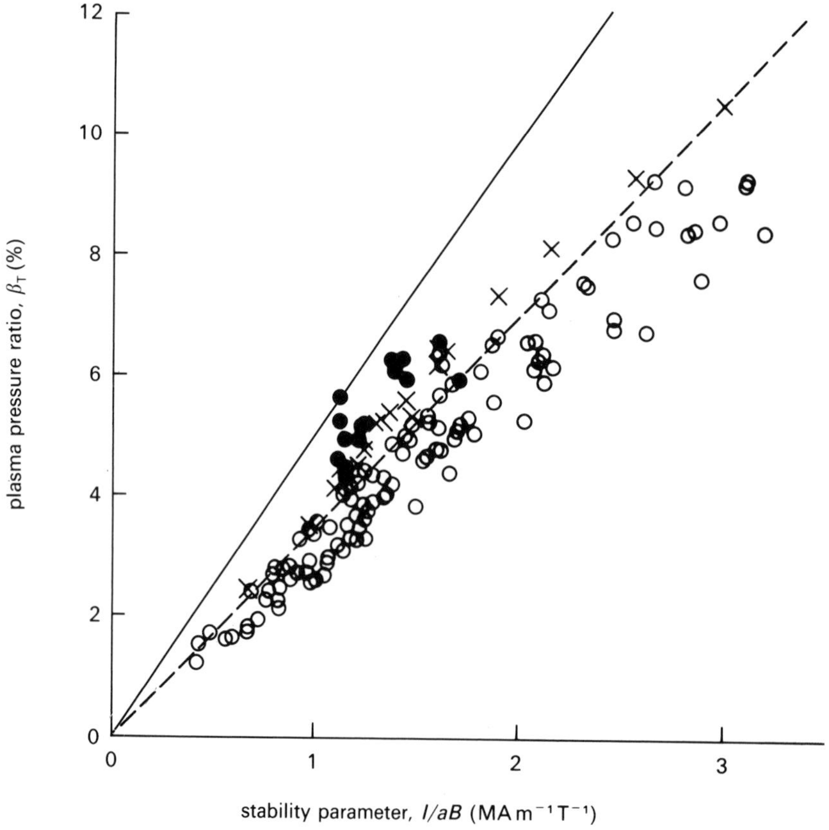

Fig. 18.16. Plasma pressure created by additional heating as a function of stability limit parameter in a DIII-D tokamak. β_T is the mean value of the ratio of plasma pressure to the toroidal magnetic field pressure, I is the toroidal current, a is the minor radius, and B is the magnetic field. Data shown represent the highest values obtained, not necessarily a limit value. Straight lines show alternate theoretical values: ———, $5.0\,I/aB$; – – – – –, $3.5I/aB$. ● $\beta_N > 4$, × $3.5 < \beta_N < 4$, ○ $2.5 < \beta_N < 3.5$. From Taylor *et al.* (1991).

$$n_e\tau_E \propto (I^2/T)(R/a)^{2.5}a^{-0.24}, \tag{18.10.1}$$

where T is the mean temperature, and R and a are the major and minor radii, respectively. The most striking feature of this empirical scaling is that the product $n_e\tau_E$ decreases with ion temperature, whereas according to Eq. (18.10.2) it should increase slowly. This discrepancy is not yet accounted for, but is thought to be due to one or other of the many instabilities proposed by theoreticians as possible causes, and which could produce small stochastic radial components of the magnetic field.

18.11 Nuclear fusion in tokamak plasmas

Thermonuclear fusion reactions in DD, D^3He and DT fuels have all been observed at powers up to 1.7 MW in tokamak plasmas. DD nuclear fusion reactions occur with power up to about 10 kW. The reaction rates are in good agreement with those expected from Eq. (18.8.1) and the measured deuteron temperatures and densities, provided that the direct interactions (between beams of ions and the plasma) resulting from the heating method are properly allowed for. Typically up to half the fusion reactions are due to thermonuclear fusion, the other half being due to direct interaction. Experiments with lean DT mixtures

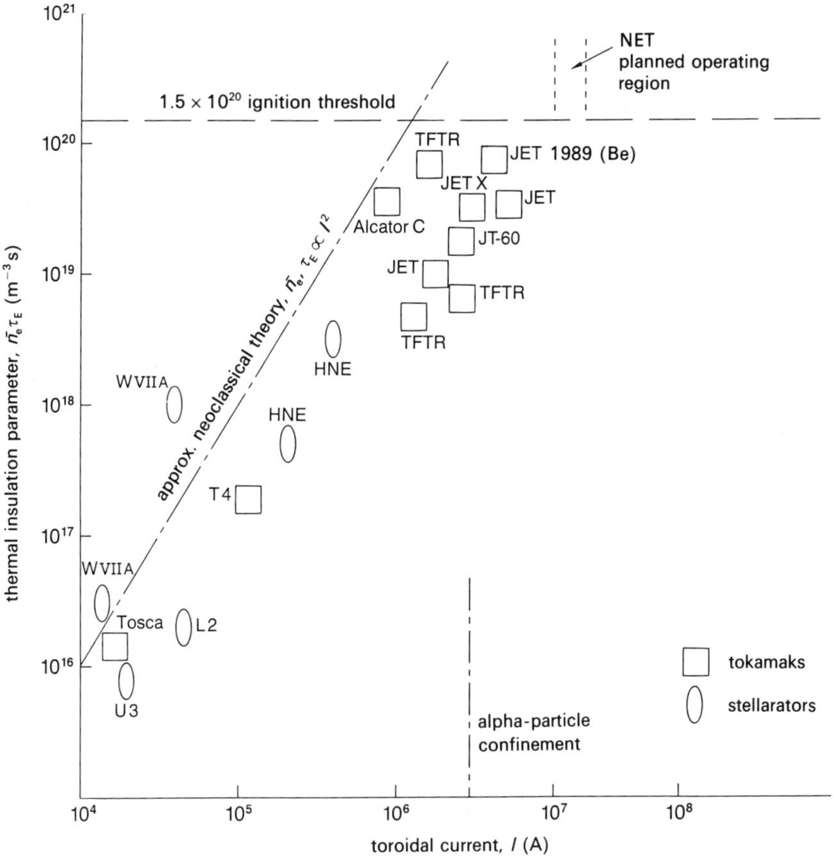

Fig. 18.17. Best values of the thermal insulation parameter $\bar{n}_e\tau_E$ observed versus I_ϕ from various tokamaks and stellarators; -----, rough ideal theory. From Pease (1987b).

have been carried out on JET at a current of 3.1 MA. The maximum fusion power output was 1.7 MW, and the peak alpha-particle power was about 300 kW. Because the additional heating power needed to sustain the plasma temperature was about 10 MW, the alpha-particle power fell short of that needed for ignition by a factor of about 30, roughly the same shortfall as suggested by Fig. 18.17. The neutron yield obtained in this experiment as a function of time is shown in Fig. 18.18, together with the theoretical estimates obtained from the measured plasma parameters. The ratio of fusion power to plasma heating power is about 100 times better than that achieved by any other controlled method, and is indeed within sight of 'break-even', $W_N = P_L$.

18.12 Stellarators

Fig. 18.17 also shows values of $n_e\tau_E$ obtained in stellarator systems, whose magnitudes are comparable with those obtained in tokamaks of around 100 kA current. In stellarators there is no toroidal current, and the confinement depends on nonzero azimuthal and radial components of the field provided by multipole external windings. A simple example is shown in Fig. 18.19. Because the currents flow entirely in external windings, the confining magnetic fields are naturally maintained for indefinitely long times. But the systems are not axisymmetric, and

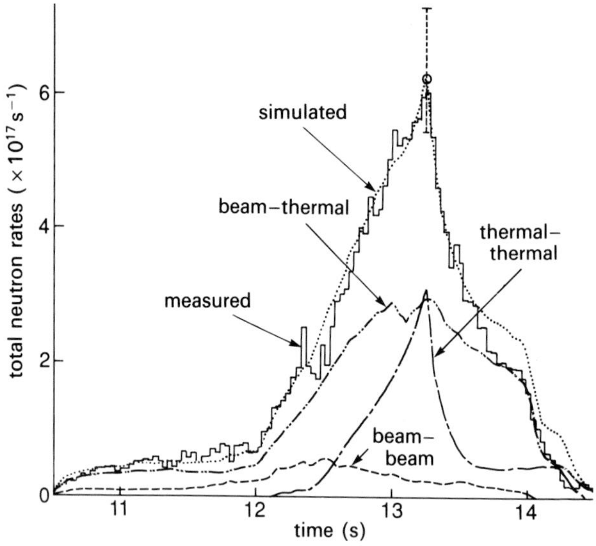

Fig. 18.18. DT nuclear fusion in JET. Observed neutron emission rates as a function of time (JET team, 1992). 6×10^{17} neutrons s^{-1} is 1.7 MW of fusion power. ········ Total theoretical value; —··— partial theoretical value from purely thermonuclear fusion.

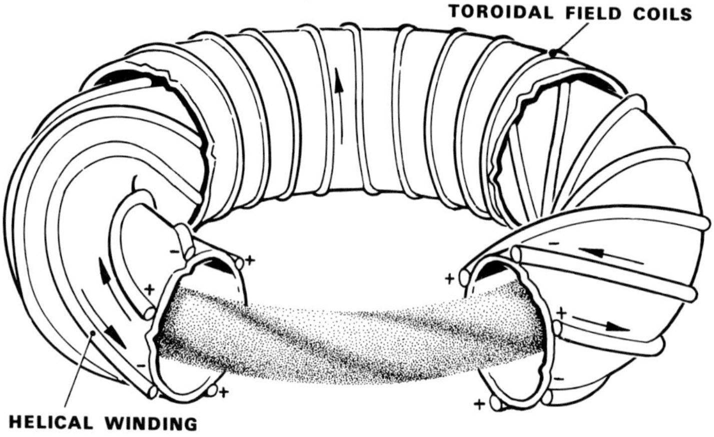

Fig. 18.19. Sketch of a stellarator geometry.

so do not generally admit of simplified discussion. Stellarators are similar to tokamaks in that the azimuthal (B_θ) and radial (B_r) components of the field are small compared with the main toroidal field B_ϕ. Moreover the mean value of B_θ averaged along the lines of force as they circle around the torus is nonzero, and gives rise to a mean helical trajectory of the lines of force. The mean value of the flux rB_θ is the mean confining flux, and can be interpreted as if a toroidal current were flowing which gives the same azimuthal flux. This equivalent current is a rough estimate of the mean confining poloidal flux, and is used in Fig. 18.17 to provide comparison of the values of $n_e\tau_e$ observed in stellarators with those of tokamaks. At the lowest equivalent currents, this comparison shows some favourable stellarator confinement parameters. But the largest of the stellarators, the Heliotron-E experiment in Kyoto, has values of $n_e\tau_E$ about those that would be expected from a tokamak of comparable mean azimuth flux. Comparison between experiment and detailed theory also shows that there is a shortfall between observed values of $n_e\tau_E$ and those ideally expected for stellarators, of the same order as that found in tokamaks (Murakami *et al.*, 1991; Obiki *et al.*, 1991; Renner *et al.*, 1991).

18.13 Concluding summary

This chapter has reviewed the basic physics of a plasma and the main results of magnetic fusion research. To answer the initial question 'What is a plasma?', the following main characteristics have been identified:

(i) plasma is an ionized gas, generally with a high degree of ionization when the temperature exceeds about one-tenth that characterized by the ionization potential;
(ii) it is often not in full thermodynamic equilibrium: it is often optically thin, and often $T_e \neq T_i$;
(iii) it is electrostatically neutral, over scale lengths more than the Debye length, λ_D;
(iv) the basic plasma approximation is $n_e\lambda_D^3 \equiv N_D \gg 1$;
(v) this theoretical basis of a plasma is experimentally established by light-scattering measurements, and by the propagation and damping of electron plasma oscillations;
(vi) the theory of the Coulomb collisions in a plasma has provided good agreement with experimental measurements of a rather wide variety of phenomena, especially of electrical resistivity, up to temperatures approaching 10^8 K;
(vii) plasma is highly conducting, so that the skin depth is often small compared with the plasma dimension. Magnetic forces are often dominant, so that MHD has to be used in place of fluid dynamics, especially in astrophysical plasmas;
(viii) magnetized plasma supports a wide variety of waves, such as Alfvén waves and whistler waves;
(xi) plasma can be confined by magnetic fields, largely in a direction normal to the lines of force.

On magnetic fusion, the main points made are:

(i) to obtain nuclear fusion power, use the D(T, n)α reaction in a thermonuclear plasma with 50/50 DT mixture;
(ii) research is seeking to achieve ignition by alpha-particle heating, which needs the approximate conditions $T_i \sim T_e \sim 25$ keV and $n_e\tau_E > 1.5 \times 10^{20}\,\mathrm{m^{-3}\,s}$;
(iii) the most researched system is the tokamak, a highly stabilized version of the toroidal pinch. Other methods include the reverse field pinch and the stellarator;
(iv) present-day large tokamak experiments have currents up to 7 MA and many tens of megawatts of additional heating. Temperatures are up to 30 keV, the confinement time of nearly 2 s provides $n\tau_e$ of $10^{20}\,\mathrm{m^{-3}\,s}$. Fusion output power has reached ~ 1.7 MW;
(v) many observed properties of tokamak plasmas – the equilibrium, their MHD stability, and the plasma resistivity – agree excellently with theory;
(vi) the major discrepancy with theory is that the observed thermal insulation, $n_e\tau_E$ is between one-hundredth to one-tenth of ideal theory. Experi-

ments indicate that a similar discrepancy also occurs in the related toroidal systems, the stellarators;

(vii) the question of which instabilities cause this discrepancy, and how the deleterious effects can be limited or reduced, is a major objective of present research;

(viii) the major target of developments is to obtain and research confined plasma in ignition conditions.

References

ALFVÉN, H. (1950). *Cosmical Electrodynamics.* Oxford University Press. See also *ibid.*, 2nd edn, by H. Alfvén and C-G. Fälthammer, 1963.

ALLEN, C. W. (1973). *Astrophysical Quantities*, 3rd edn. Athlone Press, London.

ANASHIN, A. M. *et al.* (1971). *Sov. Phys. JETP*, **33**, 1127.

ANDERSEN, L. H. *et al.* (1990). *Phys. Rev. Lett.*, **64**, 729.

ARNAUD, M. and ROTHENFLUG, R. (1985). *Astron. Astrophys. Suppl. Ser.*, **60**, 425.

ARTSIMOVICH, L. A. (1965). *Elementary Plasma Phys.* Blaisdell, New York. Chap. 1.

BARTLETT, D. V. *et al.* (1988). *Nucl. Fusion*, **28**, 73.

BECK, B. R. *et al.* (1992). *Phys. Rev. Lett.*, **68**, 317.

BENNETT, W. (1934). *Phys. Rev.*, **45**, 890.

BODIN, H. A. B. and NEWTON, A. (1969). *Phys. Fluids*, **12**, 2175.

BORNATICI, M. *et al.* (1983). *Nucl. Fusion*, **23**, 1153.

BOWLES, K. L. (1958). *Phys. Rev. Lett.*, **1**, 454.

BUTT, E. P. *et al.* (1958). *Proc. 2nd UN Conf. PUAE*, **32**, 42.

CARRUTHERS, R. and DAVENPORT, P. A. (1957). *Proc. Phys. Soc. B*, **70**, 49.

CHEN, W. J. *et al.* (1982). *J. Phys. Soc. Japan*, **51**, 1620.

COUSINS, A. and WARE, A. (1951). *Proc. Phys. Soc. B*, **64**, 1959.

DARVAS, J. *et al.* (1991). *Nucl. Fusion Suppl.*, **3**, 633.

DREICER, H. (1959). *Phys. Rev.*, **115**, 238.

FORREST, M. *et al.* (1977). Report CLM P499, UKAEA, Culham.

GAMOW, G. and CRITCHFIELD, C. (1949). *The Atomic Nucleus and Nuclear Energy Sources.* Oxford University Press.

GUROVICH, V. and SOLOV'EV, L. (1986). *Sov. Phys. JETP*, **64**, 677.

JEPHCOTT, D. F. and STOCKER, P. M. (1962). *J. Fluid Mech.*, **13**, 587.

JET team (1992). *Nucl. Fusion*, **32**, 187–203.

KRALL, N. A. and TRIVELPIECE, A. W. (1973). *Principles of Plasma Physics.* McGraw-Hill, New York.

LANDAU, L. D. (1946). *J. Phys. USSR*, **10**, 25.

LANGMUIR, I. (1928). *Proc. Natl. Acad. Sci.*, **14**, 627.

LAWSON, J. D. (1957). *Proc. Roy. Soc. B*, **70**, 6.

MALMBERG, J. H. and WHARTON, C. (1966). *Phys. Rev. Lett.*, **17**, 175.

MURAKAMI, M. *et al.* (1991). *Nucl. Fusion Suppl.*, **2**, 455.

OBIKI, T. *et al.* (1991). *Nucl. Fusion Suppl.*, **2**, 425.

PEASE, R. S. (1987a). In *The JET Project and the Prospects for Controlled Thermonuclear Fusion.* Royal Society, London, and Cambridge University Press, pp. 3–16.

PEASE, R. S. (1987b). *Plasma Phys. Control. Fusion*, **29**, 1171.

POLLOCK, J. A. and BARRACLOUGH, S. (1905). *Proc. Roy. Soc. New South Wales*, **39**, 13.

PRAGER, S. C. (1990). *Plasma Phys. Cont. Fusion*, **32**, 903.

RAMSDEN, S. A. and DAVIES, W. E. R. (1966). *Phys. Rev. Lett.*, **16**, 303.

RENNER, H. *et al.* (1991). *Nucl. Fusion Suppl.*, **2**, 439.

RIVIERE, A. C. *et al.* (1981). *Phil. Trans. Roy. Soc.*, **300**, 547.

SHEFFIELD, J. (1975). *Plasma Scattering of Electromagnetic Radiation.* Academic Press, New York.

SPITZER, L. (1962). *Physics of Fully Ionized Gases.* Interscience, New York.

STIX, T. (1962). *The Theory of Plasma Waves.* McGraw-Hill, New York.
STOREY, L. R. O. (1953). *Phil. Trans. Roy. Soc. A*, **246**, 113.
TAYLOR, J. B. (1976). In *Pulsed High Beta Plasmas* (ed. D. E. Evans). Pergamon, Oxford, p. 59.
TAYLOR, T. S. *et al.* (1991). *Nucl. Fusion Suppl.*, **1**, 177.
THOMSON, G. P. and BLACKMAN, M. (1946, 1959). British Patent 817618, 8 May 1946.
TONKS, L. and LANGMUIR, I. (1929). *Phys. Rev.*, **33**, 954.
WESSON, J. (1987). *Tokamaks.* Oxford University Press.
ZARNSTORFF, M. C. *et al.* (1990). *Phys. Fluids B*, **2**, 1852.

Index